LOGIC, LANGUAGE AND REASONING

TRENDS IN LOGIC

Studia Logica Library

VOLUME 5

SCOPE OF THE SERIES

Trends in Logic is a bookseries covering essentially the same area as the journal *Studia Logica* – that is, contemporary formal logic and its applications and relations to other disciplines. These include artificial intelligence, informatics, cognitive science, philosophy of science, and the philosophy of language. However, this list is not exhaustive, moreover, the range of applications, comparisons and sources of inspiration is open and evolves over time.

LOGIC, LANGUAGE AND REASONING

Essays in Honour of Dov Gabbay

Edited by

HANS JÜRGEN OHLBACH
King's College, London, United Kingdom

and

UWE REYLE
University of Stuttgart, Stuttgart, Germany

SPRINGER SCIENCE+BUSINESS MEDIA, B.V.

A C.I.P. Catalogue record for this book is available from the Library of Congress.

ISBN 978-94-010-5936-7 ISBN 978-94-011-4574-9 (eBook)
DOI 10.1007/978-94-011-4574-9

Printed on acid-free paper

Photograph by Elsbeth Mastenbroek
Reproduced with kind permission

Preface

This volume is dedicated to Dov Gabbay who celebrated his 50th birthday in October 1995. Dov is one of the most outstanding and most productive researchers we have ever met. He has exerted a profound influence in major fields of logic, linguistics and computer science. His contributions in the areas of logic, language and reasoning are so numerous that a comprehensive survey would already fill half of this book. Instead of summarizing his work we decided to let him speak for himself.

Sitting in a car on the way to Amsterdam airport he gave an interview to Jelle Gerbrandy and Anne-Marie Mineur. This recorded conversation with him, which is included gives a deep insight into his motivations and into his view of the world, the Almighty and, of course, the rôle of logic.

In addition, this volume contains a partially annotated bibliography of his main papers and books. The length of the bibliography and the broadness of the topics covered there speaks for itself.

The authors of the papers in this volume are, by far, not all of his close colleagues and friends. Therefore this book can only be the first in a series of books dedicated to him. Most of the articles included build on his work and present results or summarize areas where Dov has made major contributions. The fact that one cannot avoid having him as coauthor in his own festschrift confirms what he said in the interview:

> "I try to work in these areas in such a way that when, sooner or later, the roads come together, like on a roundabout, it will be Gabbay coming from this way, Gabbay coming from that way ..."

Hans Jürgen Ohlbach and Uwe Reyle

Contributions

DOV GABBAY: "I AM A LOGIC"

JELLE GERBRANDY AND ANNE-MARIE MINEUR

An Interview with Dov Gabbay[1]

Based on the assumption that the Almighty has created a coherent being, that He has sprinkled a little logic in our minds, Dov M. Gabbay is working hard on getting theories on language, logic and information to converge. With that in mind, he publishes handbooks on all kinds of logic, he is an editor of the Journal of Logic and Computation and he is involved in the International Society for Pure and Applied Logic. When the roads come together, he wants to be on every one of them. 'Like a roundabout, it will be Gabbay coming from this way, Gabbay coming from that way...' We had to accompany him to the airport to have our interview, but then again, some people deserve the 'Superstar approach'. Gabbay is now working at Imperial College in London, though officially on sabbatical to study Labelled Deductive Systems.

Biography – 'This is what I want to do'.

I was born in 1945, and I grew up in Israel. I started my university studies in '63, I studied mathematics and physics for the BSc, mathematics for the MSc – I did my Master's on many-valued logics – and then I did my PhD on non-classical logics, in 1969.

I went to an extremely religious school. Take for example the way they taught physics. The teacher came to class and said: 'When God created the world, He used these equations, and then He derived everything from that'. No experiments, nothing, it was all mathematics. They taught history only because it was necessary, teaching languages was good, and they taught (mathematically) some science. Humanities – Arts, Music – they did not

[1]This interview is also published in Ta!, the Dutch students' magazine for computational linguistics.

H.J. Ohlbach and U. Reyle (eds.), Logic, Language and Reasoning, 1–12.

take seriously. And they taught us a lot of Bible. So I naturally became one-sided, not only in what I knew, but also in my attitude. The school attitude was: 'Here is the word of God, you concentrate on that. Don't be distracted by junk'. I don't believe that you have to follow the Bible in the same way, although I believe it is good to know some things about it. But the attitude of 'this is what I want to do, don't be distracted', was ingrained in me.

At that time, this was a good attitude. A lot of my fellow students did get distracted. I didn't go to discotheques or out dancing with the girls. I made a concentrated effort. Of course, I could have gone funny at the age of forty.

A part of the religious teaching was for everyone to get married, have children, and so forth. I got married in 1970. My wife is an artist, and I learned a lot from her; the fact that I can talk about things, for instance. I remember I was going out with her, before we were married, and we were walking from one part of the university to another part. My objective was to get from A to B, she wanted to stop and look at the moon, because it looked very nice. And I thought: 'What the hell would I want to look at the moon for, when I want to go to B?' Now, of course, I will look at the moon at all times with her.

Then I went to Stanford, from 1970 to 1974, 1975. In Stanford I took up Dana Scott's position. When I worked in Stanford, I wanted to become a professor as fast as possible. I thought that if I worked only in one subject, intuitionistic logic for example, a large part of the department in Stanford would not be interacting with me. Then I saw that there was Logic and Language, and I started working on language with Julius Moravcsik. And I loved it.

At that time I also used to go to statistics seminars in the other faculty. Probably, scientifically it was right, because now we have uncertainty, probabilistic reasoning and so on, but from the effort point of view, it would have been too much to do. Then one day Kreisel came to me and said that Gödel wanted to talk to me: 'Come to my office on Sunday'. So I went to his office on Sunday, and Gödel talked to me through Kreisel. It was a very strange situation: Kreisel was sitting on the phone, talking to Gödel in German, Gödel would ask a question, Kreisel would repeat it to me in English, I would answer, or Kreisel would answer for me. This is how I talked to Gödel. Basically, what Gödel said was: 'What is this young Gabbay doing? He is doing this, he is doing that, what is this? I knew what I was doing when I was sixteen'. And Kreisel said: 'Well, he's young, he's enthusiastic'. So I dropped statistics after that, but kept the language, because I was interested in that. I will get into statistics now, probably. After that, we decided to go back to Israel. So I resigned and went back to the Bar-Ilan University, and I stayed there until 1982. In that year I went for a year to

Imperial College, to work with Bob Kowalski, and to see what all this talk about logic programming and applications of logic in computer science and so on was. I had never taken this subject seriously before. I had done some things with computer science when I was in Israel, I used to work a lot with Amir Pnueli, who developed temporal logic for computer science. We had a joint seminar. But I never took it seriously in the sense that there is a lot to be learned there. It just didn't hit me. I was thinking more in terms of expressive power and so on. Some of the results I was getting were useful to computer science, but I never thought of them in that way. At Imperial it suddenly did hit me. So I stayed there.

The Handbooks – a legacy to the next century

Imperial College is very big. Had I gone back to Israel, I would have had to read more, but at Imperial, if you want to know something, you grab someone. That saves a lot of time. The flow of information there is from logic to computer science: you show me the problem, I will solve it. The bottleneck is in understanding the problem. So at that time I decided that the best way to learn is to start the Handbook of Logic in Computer Science (Abramsky *et al*, [4]) and the Handbook of Logic in AI and Logic Programming (Gabbay *et al*, [1]). I started this in 1984 with some of my Imperial College colleagues, as a method of learning it myself and so that I would be forced to see what was happening. That is how those Handbooks started.

There are four volumes of Logic and Computer Science, and two more volumes are ready. Then we have four volumes of Logic in AI, with another volume ready. Also, we have plans now with Kluwer for the Handbook of Practical Reasoning with Hans Jürgen Ohlbach, which will be five volumes, the Handbook of Uncertainty, five volumes, the Handbook of Tableaux, one volume, and several volumes of the Handbook of Algebraic Logic. Then of course the second edition of the *Handbook of Philosophical Logic* (Gabbay *et al*, [3]), which will probably be seven or eight volumes, and a new Handbook of Mathematical Logic with Jon Barwise.

Also, we want to make a dictionary of logic. What we would like to do is make a big collection of material on logic. We want to put it on the network and allow people to log in, see the entries, maybe suggest new entries, and let it grow like this. After a while we'll have a proper dictionary, which then we will collect on a CD with all the Handbooks. So, if you want to know about Hintikka, you get the dictionary, you look up Hintikka, you get references, short descriptions and so on. Then you click on those and you might go to the relevant chapters of the handbooks, you can browse around there... I think that will take seven years to do. We are starting it now, and we will see how it goes. And I think we will do it through the IGPL (*The Interest Group in Pure and Applied Logic*).

Why am I doing all these Handbooks? We want to leave a legacy to the students of logic of the next century. It serves the community, you need these things, it helps to bring different areas together, it helps to clarify concepts. Also, the field is moving fast: you have to read a lot of papers. A Handbook is systematic: you write a chapter with a view, and it is coordinated with the other chapters. Therefore, you get an attitude. When a survey is written and coordinated with other authors, they agree on an attitude. And the attitude – a theme – is important for new problems. Sometimes you write a chapter to provide a coordinate system, so that others can relate to it. You see it in mathematics: you have several equivalent definitions of the same thing, but only one of them generalizes. That means that the others were not the right formulations of the concept.

We had this problem with the chapter on non-monotonic logic: we did not put it in the Handbook of Philosophical Logic, at that time. I think I was the only one who wanted to put it in, but all the others were against it. They said that it was not ready yet, that there was no field. But now we have a whole volume on it. So, would it have been the right move, to put such a chapter in? Maybe it would have given a view that all the AI people since then would have used or related to, maybe it would have misled them, I don't know. There was nothing on it, then. It's difficult, you could be wrong.

With the Handbooks, we tried to organize the area. And indeed, some of the chapters, like the chapter on topology, was completely new (in the *Handbook of Logic in Computer Science*), it was invented because there was a gap in how to describe the use of topology for computer science. Mike Smyth did very strong research: he simply discovered the whole thing. And there was new research generated by other questions.

I want to be safe not to do things that will disappear after ten years. The best thing is to look for more than one reason for doing what you are doing, and also look at the principles involved. I think the dictionary and the Handbooks are the things I leave for the next century.

God and logic – 'A strong image'

When you do research, I think there are two ways you can go about choosing your work, and I think the choice has to do with one's character. Some people just home in to something, others like to expand to see different points of view. So you can either choose something, like situation calculus, and work on it all the time and you can spend years and years doing that. And then if it is a hit, you have done something, and when it is not, you have gone out of history together with situation calculus, or whatever it was you were doing.

On the other hand, you cannot do everything. You must have a model, a strong image. An image such as the Americans had when they said they wanted to put a man on the moon – that is a strong image. If you go towards it, you will develop all kinds of things, you will affect the economy, affect science. My strong image is this: God created the physical universe, with some rules, and we study them, describe them. Some admire the Almighty for this, some don't, that does not matter. In this universe, He put us, with practical reasoning in our minds. There's something there that you can study and model, just like the physical universe. Analyzing what goes on in our minds may be much more difficult than studying the physical universe. It is a sort of a joke: He put the more complex thing in us. Anything that has to do with this, I go after.

You can immediately draw some conclusions from this. We are coherent, at least I think so. As practical reasoners – somebody who switches, in today's terms, from one logic to another, who takes actions, decisions – we are coherent. Then any major theory that can legitimately describe part of this activity must be connected to any other theory describing other parts of this activity.

So I started looking around: what is common? You have substructural logics on one hand, you have inheritance systems in artificial intelligence, you have categorial logics... There are those people who believe in syntax, in proof theory, people who don't want to look at semantics, there are people who believe in semantics and not in proof theory, and there are people who believe that classical logic is the only logic. And when you look around to see how logic is applied, you see you have different communities: you have temporal logic in software engineering, you have temporal logic in AI, you have different communities analyzing language, and so on.

All these theories must be connected, because they are modelling the activity of the same human, but you need a theory that makes the connection. I try to work in these areas in such a way that when sooner or later the roads come together, like on a roundabout, it will be Gabbay coming from this way, Gabbay coming from that way... There is a lot to be done, and I feel the same excitement as somebody who is pointing the telescope looking for new stars.

This is another reason for doing the Handbooks, The Journal of Logic and Computation and for the Interest Group in Pure and Applied Logic: to bring these people together, to accelerate this process. All based on the assumption that the Almighty has created a coherent being, that He has sprinkled a little logic in our minds.

Logic and reality – 'Why was that sin?'

Whether I am a sort of preacher? I am not saying that if you teach people more logic, they will behave more rationally. I do not think that. But I think that if you teach people more logic, at least they will not make mistakes in evaluating the situation. We have our vices, right? I may look at your wife and want her – I might even kill you. No amount of logic will change that: these are basic instincts, they will not change. But I may not realize that maybe I do not exactly want her, perhaps it is something else and I got my signals wrong. So if I can think more clearly, I can reason it out. It will not make me better or worse, but I will have better glasses to see reality as it is. And then I can be bad if I want to. So it will eliminate some of the problems. If I do not want to share my goodies with you, I will not. And you may not want to share your goodies with me. But maybe if we can think, we might realize that when we put a little bit together, and we watch each other without machine guns ready, we might make more money this way, and then we are cooperating. You have to reason it out.

Me, I tend to keep my options open and try and reason things through. In other words, I want to know that from x_1 I can go to x_2 and from x_2 I can go to x_3, but I am not going to x_3 because the Bible says: Don't do that – and I believe in the Bible. For example, if I have a bad colleague, I might find it nice to run him over with my car. I will not do it, because at the moment, I am not running people over with my car. But I want it clear, I don't want it fuzzy.

What I find sometimes is that there is no reality. Reality is the way we perceive things, and a part of this is representation in our minds. You might be happy as you are now. I can make you unhappy in two ways: I can stand on your toe and it will hurt – that's real. But I can also say: 'Did you know that your neighbour is getting more?' I change the modelling in your mind, and suddenly then you are not happy. So what has happened? You changed reality. A lot of reality is how you see things, not everything is real. And that part of reality, logic can affect.

Take the story of Adam and Eve and the snake. What was the big sin of the snake? What did he do? He talked. He just changed the point of view of Eve. He told her: 'Why are you not eating just because God said so?' Is that wrong? He just talked. He did not deny the facts, tell any lies, he just changed the point of view. So why was that sin? I think because points of view are very important. And point of view is representation, which is the area of logic.

You have to be very careful. If you ask a colleague: 'Why isn't your wife supportive of you?' or 'Why isn't your husband coming?' this could have the same effect as knocking them on their heads. So you should be careful in what you say to other people, because you are affecting their model; in fact, you are changing reality.

LDS – 'Perhaps this was it'

In the literature, there have been instances of where labels were used. You had, for example, Anderson and Belnap who used labels to compute relevance. But labels were used only as a side effect. It was a bit like moving all the furniture against the wall because you want to wash the floor. It is a side effect of washing the floor, not redesigning the room. So people used labels, but not as a general method.

I tried to see what happens if you put labelling into the logic, and then I saw that diverse systems begin to look similar. I thought that perhaps this was it. I gave some lectures, checked more systems, and then applied to the SERC (Science and Engineering Research Council) for a five years sabbatical, to do labelled deductive systems. I got some projects; a project on labelled tableaux, a project on consequence relations, and started working on it. The motivation was to connect all these roads in the roundabout. Fibering systems, why we move from one system to another. Because this is what we do. This is intelligence. If I say that she is a smart girl, I do not say that because she can do so many resolutions per second. I say that because she can move from one logic to another, one mode to another. It is not only power, but also the right adjustments; intelligence is a mixture of these things.

I do not believe that there is a single logic, like classical logic. I look at how people reason, and that is the logic. In order to describe this logic you would have to have notations for action, notations for mechanisms. You should not look at a theory and what follows from it, but at a theory and how it develops. I think a logical system is really what AI people call agents. The whole matter comes into it, and that's a system: evolving, maybe continuously reacting systems. The way we are; I am a logic, each one of us is a logic (Gabbay, [2]). Someone said: 'Each man is a world unto himself'. I say: 'Each man is a logic unto himself'.

Perhaps LDS could be the framework to connect these different formalisms. LDS is a very flexible formalism. For example, if you take lambda calculus and you have an application area, then you have to translate the application area into lambda calculus formulas. With LDS, you look at the application, take some of the application area, name it and use it as labels. So you are bringing the semantics into the language, you help the natural logic in there. You can go with the application.

LDS is not a single logic, it is a methodology, a framework in which you can bring things from the application area into whatever system you are doing. It means that you never necessarily have a clash between the formalism and the application. You do not have to bend the formalism to hack the application in. You don't have to do this, because you take from the application as labels and bring it in in that way.

Consider Newtonian mechanics. It does not matter for Newtonian mechanics whether you invent relativistic mechanics before or after, because it is a limit case of relativity theory for low speeds. So if you get it before or after you know relativity, that does not matter. But if you take the steam engine: you don't want to look at a steam engine if you already have diesel. The question is whether LDS, or anything you put forward, is like a steam engine – when something better comes, you don't like it anymore – or it is like Newtonian mechanics, which is a limit case or part of something bigger. I believe that I am looking for some logical principles that people will use. I hope that some of the stuff that I isolated will be kept because I isolated the principles.

I once followed a very strange course on meteorology. They had models of the atmosphere and stratosphere and how particles come from the sun and fall down again, all kinds of things like this. They had an ideal model, but they would show that it was wrong. Made a correction, and then made another correction... It looked like a suit with a lot of patches on it. And I always asked myself: 'Don't they have a new model?' But that was all there was: so-and-so's correction, and another guy's correction. Maybe we are doing the same for the time. Until we have better theories.

Computer science and logic – Honey through a pipe

Many of the problems we want to solve are not different from what Aristotle wanted: you take the human, you see how the human interacts with his environment, you try to analyze it. A lot of that is logic. In those days, there was only philosophy, there wasn't pressure. Then when computer scientists came in and they wanted a machine to help or simulate a human, there came industrial interest, there was money in it. Also, because the Japanese started to put money into it and talked about a fifth generation, everybody started to get interested. Logic became a buzz-word, and that is in itself an important service.

So there was a push to look into these problems. What happened first is that some people started building systems without thinking, just things that worked. Then other people migrated from philosophy departments and from mathematics into computer science, and they did start thinking. The interest in computer science forced the issue, and pushed logic more into the area of human activity, of human problems in thinking, and from that into processing, into theorem proving and verification systems, and so on. All these applications led to a lot of new developments.

I see computer science as a rich area of applications, and if you work there, you get ideas. Take this example: suppose the city of Utrecht decided they wanted to put a big pipe and somehow push honey into the nearby villages. Now this is going to cause a lot of technical problems, possibly will lead to new equations, they might discover new mathematical spaces. And this is what's happening with logic and computer science. You may agree or disagree on putting honey through a pipe, but that does not matter. The rise of logic programming has importance in providing a computational tool for the logician. Even if logic programming as a computer language is not as important as some others, it did its service as a tool for logicians by pushing logic up front. And now, fuzzy logic is new.

Linguistics – 'Mrs Thatcher is thin'

I think the study of language is important for logic, because a lot of our way of reasoning is reflected in the way we use language. I do not know exactly in what way logic and linguistics stand in relation to each other. There is a lot of logic in language. For example, if you say: 'John loves John', then that is ungrammatical. Let's say that is for syntactical reasons. If you say: 'John looked through the window, but he saw nothing', that is alright. But 'John looked through the window, but Mrs Thatcher is thin', that does not sound right. This is not because of the structure, but because of non-monotonic expectations of the language, so that part is logic. And I am sure that logic and linguistics interact in more complex ways. To parse a sentence, you may need a combination.

A lot of the modelling methods used in the study of language come from logic. But language also influences logic: we developed new temporal logic that came from the study of language. Like two-dimensional or three-dimensional temporal logics. Or consider quantifiers: we have quantifiers now that we didn't have in logic before.s

Fallacies – Bonzo: easy to feed your children

I think it is important to look at the phenomenon of fallacies and what the community of informal logic has to say about that. This is a very important subject, and I intend to work on it. I am planning a book, with John Woods and Andrew Irvine. A book on fallacies and informal reasoning. We are going to make a new edition and we agreed that I would participate: to analyze, using LDS, what's happening here.[2]

[2] A second edition of the book by John Woods and Douglas Walton: *Argument, the Logic of the Fallacies*, which will be done by John Woods, Douglas Walton, Dov Gabbay and Andrew Irvine.

When we reason, it is much more effective to use all these fallacies than to use proper deduction. Say you owe me one hundred pounds and you don't want to pay me back. It doesn't matter how I argue, that I say that you are a real jerk for not paying. But if I say that your wife won't like you, or your girlfriend won't like you, then that might be most effective. So real reasoning is very different from what is usual in logic. And I plan to move into it. I think it is important.

I have some examples, and I don't know what they illustrate. Suppose you take a taxi to Schiphol. It should cost 25 guilders. But you have a charter flight to America: if you miss it, you will lose a lot of money. And it is raining. Then the driver stops, and says: 'I want fifty guilders. And if you do not pay, you are going to miss your flight, even if you take another taxi'. But he will not say it like this, he'll say: 'Well, it is raining, it's more difficult than I thought, you have five pieces of luggage, your children are screaming: it's fifty'. He'll feel the need to find some excuse. I think there is something there, some rules we play by. All we need is to keep on looking, communicate with practical reasoning people, psychologists...

Let me do an experiment with you. I claim that if you give me a hundred guilders now, just this once, it will be a hundred guilders you will never use. Because I am sure you will have more than a hundred guilders, or else an overdraft with a hundred guilders more or less makes no difference. It is not a big order of magnitude. And on the day you die, you will have never used your last hundred guilders. It does not matter whether you give it to me or not. So if I have to buy a present for somebody, and it is a hundred guilders more, or if I lose a hundred guilders, I don't worry about it, because it does not matter.

Another example of how people reason. In Israel, I was teaching logic to middle-aged people, managers, housewives, teachers, who take a year of university in the middle of their lives. There is this dog food called Bonzo. It is made of meat: little round dry pieces, just like rice crispies or whatever – dry round pieces. The way you feed it to your dog: you put it in a bowl and pour water over it. It is very healthy, and not only it is good meat, but it is even kosher. You get it in big bags, and it is very easy to feed your children. In the morning, you give them cornflakes, and when they come back from school, you can give them some Bonzo and pour water over it. So I said to the class: 'Fine, do you want to use it? Look, I brought a bag'. And there was a revolution, they went crazy. Some of the mothers said: 'I am not feeding my children dog food'. And I said: 'It is not, it is kosher, it is a safe meat. The substance is alright, it is just labelled 'dog food''. I asked: 'If I take something from your fridge, and put a label on it that says 'dog food', would you then not eat it?' And they said: 'No, it is dog food now'. A lot of this kind of reasoning is not traditional logic.

Teamwork: Doing dishes

I can't do what I do if I don't think about it all the time. My wife, Lydia Rivlin is very helpful as well, she takes care of things. So in that sense I think I am lucky, the home is taken care of, I have very good teamwork in the university, with my friend and colleague, Jane Spurr, doing the Handbooks and such, and I also have very good research assistants. My policy is, as I put it, to get people who were born with a jet engine stuck to their backs. Very strong people. Some professors are afraid of very strong people, because if the guy is good, and you come up with some new theorem, he might come and say that it is rubbish. He or she will tell you that, if he or she is good. And if he or she is right, you must follow it.

I always follow the truth. I want to follow the truth, so I like very good people. I have many projects, some of them are run by other people. It is much better, that way. You have to trust them, feel that they are competent in what they are doing to the extent that you don't have to worry. Whatever they do, you accept, even though it is not exactly, because it almost never is exactly, and you do not know that if you would have done it, you would have done it better. It is a partnership. I like teamwork. It is like painting the fence with the other kids.

Usually, there are things you are better at, and there are things that are still important to whatever you are doing, but you are not as good at it. So if you team up with someone else who happens to be very well complemented, and you have similar ways of thinking, if you are compatible, one can make a terrific team this way. The best image I know of this is the following. At the beginning of this century the British were very good at building ships. They used teams of right-handed persons and left-handed persons. A right-handed person hits with his hammer this way, a left-handed person that way, and they stood next to each other, each hitting the nail, one, one, one... And if they are well-coordinated, they can hit nails in very quickly.

There are things you do not want to do. I can do things very easily that other people find very difficult. For example, I don't mind adding numbers for hours and hours, I don't mind cleaning toilets, I don't mind washing dishes, I don't mind making tea, I don't mind xeroxing for hours and hours... Because it is automatic: I can think of something else. I am sure there are lots of people who hate doing this, even though they can. It would be perfect for me to share a flat with somebody who doesn't like doing this, but who does like to pay the bills, to check whether the amounts are correct, etcetera. That is something I hate doing. It requires thinking, and thinking I keep for logic.

References

1. Chris J. Hogger, Dov Gabbay and John Alan Robinson (eds). *Handbook of Logic in Artificial Intelligence and Logic Programming (Volume 1).* Clarendon Press, 1993.
2. Dov Gabbay. *What is a Logical System?*, Chapter 'What is a Logical System', pages 179–216. Studies in Logic and Computation Series. Oxford University Press, 1994.
3. Dov Gabbay and Franz Guenthner (eds). *Handbook of Philosophical Logic* (Volume 1). D. Reidel Publishing Company, 1983.
4. Dov Gabbay, Samson Abramsky and Tom S. Maibaum (eds). *Handbook of Logic in Computer Science (Volume 1).* Clarendon Press, 1992.

RESEARCH THEMES OF DOV GABBAY

In the course of three decades of research in logic, Dov Gabbay has put forward several research themes and methodologies which have had strong influence over the pure and applied logic community. In what follows we shall list them briefly and explain the basic ideas.

1 The Decision Problem for Non-Classical Logics

In a series of publications dating back to 1969, Dov Gabbay developed two methods, one for proving decidability and one for proving undecidability for non-classical logics and theories. The method of proving decidability is by expressing a suitable semantics of the logic or theory in the monadic 2nd order theory $S\omega S$ of ω-successor functions, proved decidable by Michael O. Rabin [165].

The candidate system is manipulated in such a way that a semantics expressible in ω can be proved complete for it and thus decidability is assured. Undecidability is proved by interpreting the classical theory of reflexive and symmetric binary relation in the candidate system. The decision problem for a large class of logics and theories can be solved using these methods. See references [6, 7, 13, 14, 17, 18, 30, 36]. Gabbay's methods remain until this day the most powerful methods available for proving decidability. The traditional most widely applied method of proving decidability is through showing the finite model property, which together with axiomatizability of the system entail its decidability.

Gabbay has produced in [6, 7] a decidable finitely axiomatizable system without the finite model property, which can be shown decidable by his methods. The decision procedure for $S\omega S$ is double exponential, so further analysis is needed in each particular case, for sharper bounds.

We quote one striking example applying Gabbay's method. Consider the intuitionistic theory of Abelian groups of order m formulated by the traditional axioms for $=$, $+$ and 0, and the additional axiom $mx = 0$. This theory is decidable iff m is a product of different (non-repeating) prime numbers. The author is still active in this area. His most recent paper is [99].

H.J. Ohlbach and U. Reyle (eds.), Logic, Language and Reasoning, 13–30.

2 Goal Directed Algorithmic Proof Theory

The idea of goal directed algorithmic proof formulation of various logics originates from Dov Gabbay's desire to present human oriented (as opposed to machine oriented) theorem proving methods for classical and other logics. Implication (with its various meanings and forms) seems to be the main connective compatible with human thinking. It is also the base for Horn clause effective computation. Inspired by logic programming, Gabbay proceeded to formulate proof theory for a variety of known logics around implication and thus offer goal directed Prolog-like proof theory for them, as well as pioneering new more expressive extensions of logic programming.

The power and flavour of the method can be seen from the goal directed formulation of intuitionistic implication. A theory Δ is a set of implicational formulas of the form $B = (A_1 \to (A_2 \to \ldots \to (A_n \to q))\ldots)$ The proof rules involve remembering the history (a sequence H) of the atomic queries asked during the computation and a family of restart rules of various forms allowing for past queries to be re-asked under certain conditions.

The following are the rules for pure propositional intuitionistic implication.

1. $\Delta \vdash q; H$ if $q \in \Delta, q$ atomic.
2. $\Delta, A_1 \to (A_2 \to \ldots \to (A_n \to q)\ldots) \vdash q; H$ if for $i = 1, \ldots, n, \Delta \vdash A_i;\ H * (q)$ where $*$ is concatenation. Note that B was thrown out of the database and (q) was added to the history.
3. $\Delta \vdash (A_1 \to \ldots \to (A_n \to q)\ldots); H$ if $\Delta, A_1, \ldots, A_n \vdash q; H$
4. $\Delta \vdash q; H_1 * (q) * H_2 * (a) * H_3$ if $\Delta \vdash a; H_1 * (q) * H_2 * (a) * H_3$
   i.e. we can ask again (restart) an old query and provided it was asked after a previous instance of q.
   This rule is called bounded restart. If we allow for arbitrary unbounded restart we get classical implication.

The method is clearly powerful. The database gets smaller all the time, there is no use of cut and it is implication-based.

The goal directed implication based approach sparked the study and use of many extensions of logic lrogramming with implication in the body of clauses following the original extension of Prolog by Dov Gabbay and Uwe Reyle [55].

Gabbay, together with Olivetti is currently actively involved in formulating many logics in this manner including strict implication logics, substructural logics and fuzzy logics. We are looking forward to their book [163].

3 The Irreflexivity Rule

Traditionally axiom systems for logics involve Hilbert axioms, Modus Ponens and some rules like generalization and necessitation. In non-classical logics there is a wide correspondence between axioms and conditions on the semantics. Some conditions, however, have no corresponding axiom. The irreflexivity of the possible world accessibility relation is such a condition. In 1979, Gabbay proposed a new type of rule called IRR-rule [51], which can help axiomatise logics. In modal context the rule states

$$\vdash q \wedge \Box \sim q \rightarrow A \text{ implies } \vdash A$$

provided q is not an atom of A.

Since then, many authors have found rules like the above (referred to as Gabbay like rules) necessary for the axiomatic presentation of a wide variety of systems. In fact it is accepted now that such rules are part of the proper axiomatic presentation of any system alongside Modus Ponens.

4 Temporal Expressive Power and Execution

Gabbay has continuously been working on various aspects of temporal logics. These include axiomatisations, applications to the logical analysis of language, expressive power of temporal connectives, temporal databases and executable temporal logics. He has put forward two memorable themes in this area.

(a) Gabbay's work on expressive power was inspired by Hans Kamp's thesis of 1968 on the functional completeness of *Since* and *Until* for Dedekind complete flows of time. Gabbay observed the link between the number of distinct variable letters used (also called Henkin dimension by Gabbay) in a theory and its expressive power in 1981 [51, 52]. Modal logic semantics, when translated into classical logic requires only two bounded variables. Gabbay investigated such fragments and their connection with expressive power of modal and temporal logics. His main result in this area is the separation theorem:

Let L be a temporal propositional logic with connectives such that any wff of L can be re-written as a Boolean combination of pure past future and pure present formulas. Then L has the same expressive power as monadic classical logic over the flow of time. The separation idea gives a powerful tool for testing and finding additional connectives for increasing expressive power. Gabbay's book [107] is a classical monograph on temporal logic and its computation properties. Volume 2 is now also available in draft form.

(b) The separation theorem has led to the so-called imperative future paradigm, of viewing future wffs of temporal logic as instructions for execution. The separation theorem tells us that we can rewrite any formula of temporal logic as conjunctions of the form

$$\text{Past wffs} \wedge \text{Present wffs} \rightarrow \text{Future wff}$$

reading the past and present as a query to the history so far, we can read the future wff as an imperative wff to execute and make true. This has been developed by Gabbay and colleagues (see the book *The Imperative Future* [159]) as a logical programming language METATEM and is now an area of intense activity. (See also [167].)

5 Consequence Relations

In 1984 there was in the AI literature a multitude of proposed non-monotonic logical systems defined for a variety of reasons for a large number of applications. In an attempt to put some order in what was then a chaotic field, Gabbay asked himself what minimal properties do we require of a consequence relation $A_1, \ldots, A_n \vdash B$ in order for it to be considered as a logic. In his seminal paper [56] he proposed the following.

$$\Delta, A \vdash A \qquad \textit{Reflexivity}$$

$$\frac{\Delta \vdash A; \Delta \vdash B}{\Delta, A \vdash B} \qquad \textit{Restricted Monotonicity}$$

$$\frac{\Delta, A \vdash B; \Delta \vdash A}{\Delta \vdash B} \qquad \textit{Cut}$$

The idea is to classify non-monotonic systems by properties of their consequence relation. Kraus–Lehman–Magidor developed preferential semantics corresponding to various additional conditions on $\vdash$ and this has started the area now known as the axiomatic approach to non–monotonic logics. For a good coverage of the current state of affairs see Makinson [166].

Gabbay continued to work in this area refining the notion of consequence relation to that of structured consequence relation and that of a cut as surgical cut. His study of better and better formulations eventually led him to develop his theory of Labelled Deductive Systems.

6 Inconsistency and Negation

In the course of asking what is a logical system and refining the notion of a database and consequence relation, other traditional concepts came under scrutiny. Main among them was the notion of inconsistency. It was clear that the logical notion of inconsistency was impractical. Close observation of practical reasoning examples led Gabbay and Hunter to put forward the idea that inconsistency was a good thing (inconsistency made respectable) and that it is closely connected with a context of sequences of actions in a changing world. The paradigm

$$\text{Inconsistency} \rightarrow \text{Action}$$

was put forward, meaning it is OK to have everything inconsistent as long as we know exactly how to act, faced with consistency. This idea seems to have become a hit in certain areas of software engineering, where conflicting data and views seem to be constantly emerging. A series of papers on handling inconsistencies and conflicting revisions and updates has followed and this area is now subject to active research by colleagues and students.

7 Fibring Logics

Gabbay has always maintained that intelligence has to do more with the ability to move between different kinds of reasoning systems than with the strength and speed of any individual system. His current work on fibring systems is to develop such a methodology.

The basic idea of fibred semantics is very simple. Assume several systems S_i which can be syntactically constructed and given semantics in terms of basic atomic components a_1^i, $a_2^i, \ldots$. The expressions of each of the systems are generated from these atoms via the system constructors. The combined system can be viewed as having the union of the atoms of the components and is generated by the union family of constructors. We are thus faced with the problem of trying to understand expressions of the form $C_1(a, C_2(b))$ where a, b are atoms and C_i are constructors of S_i. The idea of fibring is to have two fibring mappings $f_{1,2}$ and $f_{2,1}$ allowing us to shuttle between the individual semantics of the components, thus creating the fibred semantics of the combined system. Since the process is essentially algorithmic, it is possible to do the fibring automatically as well as generate properties of the combined system from properties of the components.

The above idea is general but extremely intuitive and simple. When made concrete in different contexts it can be simplified and it yields astonishing results. We list the main areas of application.

- Straightforward combination of logical systems. Under this heading we find a variety of multi-modal systems, modal intuitionistic logics, systems of knowledge and belief, etc. The transfer theorems and fibring methodology has laid the foundation for 30 years of sporadic single combined systems put forward by a large number of authors. See [128].
- Bringing the meta-level into the object level.
  This is a different kind of fibring, where meta-level constructs (say a consequence relation) is brought into the object level as (say a con-

ditional). In [115] Gabbay derived the syntax and semantics for the conditional via the fibring process.

- A variety of fibring options of many systems S_i with a particularly chosen distinguished fibring partner S_O.
  - The choice of S_O can be a particular temporal logic, in which case we get a variety of options for making a general system S_i time dependent.
  - The choice of S_O as Łukasiewicz infinite valued logics yields various ways of making a system fuzzy.

  The options were investigated in [120, 130] and [133] and brought order to a chaotic literature of ad hoc methods especially in the fuzzy case.
- Self fibring of predicate logics.
  This aspect of combining allows one to fibre a logic with itself and write expressions like $A(x, \phi(y))$, where $A(x, y)$ and $\phi(y)$ are formulas and x, y terms or expressions like $x = \phi$. Such expressions are abundant in the applied literature, they occur in meta-programming, natural language analysis, selfreference and liar paradox. Other surprisingly related applications are generalized quantifiers, default theory ambivalent syntax and channel theory. See [164] for details.
- The fibring idea is similar to the notion of fibring occurring in mathematics (topology and algebra) and contact with equational theories for fibred systems has already been made through fibring constraint languages in CLP [168].

8 Labelled Deductive Systems

Gabbay was looking for a natural unifying framework for the wide variety of logics and applied logics used by the research community. He observed that such systems manipulate different kinds of information in parallel and that there is a core logic involved (more or less the familiar implication with Modus Ponens) and the rest is variation in handling and control. He therefore put forward the idea that the basic declarative unit is a pair $t : A$, A a formula and t a term from an algebra annotating the formula. Logical manipulation affects both formula and labels at the same time. Modus Ponens becomes

$$MP : \frac{t : A; s : A \to B; \phi_{MP}(s,t)}{f_{MP}(s,t) : B}$$

where t, s are labels. f_{MP} is a function giving the new label of B and ϕ_{MP} is a relation which has to hold to licence the application of the rule. Different logics can be identified through different choices of labelling, and different functions f and relations ϕ. The idea was remarkably successful in unifying a great diversity of logics and systems.

A labelled theory is a diagram of labelled formula with some special relations required on the participating labels. Notions of proof, cut, semantics inconsistency, etc. had to be developed for the new framework. These are presented in Volume 1 of Gabbay's book [122].

People have used labels before, but only as a side effect not as an essential part of the logical ontology. The labelled theory is now accepted as a major logical framework by a large cross section of the community. To see the power of such a concept, note that for intuitionistic A, $t : A$ can mean t is a λ-term inhabiting the type A or t is a possible world name in which A should hold. One is pure formula-as-type proof theory and the other is a way of bringing semantics into the syntax.

9 The SCAN Algorithm

Gabbay and Ohlbach [88, 91] put forward an algorithm for eliminating second-order existential quantifiers. This algorithm allows, under fairly general conditions, semantics to be found automatically for a given Hilbert system. In principle the following can be done:

Given a mixed specification involving a formula $\psi(P_1, P_2)$, where P_1, P_2 are predicates in languages L_1 and L_2 respectively, the formula $\exists P_1 \psi$ essentially gives the conditions on P_2 to be "linked" with P_1 through the mixed specification ψ. Eliminating $\exists$ yields the conditions in the pure L_2 language.

Other applications are in automated reasoning for set theory and the automated computation of first-order circumscription (see the chapter by Ohlbach, Nonnengart and Szałas in this volume.)

DOV GABBAY'S MAIN PAPERS AND BOOKS

1. Dov M. Gabbay. Semantic proof of the Craig interpolation theorem for intuitionistic logic and extensions, part I. In *Proceedings of the 1969 Logic Colloquium in Manchester*, pages 391–401. North-Holland Publishing Co., 1969.
2. Dov M. Gabbay. Semantic proof of the Craig interpolation theorem for intuitionistic logic and extensions, part II. In *Proceedings of the 1969 Logic Colloquium in Manchester*, pages 403–410. North-Holland Publishing Co., 1969.
   Note: The methods used to prove interpolation in the papers including [1, 2, 5, 37] seem to be general enough to be applied in categorial context, as shown by Makkai 25 years later.
3. Dov M. Gabbay. The decidability of the Kreisel–Putnam system. *Journal of Symbolic Logic*, 35:431–437, 1970.
4. Dov M. Gabbay. Selective filtration in modal logics. *Theoria*, 36:323–330, 1970.
   Note: This is part of a series of papers studying the finite model property in modal and intuitionistic logics. These methods give improved completeness theorems and can help showing decidability. Other related papers are [3, 6, 7, 12, 16, 22].
5. Dov M. Gabbay. Craig's interpolation theorem for modal logics. In W. Hodges, editor, *Proceedings of the Logic Conference, London*, pages 111–128. Springer Verlag, 1970.
6. Dov M. Gabbay. On decidable finitely axiomatizable modal and tense logics without the finite model property, part I. *Israel Journal of Mathematics*, 10:478–495, 1971.
7. Dov M. Gabbay. On decidable finitely axiomatizable modal and tense logics without the finite model property, part II. *Israel Journal of Mathematics*, 10:496–503, 1972.
8. Dov M. Gabbay. Montague type semantics for modal logics with propositional quantifiers. *Zeitschrift für Mathematische Logik und Grundlagen der Mathematik*, 17:245–249, 1971.
9. Dov M. Gabbay. Decidability results in non-classical logic III (systems with statability operators). *Israel Journal of Mathematics*, 10:135–146, 1971.
10. Dov M. Gabbay. Tense systems with discrete moments of time. *Journal of Philosophical Logic*, 1:35–44, 1972.
11. Dov M. Gabbay. Model theory for intuitionistic logic. *Zeitschrift für Mathematische Logik und Grundlagen der Mathematik*, 18:49–54, 1972.
12. Dov M. Gabbay. Applications of trees to intermediate logics I. *Journal of Symbolic Logic*, 37:135–138, 1972.
13. Dov M. Gabbay. Sufficient conditions for the undecidability of intuitionistic theories with applications. *Journal of Symbolic Logic*, 37:375–384, 1972.
    Note: This paper outlines a method for proving undecidability of many intuitionistic theories. Related papers are [18, 28, 36, 99].
14. Dov M. Gabbay. Decidability of some intuitionistic predicate theories. *Journal of Symbolic Logic*, 37:579–587, 1972.
15. Dov M. Gabbay. A general theory of the conditional in terms of a ternary operator. *Theoria*, 38:97–105, 1972.
16. Dov M. Gabbay. A general filtration method for modal logics. *Journal of Philosophical Logic*, 10:135–146, 1972.
17. Dov M. Gabbay. A survey of decidability results for modal tense and intermediate logics. In P. Suppes *et al*, editors, *Proceedings of the Fourth International Congress on Logic, Methodology and Philosophy of Science*, pages 29–43. North-Holland Publishing Co, 1973.
18. Dov M. Gabbay. The undecidability of intuitionistic theories of algebraically closed fields and real closed fields. *Journal of Symbolic Logic*, 38:86–92, 1973.
19. Dov M. Gabbay. Applications of Scott's notion of consequence to the study of general binary intensional connectives and entailment. *Journal of Philosophical Logic*, 2:340–351, 1973.

20. Dov M. Gabbay. Representation of the Montague semantics as a form of the Suppes semantics with applications to the problem of the introduction of the passive voice, the tenses, and negation as transformations. In K. J. J. Hintikka *et al*, editors, *Approaches to Natural Language*, pages 395–409. D. Reidel, 1973.
21. Dov M. Gabbay and Julius M. E. Moravcsik. Sameness and individuation. *Journal of Philosophy*, 70:513–526, 1973.
22. Dov M. Gabbay and Dick H. de Jongh. A sequence of decidable finitely axiomatizable intermediate logics with the disjunction property. *Journal of Symbolic Logic*, 39:67–79, 1974.
23. Dov M. Gabbay. On 2nd order intuitionistic propositional calculus with full comprehension. *Archiv für Mathematische Logik und Grundlagenforschung*, 16:177–186, 1974.
24. Dov M. Gabbay. A generalization of the concept of intensional semantics. *Philosophia*, 4:251–270, 1974.
25. Dov M. Gabbay and Julius M. E. Moravcsik. Branching quantifiers, English, and Montague grammar. *Theoretical Linguistics*, 1:139–157, 1974.
26. Dov M. Gabbay. Tense logics and the tenses of English. In J. M. E. Moravcsik, editor, *Readings in Logic*, pages 177–186. Mouton Publishing Co., 1974.
Note: This is part of a series of papers analyzing logical structures in natural language and English. Other papers are [21, 25, 34, 35, 39, 40, 41, 43, 44, 46, 49].
27. Dov M. Gabbay. A normal logic that is complete for neighbourhood frames but not for Kripke frames. *Theoria*, 41:145–153, 1975.
28. Dov M. Gabbay. The decision problem for finite extensions of the intuitionistic theory of abelian groups. *Studia Logica*, 34:59–67, 1975.
29. Dov M. Gabbay. Model theory of tense logics. *Annals of Mathematical Logic*, 8:185–236, 1975.
30. Dov M. Gabbay. Decidability results in non-classical logics I. *Annals of Mathematical Logic*, 8:237–295, 1975.
Note: This paper outlines a general method for proving decidability and undecidability for non-classical logical systems. The method is based on Rabin's results on $S\omega S$ and uses a variety of semantical and syntactical interpretations. It is the main, most powerful and most extensive method for solving the decision problem in the area of non-classical logics. Related papers which widely extend and develop the methods are [3, 6, 7, 13, 14, 17, 18, 22, 28] and [36].
31. Dov M. Gabbay. *Investigations in Modal and Tense Logics with Applications*, volume 92 of *Synthese*. D. Reidel, 1976.
Note: The main research thrust of this monograph is to present comprehensive methods for proving decidability and undecidability for modal and temporal systems. General theorems are proved on the one hand and new classification and semantical characterizations are given to many logics in order to show that they satisfy these general theorems. Counterexamples are constructed to show the limitations of various methods. The book also lays the mathematical and conceptual foundations for non-classical logics.
32. Dov M. Gabbay. Completeness properties of Heyting's predicate calculus with respect to RE models. *Journal of Symbolic Logic*, 41:81–95, 1976.
Note: This paper studies the possibility of providing constructive semantics for intuitionistic and non-classical logics. It shows that results depend very much on formulation. The related paper is [33].
33. Dov M. Gabbay. On Kreisel's notion of validity in Post systems. *Studia Logica*, 35:285–295, 1976.
34. Dov M. Gabbay. Two dimensional propositional tense logic. In A. Kasher, editor, *Bar-Hillel Memorial Volume*, pages 145–183. D. Reidel, 1976.
35. Dov M. Gabbay and Asa Kasher. On the semantics and pragmatics of specific and non-specific indefinite expressions. *Theoretical Linguistics*, 3:145–190, 1976.
36. Dov M. Gabbay. Undecidability of intuitionistic theories formulated with the apartness relation. *Fundamenta Mathematica*, 97:57–69, 1977.

37. Dov M. Gabbay. Craig's theorem for Intuitionistic Logic III. *Journal of Symbolic Logic*, 42:269–271, 1977.
38. Dov M. Gabbay. A new version of Beth semantics. *Journal of Symbolic Logic*, 42:306–309, 1977.
39. Dov M. Gabbay and Asa Kasher. On the quantifier there is a certain X. In *Proceedings of the International Workshop on the Cognitive Viewpoint*, pages 329–334, 1977. Appeared also in [40].
40. Asa Kasher and Dov M. Gabbay. On the quantifier there is a certain X. *Communication and Cognition*, 10:71–78, 1977.
41. Asa Kasher and Dov M. Gabbay. Improper definite descriptions: Linguistic performance and logical spaces. *Hebrew Philosophical Quarterly*, 27:74–89, 1977.
42. Dov M. Gabbay. On some new intuitionistic propositional connectives I. *Studia Logica*, 36:127–139, 1977.
43. Dov M. Gabbay and Julius M. E. Moravcsik. Negation and denial. In F. Guenthner and C. Rohrer, editors, *Studies in Formal Semantics*, pages 251–265. North Holland Pub Co, 1978.
44. Dov M. Gabbay. A tense system with split truth table. *Logique et Analyse*, 21:5–39, 1978.
45. Dov M. Gabbay. What is a classical connective? *Zeitschrift für Mathematische Logik und Grundlagen der Mathematik*, 24:37–44, 1978.
46. Dov M. Gabbay and Christian Rohrer. Relative tenses. In C. Rohrer, editor, *Papers on Tense, Aspect and Verb Classification*, pages 99–111. TBL Verlag G Narr, Tübingen, 1978.
47. Dov M. Gabbay and Christian Rohrer. Do we really need tenses other than future and past? In A. Van Stechow R. Bauerli, U. Ugli, editors, *Semantics from Different Points of View*, pages 15–21. Springer Verlag, 1979.
48. Dov M. Gabbay, Amir Pnueli, Saharon Shelah, and Jonathan Stavi. On the temporal analysis of fairness. In *Conference Record of the 7th Annual ACM Symposium on Principles of Programming Languages, Las Vegas*, pages 163–173, 1980.
Note: Outlines the way to use temporal logic as a tool in software engineering, for program specification and verification. It contains results on axiomatization and decision problems and is one of the standard quoted papers in the area.
49. Dov M. Gabbay and Julius M. E. Moravcsik. Verbs, events, and the flow of time. In C. Rohrer, editor, *Time, Tense and Quantifiers*, pages 59–83. Niemeyer, Tübingen, 1980.
50. Dov M. Gabbay. *Semantical Investigations in Heytings's Intuitionistic Logic*, volume 148 of *Synthese Library*. D. Reidel, 1981.
Note: This monograph uses semantical methods to study intuitionistic and various neighbouring systems. It develops their mathematical model theory and finite model property and studies their proof theory and interpolation properties. It further applies the methods of book 1, to obtain decidability and undecidability results for intuitionistic algebraic theories.
51. Dov M. Gabbay. An irreflexivity lemma with applications to axiomatizations of conditions on tense frames. In U. Monnich, editor, *Aspects of Philosophical Logic*, pages 67–89. D Reidel, 1981.
Note: This paper introduces the – what is now known as – Gabbay's *Irreflexivity Rule*. The idea has been taken on-board and pursued by many authors. Many systems can be formulated without this rule. Its full nature is not yet fully understood.
52. Dov M. Gabbay. Expressive functional completeness in tense logic. In U. Monnich, editor, *Aspects of Philisophical Logic*, pages 91–117. D Reidel, 1981.
Note: This paper introduces the separation methods of studying expressive power of temporal languages. Besides deep theoretical results and inter-connections with other areas of logic it gives a practical way for any software engineering user of temporal and modal logic to test and adjust the expressive power of his system. Further papers on the expressive properties of temporal logics are [44, 51] and [61].

53. Dov M. Gabbay and Franz Guenthner. A note on systems of n-dimensional tense logics. In T. Pauli, editor, *Essays Dedicated to L. Aqvist*, pages 63–71. 1982.
54. Dov M. Gabbay. Intuitionistic basis for non-monotonic logic. In D. W. Loveland, editor, *Proceedings of CADE-6, LNCS, Vol. 138*, pages 260–273. Springer-Verlag, 1982.
Note: This paper started the area now known as the intuitionistic approach to non-monotonicity. It is now a chapter in most monographs on the subject. Another paper on this topic is [62].
55. Dov M. Gabbay and Uwe Reyle. N-Prolog: An extension of prolog with hypothetical implications I. *Journal of Logic Programming*, 1:319–355, 1984.
56. Dov M. Gabbay. Theoretical foundations for non-monotonic reasoning. In K. Apt, editor, *Expert Systems, Logics and Models of Concurrent Systems*, pages 439–459. Springer-Verlag, 1985.
Note: This paper, which has had a strong following, proposes an answer to the question: what is a non-monotonic system? It gives axiomatic conditions on the notion of consequence relation, which characterizes it as a non-monotonic logic. Further papers in this area are [54, 62, 80] and [100]. This paper started the area now known as "Axiomatic" non-monotonic reasoning. Related papers asking similar "what is" questions are [45, 59, 106, 125].
57. Dov M. Gabbay. N-prolog: An extension of prolog with hypothetical implications II, logic foundations, and negation as failure. *Journal of Logic Programming*, 2:251 283, 1985.
Note: This paper is the first in a series in reformulating classical and non-classical logic in a goal directed way. It initiates the program, continued in other papers of re-evaluating the notions of logic and proof theory in the light of applications of logic in Information Technology. Further papers are [55, 59, 60, 103, 67, 94, 112] and summarized in [86].
58. Dov M. Gabbay and Marek J. Sergot. Negation as inconsistency. *Journal of Logic Programming*, 4:1–35, 1986.
59. Dov M. Gabbay. What is negation in a system? In F. R. Drake and J. K. Truss, editors, *Logic Colloquium '86*, pages 95–112. Elsevier Science Publishers (North Holland), 1986.
60. Dov M. Gabbay. Modal and temporal logic programming. In A. Galton, editor, *Temporal Logics and Their Applications*, pages 197–237. Academic Press, 1987.
Note: A basic paper showing what the Horn clause fragment of temporal logic looks like and how to identify such fragments in non-classical logics. Other related papers are [65] and [84].
61. Dov M. Gabbay and Amihud Amir. Preservation of expressive completeness in temporal models. *Information and Computation*, 72:66–83, 1987.
62. Mike Clarke and Dov M. Gabbay. An intuitionistic basis for non-monotonic reasoning. In P. Smets, editor, *Automated Reasoning for Non-standard Logic*, pages 163–179. Academic Press, 1987.
63. Dov M. Gabbay. The declarative past and imperative future. In H. Barringer, editor, *Proceedings of the Colloquium on Temporal Logic and Specifications, LNCS, Vol. 398*, pages 409–448. Springer-Verlag, 1989.
Note: Proposes temporal logic as a framework for handling time phenomena in computing. Shows that temporal logic can serve as a unifying background for the declarative and imperative paradigms in programming. The basic intuition it develops, all backed by mathematical logic, is that future statements can be read both declaratively (as describing what will happen) and imperatively as commands to go ahead and make it happen. A specific temporal logic is proposed, its mathematical properties studied and its range of applicability is demonstrated. Further papers are [65, 64, 82, 69, 71, 78, 79, 84, 87] and [90].
64. Howard Barringer, Dov M. Gabbay, Michael Fisher, Graham Gough, and Richard P. Owens. METATEM: A framework for programming in temporal logic. In *REX Workshop on Stepwise Refinement of Distributed Systems: Models, For-*

malisms, Correctness. Mook, Netherlands. LNCS Vol. 430, pages 94–129. Springer-Verlag, 1989.

65. Dov M. Gabbay. Modal and temporal logic programming II (a temporal logic programming machine). In R. P. Owens T. Dodd and S. Torrance, editors, *Logic Programming - Expanding the Horizon*, pages 82–123. Blackwells, 1990.
66. Dov M. Gabbay and Ian Hodkinson. An axiomatization of the temporal logic with until and since over the real numbers. *Journal of Logic and Computation*, 1:229–260, 1990.
67. Dov M. Gabbay and Frank Kriwaczek. A family of goal directed theorem provers, part I, based on conjunction and implications. *The Journal of Automated Reasoning*, 7:511–536, 1991.
68. Dov M. Gabbay and Anthony Hunter. Making inconsistency respectable, part I. In P. Jorrand and J. Kelemen, editors, *Fundamental of Artificial Intelligence Research (FAIR '91). Lecture Notes in Artificial Intelligence, Vol. 535*, pages 19–32. Springer Verlag, 1991.
69. Dov M. Gabbay, Ian Hodkinson, and Anthony Hunter. Using the temporal logic RDL for design specifications. In A. Yonezawa and T. Ito, editors, *Concurrency: Theory, Language and Architecture, LNCS Vol. 49*, pages 64–78. Springer-Verlag, 1991.
70. Dov M. Gabbay. Modal provability interpretation for negation by failure. In P. Schroeder-Heister, editor, *Extensions of Logic Programming. LNCS Vol. 475*, pages 179–222. Springer-Verlag, 1991.
71. Howard Barringer, Dov M. Gabbay, Michael Fisher, and Anthony Hunter. Meta reasoning in executable temporal logic. In E. Sandewall J. Allen, R. Fikes, editors, *Proceedings of KR'91*, pages 40–49. Morgan Kaufmann, 1991.
72. Dov M. Gabbay, Els Laenens, and Dirk Vermeir. Credulous vs. sceptical semantics for ordered logic programs. In E. Sandewall J. Allen, R. Fikes, editors, *Proceedings of KR'91*, pages 208–217. Morgan Kaufmann, 1991.
73. Dov M. Gabbay. Algorithmic proof with diminishing resources, part I. In E. Boerger, H. K. Buening, M. Richter, and W. Schoenefeld, editors, *Proceedings of Computer Science Logic (CSL '90), LNCS Vol. 533*, pages 156–173. Springer-Verlag, 1991.
74. Dov M. Gabbay. Abduction in labelled deductive systems; a conceptual abstract. In R. Kruse and P. Siegel, editors, *Proceedings of the European Conference on Symbolic and Quantitative Approaches for Uncertainty, 91, LNCS, Vol. 548*, pages 3–12. Springer Verlag, 1991.
75. Jim Cunningham, Dov M. Gabbay, and Hans Jürgen Ohlbach. Towards the MEDLAR framework. In *ESPRIT 91 Conference Proceedings*, pages 822–841, Directorate-General Telecommunications, Information Industries and Innovation, L-2920 Luxembourg, 1991. Comission of the European Communities.
76. Dov M. Gabbay and Ruth Kempson. Natural language content and information flow; a proof theoretic perspective – preliminary report. In P. Dekker and M. Stokhof, editors, *Proceedings of The Eigth Amsterdam Colloquium*, pages 173–196. ILLC, Amsterdam, 1991.
77. A Finkelstein, Dov M. Gabbay, Anthony Hunter, Jeff Kramer, and Bashar Nuseibeh. Inconsistency handling in multi-perspective specifications. In Axel van Lamsweerde and Alfonso Fugetta, editors, *Proceedings of the European Conference on Software Engineering, LNCS Vol. 550*, pages 569–578. Springer Verlag, 1991.
78. Dov M. Gabbay and Richard Owens. Temporal logics for real-time systems. In *Proceedings of the IMACS Symposium on the Modelling and Control of Technological Systems '91*, pages 97–103, 1991.
79. Dov M. Gabbay and Peter McBrien. Temporal logic and historical databases. In *Proceedings of the 17th International Conference on Very Large Databases, Barcelona '91*, pages 423–430. Morgan Kaufmann Publishers, Inc., 1991.
80. Dov M. Gabbay. Theoretical foundations for non-monotonic reasoning part II: Structured non-monotonic theories. In B. Mayoh, editor, *Proceedings of SCAI'91*,

pages 19–40. IOS Press, 1991.

81. Dov M. Gabbay. Temporal logic, tense or non-tense? In R. Spencer-Smith and S. Torrance, editors, *Machinations. Computational Studies of Logic, Language and Cognition*, pages 1–30. Ablex Publishing Co., 1992. Inaugural lecture at Imperial College, 17 May 1988.
82. Dov M. Gabbay and Howard Barringer. The imperative future: Past successes implies future action. A survey position paper. In Y. N. Moschovakis, editor, *Proceedings of the Logic from Computer Science*, pages 1–16. Springer-Verlag, 1992.
83. Dov M. Gabbay, Donald Gillies, Anthony Hunter, Steve Muggleton, Y Ng, and Barry Richards. The rule-based systems project: Using confirmation theory and non-monotonic logics for incremental learning. In S. Muggleton, editor, *Inductive Logic Programming*, pages 213–229. Academic Press, 1992.
84. Dov M. Gabbay. Metalevel features in the object level: Modal and temporal logic programming III. In L. Fariñas del Cerro and M. Penttonen, editors, *Non-classical Logic Programming*, pages 85–124. Oxford University Press, 1992.
85. Dov M. Gabbay and Ruy de Queiroz. Extending the Curry-Howard interpretation to linear, relevant and other resource logics. *Journal of Symbolic Logic*, 57:1319–1366, 1992.
86. Dov M. Gabbay. Elements of algorithmic proof theory. In T. Maibaum S. Abramsky, D. Gabbay, editor, *Handbook of Logic in Theoretical Computer Science, Vol 2*, pages 307–408. Oxford University Press, 1992.
87. Dov M. Gabbay and Marcelo Finger. Adding a temporal dimension to a logic system. *Journal of Logic, Language and Information*, 1:203–234, 1992.
88. Dov M. Gabbay and Hans Jürgen Ohlbach. Quantifier elimination in second-order predicate logic. In B. Nebel, C. Rich, and W. Swartout, editors, *Principles of Knowledge Representation and Reasoning (KR92)*, pages 425–435. Morgan Kaufmann, 1992. Short version of [89].
89. Dov M. Gabbay and Hans Jürgen Ohlbach. Quantifier elimination in second-order predicate logic. *South African Computer Journal*, 7:35–43, July 1992.
Note: This is a seminal paper which is now influential in the AI and the Automated Reasoning community. It provides an algorithm for eliminating second-order quantifiers. It has a wide range of applications especially in the following form: Given two specification languages L_1 and L_2 and some axioms on how they interact, the algorithm can automatically extract the projected specification on each language alone. This is strongly related to interpolation. The research is continued in [91, 101].
90. Dov M. Gabbay and Marcelo Finger. Updating atomic information in labelled database systems. In R. Hull J. Biskup, editors, *ICDT '92. Database Theory. 4th International Conference Berlin, LNCS 646*, pages 188–200. Springer-Verlag, 1992.
91. Dov M. Gabbay and Hans Jürgen Ohlbach. From a Hilbert Calculus to its model theoretic semantics. In K. Broda, editor, *Proceedings of ALPUK Logic Programming Conference, Springer LCS Series*, pages 218–252. Springer-Verlag, 1992.
92. Dov M. Gabbay. Logic made reasonable. *KI (German AI Journal*, 3:39–41, September 1992. In German, translated by Jörg Siekmann.
93. Dov M. Gabbay. How to construct a logic for your application. In H. J. Ohlbach, editor, *GWAI-92: Advances in Artificial Intelligence. Proceedings of German AI Conference, LNAI 671*, pages 1–30. Springer-Verlag, 1992.
94. Dov M. Gabbay and Uwe Reyle. Computation with run time skolemisation. *Journal of Applied Non-classical Logic*, 3:93–134, 1993.
95. Dov M. Gabbay, Ian Hodkinson, and Mark A. Reynolds. Temporal expressive completeness in the presence of gaps. In J. Vaananen and J. Oikkonen, editors, *Proceedings of Logic Colloquium '90. Lecture Notes in Logic, Vol. 2*, pages 89 – 121. Springer-Verlag, 1993.
96. Dov M. Gabbay. Labelled deductive systems: a position paper. In J. Vaananen and J. Oikkonen, editors, *Proceedings of Logic Colloquium '90, Lecture Notes in Logic, Vol. 2*, pages 66 – 88. Springer-Verlag, 1993.

Note: This paper proposes a new logic discipline for unifying the currently used classical and non-classical logical system. Since this paper was presented in Helsinki in 1990, many European researchers and projects are using this framework as a unifying theme. A manuscript of a two volume book exists presenting the results. Volume 1 is now published by Oxford University Press (see [122]) Subsequent papers are [74, 85, 90, 92, 93, 104, 105] and [111].

97. Dov M. Gabbay and Anthony Hunter. Making inconsistency respectable, part II. In S. Seraffín M. Clarke and R. Kruse, editors, *Symbolic and quantitative approaches to reasoning and uncertainty: European Conference ECSQARU '93, Granada, Spain, LNCS Vol. 747*, pages 129–136. Springer-Verlag, 1993.
Note: A first in a series of papers claiming that inconsistency is good and welcome as long as we know what to do with it. It triggers us to action. Other papers are [77, 98] and [110].
98. Dov M. Gabbay and Anthony Hunter. Restricted access logics for inconsistent information. In S. Seraffín M. Clarke and R. Kruse, editors, *Symbolic and quantitative approaches to reasoning and uncertainty: European Conference ECSQARU '93, Granada, Spain, LNCS Vol. 747*, pages 137–144. Springer-Verlag, 1993.
99. Dov M. Gabbay and Valentin B. Shehtman. Undecidability of modal and intermediate first-order logics with two individual variables. *Journal of Symbolic Logic*, 58:800–823, 1993.
100. Dov M. Gabbay. General theory of structured consequence relations. In K. Dosen and P. Schroeder-Heister, editors, *Substructural Logics, in Studies in Logic and Computation*, pages 109–151. Oxford University Press, 1993.
101. Rolf Nossum and Dov M. Gabbay. Semantical correspondence properties of some modal systems of logic. In E. Sandewall and C. Jansson, editors, *Proceedings of Scandinavian Conference on Artificial Intelligence '93*, pages 10–19. IOS Press, 1993. Prize winning paper.
102. Dov Gabbay. Labelled deductive systems and situation theory. In P. Aczel, D. Israel, Y. Katagin, and S. Peters, editors, *Situation Theory and Applications, Vol. 3*, pages 89–118. CSLI, 1993.
103. Uwe Reyle and Dov M. Gabbay. Direct deductive computation on discourse representation structures. *Linguistics and Philosophy*, 17(4):345–390, 1994.
104. Dov M. Gabbay, Ruth Kempson, and Jeremy Pitt. Labelled abduction and relevance reasoning. In R. Demolombe, editor, *Non-standard Queries and non-standard Answers*, pages 155–186. Oxford University Press, Studies in Logic and Computation Series, 1994.
105. Marcello D'Agostino and Dov M. Gabbay. A generalization of analytic deduction via labelled deductive systems, part 1: Basic substructural logics. *Journal of Automated Reasoning*, 13:243–281, 1994.
106. Dov M. Gabbay. What is a logical system. In D. Gabbay, editor, *What is a Logical System*, pages 181–215. Oxford University, 1994.
107. Dov M. Gabbay, Ian Hodkinson, and Mark A. Reynolds. *Temporal logic: mathematical foundations and computational aspects. Vol. 1.*, volume 28 of *Oxford logic guides*. Oxford University Press, Oxford, 1994.
Note: This monograph is the standard reference work in the area.
108. Ben Strulo, Dov M. Gabbay, and Peter Harrison. Temporal logic in a stochastic environment. In A. Szałaz and L. Bloc, editors, *Time and Logic*, pages 229–248. Univ. of London Press, 1994.
109. Dov M. Gabbay. Classical vs. non-classical logic. In D. Gabbay, C. J. Hogger, J. A. Robinson, and J. Siekmann, editors, *Handbook of Logic in Artificial Intelligence and Logic Programming, Vol. 2*, pages 349–489. Oxford University Press, 1994.
110. Dov M. Gabbay, Laura Giordano, Alberto Martelli, and Nicola Olivetti. Conditional logic programming. In P. van Hentenryck, editor, *Logic Programming, Proceedings of the ICLP-94*, pages 272–289. MIT press, 1994.
111. Dov M. Gabbay. Labelled deductive systems and the informal fallacies. In F. H. Van Eemeren *et al*, editors, *Proceedings of 3rd International Conference*

on Argumentation, 1994, pages 308–319. The International Center for the Study of Argumentation, 1994.
112. Klaus Schulz and Dov M. Gabbay. Logic finite automata and constraint logic finite automata. In M. Masuch L. Polos, editor, *Applied Logic: How, What and Why*, pages 237–286. Kluwer, 1995.
113. Dov M. Gabbay and Mark A. Reynolds. Towards a computational treatment of time. In D. Gabbay, C. Hogger and J. A. Alan Robinson, editors, *Handbook of logic in artificial intelligence and logic programming, Vol. 4*, pages 343–428. Oxford University Press, 1995.
114. Chris Brink, Dov M. Gabbay, and Hans Jürgen Ohlbach. Towards automating duality. *Journal of Computers and Mathematics with Applications*, 29(2):73–90, 1995.
115. Dov M. Gabbay. Conditional implications and non-monotonic consequence. In L. Fariñas del Cerro, editor, *Views on Conditional. A volume in Studies in Logic and Computation*, pages 347–369. Oxford University Press, 1995.
116. Dov M. Gabbay, Laura Giordano, Alberto Martelli, and Nicola Olivetti. Hypothetical updates, priority and inconsistency in a logic programming language. In M. Truszczynski V. W. Marek and A. Nerode, editors, *Logic Programming and Non-monotonic Reasoning, LNCS Vol. 928*, pages 203–216. Springer Verlag, 1995.
117. Ruy de Queiroz and Dov M. Gabbay. The functional interpretation of the existential quantifier. *Journal of the IGPL*, 3(2/3):243 290, 1995.
118. Howard Barringer, Michael Fisher, Dov M. Gabbay, Graham Gough, and Richard Owens. METATEM: An introduction. *Formal Aspects of Computing*, 7(5):533–549, 1995.
119. Dov M. Gabbay and Ian Hodkinson. Temporal logic in context of databases. In J. Copeland, editor, *Logic and Reality, Essays on the legacy of Arthur Prior*, pages 69–87. Oxford University Press, 1995.
120. Dov M. Gabbay. Fibred semantics and the weaving of logics,part 2: Fibring non-monotonic logics. In M. de Rijke L. Csirmaz and D. M. Gabbay, editors, *Proceedings of Logic Colloquium 92, SILLI Book Series*, pages 75–94. CSLI, 1995.
121. Dov M. Gabbay. An overview of fibred semantics and the combination of logics. In F. Baader and K. Schulz, editors, *Proceedings of FroCoS'96, Frontiers of Combining Systems*, pages 1–56. Kluwer, 1996.
122. Dov M. Gabbay. *Labelled Deductive Systems; principles and applications. Vol 1: Introduction.* Oxford University Press, 1996.
123. Dov M. Gabbay, Luigia Carlucci Aiello, Fiora Pirri, and Gianni Amati. A proof theoretical approach to default reasoning 1: tableaux for default logic. *Journal of Logic and Computation*, 6(2):205–235, 1996.
124. Derek Brough, Michael Fisher, Anthony Hunter, Richard Owens, Howard Barringer, Dov M. Gabbay, Graham Gough, Ian Hodkinson, Peter McBrien, and Mark A. Reynolds. Languages, meta-languages and METATEM, a discussion paper. *Journal of the IGPL*, 4(2):229–246, March 1996.
125. Dov M. Gabbay and Heinrich Wansing. What is negation in a system, part II. In H. Hans Rott and A. Fuhrman, editors, *Logic, Action and Information*, pages 328–350. de Gryuter, Berlin, 1996.
126. Gianni Amati, Luigia Carlucci Aiello, Dov M. Gabbay, and Fiora Pirri. A structural property on modal frames characterizing default logic. *Journal of the IGPL*, 4(1):7–22, 1996.
127. Dov M. Gabbay, Jon Barwise, and Chris Hartonas. Information flow and the Lambek calculus. In J. Seligman and D. Westerstahl, editors, *Logic, Language, and Computation, Vol 1*, volume 58 of *CSLI Lecture Notes*, pages 47–62. CSLI, 1996.
128. Dov M. Gabbay. Fibred semantics and the weaving of logics, part 1: Modal and intuitionistic logic. *Journal of Symbolic Logic*, 61:1057–1120, 1996.
Note: A revolutionary paper providing methodology for combining systems. Other related papers are [115, 120, 129, 130, 121]. There are several additional papers

forthcoming covering topics such as how to make your logic fuzzy by fibring, fibred semantics for free logic and fibred semantics for systems with self reference.

129. Jochen Dörre, Esther König, and Dov M. Gabbay. Fibred semantics for feature based grammar logic. *Journal of Logic Language and Information*, 5(3-4):387–422, 1996.
130. Marcelo Finger and Dov M. Gabbay. Combining temporal logic systems. *Notre Dame Journal of Formal Logic*, 37(2):204–232, 1996.
131. Marcello D'Agostino and Dov M. Gabbay. Fibred tableaux for multi-implicational logic. In P. Miglioli, U. Moscato, D. Mundici, and M. Ornaghi, editors, *Theorem Proving with Analytic Tableaux and Related Methods*, volume 1071 of *Lecture Notes in Artificial Intelligence*, pages 16–38. Springer Verlag, 1996.
132. Dov M. Gabbay and Odinaldo Rodrigues. A methodology for iterated theory change. In D. M. Gabbay and H. J. Ohlbach, editors, *Practical Reasoning, International Conference on Formal and Applied Practical Reasoning (FAPR'96)*, volume 1085 of *Lecture Notes in Artificial Intelligence*, pages 193–207. Springer Verlag, 1996.
133. Dov M. Gabbay. How to make your logic fuzzy (preliminary version). *Mathware and Soft Computing*, 3(1):5–16, 1996.
134. Dov M. Gabbay and Uwe Reyle. Resolution for classical and non-classical logic. *Studia Logica*, 1996. To appear in a special issue on combining logics.
135. Ruy J. G. B. de Queiroz and Dov M. Gabbay Labelled natural deduction. In H. J. Ohlbach and U. Reyle, editors, *Logic, Language and Reasoning – Essays in Honor of Dov M. Gabbay*, pages 201–281. Kluwer Academic Publishers, 1997.
136. Philippe Besnard, Luis Fariñas del Cerro, Dov M. Gabbay, and Anthony Hunter. Logical handling of default and inconsistent information. In A. Motro and P. Smets, editors, *Uncertainty Management in Information Systems – from Needs to Solutions*, pages 325–341. Kluwer, 1997.
137. Dov M. Gabbay and Ruy J. G. B. de Queiroz. The functional interpretation of modal necessity. In M. de Reyke, editor, *Advances in Intentional Logic*, pages 59–91. Kluwer, 1996.
138. Dov M. Gabbay, Laura Giordano, Alberto Martelli, and Nicola Olivetti. A language for handling hypothetical updates and inconsistency. *Journal of the IGPL*, 4(3):385–416, 1996.
139. Marcello d'Agostino, Dov M. Gabbay, and Alessandra Russo. Grafting modalities into substructural implicational logics. *Studia Logica*, 1997. To appear in a special issue on combining logics.
140. Ruth Kempson, Dov M. Gabbay, Marcelo Finger, and Roger Kibble. The LDN_NL prototype. In R. de. Queiroz, editor, *Proceedings of WOLLIC'96*, 1997. To appear.
141. Marcello d'Agostino, Dov M. Gabbay, and Alessandra Russo. Information frames, implication systems and modalities. *Mathware and Soft Computing*, 1:67–82, 1996.
142. Dov M. Gabbay. *Elementary Logic: A Procedural Perspective.* Prentice Hall, 1997/98.

Editorial Work

143. Dov M. Gabbay and Franz Guenthner, editors. *Handbook of Philosophical Logic, Vol. 1. Elements of Classical Logic*, volume 164 of *Synthese Library*. Kluwer, Dordrecht, 1983.
144. Dov M. Gabbay and Franz Guenthner, editors. *Handbook of Philosophical Logic, Vol. 2: Extensions of Classical Logic*, volume 165 of *Synthese Library*. Kluwer, Dordrecht, 1984.
145. Dov M. Gabbay and Franz Guenthner, editors. *Handbook of Philosophical Logic, Vol. 3: Alternatives to classical logic*, volume 166 of *Synthese Library*. Kluwer, Dordrecht, 1986.
146. Dov M. Gabbay and Franz Guenthner, editors. *Handbook of Philosophical Logic, Vol. 4: Topics in the Philosophy of Language*, volume 167 of *Synthese Library*. Kluwer, Dordrecht, 1989.
147. Dov M. Gabbay and Michael de Glass, editors. *WOCFAI 91, Proceedings of the First International Conference on the Foundations of Artificial Intelligence.* Angkor, 1991.
148. Samson Abramsky, Dov M. Gabbay, and Tom S. E. Maibaum, editors. *Handbook of Logic in Computer Science, Vol 1: Background: Mathematical Structures.* Oxford Univ. Press, Oxford, 1992.
149. Samson Abramsky, Dov M. Gabbay, and Tom S. E. Maibaum, editors. *Handbook of Logic in Computer Science, Vol 2: Background: Computational Structures.* Oxford Univ. Press, Oxford, 1992.
150. László Csirmaz, Dov M. Gabbay, and Maarten de Rijke, editors. *Logic Colloquium 1992.* CSLI Publications, August 1992.
151. Dov M. Gabbay, Chris J. Hogger, and John Alan Robinson, editors. *Handbook of Logic in Artificial Intelligence and Logic Programming, Vol. 1: Logical Foundations.* Oxford Univ. Press, Oxford, 1993.
152. Dov M. Gabbay, C. J. Hogger, and J. A. Robinson, editors. *Handbook of Logic in Artificial Intelligence and Logic Programming, Vol 2: Deduction methodologies.* Clarendon Press, Oxford, 1994.
153. Dov M. Gabbay, Chris J. Hogger, John Alan Robinson, and Jörg Siekmann, editors. *Handbook of Logic in Artificial Intelligence and Logic Programming, Vol 3: Nonmonotonic Reasoning and Uncertain Reasoning.* Clarendon Press, Oxford, 1994.
154. Dov M. Gabbay, editor. *What is a Logical System?* Studies in Logic and Computation. Clarendon Press, Oxford, 1. edition, 1994.
155. Dov M. Gabbay and Hans Jürgen Ohlbach, editors. *Temporal Logic. First International Conference, ICTL'94, Bonn, Germany, July 11-14, 1994. Proceedings*, volume LNAI 827 of *Lecture Notes in Computer Science.* Springer, Berlin, Heidelberg, New York, 1994.
156. Samson Abramsky, Dov M. Gabbay, and Tom S. E. Maibaum, editors. *Handbook of Logic in Computer Science, Vol 3: Semantic Structures.* Oxford Univ. Press, Oxford, 1994.
157. Samson Abramsky, Dov M. Gabbay, and Tom S. E. Maibaum, editors. *Handbook of Logic in Computer Science, Vol 4: Semantic Modelling.* Oxford Univ. Press, Oxford, 1995.
158. Dov M. Gabbay, Chris J. Hogger, and John Alan Robinson, editors. *Handbook of logic in Artificial Intelligence and Logic Programming, Vol 4: Epistemic and Temporal Reasoning.* Clarendon Press, Oxford, 1995.
159. Howard Barringer, Michael Fisher, Dov M. Gabbay, Richard Owens, and Mark A. Reynolds. *The Imperative Future.* RS Press, John Wiley, 1996.
160. Dov M. Gabbay and Hans Jürgen Ohlbach, editors. *Practical Reasoning, International Conference on Formal and Applied Practical Reasoning (FAPR'96)*, volume 1085 of *Lectures Notes in Artificial Intelligence.* Springer Verlag, 1996.
161. Dov M. Gabbay, Chris J. Hogger, and John Alan Robinson, editors. *Handbook of*

Logic in Artificial Intelligence and Logic Programming, Vol 5: Logic Programming. Clarendon Press, Oxford, 1997.

Additional References

162. Dov M. Gabbay and Nicola Olivetti. Algorithmic proof methods and cut elimination for implicational logics: part 1, modal logics, 1995. manuscript.
163. Dov M. Gabbay and Nicola Olivetti. *Goal Directed Algorithmic Proof Theory.* 1995. Draft available.
164. Dov M. Gabbay. Self fibring in predicate logics, fibred semantics and the weaving of logics, part 4, 1996. manuscript.
165. Michael O. Rabin. Decidability of second-order theories and automata on infinite trees. *Transactions of AMS*, 141:1–35, 1969.
166. David Makinson. General patterns in non-monotonic logics. In D. Gabbay, C. Hogger, and A. Robinson, editors, *Handbook of Logic in AI and Logic Programming*, volume 3, pages 35–110. Clarendon Press, 1994.
167. Dov M. Gabbay, Mark Reynolds, and Marcelo Finger. *Temporal Logic, Volume 2.* 1997. in preparation.
168. Klaus Schulz and Franz Baader, editors. *Proceedings of FroCoS'96, Frontiers of Combining Systems.* Kluwer, 1996.
169. Dov M. Gabbay. *Fibring Logics.* Book manuscript, Imperial College.

PROOFS, LABELS AND DYNAMICS IN NATURAL LANGUAGE

JOHAN VAN BENTHEM

1 Encounters with Dov Gabbay

Dov Gabbay is not just a 50-year-old person, his name also denotes a phenomenon. I have felt his and its influence for many years: which are hereby gratefully acknowledged. Two of these influences are especially relevant for what follows. The first is Dov's general view of modal logic as a theory of first-order definable operators over relational models (Gabbay [8]). The second is his work on labelled deduction as a general format for the proof theory of substructural logics with a resource-sensitive slant, be it categorial or dynamic (Gabbay [9]). This generalizes standard type theories, with their binary statements assigning types to terms, or proofs to propositions. The two themes are related. In my view, the following equation sums up much of Dov's recent work:

$$LDS = MGU[ML, TT].$$

That is, labelled deductive systems form a *most general unifier* over two systems: modal logic, with statements $w \models A$ (A is true at world w) and *type theory*, with statements $\tau : A$ (term τ has type A). This marriage reflects some natural tensions that crop up in currently emerging proof-theoretic approaches to natural language. The purpose of my brief contribution is to discuss these matters and raise some logical questions, all in a loose and informal manner. I hasten to add that my sketch of Dov's professional influence is totally incomplete. For instance, we are all living in a Handbook Era for which he is responsible. The brief span of my text prevents me from analysing his deeper motivations here. Will Dov be the Diderot of our times, or rather our Hari Seldon (Asimov [3]) making us write his 'Encyclopedia Galactica' to save the cosmos? In the latter (not unlikely) case, let us proceed just as we please: we shall be temporally executing the Gabbay Plan no matter what.

H.J. Ohlbach and U. Reyle (eds.), Logic, Language and Reasoning, 31–41.

2 Life with Labels

The research program of labelled deductive systems has the following motivation. Binary 'labelled statements' $l : A$ encode richer information than what is usually manipulated in inference, combining logical syntax (A) with explicit display of semantic indices of evaluation or other useful items (l) that do not show up in surface syntax. Thus, the regime of labels removes self-imposed artificial constraints on the expressive power of logical formalisms, allowing us to construct and reason about explicit proofs or semantic verifiers for statements. Most information passed in natural language and natural reasoning is heterogeneous, including both linguistic code and conversational or physical context – so this broader format is well-advised. This point seems convincing, and it can be motivated from many different angles. In epistemic reasoning, there is a clear advantage to keeping explicit reasons for propositions around – in categorial grammar, we need to record various syntactic and semantic resources – and in dynamic logic, we want to manipulate explicit programs moving from preconditions to postconditions. This motivation establishes a format. But it does not commit us to any particular choice of labels, or any particular calculus. Concrete examples are often the above-mentioned model-theoretic statements $w : A$ from modal logic (world w verifies statement A), or proof-theoretic ones $\tau : A$ (τ has type A, τ proves proposition A). These cases raise several logical issues.

First, there are matters of expressive power, best demonstrated with modal logic. Binary statements $w : A$ can be rendered explicitly in a first-order language over possible worlds models, via well-known translations (cf. van Benthem [26]). Thus, they form an intermediate level containing both the modal language and part of the first-order meta-language. This is even clearer with labelled deductive rules, such as

$$y : A \text{ and } Rxy \text{ imply } x : \Diamond A \qquad \text{with side condition } Rxy.$$

In principle, one can remove the labels here, and work with the first-order language directly – which is well-known and perspicuous. The rule then becomes the validity

$$(A(y) \wedge Rxy) \rightarrow \exists z\ (Rxz \wedge A(z)).$$

Sometimes, the 'mixed' labelled language is more revealing, moving back and forth between different components (cf. van Benthem [30] on the dynamics of context). Translation produces a combined theory of states or contexts and their properties, more than what may be found expressed in the surface code of natural language (as advocated in Buvač *et al.* [6] for artificial intelligence), while labelling does this in a piece-meal fashion. Practice must decide what is best for which applications.

The second point concerns options for labelled deduction. Even for the simple minimal modal logic, it is unclear what its unique canonical labelled version should be. One plausible format employs sequents $D \mid B \Rightarrow C$, where B, C consist of labelled modal statements, and D of relational side conditions. Here are two proof rules for the existential modality (modulo some technical variable conditions):

$$\frac{D, Rxy \mid y : A, B \Rightarrow C}{D \mid x : \Diamond A, B \Rightarrow C} \qquad \frac{D \mid B \Rightarrow C, y : A}{D, Rxy \mid B \Rightarrow C, x : \Diamond A} .$$

These derive the above rule, and indeed they are complete for minimal modal logic. But the calculus could be set up in other ways – and even the present one leaves it open which structural rules are to be employed (Kurtonina [16]). We know that the fragment of first-order logic consisting of only translated modal statements can be described completely by a spectrum of calculi, from a full first-order axiomatics to one which has dropped the structural rule of contraction (Andréka, van Benthem & Németi [2]). Naturally, the same options will return for labelled deductive systems.

Finally, the two motivations, modal and type-theoretic, do not suggest the same basic rules of inference for related logical notions. We demonstrate this with one example out of many. Consider combination of information. In type theory, $x : A$ and $y : B$ naturally combine to form the product $x \bullet y : A \wedge B$, whereas in model theory, $x \models A$ and $x \models B$ combine to a conjunction $x \models A \wedge B$. To bring the latter in line with the former, one must rather have a sum operation on information states, taking $x \models A$ and $y \models B$ to $x + y \models A \wedge B$. These notions return with Modus Ponens:

- $x : A \quad y : A \to B \quad \vdash \quad x + y : B$ modal
- $x : A \quad y : A \to B \quad \vdash \quad y(x) : B$ type-theoretic.

The format is the same, but the operations are quite different. Next, we state the matching introduction rules of Conditionalization (under suitable conditions on Δ):

- $\Delta, y : A \vdash x + y : B$ implies $\Delta \vdash x : A \to B$
- $\Delta, y : A \vdash \tau : B$ implies $\Delta \vdash \lambda y \bullet \tau : A \to B$.

The two rules are again different. Can we find a common perspective merging both? This would be a labelled deductive core system unifying its two main sources, solving our earlier meta-equation. We shall return to this issue later. For the moment, we wish to show this is a substantial issue. Competing modal and type-theoretic views crop up in natural language in many places. To see this, we review some recent developments.

3 Proof Theory in Natural Language

Proof theory is coming to the fore as a logical paradigm for natural language analysis. Indeed, Gabbay & Kempson [10] advocate labelled deductive systems as a general paradigm for linguistics. Now, proof theory is about syntactic structure of proofs, their transformations, and theoretical issues such as the existence of normal forms – leading, e.g., to cut elimination theorems. The roots of this field lie in meta-mathematics, where Hilbert's Program inspired formal syntactic analysis of mathematical theories (cf. Smorynski [23]). But, how can the latter be relevant to natural language? One good answer is that proofs encode constructive information about mechanisms of reasoning, closer to what we actually do when using language than classical semantic paradigms, which rather describe more abstract forms of 'correctness'. We briefly review some loci for proofs, with a focus on dynamic aspects of natural language use.

Proof and Grammatical Derivation. The traditional foothold for proof theory in natural language is grammatical analysis. It has often been observed that grammatical derivation is like logical proof ('parsing as deduction'). Notably, Categorial Grammar since Lambek employs substructural calculi of sequents for implicational logic (van Benthem [27], Moortgat [20]). Implicational formulas mirror functional categories, while other logical connectives reflect further type-forming operators. Categorial calculi have the usual logical introduction rules, including two for a left implication $\rightarrow$ seeking its arguments on the left-hand side of the functor position:

$$\frac{X \Rightarrow A \qquad\qquad Y, B, Z \Rightarrow C}{Y, X, A \rightarrow B, Z \Rightarrow C}$$

$$\frac{A, X \Rightarrow B}{X \Rightarrow A \rightarrow B\ .}$$

A clear difference with classical logical deduction is that the usual structural rules are not available (only Reflexivity and Cut remain valid). Categorial deduction treats its premises as sequences, rather than sets, in which order and multiplicity count (Buszkowski [5]). One can even remove the implicit associativity from the above sequent notation, to obtain the basic *Non-Associative Lambek Calculus* (cf. Kandulski [14], Kurtonina [16]), which distinguishes different bracketings as different ways of packaging linguistic information. This minimal system will return in what follows.

Logical proofs also show another face. They also provide *readings* expressible in a typed lambda calculus via the Curry–Howard–deBruyn isomorphism. More precisely, lambda/application terms $\tau_A(..., x_Y, ...)$ of type A encode logical proofs of a conclusion A from initial assumptions Y (given by the free variables x_Y in τ). Grammatically, assumptions stand for types of component words, and the conclusion for the type of the complex expression formed out of them. Here, left introductions are Modus Ponens steps, encoded by function applications $u_{A\to B}(v_A)$, while right introductions are conditionalization steps, encoded by lambda abstractions $\lambda x_A \bullet \sigma_B$. In the intended reading, the term τ_A serves as a constructive semantic recipe which turns denotations for lexical items in type B into denotations for complex linguistic expressions of category A. Categorial proofs are algorithms for computing semantic properties of compound expressions from components. (Proof terms for the Lambek Calculus are a 'linear fragment' of the typed lambda calculus, capturing the above structural restrictions via simple semantic procedures. Cf. van Benthem [25].)

Standard semantics, however, abstracts from proof structure. Standard completeness theorems say that a sequent is *provable* iff it is valid on somc class of models. Indeed, the Lambek Calculus is complete for the following semantics (Orłowska [22], van Benthem [30]). Categorial types may be viewed as *transition relations* over models consisting of abstract states – with atoms A as basic actions R_A, categorial product & as relational composition, and left and right implications $\to, \leftarrow$ as natural left and right inverses of composition. We call a categorial sequent $X_1, ..., X_k \Rightarrow B$ valid if, under any such interpretation, the composition of the relations R_{X_i} is contained in the relation R_C. This semantics has a clear dynamic flavour. It says that executing the input procedures sequentially is guaranteed to achieve the conclusion procedure. Its completeness for Lambek derivability was proved in Andréka & Mikulas [1]. This dynamic semantics provides an intriguing new link between categorial grammar, relational algebra and modal logic. But this very success also raises a question. *How are these two senses of dynamic modelling for natural language syntax related?* The first form of dynamics is type-theoretic, originating in the mechanics of proofs. The second form of dynamics is modal, via state-based models for dynamic logic. Okada [21] studies the parallel via constructivized versions of completeness proofs for categorial calculi. But a truly explanatory joint perspective seems lacking so far.

Proof and Discourse. Categorial grammars combine parts of sentences, a fast autonomous process. But logical proofs also model conscious reasoning at the higher level of text and discourse. This is the sense in which proof theory has been advocated as a paradigm of meaning by prominent philosophers (cf. Sundholm [24]), who replace 'truth' as a logical paradigm by 'justification'. But also linguistically, there are obvious analogies. Formal proof calculi are mathematical models of discourse, whose two-dimensional structure combines local steps (inference rules) with global dynamic mechanisms (notably, dependency on varying assumptions). It would be of interest to extend them to a discourse model capable of dealing with the subtleties of actual discourse particles and argumentative conventions (cf.

van Benthem [28]). The two aggregation levels so far give rise to an obvious question. *How does a proof theory for grammatical analysis relate to that for analysing reasoning and discourse?* In most logical theories, a certain harmony prevails – witness the popular slogan of Propositions-as-Types, which says that one system does double duty for both levels. But natural language may have a more diverse architecture. We may need categorial logic inside sentences, sensitive to occurrences of syntactic resources, whereas our discourse logic is more classical, manipulating mere sets of assumptions. Thus, validity seems more 'construction-driven' (i.e. proof-dependent) at lower levels, and more 'truth-driven' at higher ones. What is the right architecture for natural language? Which logics are involved, and how do they interconnect to pass information? There is evidence for a convergence. Recent dynamic logics of discourse show structural behaviour resembling that of categorial logics (Veltman [32], van Benthem [31], Groeneveld [12]). The above categorial completeness theorem may then be viewed as one mathematical corroboration of such analogies. Nevertheless, no serious account of language processing explains these analogies and gauges their full extent.

Dynamics of Language Use. Natural language is not just a structure, but a set of skills which are central to human cognition. Various logical paradigms bring out this dynamic structure, including game-theoretical ones (Hintikka [13]) as well as newer computationally inspired ones (Groenendijk & Stokhof [11], Muskens, van Benthem & Visser [19]). Now, as we have seen, proof theory provides another paradigm with such a dynamic flavour. Formal proofs encode constructive information that is the 'logical glue' for dynamic composition of sentence meanings, while at a discourse level, they define winning strategies for argumentation games (Lorenzen & Lorenz [17]). In addition, formal proofs exhibit key moves in the dynamics of cognition: creating worlds by assumptions, distinguishing cases, making choices, and so on (cf. Barwise & Etchemendy [4]). This again illustrates the previous question, from a slightly different perspective. The cognitive dynamics of natural language is a fact of life. In particular, categorial proofs and dynamic logics both model this phenomenon. *How are these two dynamic viewpoints related?* (E.g. in van Benthem [27], the two co-exist, but do not meaningfully interact.) The question resembles a better-known one concerning 'constructivism'. How does the proof-theoretic Brouwer–Heyting–Kolmogoroff interpretation of intuitionistic logic *really* relate to its information-style modelling by Kripke and Beth? No truly satisfactory answer to this question exists – despite various completeness theorems showing 'extensional equivalence'.

4 Unifying Proofs and Semantics

Having shown the ubiquity of our concern, we return to labelled deductive systems. What becomes of the earlier issues in this framework? Consider categorial calculi and their dynamic modelling. As with modal logic one can *translate* such systems into first-order logic, transcribing their semantics (van Benthem 1984). Namely,

primitive types A go to first-order formulas $T(A) = R_A xy$;
products go to $T(A \bullet B) = \exists z\ (T(A)(x, z) \wedge T(B)(z, y))$;
left implications go to $T(A \to B) = \forall z\ (T(A)(z, x) \to T(B)(z, y))$.

Validity of a categorial sequent $X_1, \ldots, X_k \Rightarrow B$ is equivalent to first-order validity of the corresponding implication $T(X_1 \bullet \ldots \bullet X_k)(x, y) \to T(B)(x, y)$. In this way, we can analyse categorial validity in a first-order meta-language over transition models. E.g., one can compare the earlier introduction rules for left implication with their first-order counterparts (which involve both implications and universal quantifiers). There is some slack in this translation. The basic Lambek Calculus is decidable – unlike full first-order logic. Thus, as observed before, we are dealing with first-order *fragments*. This shows in the language (translated categorial formulas need only 3 state variables in all) and in the proof calculus needed to drive the above equivalence: a decidable sublogic suffices. Here labelled deduction comes in. We can also analyse categorial reasoning via labelled statements $xy : A$ (cf. Orłowska [22]) and decidable calculi in between the Lambek Calculus and full first-order logic. There is no single system doing this job. Kurtonina [16] discusses this, showing how different labelled calculi may be complete for the same categorial logic. These are just some of many new logical questions concerning labelled deductive systems, enriching the traditional agenda of proof theory. Now, back to our main issue. Can we meaningfully merge *type-theoretic* statements $\tau : A$ and *model-theoretic* ones $w : A$? Consider the following labelled versions of Modus Ponens – coming from standard logic, its earlier modification, relational categorial semantics, and lambda calculus:

$$\begin{array}{llll} x : A & x : A \to B & \vdash & x : B \\ x : A & y : A \to B & \vdash & x + y : B \\ xy : A & yz : A \to B & \vdash & xz : B \\ x : A & y : A \to B & \vdash & y(x) : B. \end{array}$$

The most natural labelled generalization covering all these runs as follows:

$$x : A \qquad y : A \to B \qquad Rz, xy \qquad \vdash \qquad z : B$$
where Rz, xy *is some* ternary *condition relating* z, x, y.

The condition Rz, xy can be '$z = x + y$' (z is the supremum of x and y in some partially ordered Kripke model – if one exists) or 'z is the composition of the arrows x and y' (again, if one exists) or 'z is the result of applying y to x' (if defined). Keeping this analysis in mind, we now analyse the matching introduction rules of Conditionalization. The outcome is that they all exhibit the following format:

$$\begin{array}{llll} \tau : X & x : A & Rz, xy & \vdash z : B \\ \textit{implies} & \tau : X & & \vdash y : A \to B. \end{array}$$

For instance, consider the specific case of lambda abstraction:

$$\begin{array}{lll} \tau : X & x : A & \vdash \sigma : B \\ \multicolumn{3}{l}{\textit{where } x \textit{ does not occur free in the term } \tau, \textit{ implies}} \\ \tau : X & & \vdash \lambda x \bullet \sigma : A \to B. \end{array}$$

This becomes an instance of the above by reading Rz, xy as the true ternary application condition $z = \lambda x \bullet \sigma(x)(= \sigma)$, with $y = \lambda x \bullet \sigma$. In full detail:

$$\begin{array}{lllll} \tau : X & x : A & \vdash \sigma : B & & \textit{is equivalent to} \\ \tau : X & x : A & z = \lambda x \bullet \sigma(x) & \vdash z : B. & \end{array}$$

Thus, the 'most general unifier' that we were looking for turns out to be a ternary transcription of implicational logic, which reads, e.g.

$$A \to B \text{ as } \forall xz\ ((Rz, xy \wedge A(x)) \to B(z)).$$

And this is precisely the semantics of *relevant implication* (cf. the survey chapter Dunn [7]), as developed in great detail in Kurtonina [16]. This ternary semantics validates just the non-associative Lambek Calculus – which is then arguably the basic labelled calculus unifying modal logic and type theory.

Nevertheless, logical questions remain. The ternary relevant semantics is a decidable common ground for lambda calculus and dynamic logic (cf. ternary Arrow Logic: van Benthem [27], Marx & Venema [18]). But then, it has hardly any computational specifics left. What then is the computational surplus of the typed lambda calculus? One way of doing this lets the above schemata specialize to better-behaved concrete ternary relations R, satisfying additional mathematical constraints. For instance, consider the structural rule of Associativity in the Lambek Calculus, which underlies such crucial principles as Geach's Composition Rule. The latter presupposes (Kurtonina [16]) that the relation R be associative in a natural sense appropriate to ternary frames. But the application relation for the typed lambda calculus is not associative in any such sense. Now, the justification for, e.g. the Geach Rule in a typed lambda calculus is somewhat different. Validity on the proof-theoretic reading of sequents says that, given verifiers for the premises, there exists some *construction out of these* verifying the conclusion (as is indeed the case for function composition.) In this sense, stronger frame conditions on R may enrich the logic produced, by producing further witnesses to conclusions than those available in the base calculus. We conclude with another approach. Modal logic seems to have a *binary* semantic format rather than the above ternary one, and so does intuitionistic logic. That is, its rules for implication rather have the following simplified shape:

$x : A \quad y : A \to B \quad Ry, x \ \vdash \ x : B$
where Rxy *is some* binary *condition relating* x, y

$\tau : X \quad x : A \qquad Ry, x \ \vdash \ x : B$
implies $\tau : X \qquad \vdash \qquad y : A \to B.$

Again, the typed lambda calculus should match up, as it also validates intuitionistic implicational logic. How can this be explained? The answer provides a common perspective for the dynamics of proofs and that for information-based Kripke models. We may read the lambda calculus rules as specializing the above schema as follows (using upward heredity of intuitionistic formulas along the information ordering $\subseteq$):

- Ry, x *becomes* $y(x) \subseteq x$ *in the partial order of information extension;*
- *the rule of lambda abstraction involves the premise* $\tau : X \ \ x : A \vdash \sigma : B$
  which implies $\tau : X \ \ x : A \ \sigma \subseteq x \ \ \vdash x : B$
  or equivalently $\tau : X \ \ x : A \ \lambda x \bullet \sigma(x) \subseteq x \ \ \vdash x : B$
  which implies $\tau : X \ \ \vdash \lambda x \bullet \sigma : A \to B.$

So, we have found at least one possible way in which the format of labelled deduction provides a unified dynamics for natural language.

Finally, a question of conscience emerges. At this level of logical generality, is there any bite left to the original claim that natural language can be understood by proof-theoretic paradigms? The most general labelled rule format reads as follows:

from $x_1 : A_1$ *and... and* $x_k : A_k$*, infer* $y : B$*,*
allowing side conditions on all items involved.

Now, this is the format of *unrestricted rewrite rules*, which buys universal power at the price of emptiness. Our answer to the question is this. Proof Theory has proved itself as a powerful logical paradigm for linguistic analysis. But there is indeed a real challenge in understanding just how it works, and what its success means.

University of Amsterdam, The Netherlands.

References

1. Hajnal Andréka and Szabolcs Mikulas. Lambek calculus and its relational semantics: Completeness and incompleteness. *Journal of Logic, Language and Information*, 3:1–37, 1994.
2. Hajnal Andréka, Johan van Benthem, and Istvan Németi. Back and forth between modal logic and classical logic. *Bulletin of the IGPL*, 3:685–720, 1995. Revised version appeared in Journal of Philosophical Logic 27:3, 1998, 217-274.
3. Isaac Asimov. *Foundation Trilogy*. Granada Publishing Company, London, 1960.
4. Jon Barwise and John Etchemendy. *Hyperproof.* CSLI Publications, Stanford, 1994.
5. Wojciech Buszkowski. Mathematical linguistics and proof theory. In J. van Benthem and A. ter Meulen, editors, *Handbook of Logic and Language*. Elsevier Science Publishers, pages 683 - 736, Amsterdam, 1996.
6. Sača Buvač and Richard Fikes. Formalizing context. In *Working Notes for AAAI-5 Fall Symposium Series*. Cambridge (Mass.), 1995.
7. Michael Dunn. Relevance logic and entailment. In D. M. Gabbay and F. Guenthner, editors, *Handbook of Philosophical Logic. Vol. III*, pages 117–224. Reidel, Dordrecht, 1984.
8. Dov M. Gabbay. Expressive functional completeness in tense logic. In U. Mönnich, editor, *Aspects of Philosophical Logic*, pages 91–117. Reidel, Dordrecht, 1981.
9. Dov M. Gabbay. *Labelled Deductive Systems; Principles and Applications, Vol. 1: Basic Principles*. Oxford University Press, 1996.
10. Dov M. Gabbay and Ruth Kempson. Natural-language content: A proof-theoretic perspective. In P. Dekker and M. Stokhof, editors, *Proceedings Eighth Amsterdam Colloquium*, pages 173–196. Department of Philosophy, Univesrity of Amsterdam, 1991.
11. Jeroen Groenendijk and Martin Stokhof. Dynamic predicate logic. *Linguistics and Philosophy*, 4:39–100, 1991.
12. Willem Groeneveld. *Logical Investigations into Dynamic Semantics*. PhD thesis, Institute for Logic, Language and Computation, University of Amsterdam, 1995.
13. Jaakko Hintikka. *Logic, Language Games and Information*. Clarendon Press, Oxford, 1973.
14. Maciej Kandulski. The non-associative Lambek calculus. In W. Marciszewski W. Buszkowski and J. van Benthem, editors, *Categorial Grammar*, pages 141–151. John Benjamin, Amsterdam, 1988.
15. Ruth Kempson, editor. *Deduction and Language, special issue, Bulletin of the Interest group in Pure and Applied Logics, Vol 3:2/3*. Max-Planck-Institut, Saarbrücken, 1995.
16. Natasha Kurtonina. *Frames and Labels. A Modal Analysis of Categorial Deduction*. PhD thesis, Onderzoeksinstituut voor Taal en Spraak, University of Utrecht and Institute for Logic, Language and Computation, University of Amsterdam, 1995.
17. Peter Lorenzen and Karl Lorenz. *Dialogische Logik*. Wissenschaftliche Buchgesellschaft, Darmstadt, 1979.
18. Maarten Marx and Yde Venema. *Many-Dimensional Modal Logic and Arrow Logic*. Oxford University Press, 1996.
19. Reinhard A. Muskens, Johan van Benthem and Albert Visser, *Dynamics*. In J. van Benthem and A. ter Meulen, editors, *Handbook of Logic and Language*. Elsevier Science Publishers, Amsterdam, pp. 587–648, 1997.
20. Michael Moortgat. Type-logical grammars. In J. van Benthem and A. ter Meulen, editors, *Handbook of Logic and Language*. Elsevier Science Publishers, Amsterdam, 1997.
21. Mitsuhiro Okada. A uniform phase-semantic proof of completeness, cut-elimination and strong normalization for polymorphic Lambek calculus. Technical report, Department of Computer Science, Keijo University, Tokyo, 1995.
22. Ewa Orłowska. Relational interpretation of modal logics. In J. Monk, H. Andréka and I. Németi, editors, *Algebraic Logic, Colloq. Math. Soc. J. Bolyai*, pages 443–471.

North-Holland, Amsterdam, 1991.
23. Craig Smorynski. The incompleteness theorems. In *Handbook of Mathematical Logic*, pages 821–865. North-Holland, Amsterdam, 1977.
24. Göran Sundholm. Proof theory and meaning. In D. M. Gabbay and F. Guenthner, editors, *Handbook of Philosophical Logic. Vol. III*, pages 471–506. Reidel, Dordrecht, 1986.
25. Johan van Benthem. The semantics of variety in categorial grammar. Technical Report Report 83-26, Department of Mathematics, Simon Fraser University, Burnaby (B.C.)., 1983.
26. Johan van Benthem. Correspondence theory. In D. M. Gabbay and F. Guenthner, editors, *Handbook of Philosophical Logic. Vol. II*, pages 167–247. Reidel, Dordrecht, 1984.
27. Johan van Benthem. *Language in Action. Categories, Lambdas and Dynamic Logic.* North-Holland, Amsterdam, 1991.
28. Johan van Benthem. Logic and argumentation theory. Technical Report Report X-94-05, Institute for Logic, Language and Computation, University of Amsterdam, 1994. TAppeared in F. van Eemeren, R. Grootendorst and F. Veltman, eds., *Proceedings Academy Colloquium on Logic and Argumentation*, pages 27-41, Royal Dutch Aacdemy of Sciences, Amsterdam.
29. Johan van Benthem. Modal foundations for predicate logic. Technical Report Report ML–95–07, Institute for Logic, Language and Computation, University of Amsterdam, 1995. First version appeared in Bulletin of the IGPL 5:2, 259-286, London and Saarbruecken. Second version to appear in E. Orlowska, ed., Memorial Volume for Elena Rasiowa, Springer Verlag, Berlin.
30. Johan van Benthem. Changing contexts and shifting assertions. In R. van Glabbeek A. Aliseda and D. Westerståhl, editors, *Proceedings 4th CSLI Workshop in Logic, Language and Computation.* CSLI Publications, Stanford, pages 51 - 65, 1996.
31. Johan van Benthem. *Exploring Logical Dynamics.* Studies in Logic, Language and Information, CSLI Publications, Stanford, 1996.
32. Frank Veltman. Defaults in update semantics. Technical Report Report LP–91–02, Institute for Logic, Language and Computation, University of Amsterdam, 1991. Appeared in the *Journal of Philosophical Logic*, 25 (1996), 221 - 261.

WHAT A LINGUIST MIGHT WANT FROM A LOGIC OF MOST AND OTHER GENERALIZED QUANTIFIERS

HANS KAMP

When Dov and I received our logical education – Dov is quite a bit younger than I am, still we got our education at more or less the same time – the overall picture of what logic was, seemed comfortably clear. There were four main branches of mathematical logic – model theory, set theory, recursion theory and proof theory. Underlying this clear and simple picture were a number of widely shared assumptions, some of them to the effect that certain basic problems of logic had essentially been solved. Of central importance among these were: the belief that one had, through the work of Peano, Frege, Peirce, Russell, Hilbert, Gentzen and others, a definitive formal analysis of the notion of *logical deduction* (or *logical proof*); the belief that the conceptual problem of defining *logical consequence* and *logical truth*, and of explicating the relationship between those concepts and the concepts of *truth*, *reference* and *satisfaction* on one hand, and their relationship with the concept of a formal deduction on the other, had found a definitive solution in the work of Gödel and Tarski; and, finally, the conviction that with the characterizations of recursive functions proposed by Gödel, Turing and Church, one had uncovered what had to be the right concept of computability. With regard to set theory the situation was perhaps a little different; then as now, one could not help feeling that each of the available systems of set theory (the most popular ones, Z(ermelo–)F(raenkel) and G(ödel–)B(ernays), among them) embodied an element of arbitrariness. Nevertheless, for better or worse even in this domain a certain consensus had established itself which heavily favoured GB and ZF.

True, the picture wasn't really quite as simple as that. At the fringes hovered logical alternatives such as intuitionistic and other constructive logics; the basic concepts of set theory were challenged by the mereological logics; the spectre of undefinedness had produced, in the course of several decades, a still modest, but steadily growing literature on many-valued, probabilistic and partial logics; and the need for new logical tools for philo-

H.J. Ohlbach and U. Reyle (eds.), Logic, Language and Reasoning, 43–72.

sophical analysis was beginning to give rise to a variety of new logical formalisms and to new and better meta-mathematical foundations for the formalisms already in existence. Decisive in this connection was Kripke's work on the semantics of modal and intuitionistic logic, which more than anything gave the impetus to what has developed into the vast and still growing field of modal logic in its comprehensive sense (encompassing such disciplines as tense logic, deontic logic, counterfactual logic, etc.) and which not only contributed to our conceptual understanding of those systems, but also established the foundations for their mathematical investigation.

Still, there was a strong tendency then to see all these alternatives as marginal. The core of logic remained – in the eyes of most, and certainly in the eyes of almost everyone who seemed to count – the four branches mentioned above; and one feature that those four branches shared was a primary, almost exclusive preoccupation with the new Characteristica Universalis, the predicate calculus – in the first place its first-order fragment, but, to a lesser extent, also parts of higher-order logic, or alternative extensions of first-order logic such as the infinitary logics.

If since that time the picture has changed dramatically, Dov Gabbay certainly has been foremost among those to whom that change is due. Already in the days when modal logic was only beginning to develop into the sophisticated field it has become, he made substantial contributions to it, many of which have become so much part of the logician's intellectual equipment that many who have joined the field in the course of the past three decades and who now make up the clear majority of its active representatives, aren't even aware that they owe these tools and insights to him. Yet emphasizing solely the important work that Dov has done – over so many years – on modal and related logics would seriously understate the influence he has had on our general understanding of what logic is and ought to be, an influence which continues to be as strong as it ever was.

It is important in this connection to note in what ways the general conception of logic has changed, and what have been the forces that have led to that change. As it appears to me, the central difference between the views of logic that are held by many today and the earlier one I sketched in the opening paragraphs, is that in the meantime we have arrived at a much more abstract, and, therewith, a more comprehensive, perception of what logic is about – a much more abstract perspective on what can qualify as a formal analysis of reasoning and what counts as a logical formalism (or 'logical language') suitable for the representation and manipulation of information. Pressure towards such a more liberal perspective has come from many different directions – philosophy, computer science, artificial intelligence, linguistics and (inasmuch as that is a discipline in its own right) computational linguistics. Of course, the strongest personal influence on this change has come from those at home in those neighbouring disciplines as well as in the classical branches of symbolic logic itself, and most of all from those, if any, who were conversant in all these fields at the same time. It needs no comment that such individuals are few and far between. Still, their class is not empty; though it may well be that it equals {Gabbay}.

To the needs coming from those neighbouring disciplines – for formalisms providing novel means of expression, new ways of defining the semantic values of expressions, new ways of defining inference or computing inferences – the logical community has not only responded with a forever expanding panopticum of different logical systems; it has also reacted by rethinking its own credo, and tried to come up with abstract, meta-logical analyses of what the central concepts of logic, those which anything deserving the predicate 'logic' should instantiate, might be. And here again, Dov has played a pivotal role, for instance through his work on the question: what is a relation of logical inference? Or, more recently, through his development of the concept of Labelled Deduction.

Labelled Deduction is perhaps the most promising of a number of current proposals of frameworks in which (hopefully) the entire spectrum of logical alternatives which offer themselves to the bewildered observer today can be compared and helpfully classified, at least when this task is seen from a proof-theoretical perspective. Thus it promises to fill the increasingly felt need for a conceptually sound and accessible map through the labyrinthine landscape of contemporary formal logic, in which the potential customer, whether from philosophy, linguistics or computer science, is in danger of getting lost almost as soon as he makes an attempt to enter.

The present paper pursues by no means so lofty a purpose as this. Rather than concerning itself with the labyrinth of logics as a whole, it looks at one little corner of what is itself only a (somewhat larger) corner of that labyrinth. Still, it seems to me that the lesson which can be gleaned from the little exercise we will go through is applicable to all or most of the larger corner as a whole, and even that it throws some light on the larger question that concerns the relationship between logic and one of its domains of application: the semantics of natural language.

As its title makes clear, the paper is about the quantifier *most*. More generally, it tries to address the question what can and should be expected from a logic of generalized quantifiers. The motivation comes from the semantics of natural language and has an eye not only on the correct logical representation of the quantificational devices that natural languages employ, but also on the computability of those representations and their logical properties. I must add emphatically that from the perspective of mathematical logic the paper offers hardly anything that is really new. All the facts discussed in Sections 1 and 2 – they are presented as lore, and rightly so, for most people who are reasonably familiar with the metamathematics of generalized quantifiers have known about these facts for quite a long time, and any competent logician who hasn't actually seen a proof of them should have little trouble concocting one himself – and most of those of Section 3, can be found explicitly or implicitly in the existing literature. See for instance [10], esp. Section 1.7.

To my knowledge, the paper does not relate in any direct way to Dov's own work. My excuse is that it is surely much harder to find a topic which does not directly relate to any of his work than to find one which does. What better way could there be to pay homage to this œuvre than by finding one of the few logical niches which it has left untouched? But then, probably I have failed anyway and all I am going to say, and more, is hidden somewhere in some paper of Dov's that I have missed.

1 Some Established Views on 'most' and Other Generalized Quantifiers

I regard as uncontroversial that nominal quantification in natural languages such as English has the logical form of what have come to be called Generalized Quantifiers: operators which take a pair of formulas as arguments and return a new formula, while binding a variable.[1] In fact, this is as true of the standard quantifiers *every* and *some* as it is of others (such as *many* or *most*); and it is a simple exercise to develop a version of first-order logic, straightforwardly inter-translatable with its standard versions, in which the universal and existential quantifier are formally treated as generalized (i.e. two-place, not one-place) quantifiers.[2] In a way, in the context of this paper such a version, in which even the standard quantifiers are two-place, would make for greater uniformity. But I believe the presentation will be more perspicuous if first-order predicate logic is kept in the form in which most of us are familiar with it. So I will assume, as 'basis logic', a first-order language L_0 with an infinite set of individual variables $x_1, x_2, x_3, ...$; infinitely many predicate constants $P_1^n, P_2^n, P_3^n, ...$ for each arity n; the connectives $\neg, \wedge, \vee, \rightarrow$ and $\leftrightarrow$, the quantifiers $\forall$ and $\exists$, and the identity $=$. x, y and z are the first three variables x_1, x_2, x_3 and P and Q the first two 1-place predicate constants P_1^1 and P_2^1.

It was one of Frege's insights, which led to the predicate calculus as we now have it, that the universal and existential quantifier *can* be treated as one-place operators. That from the point of view of the grammar of English

[1]Recent work on natural language quantification, especially that of [7] and [8], has shown convincingly that the quantificational possibilities in English and other natural languages go well beyond this – there are expressions that must be analyzed as operators taking more than two formulas as arguments and/or binding more than one variable. Such constructs will play no role in this paper.

[2]To prove the point (if a proof is wanted) see [5], footnote 1.

(or, historically more accurately, German) they rather behave like two-place operators (i.e. as generalized quantifiers) than as the quantificational devices he adopted in his Begriffsschrift, is something of which he was as much aware as anyone. But he noted that for both these quantifiers the contributions made by the two arguments can be contracted into one – by forming material conditionals in the one case and conjunctions in the other; and, for reasons we need not go into here, these are the devices that have remained with us ever since.

It has long been part of the general lore surrounding natural language semantics that *every* and *some* are quite special in this respect. In general such a Boolean reduction of a two-place to a one-place quantifier is not possible. I have called this part of semantic lore since it is a conviction that many take for granted even though it is not established by actual proof. The principal reason for this is that a proof presupposes a well-defined semantics for the quantifier that is to be shown irreducible, and such a semantics is rarely available. A notorious exception – perhaps one should say: *the* notorious exception – is the quantifier *most*.

There is a fairly general consensus that 'Most As are Bs' is true provided the cardinality of the set of As that are Bs exceeds that of the remaining As, or at least that this is so, provided the number of As is finite. Since these two conditions will play a central part in the paper, let us give them a label right away:

(MOST) 'Most As are Bs' is true iff $|A \cap B| > |A \setminus B|$

(MOST^{FIN}) If A is finite, then
'Most As are Bs' is true iff $|A \cap B| > |A \setminus B|$.

This second, weaker assumption suffices to show that *most* is not reducible to a 1-place operator – or, to put it differently, we can show the slightly stronger result that such a reduction isn't possible within the Theory of Finite Models. More precisely we can show Fact 1.

Fact 1 There is no combination of (i) a function F from finite sets U to sets of subsets of U and (ii) a first-order formula $\Phi(P, Q; x)$, built up from the predicate constants P, Q, variables and logical constants, in which at most x occurs free, such that for every finite first-order model $M = \langle U, I \rangle$:

$$|I(P) \cap I(Q)| > |I(P) \setminus I(Q)| \text{ iff } \{u \in U : M \models \Phi(P, Q; x)[u]\} \in F(U).$$

To see that Fact 1 says what it ought to, first observe that a one-place variable binding operator O turns, when it binds, say, the variable x, a formula that has only x free into a sentence. Semantically this means that O maps the satisfaction set of any such argument formula to a truth value. More specifically, if M is any model, the interpretation O_M of O in M must be a function that maps for any such formula ϕ the set of individuals of M which satisfy ϕ in M to one of 0 and 1. Thus O_M must be (the characteristic function of) a set of such satisfaction sets. If we make the additional (highly plausible and generally endorsed) assumption that O_M ought not to depend on the interpretation of any non-logical constant in M and thus that it depends exclusively on the universe U of M, it follows that the meaning of O can be given as a function F from sets U to sets of subsets of U. The interpretation of O in any model M will then be the value $F(U_M)$ which F assigns to the universe of M.

Second, a reduction of *most* of the kind in question will involve a way of combining its argument formulas $A(x)$ and $B(x)$ into a single compound formula $\Phi(A(x), B(x))$ such that the generalized quantifier relation $MOST$ holds between the satisfaction sets of A and B if and only if the satisfaction set of $\Phi(A(x), B(x))$ belongs to the interpretation of the operator O. This will have to be so in particular in cases where the arguments of *most* are the formulas $P(x)$ and $Q(x)$ and in models M in which all non-logical constants other than P and Q are given a trivial interpretation (e.g. every n-place predicate is interpreted as the empty n-place relation.) In such cases $\Phi(A(x), B(x))$ reduces to a formula $\Phi(P, Q; x)$ of the sort mentioned in the statement of Fact 1. Thus Fact 1 entails the irreducibility of *most.*

N.B. the statement made by Fact 1 goes beyond what I promised insofar as the formula $\Phi(P, Q; x)$ may involve (standard first-order) quantification as well as Boolean connectives. In this regard the result is more general than a strict analogue to the reducibility of the generalized quantifiers *every* and *some*, where the combination of the two argument formulas requires only the sentential connectives $\rightarrow$ and $\wedge$, respectively.

The proof of Fact 1 rests on long known facts about monadic first-order logic and would hardly be worth a looking into if it didn't provide some insight into the question what is likely to be needed to obtain similar irreducibility results for other quantifiers than *most.* It is with this purpose in mind that I will take a little time to remind the reader of how the argument might go.[3]

[3]As matters have turned out, no further use of the proof is made in the present paper. However, in more comprehensive joint work with Tim Fernando, in which we investigate other non-standard quantifiers besides *most* and *more*, we intend to exploit this possibility. (See [2].) In retrospect, and thanks to critical remarks by Johan van Benthem, I now feel that this first section should have been written quite differently, and that a much more compact presentation would have served the purpose better.
Another defect of the section is that it does not relate the notions of definability and reducibility for quantifiers sufficiently to those that can be found in the literature on this subject. So to those familiar with this literature the section will appear rather amateurish. And for anyone familiar with the standard techniques for proving results in this domain – such as, in particular, those using Ehrenfeucht games or the notion of partial isomorphism – the old-fashioned, 'syntactic' kind of argumentation I have used will undoubtedly reinforce that impression. This is another reason why the section should have been rewritten. But unfortunately, time prevented me from doing the necessary replace-

Proof of Fact 1. (Sketch) I will state, in a form convenient to the present purpose, the facts about monadic logic which we will need. As said, these facts are standard; they, or something very much like them, is involved in familiar proofs that monadic logic has the finite model property; and they can be established by a well-known quantifier elimination argument. Let $L(P, Q)$ be the language of first-order logic with identity whose only non-logical constants are P and Q. There are sentences of $L(P, Q)$ which express the following properties of models $M = \langle U, I \rangle$ for $L(P, Q)$:

1. For $n \geq 1$ and natural numbers $m(P, Q)$, $m(P, \neg Q)$, $m(\neg P, Q)$ such that $(m(P, Q) + m(P, \neg Q) + m(\neg P, Q)) \leq n$ the proposition that
   (a) $|U| = n$,
   (b) the number of individuals in M satisfying both P and Q is $m(P, Q)$
   (c) the number of individuals satisfying P but not Q is $m(P, \neg Q)$, and
   (d) the number of individuals satisfying Q but not P is $m(\neg P, Q)$.

   (We will refer to these sentences as $\Psi_{n;m(P,Q),m(P,\neg Q),m(\neg P,Q)}$.)
2. For $n \geq 1$ and natural numbers $m(P, Q)$, $m(P, \neg Q)$, $m(\neg P, Q)$, $m(\neg P, \neg Q) < n$, and such that $(m(P, Q) + m(P, \neg Q) + m(\neg P, Q) + m(\neg P, \neg Q)) \geq n$ the proposition that
   (a) $|U| > n$,
   (b) the number of individuals in M satisfying both P and Q is $m(P, Q)$
   (c) the number of individuals satisfying P but not Q is $m(P, \neg Q)$, and
   (d) the number of individuals satisfying Q but not P is $m(\neg P, Q)$.
   (e) the number of individuals satisfying neither P nor Q is $m(\neg P, \neg Q)$.

   (We will refer to these sentences as $\Psi_{>n;m(P,Q),m(P,\neg Q),m(\neg P,Q),m(\neg P,\neg Q)}$.)
3. For $n \geq 1$ and natural numbers $m(P, Q)$, $m(P, \neg Q)$, $m(\neg P, Q) \leq n$ the proposition that
   (a) $|U| > n$,
   (b) the number of individuals that are P and Q, the number of those that are P but not Q and the number of those that are Q but not P are $m(P, Q)$, $m(P, \neg Q)$, $m(\neg P, Q)$, respectively, and
   (c) the number of elements that are neither P nor Q is $> n$;
   this sentence is denoted as $\Psi_{>n;m(P,Q),m(P,\neg Q),m(\neg P,Q)}$; analogously there are sentences $\Psi_{>n;m(P,Q),m(P,\neg Q),m(\neg P,\neg Q)}$; $\Psi_{>n;m(P,Q),m(\neg P,Q),m(\neg P,\neg Q)}$; $\Psi_{>n;m(P,\neg Q),m(\neg P,Q),m(\neg P,\neg Q)}$;

ment job in the way in which it should be done. A humble request to the cognoscenti: Please skip this section!

the first of these says that $|U| > n$, that there are $m(P,Q)$ elements that are P and Q, $m(P,\neg Q)$ that are P but not Q and $m(\neg P,\neg Q)$ that are neither P nor Q, while the number of elements that are Q but not P is $> n$; similarly for the other three.

4. In analogy with the sentences mentioned under 3, there are those which say of two of the four categories that there are $\leq n$ individuals of that category and say exactly how many there are, while of the remaining two categories there are $> n$ – these sentences are denoted as $\Psi_{>n;m(P,Q),m(P,\neg Q)}$, $\Psi_{>n;m(P,Q),m(\neg P,Q)}$ etc. – and there are sentences $\Psi_{>n;m(P,Q)}$; $\Psi_{>n;m(P,\neg Q)}$; $\Psi_{>n;m(\neg P,Q)}$; $\Psi_{>n;m(\neg P,\neg Q)}$, saying of just one category that there is some particular number $m \leq n$ of elements of that category, whereas for each of the other three there are more than n; and finally there is a sentence $\Psi_{>n;>}$ which says that there are more than n elements of each of the four categories.
5. Corresponding to each of the sentences $\Psi_{n;m(P,Q),m(P,\neg Q),m(\neg P,Q)}$ for which $(m(P,Q)+m(P,\neg Q)+m(\neg P,Q)) < n$ there are four $L(P,Q)$ formulae with x as only free variable, to which we will refer as $\Psi_{n;m(P,Q),m(P,\neg Q),m(\neg P,Q)}(P,Q;x)$, $\Psi_{n;m(P,Q),m(P,\neg Q),m(\neg P,Q)}(P,\neg Q;x)$, $\Psi_{n;m(P,Q),m(P,\neg Q),m(\neg P,Q)}(\neg P,Q;x)$ and $\Psi_{n;m(P,Q),m(P,\neg Q),m(\neg P,Q)}(\neg P,\neg Q;x)$. $\Psi_{n;m(P,Q),m(P,\neg Q),m(\neg P,Q)}(P,Q;x)$ is satisfied by $u \in U$ if $|U| = n$, there are $m(P,Q)$ individuals *other than* u which are both P and Q, $m(P,\neg Q)$ individuals *other than* u which are P but not Q, $m(\neg P,Q)$ individuals *other than* u which are Q but not P, while moreover u is a P as well as a Q; $\Psi_{n;m(P,Q),m(P,\neg Q),m(\neg P,Q)}(P,\neg Q;x)$ is satisfied by u if the same conditions obtain except that u is a P but not a Q; and similarly for the remaining two formulas.
6. Similarly there are four formulas with free x for each of the sentences described in 2, 3 and 4. (Thus, to take just one example, there is a formula $\Psi_{>n;>}(P,Q;x)$ which is satisfied by u iff there are more than n individuals, there are more than n individuals different from u which are both P and Q,..., and u itself is both P and Q.)
7. For each formula $\phi(x)$ of $L(P,Q)$ in which only x occurs free there is a number n_ϕ such that $\phi(x)$ is logically equivalent to a disjunction of formulas of the types described under 5 with $n \leq n_\phi$ and those in 6 with $n = n_\phi$.

7 gives us the result we are trying to establish (i.e. Fact 1) fairly straightforwardly. For suppose there was a formula $\Phi(P,Q;x)$ as described in the statement of Fact 1. Then there would be a number n_Φ as described in 7 such that $\Phi(P,Q;x)$ is equivalent to a disjunction D of the indicated kind. Now consider any model $M = \langle U,I\rangle$ such that $|U| = 8 \times n_\Phi$ and in which there are more than $n_\Phi + 1$ individuals that are both P and Q, more than $n_\Phi + 1$ individuals which are P but not Q, etc. It is clear that the set D_M

of those $u \in U$ which satisfy D in M will consist of the union over some subset (possibly empty) of the following four disjoint sets: (i) the set of individuals that are both P and Q in M, (ii) the set of those that are P but not Q, and (iii, iv) similarly for the other two combinations, Q but not P and neither P nor Q. Whether or not the first of these sets is part of D_M depends on whether D contains as one of its disjuncts the formula $\Psi_{>n;>}(P, Q; x)$. For any other possible disjunct of D will fail to be satisfied by a u that is both a P and a Q in M either because what it says about the size of U or else because of what it says about the number of individuals that are P and Q, P but not Q, Q but not P, or neither P nor Q; or, finally, because it requires u to be not a P or not a Q. Similarly, the second set is part of X_1 iff D contains the disjunct $\Psi_{>n;>}(P, \neg Q; x)$ and likewise for the remaining two. This gives us a small, finite number of possibilities for D_M: the empty set $\{\}$, the set of u which are both P and Q, the set of u which are P but not Q, the union of these two sets, i.e. the set of u which are P, etc. with as largest possibility the set U itself. It is tedious, but not hard, to construct for each of those possibilities a pair of models $M_1 = \langle U, I_1 \rangle$ and $M_2 = \langle U, I_2 \rangle$ which satisfy the above conditions for M and which are such that

1. according to our adequacy criteria (MOST^{FIN}) for MOST most Ps are Qs in M_1 but not in M_2, and
2. $D_{M_1} = D_{M_2}$.

We will consider just two cases, that where D_M is the set $I(P) \cap I(Q)$ and that where it is $(I(P) \setminus I(Q)) \cup (I(Q) \setminus I(P))$. In the first case let $I_1(P) \cap I_1(Q) = I_2(P) \cap I_2(Q) = I_1(P)$ be a subset of U of $n_\Phi + 2$ elements and let $I_2(P)$ be U. Then evidently both 1 and 2 are satisfied. For the second case let M_2 be as in the preceding case and let M_1 be like M_2 but with the interpretations of P and Q reversed. Since in the present case D_M is symmetric in P and Q, 2 is satisfied again. Moreover, it should be clear that most Ps are Qs in M_1, as there the Ps are included in the Qs while, as in the first case, only a minority of the Ps are Qs in M_2. ■

The reader will no doubt agree that this proof is every bit as unenchanting as I promised it would be. The point of presenting it nevertheless is, as I said before embarking upon it, that very similar arguments may well be usable to show the irreducibility of other quantifiers, such as, say, *many*, and that this may require comparatively weak assumptions about the semantics of such a quantifier. For instance, it would be enough to assume that (if necessary, only under certain conditions, provided these are compatible with the set of As and the set of Bs being of arbitrarily large finite size) the truth of 'many As are Bs' requires that some suitable proportion of the As are Bs.)

There is a second point to be made, this one not so much about the proof of Fact 1, but rather about what the Fact asserts. What it asserts is *not* that the quantifier MOST is not first-order definable. By the first-order definability of a generalized quantifier we mean the following. First, by a *generalized quantifier relation* we understand a function from sets A to sets of pairs of subsets of A. (Generalized quantifier relations are the kinds of objects that are to serve as meanings of binary generalized quantifiers. The motivation for the definition is the same as the one given above for the meaning of a one-place quantifier as a function from sets to sets of subsets of those sets.) Suppose R is such a relation and that $\Psi_R(P, Q)$ is a sentence of $L(P, Q)$. Then we say that R is *first-order defined by* $\Psi_R(P, Q)$ iff for any model $M = \langle U, I \rangle$ for $L(P, Q)$:

$$\langle I(P), I(Q) \rangle \in R(U) \text{ iff } M \models \Psi_R(P, Q); \tag{1}$$

and R is said to be *first-order definable* iff there exists such a sentence. Similarly, R is said to be *first-order defined by* $\Psi_R(P, Q)$ *in the Theory of Finite Models* iff (1) holds for all finite models.

The point of these notions should be clear: if a generalized quantifier Qu has as its meaning a generalized quantifier relation R which is first-order defined by a formula $\Psi_R(P, Q)$, then any sentence δ containing occurrences of Qu will be equivalent to a sentence η in which Qu does not occur; η is obtained by replacing, going from the inside out, every subformula $Qu_v(\phi, \psi)$ of δ by a formula $\Psi'_R(\phi, \psi)$ which we get by (a) taking an alphabetic variant $\Psi'_R(P, Q)$ of $\Psi_R(P, Q)$ such that the variables of $\Psi'_R(P, Q)$ are disjoint from the free variables of $Quv(\phi, \psi)$ and (b) replacing in $\Psi'_R(P, Q)$ every subformula $P(w)$ by $\phi(w/v)$ and every subformula $Q(w)$ by $\psi(w/v)$.

First order definability is clearly a different concept from the notion of reducibility which was used in Fact 1, and which in general terms can be characterized as follows:

A generalized quantifier R is *reduced to* a one-place quantifier meaning F (i.e. a function from sets U to sets of subsets of U) *by* a formula $\Psi_R(P, Q; x)$ iff for each model $M = \langle U, I \rangle$ for $L(P, Q)$

$$\langle I(P), I(Q) \rangle \in R(U) \text{ iff } \{u \in U : M \models \Psi_R(P, Q; x)[u]\} \in F(U). \tag{2}$$

Again, we say that R *is reduced* to a one-place operator *in the Theory of Finite Models* iff the above condition holds for all finite models for $L(P, Q)$.

It is easy to see that first-order definability entails reducibility to a one-place operator. For suppose that R is first-order definable by $\Psi_R(P, Q)$. Then the formula $x = x \wedge \Psi_R(P, Q)$ will (trivially) reduce R to the one-place operator which maps each set U onto $\{U\}$. Of course, the converse entailment does not hold: there are uncountably many one-place quantifiers F which are not first-order definable, in the sense that there is no sentence

$\Psi_F(P)$ of the language $L(P)$ such that for all $M = \langle U, I\rangle$, $I(P) \in F(U)$ iff $M \models \Psi_F(P)$. For each such quantifier F we can make up any number of 2-place quantifiers reducible to it; consider for instance the generalized quantifier relation R_F defined by the condition that for any set U and subsets A, B of U, $\langle A, B\rangle \in R_F(U)$ iff $A \in F(U)$. This relation is reduced to F by the formula $P(x)$. And it is easy to see that any first-order definition for R_F would yield a first-order definition for F in the sense just given. For suppose that $\Psi_R(P, Q)$ were a first-order definition of R_F. Then the sentence $\Psi'_R(P)$ (= '$\Psi_R(P, \top)$'), which we obtain by replacing in $\Psi_R(P, Q)$ each subformula $Q(v)$ by the corresponding formula $v = v$, would be a first-order definition of F. Thus R_F cannot be first-order definable.

2 Another Piece of 'most'-lore: Non-axiomatizability

The next bit of lore about *most* I must mention is that adding it to first-order logic leads to non-axiomatizability.[4] What is meant is this. Suppose we extend our first-order language L with a generalized quantifier symbol Mo, subject to the syntactic rule that

> if ϕ and ψ are formulas of the new language $L(Mo)$ and v is any variable, then $Mo_v(\phi, \psi)$ is a formula; (3)

and the accompanying semantic principle that for any model $M = \langle U, I\rangle$

$$M \models_a Mo_v(\phi, \psi) \text{ iff } \langle \{u \in U : M \models_{a[u/v]} \phi\}, \{u \in U : M \models_{a[u/v]} \phi\}\rangle \in \text{MOST}(U), \quad (4)$$

where MOST is the binary generalized quantifier we choose to interpret Mo. Together with the familiar clauses of the truth definition for first-order logic (4) provides us with the usual characterizations of logical consequence (as preservation of truth in all models) and of logical truth (as truth in all models). Then, as lore has it, neither the consequence relation nor the set of logical truths of the resulting language $L(Mo)$ is recursively enumerable.

Whether the claim is true depends of course on exactly what the generalized quantifier MOST is taken to be; and here for the first time the distinction between the strong version (MOST) and the weak version (MOST^{FIN}) of our intuitive meaning constraint for the quantifier *most* becomes important. For it is only when we adopt the strong version that the claim holds

[4] Proofs of this fact seem to be ten to the gallon and have been around for (probably) at least two decades. For instance, a slightly different demonstration can be found in [10], leading to a more informative result than will be given here – but one which is for our present aims is not needed in its full strength.

true. This constraint fixes the generalized quantifier relation MOST completely. For now and later reference we repeat the definition:

Definition MOST is the function which maps each set U onto the set of all pairs $\langle V, W\rangle$ such that $V, W \subseteq U$ and $|V \cap W| > |V \setminus W|$.

We state the strongest part of the claim just made, the non-recursive enumerability of the set of logical truths, as Fact 2:

Fact 2 Let $L(Mo)$ be the language defined above, through the clauses (3) and (4). Then the set of all logical truths of $L(Mo)$ is not recursively enumerable.

Proof. Here is a simple proof of this fact. Let L_{Ar} be a sublanguage of L suitable for the formulation of arithmetic (I assume that the operations of successor, plus and times are represented by corresponding predicates) and let T_{Ar} be some finite axiomatization of first-order arithmetic strong enough to yield Gödel's incompleteness theorem and to prove that every model has an initial segment isomorphic to the standard model of arithmetic. Suppose we add the quantifier Mo to L_{Ar}, thus obtaining the language $L_{Ar}(Mo)$ and extend T_{Ar} with a single axiom of the following form (modulo some straightforward definitions)

$$(\forall y)(Mo_x(x \leq y, (\exists z)(x = z + z)) \vee Mo_x(x \leq y + 1, (\exists z)(x = z + z))). \quad (5)$$

Given (4) and our identification of MOST, (5) says that for any number y (finite or transfinite) either the cardinality of the even numbers $\leq y$ exceeds that of the set of the remaining numbers $\leq y$, or else the cardinality of the even numbers $\leq y + 1$ exceeds that of the set of the remaining numbers $\leq y + 1$. It is clear that this condition is satisfied for every finite number y (the first disjunct is true when y is even, the second when y is odd) but that it fails for any transfinite number (for then the sets that are being compared are all denumerably infinite and thus of the same cardinality). Thus the only model of the theory T_{Ar} + (5) (up to isomorphism) is the standard model of arithmetic. But then, if Ψ is the conjunction of the axioms of T_{Ar} + (5), we have that for any sentence ϕ of L_{Ar} ϕ is true in the standard model of arithmetic iff the sentence $\Psi \rightarrow \phi$ is a logical truth of L_{Ar}. ∎

It is important to note that this proof depends crucially on the assumption that the semantics for Mo satisfies the condition (MOST) of the preceding section also for infinite sets A and not only for finite ones. Indeed, we will see in the next section that if we weaken the assumptions of Fact 2 in that we replace (MOST) by (MOST^{FIN}) the assertion it makes is no longer true.

3 An Axiomatizable Logic for 'most'

When reflecting on the implications of Fact 2, we do well to ask once more what and how good is the intuitive justification for conditions such as (MOST) and (MOST^{FIN}). In Section 1 I ventured the observation that there is a firmer consensus concerning (MOST^{FIN}) than there is concerning the more comprehensive condition (MOST). Perhaps this claim is more a reflection of my own preferences than the true description of an actual distribution of opinion. In any case, I have my preferences and this is the place to try and account for them.

It seems to me that when the set A is finite, counting the set of As that are Bs and the set of As that are not Bs and finding there are more things in the first set than there are in the second amounts to a conclusive demonstration that most As are Bs. This is connected with the circumstance that counting a set seems to be *the* criterion for determining its size as long as the set is finite – an intuition that is reflected in the set-theoretic fact that for the finite sets the concepts of cardinal and of ordinal coincide. For infinite sets, in contrast, there is no clear pretheoretic conception of how their size should be assessed, and it seems that precisely for this reason our intuitions about when sentences of the form 'Most As are Bs' are true become uncertain too. The concept of cardinality as a measure of set size was a profound discovery when it was made and since then it has become central to the ways in which we deal with the infinite in mathematics. But cardinality remains a term of art, which has no more than a tenuous connection with the intuitions of the ordinary speakers of natural languages.

As far as those intuitions are concerned, it seems rather that when infinite sets come into play, the concept of 'majority' that one fastens on to form a judgement about the truth or falsity of a *most*-sentence varies with context, and may take factors into account that fall outside the conception of generalized quantifier meaning which has guided us so far. The stock examples

a. Most natural numbers are prime. (6)
b. Most natural numbers are not prime.

remain good illustrations of the point at issue. The tendency to judge the first sentence as false and the second as true – or at any rate, to find it much more plausible that the second should be true and the first one false than the other way round – surely reflects our inclination to think of the *rates* with which we are likely to encounter prime or non-prime numbers when going through the numbers in some special order (e.g. going up the standard ordering) or, alternatively, at random. Indeed, there exists a cluster of number-theoretic theorems which confirm these intuitions: for a vast family of ways to sample the numbers in some order the rate with which one encounters non-primes tends towards 100% while the rate with which one encounters primes tends to 0%.

What morals is the natural language semanticist to draw from these considerations? I do not know of any consensus on this point. But let me put forward my own assessment. First, a realistic semantics should respect speakers' intuitions as much as possible, and this should include cases where speakers' intuitions are unstable or simply missing; in these cases semantic theory should withhold judgement too, or it should try to identify the different conflicting strains of conceptualization that are responsible for the instability. For the case at hand – *most* applied to infinite sets – these recommendations should, I reckon, come to something like this:

(a) Eventually, the different conceptual elements that typically enter into speakers' judgements about sentences such as (6.a) and (6.b) and the ways in which they shape those judgements will have to be identified. This will evidently lead to an analysis of *most* according to which its meaning is something other (and more complicated) than the generalized quantifier relations considered hitherto. As far as I know, this is a research topic on which some work has been done (see the remarks on Colban below), but where there is much to be done still. It is a topic, however, which will not be explored here.

(b) Short of engaging in the kind of investigation advocated under (a), a semantics of *most* should remain agnostic in those cases where speakers' judgements depend on factors which are outside of the conceptual apparatus provided by quantifier meanings in the narrow sense. For a model-theoretic analysis this may have two different implications. First, that of a partial model theory in which sentences need not get a definite truth value in every model. (In particular sentences of the form 'most As are Bs' may fail to be either true or false in models where the number of individuals satisfying A is infinite.) Alternatively, one may adopt a model theory in which every model determines a truth value for all sentences, but where, intuitively speaking, several non-equivalent models may correspond to one and the same possible state of affairs, *viz.* by providing different interpretations for the generalized quantifier. (Intuitively: whenever the judgement about truth or falsity of a *most*-sentence with respect to a given state of affairs depends on such factors, some of the models compatible with that state of affairs may assign the sentence the value true while other such models assign it the value false.)[5]

[5]The difference between these two options – a partial model theory or a non-partial model theory which allows for different models corresponding to a single state of affairs – need not be all that different from each other in the end. This is one of the main lessons

These recommendations seem to me to be in the spirit of a paper by Colban [1], which has played an important part in shaping the thoughts on which the present contribution reports. When compared with the way we have been approaching the problems posed by *most*, Colban's approach could be said to start at the opposite end. Rather than trying to determine of some particular natural language quantifier, such as *most*, exactly what its meaning is and then investigating the logic that is generated by the meaning one has fastened upon, Colban begins by having a look at so-called *weak logic*, the logic for the extension $L(Qu)$ with one new binary quantifier symbol that is generated by the class of all models M for $L(Qu)$ in which the new quantifier is interpreted by any relation between subsets of U_M whatever. (In other words, this is the logic of the concept of a generalized quantifier in its full generality, in which properties that differentiate between such quantifiers are entirely ignored. The idea of weak logic appears to be quite old; one finds it for instance already in one of the mile stones in the history of generalized quantifier theory, Keisler [9], except that Keisler is concerned with a one-place quantifier – 'there are uncountably many' – rather than with the two-place quantifiers considered here and in Colban's work; a discussion of the weak logic of binary quantifiers can also be found in Appendix B of [10].) Once an axiomatization for weak logic is in place, one can then proceed, as Westerstahl and Colban do, to impose conditions on the admissible quantifier meanings and extend the axiomatization of weak logic accordingly. Those interested in the logic of some particular quantifier, semantically given by some particular generalized quantifier relation R, might wish to use this strategy to whittle down the class of permitted quantifier relations step by step until one reaches the singleton class consisting solely of R. But of course, one should be prepared for the contingency that this is too much to hope for: perhaps that no matter how the strategy is applied the resulting class will always contain some relations besides R.

However, in the light of our reflections earlier in this section reducing the class to a singleton set may not be the right goal anyway. In particular, I suggested, the best account of *most* as a generalized quantifier might well be one that admits a variety of quantifier relations, which may yield incompatible predictions about the truth of certain *most*-sentences concerned with infinite sets, while harmonizing in their predictions about sentences speaking of finite sets. Indeed, it is just such an account which I shall present here.

As a basis for our further explanations we need an axiomatization of weak logic for the language $L(Mo)$ (where Mo is, as before, a binary quan-

---

of the so-called supervaluation approach to problems of semantic underspecification. See e.g. [3], [4] or [6].

tifier symbol).[6] As can be shown by a largely standard Henkin argument, addition of the universal closures of all instances of the following schemata to a complete axiomatization of first-order logic (with the rules of Modus Ponens and Universal Generalization) is complete for this logic:

WQL.1 $(\forall v_i)(\phi \leftrightarrow \psi) \rightarrow (Mo_{v_i}(\phi, \chi) \leftrightarrow Mo_{v_i}(\psi, \chi))$

WQL.2 $(\forall v_i)(\phi \leftrightarrow \psi) \rightarrow (Mo_{v_i}(\chi, \phi) \leftrightarrow Mo_{v_i}(\chi, \psi))$

WQL.3 $Mo_{v_i}(\phi, \psi) \rightarrow Mo_{v_j}(\phi', \psi')$,
if $Mo_{v_i}(\phi, \psi)$ and $Mo_{v_j}(\phi', \psi')$ are alphabetic variants.

But where do we go from here? First a decision of convenience. In the remainder of this section I will follow Colban in pursuing an axiomatization not of the quantifier *most*, but instead for the quantifier usually referred to as *more*, which relates its arguments A and B in a way that can be paraphrased as 'there are more As than Bs'. Thus, corresponding to the 'standard semantics' for *most*, which is given by the truth condition

$$\begin{aligned} &M \models_a Most_v(\phi, \psi) \text{ iff } |V \cap W| > |V \setminus W|, \\ &\text{where } V = \{u \in U_M : M \models_{a[u/v]} \phi\}, \\ &\text{and } W = \{u \in U_M : M \models_{a[u/v]} \psi\}, \end{aligned} \tag{7}$$

we have standard semantics for more given by

$$M \models_a More_v(\phi, \psi) \quad \text{iff} \quad |V| > |W|, \quad \text{where } V, W \text{ as in (7).} \tag{8}$$

As shown in [10], on the standard semantics the language with *more* is more expressive than that with *most*. On the one hand, $Most_v(\phi, \psi)$ can evidently be expressed in the language of *more* as $More_v(\phi \wedge \psi, \phi \wedge \neg\psi)$. On the other hand, in the language of *more* we can also express the unary quantifier 'there are infinitely many ϕs' viz. as $(\exists y)(\phi(y/v) \wedge \neg More_v(\phi, \phi \wedge v \neq y))$, where y is a variable not occurring in ϕ. This quantifier cannot be expressed in the language of *most* with its standard semantics. (This is something which will not be shown here, but again, see [10].) This relationship between the two languages with *more* and *most* remains true when the standard semantics is replaced the weaker semantics which I will propose below. For although the above definition of 'there are infinitely many' no longer works in that more liberal semantic setting, the definition of *most* in terms of *more* remains valid; on the other hand there is no hope of defining *more* in terms of *most*, for such a definition, if correct, would be correct a fortiori for the standard semantics; but that is something which we just saw is impossible.

[6]See [10]. Colban presents proof theories in the Gentzen sequence calculus format, which I personally find somewhat more difficult to read and handle than the axiomatic approach we will follow.

So the axiomatizations proposed here leave open the question of an intrinsic axiomatization of *most* for the new semantics (i.e. within the language $L(most)$ rather than $L(more)$).[7] From the linguist's point of view, however, this gap is of little importance. For a satisfactory logic for *more* is as important an item on his wish list as one for *most*, and since the first will automatically give us the second, we may as well concentrate on the first.

From now on we will read the quantifier symbol Mo as short for *more* and we proceed with the question how the weak logic of WQL.1-3. may be extended to one which is a credible reflection of our intuitions about the meaning of *more*.

There are two aspects to this problem. The first concerns the behaviour of *more* on the finite sets. Here, as I have been arguing in relation to *most*, the cardinality principle – there are more As than Bs iff the cardinality of the set of As is greater than that of the set of Bs – seems intuitively right. But then, for the finite sets this principle can be fully axiomatized, albeit by an infinite set of axioms. Note that in view of WQL.1 and WQL.2 it is enough to state, for each $n \geq 0$, that for any pair of sets A, B such that B has at most n members and A has $n+1$ members $more(A, B)$ holds and for any pair A, B such that A has at most n members and B has n members $more(A, B)$ does not hold. The axioms WQL.4^n and WQL.5^n express this:

WQL.4^n $(\forall v_1)...(\forall v_n)(\forall v_{n+1})(\forall w_1)...(\forall w_n)$
$(\bigwedge_{i \neq j} v_i \neq v_j \rightarrow Mo_x(\bigvee_i(x = v_i), \bigvee_i(x = w_i))$.

WQL.5^n $(\forall v_1)...(\forall v_n)(\forall w_1)...(\forall w_n)$
$(\bigwedge_{i \neq j} w_i \neq w_j \rightarrow \neg Mo_x(\bigvee_i(x = v_i), \bigvee_i(x = w_i))$.

(In both WQL.4^n and WQL.5^n the variables $v_1, .., v_n, v_{n+1}, w_1, .., w_n, x$ are all distinct.)

The truth of the axioms WQL.4^n and WQL.5^n in a model M for $L(Mo)$ entails that the interpretation $R(U_M)$ of Mo in M has the property that for *any* two finite subsets A, B of U_M, $\langle A, B\rangle \in R(U_M)$ iff $|A| > |B|$.

The second aspect of the problem concerns the infinite sets A. As we have seen, this appears to be a more difficult matter, conceptually as well as formally. I have already expressed my doubts about the strong logic for $L(more)$ which adopts (8) for infinite as well as finite sets. Still, there surely are some principles which ought to hold also in the case where infinite sets are involved. Arguably the most unequivocal one is that when A is infinite and B finite, then '$more(A, B)$' must be true and '$more(B, A)$' must be false. But there are a number of other plausible candidate principles as well. For instance that if '$more(A, B)$' is true, then '$more(B, A)$' must be

[7] I have not looked at the problem of axiomatizing the logic of *most* in its own terms, i.e. in the language $L(most)$.

false, or that when '$more(A, B)$' and '$more(B, C)$' are both true then so is '$more(A, C)$'; or that when $A \subseteq B$, then '$more(A, B)$' cannot be true. Colban has argued for all these principles as part of what governs our intuitions about the meaning of *more* in the infinite as well as the finite domain. He shows that any set relation satisfying these conditions can be represented as the quasi-ordering induced by a *naive measure*, a function ν on $\wp(U_M)$ with the property that its range is some linear ordering $<$ with a smallest element 0 and a largest element ∞ such that $A \subseteq B$ entails $\neg(\nu(B) < \nu(A))$. With respect to such a naive measure '$more(A, B)$' is interpreted as $(\nu(B) < \nu(A))$. Note that the properties of R that are at issue here are second-order properties, as they involve quantification over all subsets of the given set U_M. For instance, transitivity of R takes the form:

$$(\forall X)(\forall Y)(\forall Z)(XRY \wedge YRZ \rightarrow XRZ) \tag{9}$$

where X, Y and Z are second-order variables. The full force of such a sentence cannot be captured within the language $L(Mo)$ as that language only has individual variables. To express (9) we would have to add second-order variables to $L(Mo)$; then (9) could be expressed as

$$\begin{aligned}&(\forall X)(\forall Y)(\forall Z)(Mo_v(v \in X, v \in Y)\\ &\wedge Mo_v(v \in Y, v \in Z) \rightarrow Mo_v(v \in X, v \in Z)).\end{aligned} \tag{10}$$

In the 'first-order' language $L(Mo)$ the force of (12) can only be approximated through the infinite set of sentences which we obtain by dropping the initial second-order quantifiers from (12), replacing the atomic subformulae '$v \in X$', '$v \in Y$', '$v \in Z$' uniformly by formulae ϕ, ψ, χ of $L(Mo)$ (and forming universal closures when the resulting formula is not a sentence). The truth of all these sentences in a model M guarantees that the interpretation R_M of the quantifier satisfies the given property (*viz.* transitivity) with respect to the subset of $\wp(U_M)$ consisting of all the $L(Mo)$-definable sets. But there is no guarantee that the property is satisfied 'absolutely', i.e. with regard to all of $\wp(U_M)$. The problem of transforming a model M in which the property is known to hold only relative to definable subsets into an equivalent model M' in which the property holds absolutely is non-trivial and varies with the property in question. But as Colban has shown, it can be solved for the property under consideration, that of being an asymmetric, transitive relation which respects set inclusion (in the sense that if $A \subseteq B$ then not $more(A, B)$). Moreover, the transformation can be carried out in such a way that the first-order reductions of M and M' (i.e. the models obtained by throwing away the interpretations of Mo) are identical and such that the interpretation $R_{M'}$ of Mo in M' coincides with R_M on the set of definable subsets of M.

This means that if we add to weak logic (i.e. to WQL.1-3) all axioms of the forms:

WQL.6 $Mo_v(\phi, \psi) \rightarrow \neg Mo_v(\psi, \phi)$
WQL.7 $Mo_v(\phi, \psi) \wedge \neg Mo_v(\chi, \psi) \rightarrow Mo_v(\phi, \chi)$
WQL.8 $(\forall v)(\phi \rightarrow \psi) \rightarrow \neg Mo_v(\phi, \psi)$

then we obtain an axiom system that is complete with respect to the class of all models M for $L(Mo)$ in which the interpretation R_M of Mo is a relation that is asymmetric and transitive and respects inclusion on all of $\wp(U_M)$. If we include moreover the axioms WQL.4 and WQL.5, then R_M will coincide with the relation $\{\langle A, B\rangle : |A| > |B|\}$ on the finite subsets of U_M. It should also be clear that transitivity and WQL.4 jointly guarantee that $\langle A, B\rangle \in R_M$ whenever A infinite and B finite.

Is this the axiomatization we want? It comes, I think, pretty close. Still, we can, if we want to, pin the interpretation of *more* for infinite domains down further in various ways and strengthen the logic accordingly. One natural strengthening of the logic, to which my attention was drawn by Johan van Benthem, involves the following principle:

> Suppose that '*more*(A, B)' and '*more*(C, D)' and that A and C are disjoint. Then it should also be the case that '*more*$(A \cup C, B \cup D)$'. (11)

This principle has a very strong intuitive appeal, and we may well want to add the corresponding schema WQL.9′ to our axiomatization.

WQL.9′ $Mo_v(\phi, \psi) \wedge Mo(\eta, \theta) \wedge (\forall y)(\phi(y) \rightarrow \neg\eta(y)) \rightarrow Mo_v(\phi \vee \eta, \psi \vee \theta)$.

It is not as straightforward, however, to modify the given semantics, based on Colban's notion of a naive measure, in such a way that WQL.9 is verified in a natural way. Intuitively, WQL.9 is an additivity principle, and so one might want it to come out valid in virtue of an operation + of 'addition' on the sizes $\nu(A)$ which the naive measure assigns to subsets A of the universe of any model for $L(Mo)$. + ought to have, in particular, the property that when A and C are disjoint, then $\nu(A \cup C) = \nu(A) + \nu(C)$ (in addition, to the usual properties of commutativity, associativity, and monotonicity w.r.t. the order on the range of ν). At present I do not see how to prove completeness for the axiom system WQL.1-9′ with respect to models in which an operation of addition with these properties is defined on the range of ν; though there may well be some way to do this.

Other possible strengthenings have to do with what happens when a finite set is added to an infinite set. For instance, we can add a schema to the effect that if y does not belong to the extension E_ϕ of ϕ, then there are more elements in $E_\phi \cup \{y\}$ than there are in E_ϕ: and, moreover, that when z is another such element, then neither of the sets $E_\phi \cup \{y\}$ and $E_\phi \cup \{z\}$ has more elements than the other:

WQL.9 $(\forall y)(\neg\phi[y/w] \rightarrow Mo_w(\phi \vee w = y, \phi))$

WLQ.10 $(\forall y)(\forall z)(\neg\phi[y/w] \wedge \neg\phi[z/w] \rightarrow \neg Mo_w(\phi \vee w = y, \phi \vee w = z))$.

(Again, to be precise, WQL.9 and WQL.10 represent the sets of all sentences which are obtained by universally closing any formula of the respective forms displayed; it is assumed that y and z are not among the free variables of ϕ.)

That WQL.9 and WQL.10 can be added consistently to WQL.1-8 will be shown in Appendix A. Of course, the circumstance that these axioms can be added consistently is no compelling reason for taking them on board. In fact, while there seems to be nothing that speaks against adopting WQL.10, WQL.9 is very dubious. If perhaps at first sight it looks like a natural generalization of WQL.4, this impression can hardly stand up to scrutiny. It is not so much that the axiom contradicts the cardinality principle adopted by the standard semantics – it would be odd for me to put this forward as a serious objection against it, after my earlier protests that the standard semantics isn't really what we want. More significant, it seems to me, is that WQL.9 is incompatible with any interpretation of *more* in its application to infinite sets that is based on converging frequency on finite samples. For it is quite clear that the limiting frequencies for two infinite sets which differ by one element only must be the same if they exist at all.

Let us be a little more explicit. Suppose that M is a denumerable model for L and that $\mathcal{S}$ is a nest of finite subsets of U_M the union of which equals U_M (we think of $\mathcal{S}$ as the 'sample sequence'). For arbitrary infinite subsets D of U_M we define the *rate of D on $\mathcal{S}$* to be $lim_{S\in\mathcal{S},|S|\rightarrow\infty}\frac{|D\cap S|}{|S|}$, in case this limit exists, and to be undefined otherwise. Then, if A is an infinite subset of U_M and $B = A \cup \{b\}$ for some element b from U_M that is not in A and the rate of A on $\mathcal{S}$ exists, then the rate of B on $\mathcal{S}$ exists also and is equal to the rate of A. Thus if we interpret 'there are more As than Bs' as true when the rates of A and B on $\mathcal{S}$ both exist and the former is bigger than the latter, then 'there are more As than Bs' will necessarily be false (if it is defined at all) for the sets A and B in question. So WQL.9 could never be true for a ϕ with an infinite extension.

As I have said already, I cannot see anything amiss with WQL.10. Note that WQL.10 is validated both by the standard semantics and by the converging frequency interpretation just sketched. Indeed, WQL.10 seems a natural candidate for a further strengthening of our theory, even if it is not immediately clear how to give a simple and natural characterization of a class of models with respect to which the logic given by WQL.1-8 + WQL.10 would be complete.

This problem, of finding a natural semantics with respect to which the new theory is complete, brings me back to my earlier plea: to investigate additional concepts in terms of which the meanings of quantifiers like *most* and *more* can be given more life-like analyses than is possible with the purely set-theoretical tools to which generalized quantifier theory has for the most part confined itself in the past. Let me, in this connection, return once more to the frequency interpretation. What I have said about this interpretation so far seems to have the draw-back that, for all we know, the frequency limits in terms of which the truth conditions of $Mo(\phi, \psi)$ are given may fail to be defined, so that models in which Mo is given a frequency interpretation will in general be partial. However, so long as the aim of a model-theoretic semantics is that of defining logical validity, partiality is no serious obstacle. One way to circumvent it is to define ϕ to be a logical consequence of Γ iff for every model in which all sentences in Γ are (defined and) true, so is ϕ.

Someone for whom this analysis of the meaning of *most* and *more* has intuitive plausibility, will want an answer to the following question: For any denumerable model M for L let, as above, a *sample sequence* for M be a chain $\mathcal{S}$ of finite subsets of U_M such that $\bigcup \mathcal{S} = U_M$ and call a *frequency model* for $L(Mo)$ any pair $\langle M, \mathcal{S}\rangle$ such that M is a denumerable model for L and $\mathcal{S}$ is a sample sequence for M. If $\mathcal{M} = \langle M, \mathcal{S}\rangle$ is a frequency model, then $Mo(\phi, \psi)$ is true in $\mathcal{M}$ iff either (i) $\{u \in U_M : \mathcal{M} \models \phi[u]\}$ is finite, and $|\{u \in U_M : \mathcal{M} \models \phi[u]\}| > |\{u \in U_M : \mathcal{M} \models \psi[u]\}|$ or (ii) $\{u \in U_M : \mathcal{M} \models \phi[u]\}$ is infinite, the rates of $\{u \in U_M : \mathcal{M} \models \phi[u]\}$ and $\{u \in U_M : \mathcal{M} \models \psi[u]\}$ on $\mathcal{S}$ are both defined and the former is bigger than the latter. For any sentence of $L(Mo)$ and frequency model $\mathcal{M}$ take $\mathcal{M} \models \phi$ to mean that the truthvalue of ϕ in $\mathcal{M}$ is defined and, moreover, ϕ is true in $\mathcal{M}$. Suppose we define the consequence relation for $L(Mo)$ as in (12).

$$\Gamma \models \phi \text{ iff for any frequency model } \mathcal{M} \text{ iff for all } \psi \in \Gamma, \mathcal{M} \models \psi, \text{ then } \mathcal{M} \models \phi. \qquad (12)$$

Question 1: Is this consequence relation axiomatizable? Question 2: If the answer to Question 1 is *yes*, what is a (nice) axiomatization for this relation?

To repeat, it is questions of this general sort to which I believe quantifier theory should increasingly turn its attention.

4 Conclusion

Let me briefly summarize the principal points and concerns of this paper. I began by rehearsing some well-known facts about the quantifier *most*: its essentially binary character, its undefinability in terms of the classical quantifiers 'for all' and 'there is', and the non-axiomatizability of first-order logic extended with *most* on the standard semantics for it (for all A, B '$most(A, B)$' is true iff $|A \cap B| > |A \setminus B|$). I then argued that the condition $|A \cap B| > |A \setminus B|$ is in agreement with our intuitions about the meaning of 'most As are Bs' only in the case where A is finite. So a more realistic semantics is obtained when we adopt this condition only for the finite case, while treating the infinite case in some other way. Since the restriction of the cardinality condition to the finite case can be axiomatized straightforwardly, axiomatizability is now again within our grasp, although whether we get it, and what an axiomatization will be like, if it can be had at all, will of course depend on what the new semantics will stipulate about the infinite case.

How then should the infinite case be treated? On this score my proposals have been incomplete. I have proposed a number of principles (WQL.6-8) to be adopted universally – for the finite case these are entailed by the axioms reflecting the cardinality condition – as a first approximation and mentioned that completeness can be obtained for the resulting system with respect to a semantics based on Colban's notion of naive measure. But clearly that is not the end of the story. I mentioned one further plausible principle (WQL.10) whose addition presents no difficulties (completeness along essentially the same lines can still be obtained as before), as well as another, (WQL.9′), suggested to me by van Benthem, for which a satisfactory semantics plus completeness is still outstanding.

But will these be enough? What *is* enough? That is, I have tried to argue, a difficult question, which is likely to involve much that goes beyond what can be found within the current model-theoretic toolkit of formal quantifier theory. In particular, the familiar arguments against adopting the cardinality condition for the infinite case suggest that our judgements about *most*-sentences with infinite A and B often involve some notion of rate, or frequency. So, I suggested, to make further progress with the question what logic governs the use of *most* with infinite sets, we should explore a semantics based on such a notion. One option, suggested towards the end of Section 4, would be a semantics which deals with the finite cases by way of cardinality and with the infinite ones in terms of frequency. An implementation of that option will have to make a number of further decisions, possibly with diverging consequences for the resulting logic. So this option alone may yield a spectrum of alternative logics, between which it may be difficult to choose. Moreover, it is possible that whichever way we go, we will have to cope with problems quite unlike those that arise for the comparatively simple model theory which has been used here. (One of the contingencies, I observed, with which a frequency-based semantics must be prepared to deal, is partiality: Some *most*-sentences may come out as lacking truth values in some models.)

In addition, frequency need not be the only conception behind our judgements about *most*-sentences involving infinite sets. Careful thought will have to be devoted to the question whether alternative conceptions might come into such judgements and what these might be like. Pursuing this question may well induce us to look into yet other model theories for *most*.

So, a potentially wide field of possible choices, and corresponding axiomatization problems, opens up to those who accept the need of probing further in these directions.

As far as the present paper is concerned, all this has been no more than a plea. In fact, I have only just begun to look into some of these options. But I am resolved to carry on, and I can only hope that I won't be all alone.

Appendix A

We show that WQL.9 and WQL.10 are consistent with WQL.1-8. As a matter of fact we will prove something slightly stronger that the consistency of WQL.1-10, *viz.* that every consistent set Σ of sentences of L is consistent with all instances of WQL.1-10. It follows from this via the completeness theorem for weak logic (see, e.g. [1], or [9]) that there is an $L(Mo)$ model in which Σ and all instances of WQL.1-10 hold. By the methods of [1] this model can then, if one wants, be turned into an equivalent one in which Mo is interpreted by a naive measure.

Let Σ be any consistent theory of L. Let S be a finite set of instances of WQL.9 and WQL.10. Let M be an at most denumerable model of Σ. We show that M can be turned into an $L(Mo)$ model M' in which Mo is interpreted by a naive measure which verifies all sentences WQL^n.4 and WQL^n.5 as well as the sentences in S.[8] For each of the finitely many ϕ which occur in WQL.9 instances or WQL.10 instances in S let E_ϕ be the set of all $u \in U_M$ that satisfy ϕ in M, and let $Umb(\phi)$ be the set $\{E_\phi\} \cup \{E_\phi \cup \{u\} : u \in U_M \setminus E_\phi\}$. We call $Umb(\phi)$ the *umbrella defined* by ϕ (in M) (thinking of E_ϕ as the *handle* of $Umb(\phi)$ and of the sets $E_\phi \cup \{u\}$ as the *spokes* of $Umb(\phi)$). Umb will be the union of the (finitely many) umbrellas $Umb(\phi)$ with ϕ occurring in S. Evidently a naive measure ν will verify all sentences in S iff it assigns the same value to all spokes of any umbrella $Umb(\phi)$ for ϕ occurring in S and assigns a smaller value to the umbrella's handle. Let $\equiv$ be the relation which holds between two subsets A and B of U_M iff their symmetric difference is finite. It is well-known that this is an equivalence relation. Furthermore, for any two sets A and B such that $A \equiv B$ let *the distance from A to B*, $d(A,B)$, be the integer $|A\setminus B| - -|B\setminus A|$. It is not hard to check that if $A \equiv B$, then $d(B,A) = -d(A,B)$ and that for $A \equiv B \equiv C$, $d(A,C) = d(A,B) + d(B,C)$. It is also clear that if A and B both belong to

[8]In case M is finite, we can directly interpret Mo by the relation which holds betweensubsets A and B of U_M iff $|A| > |B|$. This will then be a naive measure satisfying all the schemata WQL.1 - WQL.10. So we could assume at this point that M is denumerably infinite. As this assumption doesn't seem to simplify the proof, I haven't made it. However, it may help to understand the construction below to think of M as infinite and in particular of the 'umbrellas' $Umb(\phi)$ (which will be defined directly) as (for the most part) infinite.

$Umb(\phi)$ for the same ϕ, then $A \equiv B$ and, moreover, that $d(A, B) = 1$ if A is the handle of $Umb(\phi)$ and B one of its spokes; and $d(A, B) = 0$ if both A and B are spokes of $Umb(\phi)$. Also, if $A \in Umb(\phi)$, $B \in Umb(\psi)$ and $A \equiv B$, then for any other $C \in Umb(\phi)$, $D \in Umb(\psi)$, $C \equiv D$. So $\equiv$ collects the umbrellas $Umb(\phi)$ into equivalence classes. Since any equivalence class contains the members of only a finite number of umbrellas (obviously, as there are only finitely many umbrellas that are being considered altogether), it should be clear from what has been said that for each such class C there is a natural number n_C such that for all $A, B \in C$, $|d(A, B)| < n_C$. Also there will be some member $A_0(C)$ of C (not necessarily uniquely determined) such that $d(A_0(C), B) \geq 0$ for all $B \in C$.

Any two distinct equivalence classes C_1, C_2 consisting of (members of) umbrellas can stand in one of three relations; either (i) there are $A \in C_1$ and $B \in C_2$ such that $B \setminus A$ is infinite and $A \setminus B$ is finite, or (ii) there are $A \in C_1$ and $B \in C_2$ such that $B \setminus A$ is finite and $A \setminus B$ is infinite, or (iii) there are $A \in C_1$ and $B \in C_2$ such that both $B \setminus A$ and $A \setminus B$ are infinite. It is easily seen that in case (i) the same relation, $C \setminus D$ infinite and $D \setminus C$ finite, holds for any other $C \in C_1$ and $D \in C_2$, and similarly for cases (ii) and (iii). So, if we define the following relation $\prec$ between equivalence classes: $C_1 \prec C_2$ iff for some $A \in C_1$ and $B \in C_2$ $B \setminus A$ is infinite and $A \setminus B$ is finite, then (a) this definition does not depend on the choice of A and B, and (b) $\prec$ is a strict partial order on the set of equivalence classes. Since $\prec$ is finite, we can assign to each equivalence class C a degree $deg(C)$ by induction: if C has no predecessors in the sense of $\prec$, then $deg(C) = 1$; otherwise $deg(C) = max\{deg(C') : C' \prec C\} + 1$. Now we define a naive measure ν on the power set of U_M as follows:

(i) $\nu(A) = |A|$, if A is finite;
(ii) $\nu(A) = w \cdot deg(C) + d(A_0(C), A)$, if A is infinite and A belongs to the union Umb of the finitely many umbrellas under consideration;
(iii) $\nu(A) = max\{\nu(B) : B \in Umb \wedge B \subseteq A\}$, if A is infinite but not $A \in Umb$.

It is not difficult to verify that ν is indeed a naive measure (the only condition that needs a little care in checking is that $\nu(A) \leq \nu(B)$ whenever $A \subseteq B$) and that when Mo is interpreted in terms of it, then the sentences in Σ all come out true; that the interpretation also verifies WQL.4 and WQL.5 is obvious and that WQL.6-8 are satisfied follows from the results of [1]. ■

The consistency of WQL.9 and WQL.10 with any first-order extension of WQL.1-8 is only one of an indefinite number of similar results that one may try to obtain. I have presented the argument in the hope that many such results could be established by similar means, though I do not, at the present time, have a clear conception of how far these methods might carry us.

Appendix B

In Section 2 we noted that $L(more)$ is strictly more expressive than $L(most)$. As the proof of this fact in [10] makes clear, the reason for this is that the size comparisons involved in the evaluation of *most*-sentences are always between disjoint sets, whereas *more* permits the comparison of arbitrary sets. It is not clear, however, that this difference – *most* has less expressive power than *more* – remains, when we develop a logic of *most* which covers the full spectrum of uses of the word *most* in a language like English. English has sentences in which *most* requires the comparison of sets that overlap.

For instance, with respect to a situation in which a test was taken by Susan, Fred and Naomi we can say

Susan solved most problems on the test. (13)

to mean that the number of problems that Susan solved was larger than the number of problems solved by either of the others. There is no presupposition that the sets of problems each of them solved are pairwise disjoint – for instance, for all that (13) implies, the set of problems solved by Fred might be a proper subset of the set of problems solved by Susan.[9,10]

The comparison class – here {Susan, Fred, Naomi} – can also be made explicit in the sentence itself, as in

As between Susan, Fred and Naomi, Susan solved most problems on the test. (15)

[9]My attention was drawn to this use of *most* by a remark of Ruth Kempson.

[10]In English it seems that the use of *most* in contexts such as (13) is restricted to comparison classes whose cardinality is at least three; if the comparison is between two cases only, the proper word is not *most* but *more*. It is my impression that in certain other languages this constraint is not as strong as it is in English. For instance, I personally do not feel much resistance (if any) against the use of the Dutch equivalent *de meeste* in comparison between two classes. Thus I can say

Susan en Fred hebben allebei genoeg problemen opglost om voor het examen te slagen. Maar Susan heeft de meeste opgelost, en krijgt dus ook het hoogste cijfer.
(Susan and Fred both solved enough problems to pass the exam. But Susan solved more (literally: 'the most') problems and thus gets the better (literally: 'the highest') mark.) (14)

This issue is of some importance for the present discussion insofar as in languages for which the given constraint (i.e. that the comparison class must consist of at least three elements) does not hold, the question of how *more* could be reduced to *most* can be addressed without the slight complication that the constraint produces.

The presence in (15) of the adjunct *as* between Susan, Fred and Naomi, which makes the comparison class explicit rather than leaving it to be recovered from context, renders (15) unambiguous in a sense in which (13) is not. (13) has besides the reading we have just discussed also one which conforms to the analysis of *most* we have been assuming so far – the reading according to which the number of problems Susan solved was more than half the number of problems on the test altogether. As we will see below, the difference between these two readings is, in a certain sense, a matter of scope.

Before we pursue the semantics of sentences such as (13) further, first a brief remark on how this matter affects the question whether *most* is less expressive than *more*. Speaking somewhat loosely, 'there are more As than Bs' can be expressed by a sentence of the form exemplified by (13), provided we can find

(i) a binary relation R that is expressible as a simple or complex transitive verb,
(ii) a set X of three or more individuals, and
(iii) an individual a in X,

such that

(a) the As are the entities y such that a stands in the relation R to y,
(b) for some b in X with $b \neq a$ the Bs are the entities y such that b stands in the relation R to y, while
(c) for every other element c of X, the set of y such that c stands in the relation R to y forms a subset of the set of Bs.

For we can then paraphrase the statement 'there are more As than Bs' by a sentence of the form

As regards the individuals in X, a (is the one who) Rs most things. (16)

(or something in this vein that obeys the rules of English grammar and doesn't offend the English speaker's sensibilities in other ways).

It is not hard to see what it is about English that enables it to express not only those uses of *most* that can be analyzed correctly by treating *most* as a simple generalized quantifier, but also uses of the sort exemplified by (13). Roughly speaking, an NP the determiner of which is *most* can occur in any of the positions in an English clause that are open to NPs generally. Typical examples of the use of *most* which conforms to its analysis as a generalized quantifier are sentences in which the *most*-NP is the subject and in which the VP acts as a 1-place predicate whose only argument position is that subject. Among these sentences there are in particular those in which the VP consists of the copula *be* followed by a nominal or adjectival predicate – sentences such as 'Most trees in Scandinavia are conifers'. or 'Most Americans are white'. Such sentences fit the schematic paraphrase 'Most *A*s are *B*s' almost to perfection. But other sentences with *most*-NPs as subjects – such as, say, 'Most French businessmen smoke' or 'Most American families own a car'. – can, for the purposes of the present investigation, be considered to be of this form too.

Uses of *most* which display the semantic complication we observed in connection with (13) arise when the *most*-NP occurs as argument to a verb or verb phrase which has other arguments as well, and where, moreover, the *most*-NP can be interpreted as being 'within the scope' of one or more of those other NPs. Typical instances of this are clauses with transitive verbs in which the *most*-NP is the direct object; (13) is a case in point. But it is important to note that these are not the only ones. (17), for example,

Most letters were written by Susan to Fred. (17)

can be used to say that within a certain set of author-recipient pairs (containing three pairs or more) the pair Susan-Fred was involved in the writing and receiving of a larger number of letters than were any of the other pairs.

How should these uses of *most* which we have been ignoring hitherto be formally represented? It takes little reflection to see that what is needed is not some generalized quantifier – in the narrow sense of the term, that of an operator which takes two formulas as arguments, produces a formula as output and binds one variable – other than those which we have explored in the body of the present paper. The *most* that concerns us now diverges from the determiners which we have been looking at so far primarily in that it has a very different 'logical grammar'. Take for instance the occurrence of *most* in (15). Its semantic effect is to establish a certain relation between (i) the comparison class {Susan, Fred, Naomi} given by the *as between* phrase; (ii) the individual Susan given by the subject NP; and (iii) the relation 'u solved problem v' given by the VP. This effect is captured in the following clause:

(15) is true iff $(\forall u)(u \in \{Susan, Fred, Naomi\} \wedge u \neq Susan \rightarrow$ Susan solved more problems than u). (18)

If we insist on capturing this semantic relationship while treating *most* as a variable binding operator, the apparent type of this operator is that of one which (a) takes as input one term and two formulas, (corresponding to the subject, the *as between* phrase and the VP, respectively, in (15)); and (b) binds two variables, the first of which represents the relevant member of the comparison class and the subject argument of the VP, while the second represents the object argument of the VP. Thus (15) gets the logical form

$$Most^2_{u,v}(\tau, \rho(u), \chi(u,v)), \quad (19)$$

where t is the term 'Susan', $\rho(u)$ is short for for all 'u $\in$\{Susan, Fred, Naomi\}' and $\chi(u,v)$ for 'u solved problem v on the test'. The truth conditions of (19) are given in (20)

$$\begin{aligned}&Most^2_{u,v}(\tau,\rho(u),\chi(u,v)) \text{ is true iff}\\&(\forall u)(\rho(u)\wedge u\neq\tau\rightarrow MORE(\{v:\chi(\tau,v)\},\{v:\chi(u,v)\}))\end{aligned}\qquad(20)$$

where MORE is the generalized quantifier (i.e. relation between sets) expressed by *more*; or, alternatively, using the generalized quantifier *Mo* (with interpretation MORE) which we have investigated in Section 3:

$$\begin{aligned}&Most^2_{u,v}(\tau,\rho(u),\chi(u,v)) \text{ is true iff}\\&(\forall u)(\rho(u)\wedge u\neq t\rightarrow Mo_v(\chi(\tau,v),\chi(u,v))\end{aligned}\qquad(21)$$

As (21) shows, $Most^2$ is definable in terms of the old Mo. Can we define, conversely, *Mo* in terms of $Most^2$? Almost. All we need is an antecedent assumption that there are enough things to form at least one proper comparison class; if we stick to the intuitions I mentioned about the use of *most* in sentences like (13) in English, this means that the universe must contain at least three things. So let us assume that there are three distinct objects x, y and z. Consider the formula $Mo_v(\phi(v),\psi(v))$. Let $\chi(u,v)$ be the formula $(u=x\wedge\phi(u))\vee(u=y\wedge\psi(u))\vee(u=z\wedge\psi(u))$ and let $\rho(u)$ be the formula $u=x\vee u=y\vee u=z$. Then $Mo_v(\phi(v),\psi(v))$ is clearly equivalent to $Most^2_{u,v}(x,\rho(u),\chi(u,v))$. Thus we have the following conditional definition of *Mo* in terms of $Most^2$:

$$\begin{aligned}&(\exists x)(\exists y)(\exists z)(x\neq y\wedge x\neq z\wedge y\neq z)\rightarrow(Mo_v(\phi(v),\psi(v))\leftrightarrow\\&(\exists x)(\exists y)(\exists z)(x\neq y\wedge x\neq z\wedge y\neq z\wedge Most^2_{u,v}(x,\rho(u),\chi(u,v)))).\end{aligned}\qquad(22)$$

Since the operator $Most^2$ is definable in terms of Mo, its introduction does not introduce any fundamentally new axiomatization problems. One could still pose the question whether there is a direct, natural and elegant axiomatization for the new $Most^2$. This is a question that I have not explored.

The operator $Most^2$ we have just been discussing arose out of a reflection on the meaning of (15). The need to formalize (15) by means of an operator which binds not one but two variables, one variable for the problem solved and one for the one who solved it, arose from the circumstance that the different sets of solved problems which the sentence asks us to compare depend on who in each case is the solver. By analogy, formalization of a sentence like (17) will require an operator binding three variables, one variable for the letter written, one for the person who wrote it and one for the person to whom it was written. The comparison class is now, as we have seen, a set of pairs; in the setting of variable binding this comes down to a two-free-variable-formula $\rho(u,w)$. And instead of the binary relation expressed by the transitive verb 'solved' in (15) we now have the ternary relation expressed by 'u wrote v to w'; in terms of the operator treatment this amounts to a formula $\xi(u,v,w)$ with free variables u, v and w. These considerations suggest an operator $Most^3$ which binds 3 variables and takes as inputs two formulas (the ρ and ξ just mentioned) as well as two terms – in (17) these are given by the subject and the *to*-PP. Using such an operator, (17) can be represented as.

$$Most^3_{u,v,w}(\tau, \sigma, \rho(u,w), \xi(u,v,w)), \tag{23}$$

where τ is the term 'Susan', σ is the term 'Fred', $\rho(u,w)$ is short for '$\langle u, w\rangle \in C$' with C the relevant class of pairs that acts as comparison class, and $\xi(u,v,w)$ for 'u wrote letter v to w'.

I take it that the meaning of (17) is correctly captured by the following truth clause for $Most^3$:

$$\begin{array}{l} Most^3_{u,v,w}(\tau, \sigma, \rho(u,w), \xi(u,v,w)) \quad \text{iff} \\ (\forall u)(\forall w)((\rho(u,w) \wedge (u \neq t \vee w \neq \sigma)) \to Mo_v(\xi(\tau, v, \sigma), \xi(u,v,w)). \end{array} \tag{24}$$

Thus $Most^3$ is, just like $Most^2$, definable in terms of Mo.
Of course this is not the end of it. Formalization of a sentence such as

Most letters were written by Susan from Ithaca to Fred. (25)

which may report on a comparison of the number of letters which Susan wrote to Fred from Ithaca with the number of letters which Carla wrote to Algie from Corfu, the number of letters that Carla wrote to Fred from Corfu, the number of letters that Susan wrote to Fred from Athens, etc., would require for its formalization an operator binding four variables; and so forth. Operators binding even more variables would be needed to represent sentences in which the sets defined by the *most*-NP depend on four, five,... other arguments (obligatory or optional) to the main verb. Thus, the number of operators needed to formalize arbitrary sentences of this pattern will be finite only if there is an upper bound to the number of optional arguments to any given verb that can be incorporated into a single clause. Those who feel that such an upper bound would, even if it could be argued to exist, testify to an idiosyncrasy of natural language grammar to which the design of logical representation formalisms should be pay no attention, may want to adopt the entire infinite sequence of operators in any case.

From a logical perspective there exists an obvious alternative. Semantically, each of these infinitely many operators is definable in terms of Mo. So the language $L(Mo)$ is all we need in order to capture the truth conditions of any of the sentences that can be represented in the language $L(\{Most^n\}_{n \in N})$. But to what extent is this alternative acceptable linguistically? What the linguist wants is not just a formalism in which the truth conditions of natural language sentences can be stated accurately; he also wants a systematic procedure that gets him, for any one of the sentences of his concern, to a statement of its truth conditions while starting from its syntactic form – a procedure which somehow 'explains' why a sentence of *this* syntactic form has *this* meaning. I find it hard to see, however, how it might be possible to define a systematic transition from syntactic to logical representation for the sentences in question which did not pass via a representation that involves in some form or other the relevant operator $Most^n$.

But this is a matter that will have to be explored in another context than this.

Acknowledgements

Many thanks to Johan van Benthem for a number of important comments and suggestions. Unfortunately it was not possible for me to deal with his criticisms in the way they deserved. All I have been able to do here is to add a few last minute adjustments, but I hope to make better use of his observations in further projected work on the logic of non-standard quantifiers. Many thanks also to Uwe Reyle, whose help in getting this paper into a form suitable for appearance in this volume much exceeded what an author may reasonably expect from an editor.

Universität Stuttgart, Germany.

References

1. Erik A. Colban. Generalized quantifiers in sequent calculus. COSMOS-Report 18, Department of Mathematics, University of Oslo, March 1991.
2. Tim Fernando and Hans Kamp. 'Most', 'more', 'many', and more. SFB-Report, IMS, University of Stuttgart, 1996.
3. Kit Fine. Vagueness, truth and logic. *Synthese*, 30:265–300, 1975.
4. Hans Kamp. Two theories about adjectives. In E. Keenan, editor, *Formal Semantics of Natural Language*. Cambridge University Press, 1975.
5. Hans Kamp. Conditionals in DR theory. In J. Hoepelman, editor, *Representation and Reasoning*. Niemeyer, 1986.
6. Hans Kamp and Barbara Partee. Prototype theory and compositionality. *Cognition*, 57:129–191, 1995.
7. Ed Keenan. Beyond the Frege boundary. *Linguistics and Philosophy*, 15, 1992.
8. Ed Keenan. Natural language, sortal reducibility and generalized quantifiers. *Journal of Symbolic Logic*, 58(1), 1993.
9. H. Jerome Keisler. Logic with the quantifier 'there exists uncountably many'. *Annals of Mathematical Logic*, 1:1–93, 1970.
10. Dag Westerstahl. Quantifiers in formal and natural languages. In D. M. Gabbay and F. Guenther, editors, *Handbook of Philosophical Logic*, volume IV, chapter 1, pages 1–131. Reidel, 1989.

IMPERATIVE HISTORY: TWO-DIMENSIONAL EXECUTABLE TEMPORAL LOGIC

MARCELO FINGER AND MARK REYNOLDS

1 Introduction

In this chapter we combine two interesting and useful recently proposed ideas within applied temporal logic which were both initially developed by Dov Gabbay (amongst others). We coin the term 'Imperative History' for the two-dimensional executable temporal logic which results from combining the 'Imperative Future' idea of an executable temporal logic (proposed in [10] and described more fully in [3]) with the idea of using a two-dimensional temporal logic to describe the evolution of temporal databases (an idea proposed in [6] but closely related to the work in [4]). We demonstrate that this combination leads to a powerful declarative approach to handling time in databases.

Temporal logic has become one of the most important formalisms for describing, specifying, controlling and reasoning about systems which exhibit some kind of on-going interaction with their environment. The formal language with its proof-theory, decision algorithms and associated methods of practical application has found many uses in dealing with programs, complex reactive systems, databases and artificial intelligent systems: the interested reader is referred to [9] for a fuller description of these applications. In this paper we extend two different applications.

In [10] it was suggested that the formal temporal language for describing the development of a reactive system could be used, in a restricted form, to actually write the programs which control the behaviour of the system. Thus we use temporal logic as a declarative programming language: the logic becomes executable. All the well-known advantages of declarative programming languages apply: they are quick to write, easy to understand and anyone interested in formal verification has a head start.

In the executable temporal logic of [10], the simple restricted format for the formulas of the temporal language which become program rules is

H.J. Ohlbach and U. Reyle (eds.), Logic, Language and Reasoning, 73–98.

summarized as *Past* implies *Future.* The procedural effect of such a rule is that some condition on the observed past behaviour of the system (and/or its environment) controls whether the system brings about some future situation. Thus this idea is rendered as *Declarative Past* implies *Imperative Future.* There is an ever increasing body of useful work developing from this proposal and related work. The interested reader can find descriptions of first-order versions, efficient implementations and generalizations to concurrency amongst other recent developments in [3].

Another very important use of temporal logic is in dealing with databases which make use of time. We call these *temporal databases.* Time can be relevant to a database in one or both of two different ways. Each change to the contents of the database will be made at some time: we refer to this as the *transaction* time of the database update. Databases often also store information about the time of events: we refer to the actual time of occurrence of an event as its *valid* time. Depending on which of these uses is made of time or on whether both approaches have a role to play, we can identify several different types of temporal databases but what is common to all, as with all systems which change over time, is that describing or reasoning about their evolution is very conveniently done with temporal logic.

With both the forms of temporal information involved, it was thus suggested in [6], that describing the evolution of a temporal database is best done with two-dimensional temporal logic. This is because, for example, at a certain transaction time today, say, we might realize that our database has not been kept up-to-date and we may add some data about an event which occurred (at a valid time) last week. Thus a one-dimensional model which represents this morning's view of the history of the recorded world, is changed, by the afternoon, into a new one-dimensional model by having the state of its view about last week altered. A series of one-dimensional models arranged from one day to the next is clearly a structure for a two-dimensional temporal logic. Other applications of two-dimensional temporal logic exist – for example in dealing with intervals of time [1] – but the logic is generally quite difficult to reason with (see [22]). However, it has recently been shown [4] that the kind of logic needed for database applications is much more amenable.

Managing databases is not just about collecting facts. There are many uses for more general rules. For example, we often need integrity constraints, derived properties, conditional updates, side-effects and systematic corrections. All such rules must be expressed in some sort of database-control/programming language.

In this paper we suggest using a two-dimensional executable temporal logic as a declarative language for expressing rules for temporal database management. The most common form for these rules will be a formula which expresses a condition on the one-dimensional historical model at a certain time controlling a condition on the new one-dimensional historical model which should hold after the next transaction. This may necessitate an update to recorded history (about some valid times in the past, present or future). We thus call this executable temporal logic 'Imperative History'.

The paper is structured as follows. In the next section, we define propositional and predicate one-dimensional temporal logics: their languages form the basis of existing executable temporal logics and our two-dimensional temporal logic. Also in this section, we describe the existing (one-dimensional) executable temporal logic METATEM and its variations. In Section 3, we describe two-dimensional logic as it is applied to temporal databases. We also briefly describe the idea of temporal databases and their various types. In Section 4, we introduce the idea of an executable two-dimensional logics and describe how it could be used in database management. In Section 5, we provide a simple example of the idea in action in the intensive care ward of a hospital: this example develops, some previous applications of executable temporal logic. In Section 6, we give a possible extension of the technique to database triggers before summarizing our work.

2 Executable Temporal Logic

2.1 Temporal Logic

We are going to be concerned with the behaviour of processes over time. Two very useful formal languages for describing such behaviour are the propositional temporal logic **PTL** and the first-order temporal logic **FTL** based on the temporal connectives until U and since S introduced by Kamp in [12]. The simpler propositional language allows us to express less and so is easier to deal with.

A crucial point in the executable temporal logic paradigm is that the same languages are used to specify the desired behaviour of a program and to actually write the program to satisfy the specification. In fact, in the ideal case, the specification and the program are the same thing.

In any case, amongst many other advantages, using the same language for specification and implementation gives us a head start in proving correctness of programs.

2.2 Temporal Structures

The languages **FTL** and **PTL** are used to describe the behaviour of processes over time. In this paper, we will take the underlying flow of time to be either the natural numbers – equivalently some sequence $s_0, s_1, s_2, \ldots$ of states – or the integers. In general, such temporal languages can describe changes over any linear order $(T, <)$ of time points.

In the propositional case the *state* at each time is just given by the truth values of a set $\mathcal{L}_P$ of atomic propositions or atoms. The behaviour we are describing is just the way the various atoms become true or false over time. To formalize this we use a map $\pi_P : T \times \mathcal{L}_P \to \{\top, \bot\}$ where $\pi_P(t,q) = \top$ iff the atom q is true at time t.

In first-order temporal structures the state at each time is a whole first-order structure with a domain of objects on which are interpreted constant symbols, function symbols and predicate symbols. Without any restrictions such situations would be too messy to describe formally so we make some assumptions. As described in [18] there are many sets of simplifying assumptions which can be made but the ones we make here are comfortable to work with and, at the same time, so general that other approaches can be easily coded in.

For a start we assume that each state is a first-order structure in the same language. So suppose that $\mathcal{L}_P$ is a set of predicate symbols and $\mathcal{L}_F$ is a set of function symbols. We divide up $\mathcal{L}_P$ into a set $\mathcal{L}_P^n$ for each $n \geq 0$ being the n-ary predicate symbols. We also divide up $\mathcal{L}_F$ into a set $\mathcal{L}_F^n$ for each $n \geq 0$ being the n-ary predicate symbols. The 0-ary function symbols are just constants.

We assume a constant domain $\mathcal{D}$ of objects but, over time, the extensions of the predicates change. To formalize this we use a map $\pi_P = \pi_P^0 : T \times \mathcal{L}_P^0 \to \{\top, \bot\}$ and a map $\pi_P^n : T \times \mathcal{L}_P^n \to \mathcal{D}^n$ for each $n = 1, 2, \ldots$ The interpretations of the functions are constant: we use maps $\pi_F^n : \mathcal{L}_F^n \to (\mathcal{D}^n \to \mathcal{D})$.

In many of the definitions below we can include the propositional case as a special case of the first-order one by equating $\mathcal{L}_P$ with $\mathcal{L}_P^0$ and π_P with π_P^0.

2.3 Syntax

As well as $\mathcal{L}_P^n$ and $\mathcal{L}_F$, we also use a countable set $\mathcal{L}_V$ of variable symbols. The terms of **FTL** are built in the usual way from $\mathcal{L}_F$ and $\mathcal{L}_V$.

The set of formulas of **FTL** is defined by:

- if $t_1, \ldots$ are terms and p is an n-ary predicate symbol then $p(t_1, \ldots$ is a formula,
- if α and β are formulas then so are $\top$, $\neg\alpha$, $\alpha \wedge \beta$, $\forall x \alpha$, $\mathcal{U}(\alpha, \beta)$ and $\mathcal{S}(\alpha, \beta)$.

We have the usual idea of free and bound variable symbols in a formula and so the usual idea of a sentence – i.e. a formula with no free variables. The class of formulas which do not have any variable symbols or constants form the well-formed formulas of the propositional language **PTL**. In **PTL** we only use 0-ary predicate symbols which are just propositions.

A formula of the form $p(\overline{u})$ is called a *positive literal*. A formula of the form $\neg p(\overline{u})$ is called a *negative literal*. A *literal* is either a positive one or a negative one. A literal is *ground* if it is also a sentence.

2.4 Semantics

A variable assignment is a mapping from $\mathcal{L}_V$ into $\mathcal{D}$. Given such a variable assignment V we assume it extends to all terms by recursively defining $V(f(u_1, \ldots, u_n) = \pi_F^n(f)(V(u_1), \ldots, V(u_n))$ for any $f \in \mathcal{L}_F^n$.

For a temporal structure $\mathcal{M} = (T, <, \mathcal{D}, \{\pi_P^n\}_{n\geq 0}, \{\pi_F^n\}_{n\geq 0})$ a time point $t \in T$, a formula φ, and a variable assignment V, we define whether (or not resp.) $\varphi(\overline{d})$ under V is true at t in $\mathcal{M}$, written $\mathcal{M}, t, V \models \varphi$ (or $\mathcal{M}, t, V \not\models \varphi$ resp.) by induction on the construction of φ.

- $\mathcal{M}, t, V \models$ iff $\top$.
- $\mathcal{M}, t, V \models q$ iff $\pi_P^0(t, q) = \top$ for a proposition q.
- $\mathcal{M}, t, V \models p(\overline{u})$ iff $(V(u_1), \ldots, V(u_n)) \in \pi_P^n(t, p)$ for an n-ary predicate p and n-tuple $\overline{u}$ of terms.
- $\mathcal{M}, t, V \models \neg\chi$ iff $\mathcal{M}, t, V \not\models \chi$.
- $\mathcal{M}, t, V \models \chi \wedge \psi$ iff $\mathcal{M}, t, V \models \chi$ and $\mathcal{M}, t, V \models \psi$.
- $\mathcal{M}, t, V \models \mathcal{U}(\psi, \chi)$ iff there is $s > t$ in T such that $\mathcal{M}, s, V \models \psi$ and for all $r \in T$ such that $t < r < s$, $\mathcal{M}, r, V \models \chi$.
- $\mathcal{M}, t, V \models \mathcal{S}(\psi, \chi)$ iff there is $s < t$ in T such that $\mathcal{M}, s, V \models \psi$ and for all $r \in T$ such that $s < r < t$, $\mathcal{M}, r, V \models \chi$.
- $\mathcal{M}, t, V \models \forall x \chi$ iff for all $d \in \mathcal{D}$ $\mathcal{M}, t, W \models \chi$ where $x \in \mathcal{L}_V$ and W is the variable assignment given by

$$W(y) = \begin{cases} V(y) & y \neq x \\ d & y = x. \end{cases}$$

It is easy to prove that the truth of a formula at a point in a structure does not depend on assignments to variables which do not appear free in it. So we can write $\mathcal{M}, t, v \models \varphi$ where v is a partial assignment provided its domain does include the free variables of φ. When σ is a sentence – or a **PTL** formula – we also write $\mathcal{M}, t \models \sigma$ iff $\mathcal{M}, t, \varnothing \models \sigma$ where $\varnothing$ is the empty map.

2.5 Models

Say that temporal structure $\mathcal{M}$ is a *model* of a sentence σ iff $\mathcal{M}, 0 \models \sigma$. A sentence is *satisfiable* iff it has such a model. A sentence σ is *valid* iff $\mathcal{M}, t \models \sigma$ for all structures $\mathcal{M}$ and for all time points t in $\mathcal{M}$.

2.6 Abbreviations

We read $\mathcal{U}(\psi, \xi)$ as 'ξ until ψ' and similarly for since. Note that our $\mathcal{U}$ is *strict* in the sense that $\mathcal{U}(q, p)$ being true says nothing about what is true now. In some presentations of temporal logic, until is defined to be non-strict. We can introduce an abbreviation $\mathcal{U}^+$ for non-strict until: $\mathcal{U}^+(\psi, \xi)$ iff $\psi \vee (\xi \wedge (\mathcal{U}(\psi, \xi)))$.

As well as the classical abbreviations $\bot$, $\vee$, $\rightarrow$, $\leftrightarrow$ and $\exists$ we also have many temporal ones. The only ones that we need in this paper are:

$\bigcirc\varphi$	'φ is true in the next state'	$\mathcal{U}(\varphi, \bot)$
$\bullet\varphi$	'there was a last state and φ was true in this state'	$\mathcal{S}(\varphi, \bot)$
$\Diamond\varphi$	'φ will be true in some future state'	$\mathcal{U}(\varphi, \top)$
$\Box\varphi$	'φ will be true in all future states'	$\neg\Diamond\neg\varphi$
$\blacklozenge\varphi$	'φ was true in the past'	$\mathcal{S}(\varphi, \top)$
$\blacksquare$	'φ has always been true in the past'	$\neg\blacklozenge(\neg\varphi)$
start	'it is now the start of time'	$\neg(\mathcal{S}(\top, \top))$

2.7 Separation

As one would expect from the declarative past / imperative future motivation, one distinction which plays an important role in METATEM is that between formulas which refer to the past and those which refer to the future. Let us make this precise.

A formula φ is a *not necessarily strict future time formula* iff it is built without $\mathcal{S}$. The class of *strict future time* formulas include only

- $\mathcal{U}(\chi, \psi)$ where ψ and χ are both not necessarily strict future time formulas and
- $\neg\varphi$, $\varphi \wedge \psi$ and $\forall x\varphi$ where φ and ψ are both strict future time formulas.

Dually we have strict and not necessarily strict past time formulas.

It is clear that a strict past time formula only depends on the past for its truth. This classification of formulas is the basis for Gabbay's separation property and separation theorem which is itself useful for establishing the expressive power of the METATEM language. See [9] for details which also include a discussion of the proof-theory of the temporal logics mentioned above.

2.8 Explicit Time

In using executable temporal logic it is often useful to be able to refer explicitly to the time of an event as measured by some clock or calendar. This is especially so when we come to use the logic to reason about temporal databases. To support this feature we will suppose that our logic includes a special 1-ary predicate `time` which at any time t is only true of some syntactic representation of t. That is, there are enough constants and function symbols in the language to allow us to write (the name of) time t and all temporal structures $\mathcal{M} = (T, <, \mathcal{D}, \{\pi_P^n\}_{n\geq 0}, \{\pi_F^n\}_{n\geq 0})$ mentioned are supposed to have the property that $\pi_P^1(t, \mathtt{time}) = \{t\}$ for each $t \in T$.

Note that with this assumption on our structures, the temporal language **FTL** can easily be shown to be as expressive as a two-sorted first-order language.

2.9 METATEM Programming Language

METATEM is really a paradigm for programming languages rather than one particular language. The bases are three:

- programs should be expressed in a temporal language;
- programs should be able to be read declaratively;
- the operation of the program should be interpretative with individual program clauses operating according to the 'declarative past implies imperative future' idea.

Most versions of METATEM use the temporal languages **PTL** and **FTL** with until and since.

The basic idea of declarative languages is that a program should be able to be read as a specification of a problem in some formal language and that running the program should solve that problem. Thus we will see that a METATEM program can easily be read as a temporal sentence and that running the program *should* produce a model of that sentence.

The task of the METATEM program is to build a model satisfying the declared specification. This can sometimes be done by a machine following some arcane, highly complex procedure which eventually emerges with the description of the model (see [16]). That would *not* be the METATEM approach. Because we are describing a programming language, transparency of control is crucial. It should be easy to follow and predict the program's behaviour and the contribution of the individual clauses must be straightforward.

Fortunately, these various disparate aims can be very nicely satisfied by the intuitively appealing 'declarative past implies imperative future' idea of [10]. The METATEM program *rule* is of the form $P \Rightarrow F$ where P is a strict past-time formula and F is a not necessarily strict future-time formula. The idea is that on the basis of the declarative truth of the past time P the program should go on to 'do' F.

In the case of a closed system, a METATEM program is a list $\{P_i \to F_i \mid i = 1, \ldots, n\}$ of such rules and, at least in the propositional case, it represents the **PTL** formula

$$\Box \bigwedge_{i=1}^{n} (P_i \rightarrow F_i).$$

The program is read declaratively as a specification: the execution mechanism should deliver a model of this formula. To do so it will indicate which propositions are true and which are false at time 0, then at time 1, then at time 2, etc. It does this by going through the whole list of rules $P_i \rightarrow F_i$ at each successive stage and make sure that F_i gets made true whenever P_i is. This is called *forward chaining*. For details of the way it does this see [2].

In the first-order case there are several slightly different versions of METATEM. The simplest involves allows only clauses in the forms:

$$\textbf{start} \Rightarrow p(\overline{c})$$

$$\textbf{start} \Rightarrow \Diamond q(\overline{c})$$

$$\forall \overline{X}.[\bullet \bigwedge_{i=1}^{h} k_i(\overline{X}) \Rightarrow p(\overline{X})]$$

$$\forall \overline{X}.[\bullet \bigwedge_{i=1}^{h} k_i(\overline{X}) \Rightarrow \Diamond q(\overline{X})]$$

where each k_i is a literal, p and q are predicates and $\overline{c}$ is a tuple of constants. These constraints enable us to implement the program in a direct way.

3 Temporal Updates

We define a *temporal database* as a *finite* temporal structure, i.e. a temporal structure $\mathcal{M} = (T, <, \mathcal{D}, \{\pi_P^n\}_{n\geq 0}, \{\pi_F^n\}_{n\geq 0})$ obeying the following constraints:

- The set of predicate symbols is finite.
- The interpretation of each predicate is finite, i.e. for every predicate symbol p there are only finitely many tuples $\langle a_1, \ldots, a_n\rangle$ for which there exists a $t \in T$ such that

$$\pi_P^n(t, p(a_1, \ldots, a_n)) = \top.$$

- It is usual for databases that $\mathcal{M}$ be a *Herbrand model*, i.e. the domain $\mathcal{D}$ is identical to the set of constant symbols and every ground term is interpreted into itself.

- We further assume that the set of points where an atomic formula $p(a_1, \ldots, a_n)$ is true must be representable by a *temporal element*, i.e. a finite union of intervals over T. For example, if $T = \mathbb{N}$, then $[0, 10] \cup [20, 30]$ is a temporal element, but the set of all even numbers is not.[1]

These conditions guarantee that temporal databases can be finitely represented as a set of *labelled formulas*, $i : p(a_1, \ldots, a_n)$, where $p(a_1, \ldots, a_n)$ is an atomic formula and i is a temporal element over T; in this context, a labelled formula $i : p(a_1, \ldots, a_n)$ represents a partial temporal structure.

In traditional databases, the update of data means the replacement of the current value of the data by a new one. In temporal databases one is presented with the extra possibility of changing the past, the present and the future. In a nutshell, a temporal update is a 'change in history'. Note the double reference to time in such an expression: *change* relates to the temporal evolution of data, while *history* refers to the temporal record. These two notions of time are independent and coexistent. In analysing updates in temporal databases we have to be able to cope simultaneously with those two notions of time. For that, we present next a *two-dimensional temporal logic*, a formalism that will allow for the simultaneous handling of two references of time. We will stick to the propositional case, for the updates we are concerned with are only atomic updates that may be modelled as propositional atoms.

3.1 Propositional Two-dimensional Temporal Logic

There are several modal and temporal logic systems in the literature which are called *two-dimensional*; all of them provide some sort of double reference to an underlying modal or temporal structure. More systematically, two-dimensional systems have been studied as the result of combining two one-dimensional logic systems [4, 5]. In [5] two criteria were presented to classify a logical system as two-dimensional:

- *The connective approach:* a temporal logic system is two-dimensional if it contains two sets of connectives, each set referring to a distinct flow of time.
- *The semantic approach:* a temporal logic system is two-dimensional if the truth value of a formulas is evaluated with respect to two time points.

[1] This condition basically tells that there is a limit to what temporal data can be represented. The condition itself can be relaxed if data expressivity is enhanced, but some time-stamps will always remain unrepresentable. E.g. if deductive rules are added to the database, periodic sets, like the even numbers, become representable; but since there are uncountably many subsets of $\mathbb{N}$, it will never be possible to represent all of them.

The two criteria are independent and there are examples of systems satisfying each criterion alone, or both. For the purposes of this work, the two-dimensional temporal logic satisfies both criteria, and is thus a *broadly two-dimensional logic*. Both flows of time are assumed to be discrete ($\mathbb{Z}$). In databases it is usual to use $\mathbb{Z}$ instead of $\mathbb{N}$ as the underlying flow of time.

So let $\mathcal{L}$ be a countable set of propositional atoms. Besides the Boolean connectives, we consider two sets of temporal operators. The *horizontal operators* are the usual 'since' ($\mathcal{S}$) and 'until' ($\mathcal{U}$) two place operators, together with all the usual derived operators; the horizontal dimension will be used to represent *valid time* temporal information. The *vertical dimension* is assumed to be a $\mathbb{Z}$-like flow and the operators over such dimensions are the two-place operators 'since vertical' ($\bar{\mathcal{S}}$) and the 'until vertical' ($\bar{\mathcal{U}}$); in general, we use barred symbols when they refer to the vertical dimension. The vertical dimension will be used to represent *transaction time* information. Two-dimensional formulas are inductively defined as:

- every propositional atom is a two-dimensional formula;
- if A and B are two-dimensional formulas, so are $\neg A$ and $A \wedge B$, $\mathcal{S}(A, B)$ and $\mathcal{U}(A, B)$, $\bar{\mathcal{S}}(A, B)$ and $\bar{\mathcal{U}}(A, B)$.

On the semantic side, we consider two flows of time: the horizontal one $(T, <)$ and the vertical one $(\overline{T}, \overline{<})$. Two-dimensional formulas are evaluated with respect to two dimensions, typically a time point $t \in T$ and a time point $\bar{t} \in \overline{T}$, so that a *two-dimensional plane model* is a structure based on two flows of time $\mathcal{M} = (T, <, \overline{T}, \overline{<}, \pi)$. The *two-dimensional assignment* π maps every triple $(t, \bar{t}, p)$ into $\{\top, \bot\}$. The model structure can be seen as a two-dimensional plane, where every point is identified by a pair of coordinates, one for each flow of time (there are other, non-standard models of two-dimensional logics which are not planar; see [5]).

The fact that a formula A is true in the two-dimensional plane model $\mathcal{M}$ at point $(t, \bar{t})$ is represented by $\mathcal{M}, t, \bar{t} \models A$ and is defined inductively as:

$\mathcal{M}, t, \bar{t} \models p$	iff $\pi(t, \bar{t}, p) = \top$.
$\mathcal{M}, t, \bar{t} \models \neg A$	iff it is not the case that $\mathcal{M}, t, \bar{t} \models A$.
$\mathcal{M}, t, \bar{t} \models A \wedge B$	iff $\mathcal{M}, t, \bar{t} \models A$ and $\mathcal{M}, t, \bar{t} \models B$.
$\mathcal{M}, t, \bar{t} \models \mathcal{S}(A, B)$	iff there exists a $t' \in T$ with $t' < t$ and $\mathcal{M}, t', \bar{t} \models A$ and for every $t' \in T$, whenever $t' < t'' < t$ then $\mathcal{M}, t'', \bar{t} \models B$.
$\mathcal{M}, t, \bar{t} \models \mathcal{U}(A, B)$	iff there exists an $t' \in T$ with $t < t'$ and $\mathcal{M}, t', \bar{t} \models A$ and for every $t'' \in T$, whenever $t < t'' < t'$ then $\mathcal{M}, t', \bar{t} \models B$.

$\mathcal{M},t,\bar{t} \models \bar{\mathcal{S}}(A,B)$ iff there exists a $\bar{t}' \in T$ with $\bar{t}' < \bar{t}$ and $\mathcal{M},t,\bar{t}' \models A$ and for every $\bar{t}'' \in T$, whenever $\bar{t}' < \bar{t}' < \bar{t}$ then $\mathcal{M},t,\bar{t}'' \models B$.

$\mathcal{M},t,\bar{t} \models \bar{\mathcal{U}}(A,B)$ iff there exists a $\bar{t}' \in T$ with $\bar{t} < \bar{t}'$ and $\mathcal{M},t,\bar{t}' \models A$ and for every $\bar{t}' \in T$, whenever $\bar{t} < \bar{t}'' < \bar{t}'$ then $\mathcal{M},t,\bar{t}'' \models B$.

Note that the semantics of horizontal and vertical operators are totally independent from each other, i.e. the horizontal operators have no effect on the vertical dimension and similarly for the vertical operators. If we consider the formula without the vertical operators, we have a one-dimensional horizontal $\mathcal{U}\,\mathcal{S}$-temporal logic: similarly for the vertical temporal logic. Unary temporal predicates can be defined for both dimensions in the usual way, so we get $\Box$, $\blacksquare$, $\Diamond$, $\blacklozenge$, etc. for the horizontal dimension and $\overline{\Box}$, $\overline{\blacksquare}$, $\overline{\Diamond}$, $\overline{\blacklozenge}$, etc, for the vertical one.

3.2 Two-dimensional Separation

The separation result does not hold for two-dimensional temporal logic above; i.e. given a two-dimensional formula, it is not guaranteed that there exists an equivalent formula that is a conjunction of formulas of the form

$$Past \vee Present \vee Future,$$

where *Present* contains no temporal operators, *Past* contains no future (i.e. $\mathcal{U}$ and $\bar{\mathcal{U}}$) operators and *Future* contains no past (i.e. $\mathcal{S}$ and $\bar{\mathcal{S}}$) operators. For example, the formula $\Diamond\overline{\blacklozenge}p$ cannot be separated.

However, for a very useful class of formulas called *temporalized formulas*, we obtain a restricted notion of separation, called *vertical separation*, which is strong enough for our purposes here.

A *temporalized formula* is a two-dimensional formula in which no vertical temporal operator appears inside the scope of a horizontal temporal operator. E.g. $\overline{\Diamond}\Diamond A$ is a temporalized formula, but $\Diamond\overline{\Diamond}A$ is not. Temporalized formulas can be seen as the result of having the vertical temporal dimension applied *externally* to a one-dimensional temporal logic, as discussed in [4].

A formula is *vertically separable* if it is equivalent to a formula that is a conjunction of formulas of the form

$$v\text{-}Past \vee v\text{-}Present \vee v\text{-}Future,$$

where *v-Past* contains no vertical future operators (i.e. $\bar{\mathcal{U}}$ and its derived operators), *v-Future* contains no vertical past (i.e. $\bar{\mathcal{S}}$ and its derived operators) and no vertical operator occurs in *v-Present*. The following result was proved in [4].

THEOREM 3.1 *If A is a temporalized formula then A is vertically separable.*

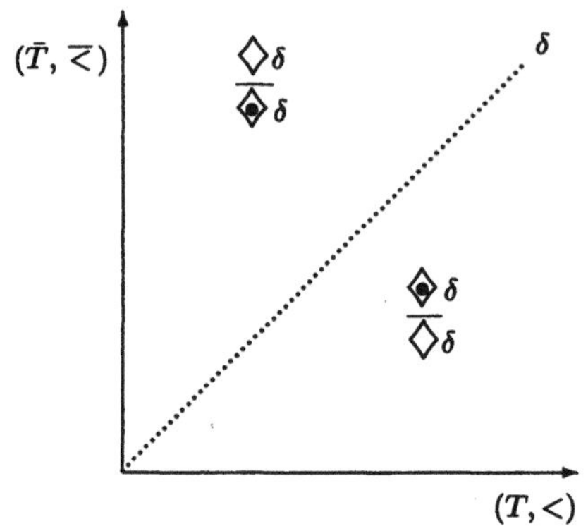

Figure 1. The two-dimensional diagonal.

Of course, a totally analogous horizontal separation can be obtained for formulas where the horizontal dimension is applied externally to the vertical one. In this work, vertical separation is emphasized because in our modelling of temporal database evolution (see Section 3.4), the horizontal dimension represents the state of the database, while the vertical dimension represents the evolution of such temporal database; hence it is the vertical dimension that is *external* to the database.

In this context, the horizontal dimension representing the temporal database state is called the *valid time* dimension. The vertical dimension representing the evolution of temporal database states is called the *transaction time* dimension.

3.3 The Two-dimensional Diagonal

We now examine some properties of the diagonal in two-dimensional plane models. The diagonal is a privileged line in the two-dimensional model intended to represent the sequence of time points we call 'now', i.e. the time points which an historical observer is expected to traverse. The observer is on the diagonal when he or she poses a query (i.e. evaluates the truth value of a formula) on a two-dimensional model.

So let δ be a special atom that denotes the points of the diagonal, which is characterized by the following property: for every $t \in T$ and every $\bar{t} \in \mathbb{Z}$:

$$\mathcal{M}, t, \bar{t} \models \delta \quad \text{iff} \quad t = \bar{t}.$$

The diagonal is illustrated in Figure 1.

The following formulas are true at all points of the two-dimensional plane model:

$$\blacklozenge\delta \vee \delta \vee \Diamond\delta \qquad \blacklozenge\delta \leftrightarrow \overline{\Diamond}\delta$$
$$\delta \leftrightarrow (\Box\neg\delta \wedge \blacksquare\neg\delta \wedge \overline{\Box}\neg\delta \wedge \overline{\blacksquare}\neg\delta) \qquad \overline{\blacklozenge}\delta \leftrightarrow \Diamond\delta.$$

The diagonal divides the two-dimensional plane in two semi-planes. The semi-plane that is to the (horizontal) left of the diagonal is 'the past', and the formulas $\Diamond\delta$ and $\overline{\blacklozenge}\delta$ are true at all points of this semi-plane. Similarly, the semi-plane that is to the (horizontal) right of the diagonal is 'the future', and the formulas $\blacklozenge\delta$ and $\overline{\Diamond}\delta$ are true at all points of this semi-plane. Figure 1 puts this fact in evidence.

The propositional approach of this section differs from the first-order treatment of temporal features in the previous section. To reconcile these two different approaches a *propositional abstraction of database manipulations* can be developed. However, for space reasons, we omit such presentation, referring the reader to [7] for details.

3.4 Temporal Database Evolution

In describing the evolution of a temporal database, we have to distinguish the database evolution from the evolution of the world it describes. The 'world', also called the *Universe of Discourse*, is understood to be any particular set of objects and relations between them in a certain environment that we may wish to describe. The database, in its turn, contains a description of the world. Conceptually, we have to bear in mind two distinct types of evolution, as introduced in [6]:

- the *evolution of the modelled world* is the result of changes in the world that occur independently of the database;
- a temporal database contains a description of the history of the modelled world that is also constantly changing due to database updates, generating a sequence of database states. This *evolution of the temporal description* does not depend only on what is happening at the present; changes in the way the past is viewed also alter this historical description; moreover, changes in expectations about the future, if those expectations are recorded in the database, also generate an alteration of the historical description. This process is also called *historical revision*.

These two distinct concepts of evolution are reflected by a distinction between two kinds of flows of time, whether their time points refer to a moment in the history of the world, or whether they are associated with a moment in time at which a historical description is in the database.

Several different names are found in the literature for these two time concepts. The former is called *evaluation time* [13, 9], *historical time*

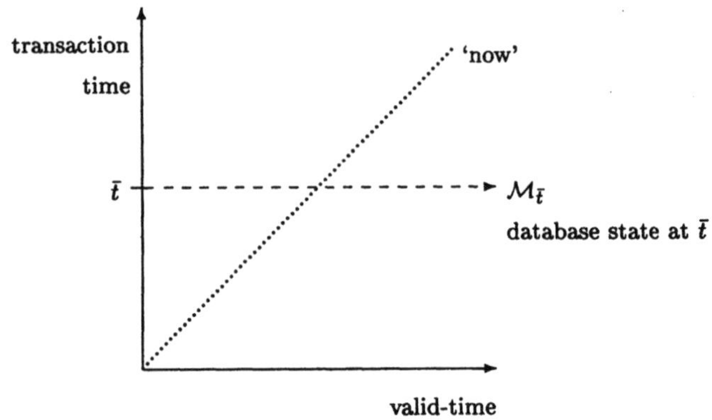

Figure 2. Two-dimensional database evolution.

[6], *valid time* [19] and *event time* [14]. The latter time concept is called *utterance time* [13], *reference time* [9], *transaction time* [6, 19] and *belief time* [20]. In this presentation we chose to follow a glossary of temporal database concepts proposed in [11], calling the former valid time, which is associated with the horizontal dimension in our two-dimensional model, and calling the latter transaction time, which is associated with the vertical dimension. So we use the two-dimensional plane model to simultaneously cope with the two notions of time in the description of the evolution of a temporal database, as illustrated in Figure 2.

Let $\mathcal{M} = (T, <, \overline{T}, \overline{<}, \pi)$ be a two-dimensional plane model; its *horizontal projection* with respect to the vertical point $\bar{t} \in \overline{T}$ is the one-dimensional temporal model

$$\mathcal{M}_{\bar{t}} = (T, <, \pi_{\bar{t}}),$$

such that, for every propositional atom q, time points $t \in T$ and $\bar{t} \in \overline{T}$,

$$\pi_{\bar{t}}(t, q) = \top \quad \text{iff} \quad \pi(t, \bar{t}, q) = \top.$$

It follows that for every horizontal $\mathcal{U}\,\mathcal{S}$-formula A and for every $t \in T$ and $\bar{t} \in \overline{T}$, $\mathcal{M}_{\bar{t}}, t \models A \quad \text{iff} \quad \mathcal{M}, t, \bar{t} \models A$. The horizontal projection represents a state of a temporal database.

Updating temporal databases requires that, besides specifying the atom to be inserted or deleted, we specify its valid time. For that reason, it is convenient to use the notation of time-stamped atoms to represent the data being inserted and deleted. As a result, infinite updates are possible as long as such update is representable by a finite set of labelled formulas, where the label (time-stamp) is a temporal element. For example, $\Theta_- = \{[-\infty, +\infty] : p\}$ deletes the atom p for all times.

An *update pair* (θ_+, θ_-) consists of two finite disjoint sets of time-stamped atoms, where θ_+ is the *insertion set* and θ_- is the *deletion set*;

by disjoint sets, in the context of data representation, it is meant that it is not the case that $\tau_+ : p \in \theta_+$ and $\tau_- : p \in \theta_-$ such that $\tau_+ \cap \tau_- \neq \varnothing$. We say that an update pair determines or characterizes a *database update* $\Theta_{\bar{t}}$ occurring at transaction time $\bar{t} \in \mathbb{Z}$ if the application of the update function $\Theta_{\bar{t}}$ to the database state $\mathcal{M}_{\bar{t}} = (T, <, \pi_{\bar{t}})$ generates a database state $\Theta_{\bar{t}}(\mathcal{M}_{\bar{t}}) = (T, <, \Theta_{\bar{t}}(\pi_{\bar{t}}))$ satisfying, for every propositional atom q and every time point $t \in T$,

- if $\tau : q \in \theta_+$, then $\Theta_{\bar{t}}(\pi_{\bar{t}})(t, q) = \top$ for every $t \in \tau$;
- if $\tau : q \in \theta_-$, then $\Theta_{\bar{t}}(\pi_{\bar{t}})(t, q) = \bot$ for every $t \in \tau$;
- if neither $t : q \in \theta_+$ nor $t : q \in \theta_-$, then $\Theta_{\bar{t}}(\pi_{\bar{t}})(t, q) = \pi_{\bar{t}}(t, q)$.

The first item corresponds to the insertion of atomic information, the second one corresponds to the deletion of atomic information, and the third one corresponds to the persistency of the unaffected atoms in the database. Note that the disjoint sets θ_+ and θ_- are represented in the same way that the underlying database, so that we can represent a temporal database update schematically as:

$$\Theta_{\bar{t}}(\mathcal{M}_{\bar{t}}) = \mathcal{M}_{\bar{t}} \cup \theta_+ - -\theta_-.$$

When the sets θ_+ and θ_- are not disjoint, i.e. the update it is trying to insert and remove the same information, the situation is undetermined; typically this would mean that the transaction in which the update was generated should be rolled back. The update $\Theta_{\bar{t}}$ is a database state transformation function. An update may be empty ($\theta_+ = \theta_- = \varnothing$), in which case the transformation function is just the identity and the database state remains the same.

4 Imperative History

The two-dimensional temporal model can be applied in two distinct situations, namely:

- in the context of standard, one-dimensional temporal databases, the two-dimensional model is used to represent the evolution of the database. The current state of the database is constantly modified, i.e. history is constantly been rewritten, and past states of the database are not recorded;
- in the context of *bitemporal* databases, both dimensions are stored, so the two-dimensional model can be seen as modelling a state of the database.

In each of these two situations, the two-dimensional model allows us to lift the restrictions imposed by the imperative future to METATEM rules of the form past $\wedge$present $\rightarrow$future

in a distinct way. In the context of bitemporal databases, this may lead to a two-dimensional imperative history over bitemporal databases, which is quite outside the scope of this presentation and is left as future work. So we concentrate on the first option, which is purely one dimensional.

4.1 Imperative History for a One-dimensional Database

When only a single state representing the history of the world is stored in the database, an *imperative history* rule has the general format of

history $\rightsquigarrow$ history

which, in formal terms is a formula of the form

$$\varphi \rightsquigarrow \psi$$

where both φ and ψ are any **FTL**-formulas that may refer to the (one-dimensional) present, past or future; φ and ψ are thus called *historical formulas*.

The meaning of an imperative history rule is, however, given in terms of the two-dimensional model. Therefore, the rule above is seen as representing the two-dimensional formula

$$\forall X \forall Y (\varphi \rightarrow \bar{\circ}\, \psi)$$

which clearly is a vertically separated formula. In this paper we will use such rules to specify that at every transaction time t, the corresponding formula holds at (t, t), i.e. on the diagonal.

This new view of **FTL**-formulas consists of the following reading of a temporal database evolution: 'for all substitutions of the free variables that makes (the temporal query) φ true at the current time (i.e. on the diagonal point of the current state) force ψ to be true at this valid time in the next state'. The formula φ is called the *condition* or *query* part of the rule, and ψ is called the *action* part.

Several restrictions are imposed on the format of those rules. First, it is required that the set of free variables of ψ be free in φ so as to avoid undetermined actions. The complete rendering of an imperative history rule becomes:

$$\forall X \forall Y (\varphi(X, Y) \rightarrow \bar{\circ}\, \psi(X)).$$

Second, it is required of the query part of the rule to be *range restricted.* This condition guarantees that queries have at most finitely many answers, and hence only finitely many actions to be executed. Range restrictedness demands that all the free variables occurring in the query formula should occur in a positive literal of the formula.

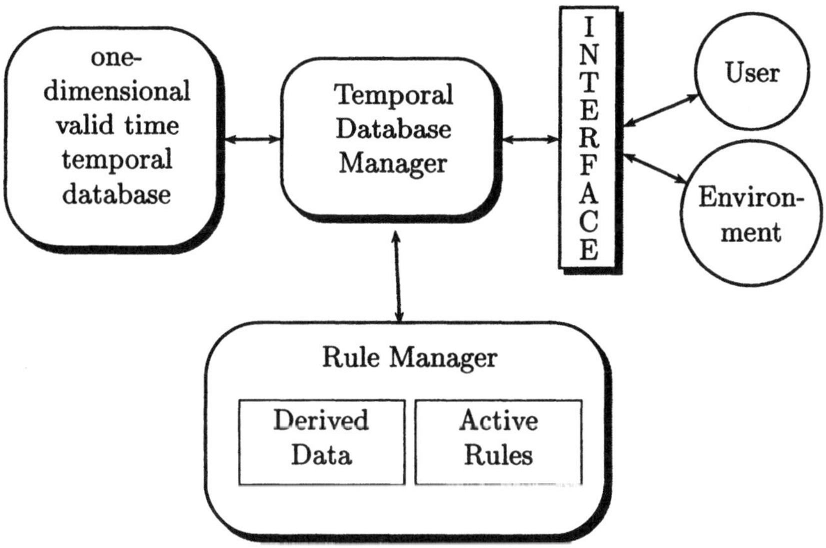

Figure 3. Possible architecture that supports IMPHIST rules.

Finally, it is convenient to require that the action part of the rule be *deterministic* so that the rule management system will always know how to execute the rule. One way to avoid non-determinism is to constrain the format of the action formula to a conjunction of positive and negative literals, possibly preceded by a string of $\circ$s or $\boldsymbol{\odot}$s. The following is a deterministic action formula:

$$\boldsymbol{\odot}\texttt{clear\_top(X)} \wedge \neg\texttt{occupied(X)} \wedge \circ\circ\texttt{place(Obj, X)}.$$

Note that in imperative history rules forcing ψ to be true may falsify φ. For instance, the simple rule $p \leadsto \neg p$ is legitimate, for it is understood as $p \rightarrow \bar{\circ}\, \neg p$, i.e. whenever $t : p$ is true in the current state at transaction time t, $t : \neg p$ will hold at the next database state, not causing any inconsistency.

4.2 A System's Architecture to Support Imperative History

Conceptually, the support of imperative history rules is not much different from the support of standard rules found in non-temporal databases. As shown in Figure 3, an architecture of such a system is centered on the temporal database manager (TDM), which performs all the tasks of a normal database manager, plus the manipulation of time.

Imperative history rules are stored in the rule manager (RM), and can be precompiled and optimized. The TDM activates the RM which, by selecting which rules to execute (see Section 6 on triggers), submit queries to the database via the TDM. The TDM then sends the answers back to the RM, and at the end of the rule execution cycle receives from it the actions that are to be performed. Some of these actions are temporal database updates which are sent to the database, others are messages sent to user or some kind of interaction with the temporal database's environment, such as an order to print a cheque or to signal an alarm, which are sent by the TDM via the environment interface module; others, still, may be triggering actions that will come back to the RM.

This apparently excessive traffic through the TDM in the processing of rules may be diminished if the RM is allowed to interact separately with the constituent parts of the TDM. In this way, the RM may interact directly with the database access module for querying and updating, receive triggers from and send triggers to the transaction selector, and send messages and requests to the environment directly to the interface module, avoiding the TDM in those cases. Of course, this is just a conceptual architecture, and each systems may adapt it differently to its own design philosophy.

5 Example

In this section we see the 'Imperative History' ideas in action in a very simple example. The example concerns a patient monitoring system (PMS) for use in an intensive care ward of a hospital. It is based on the PMS system described in a software engineering context in [21] which has been implemented in a prototype version by an executable temporal logic language in [17]. The description and these implementations all concern a distributed system problem focusing on formal specifications of the communication between separate modules. Here we will introduce a historical dimension by requiring patient data to be recorded.

So suppose that we have a central nurse console (NC) interacting with several distributed patient monitors. The responsibilities of the NC will include:

- recording information about a constant stream of patient data – here just heart rates – from the patient modules;
- recording data manually input by the nurse about changes in occupants of beds;
- notifying the nurse of alarming events (heart rate out of safety range etc);
- answering queries about beds and hearts in the present and the past;
- and correcting incorrectly recorded information.

A fully comprehensive database for this task would need to be two-dimensional so that it could record information about the correction of past mistakes. Such information would be needed to explain actions which were taken on the grounds of information subsequently corrected. However, we will consider a simpler, one-dimensional temporal database adequate to hold a representation of the most up to date account of the history of the situation in the intensive care ward. Changes to data about the past will be allowed (as mistakes in recording do occur) but we will not necessarily be able to reconstruct the superseded model of the history of the ward.

The users will be primarily interested in the following two predicates:

`heart_rate(Patient,Rate)`	the heart rate of `Patient` is `Rate`
`occ(Bed,Patient)`	`Bed` is occupied by `Patient`.

These are the most important predicates which vary in time in the world which the database will try to model. However it is easier to actually record some different predicates in the database. We introduce:

`hrm(Patient, Rate)`	the heart rate of `Patient` is measured as `Rate`
`chpat(Bed,Old,New)`	`Bed` is vacated by patient `Old` and occupied by patient `New`.

Empty beds can easily be handled within this formalism.

The simple (one-dimensional) temporal formula

$$\forall \mathtt{p}.\ \forall \mathtt{r}.\ \overline{\Box}\Box(\mathtt{heart\_rate(p,r)} \Leftrightarrow \mathcal{S}(\mathtt{hrm(p,r)}, \neg\exists \mathtt{r2}.\ \mathtt{hrm(p,r2)}))$$

specifies how to interpolate the most recent heart rate reading to any time. Below we will show how to render this in an imperative history rule.

Similarly, `chpat` and `occ` are related by

$$\forall \mathtt{b}.\forall \mathtt{p}.\overline{\Box}\Box(\mathtt{occ(b,p)} \Leftrightarrow \mathcal{S}(\exists \mathtt{q}.\mathtt{chpat(b,q,p)}, \neg\exists \mathtt{q'}.\ \mathtt{chpat(b,p,q')})).$$

In general with deductive databases, whether temporal or otherwise, there is a distinction between basic and derived predicates. The database manager must be told which predicates are to have their history recorded. Other predicates play subsidiary roles: either being able to be derived from the recorded ones or appearing temporarily to produce, in combination with Imperative History rules, systematic changes in the database. In our example the database will only record the history of the predicates `hrm` and `chpat`.

Thus, the database manager will be responsible for

- translating automatically recorded heart rate measurements into database updates;
- doing the same for information entered by the nurse;
- answering queries which may be expressed in terms of derived predicates;
- producing the side-effect of an alarm which is described in terms of derived predicates;

- allowing error correction (including corrections of corrections etc.) and disallowing non-sensical attempts at error correction.

Although our database is one dimensional, we use two dimensions of time to describe the way it changes over time. In this two-dimensional approach, the predicates we have introduced take on a slightly expanded semantics: for example, **heart_rate(Patient,Rate)** holding at valid time t and transaction time $\bar{t}$ will mean that at transaction time $\bar{t}$, the database's model of the world contained the information that at valid time t, the heart rate of **Patient** is **Rate**.

Let us now examine some of the properties of this two-dimensional account and see which need explicit statement.

We suppose that automatic measurements arrive in the form of atoms t : **bhrm(b,r)** labelled with time instants. Such an atom indicates a heart rate measurement of **r** on the patient in bed **b** was made at the instant t. An atom like this may arrive at the manager at any time after t and so its effects will be recorded at some even later transaction time.

The effect of the arrival of such an atom is given by

$$\blacklozenge\, (\mathtt{time(t)} \wedge \mathtt{bhrm(b,r)} \wedge \mathtt{occ(b,p)}) \rightsquigarrow \blacklozenge\, (\mathtt{time(t)} \wedge \mathtt{hrm(p,r)})$$

and the procedural effect of this rule is as follows. Suppose that at transaction time $\overline{t_1}$ the labelled atom t_2 : **bhrm(b,r)** is true. Then, the database manager will check, from the rule manager, whether t_2 : occ(b, p) holds for any p. If so, then by transaction time $t_1 + 1$, the database will make t_2 : **hrm**(p, r) hold.

Information entered by the nurse has a slightly more complicated effect. Suppose that the nurse enters the information that at time t, patient p vacates bed b and patient q replaces her or him. We use the atom **nurse_requests_chpat_add(b,p,q,t)** to indicate this. One of the effects of this information being entered is often the transaction

$$\blacklozenge\, \mathtt{nurse\_requests\_chpat\_add(b,p,q,t)} \rightsquigarrow \blacklozenge\, (\mathtt{time(t)} \wedge \mathtt{chpat(b,p,q)}).$$

However we might want to disallow this direct update if the database currently shows that occ(b,p) is not true at that valid time. Thus we use

$$\mathtt{nurse\_requests\_chpat\_add(b,p,q,t)} \wedge \mathtt{occ(b,p)}$$
$$\rightsquigarrow \blacklozenge\, (\mathtt{time(t)} \wedge \mathtt{chpat(b,p,q)})$$

instead.

A systematic change to history

There is also another effect of the nurse entering a bed-change fact. Say that the nurse enters the information that at time t, the patient p is replaced by patient q in bed b. Any heart rate measurements already recorded from bed b but with later valid times than t are recorded as being about patient p but are now known to refer to patient q instead. Thus we have

$$\texttt{nurse\_requests\_chpat\_add(b,old,new,t)} \wedge \\ \blacklozenge\,(\texttt{time(t)} \wedge \neg\texttt{chpat(b,old,new)}) \rightsquigarrow \blacklozenge\,(\texttt{time(t)} \wedge \texttt{chpat(b,old,new)}).$$

This rule is a very good example of the power of two-dimensional temporal logic. It is a very clear case of a systematic change to history.

Queries and side-effects

Queries are very straight forward to deal with. They are likely to be expressed in terms of occ and heart_rate and so can be answered using the derivation laws for these two predicates.

The syntax for the first of these rules is

$$\mathcal{S}(\texttt{hrm(p,r)}, \neg\exists \texttt{r2.\,hrm(p,r2)}) \rightsquigarrow \texttt{heart\_rate(p,r)}.$$

Note that there is no interesting two-dimensional character to this rule.

The side-effect of an alarm sounding for dangerous heart rate measurements can be produced by combining a derivation

$$\neg\,\mathcal{S}\,(\texttt{heart\_rate(p,r)} \wedge (\texttt{r} > \texttt{maxrate}), \texttt{alarm\_off(p)}) \rightsquigarrow \texttt{alarm!(p)}$$

with the hard-wiring of the current truth of predicate alarm! to the alarm bell. We associate an alarm sounding at time t with the predicate alarm! being true at time t in the model of the ward kept at transaction time t. The introduction of a nurse request to turn the alarm off and the corresponding predicate alarm_off is to prevent historical situations causing alarms.

Corrections

Corrections can not be handled by allowing the nurse to enter in directly the falsity of chpat for a particular historical time. This is because we must check for attempts at silly corrections. Thus we separate the fact of the nurse requesting a correction from the act of updating in accordance with that correction.

As well as corrections involving removal of recorded facts we also need to be able to add facts about the past. The nurse's requests are formalized in the predicate

nurse_requests_hrm_add(p,r,t)

which means that the nurse requests that hrm(p,r) be added for valid time t and similar predicates nurse_requests_hrm_remove(p,r,t), nurse_requests_chpat_add(b,p,q,t) and nurse_requests_chpat_remove(b,p,q,t).

We introduce a predicate `error` whose truth at particular time on the diagonal has the side-effect of notifying the nurse that she or he has just requested a silly correction.

The rules which define the truth of `error` look like the following:

$$\texttt{nurse\_requests\_hrm\_add(p,r,t)} \wedge \blacklozenge\,(\texttt{time(t)} \wedge \texttt{hrm(p,r)}) \rightsquigarrow \bigcirc(\texttt{error}).$$

When `error` does not hold, then we use a set of Imperative History formulas to bring about the correction:

$$\begin{array}{l}\texttt{nurse\_requests\_chpat\_remove(b,old,new,t)}\wedge \\ \blacklozenge\,(\texttt{time(t)} \wedge \texttt{chpat(b,old,new)}) \rightsquigarrow \blacklozenge\,(\texttt{time(t)} \wedge \neg\texttt{chpat(b,old,new)}) \\ \texttt{nurse\_requests\_chpat\_add(b,old,new,t)}\wedge \\ \blacklozenge\,(\texttt{time(t)} \wedge \neg\texttt{chpat(b,old,new)}) \rightsquigarrow \blacklozenge\,(\texttt{time(t)} \wedge \texttt{chpat(b,old,new)})\end{array}$$

and similarly for `hrm`.

Of course, just as with the nurse entering new information about bed changes, there is a consequence for heart rate measurements data from the correction of bed occupant data. We have

$$\begin{array}{l}\texttt{nurse\_requests\_chpat\_add(b,old,new,t1)} \\ \wedge\blacklozenge\,(\texttt{time(t2)} \wedge \texttt{hrm(old,r)} \wedge \blacklozenge\,(\texttt{time(t1)} \wedge \neg\texttt{chpat(b,old,new)}) \\ \rightsquigarrow \blacklozenge\,(\texttt{time(t2)} \wedge \texttt{hrm(new,r)} \wedge \neg\texttt{hrm(old,r)}) \\ \quad \text{and} \\ \blacklozenge\,(\texttt{nurse\_requests\_chpat\_remove(b,old,new,t1)} \\ \wedge\blacklozenge\,(\texttt{time(t2)} \wedge \texttt{hrm(new,r)} \wedge \blacklozenge\,(\texttt{time(t1)} \wedge \texttt{chpat(b,old,new)})) \\ \rightsquigarrow \blacklozenge\,(\texttt{time(t1)} \wedge \texttt{hrm(old,r)} \wedge \neg\texttt{hrm(new,r)}).\end{array}$$

Correcting corrections turns out to be just the same as entering in new information.

Persistence

By requiring the database manager to follow the specification above – i.e. the explicit rules and the informal description of its operation – we actually end up abstractly building a two-dimensional temporal structure. A ground atomic formula is true in this structure at times $(t, \bar{t})$ if and only if the database claims at transaction time $\bar{t}$ that the formula is true at valid time t. It is clear that this structure is a model of the formulas explicitly mentioned above as rules.

However, the structure is also a model of many other formulas. For example, because of the operation of the database manager, only changing stored information in response to new data or entered requests, there are several *persistence* axioms. The one below indicates that heart rate measurements persist from an automatic measurement unless a request for change arrives:

$$\overline{\Box}\Box\forall \mathtt{p}.\ \forall \mathtt{b}.\ \forall \mathtt{r}.\ \big[\mathtt{hrm(p,r)} \Leftrightarrow \overline{\mathcal{S}}((\mathtt{bhrm(b,r)} \land \mathtt{occ(b,p)}) \lor \mathtt{nurse\_requests\_hrm\_add(p,r)}, \neg\mathtt{nurse\_requests\_hrm\_remove(p,r)})\big].$$

It is important to note that these two-dimensional rules describe the abstract model which is built by the database manager but they do not have to be explicitly programmed by the user.

Summary

In summary then, we can think of the running of the NC as the gradual construction of a two-dimensional temporal structure. Its language involves many predicates which have a variety of connections with the real world:

- `heart_rate` and `occ` affect the truth of queries;
- `hrm` and `chpat` represent the data stored;
- `alarm` and `error` produce side-effects;
- `nurse_requests_hrm_add` etc. is true when the nurse makes a request to correct data via the NC;
- and `bhrm` is true when data arrives from a patient monitor.

There are also a variety of explicit or implicit properties exhibited by the structure:

- persistence arises from the inertia of the database;
- derivation rules define some predicates in terms of others;
- and Imperative History rules describe how various predicates require changes to the history recorded in the database.

6 Triggers

In this section we will enhance imperative history rules with *triggers*. Informally, a trigger is a mechanism that enables the processing of a rule. Only when a rule is enabled (i.e. when its trigger is fired) is its antecedent checked against the temporal database and the corresponding actions are executed; otherwise, when the trigger of a rule is not fired, the rule remains disabled and is ignored.

A trigger will be represented by a guard (i.e. a label) placed in front of the rule:

$$\textit{trigger} : \mathsf{Condition} \leadsto \mathsf{Action}$$

and it is read as

when *trigger* is fired **if** Condition holds **then** execute Action.

A rule without a trigger can be thought of as being labelled by truth, $\top$, so it is always enabled.

One of the main reasons to include triggers in rules is to increase the system's efficiency. Without triggers, *all* rules have their antecedents checked against the database at every rule evaluation cycle – which, according to the two-dimensional model, means at every transaction time. When triggers are present, only the subset of enabled rules have their antecedent evaluated. We also say that the rule is *fired* when its trigger is.

Another good reason to include triggers is that they may be treated as channels through which the outside world communicates with the system, in the sense of [15]. For instance, in the rule sketched below:

$$alarm\_ringing : \mathsf{door\_open} \leadsto \mathsf{close\_door}$$

the fact that the alarm is ringing is *external* to the system. This rule is activated only when it is communicated to the system that the alarm is ringing; therefore the trigger is called *external*. Only then is it checked whether the door is open.

Triggers may be also used in connection with an event *internal* to the database, such as the insertion (+) or deletion (−) of a fact. For example, when a door is recorded closed, we may wish to issue several warnings:

$$+door\_closed : \mathsf{emergency\_mode} \leadsto (\mathsf{lock\_door} \wedge trigger\_lights).$$

As also seen in the example above, the Action-part of the rule can also fire an *internal* trigger (*trigger_lights*), which on its turn will activate another rule at the next transaction time. Therefore a *chain of control* of rule execution may be created through the rules.

Finally, there may be triggers related to events associated to a resource managed by the system, such as the system clock, file access, etc. For example:

$$time\_is(8am) : \mathsf{user\_logged\_in(User)} \leadsto \mathsf{good\_morning\_to(User)}.$$

As we see, triggers can also carry parameters and have the same 'appearance' as a normal database predicate.

There must be a mechanism for linking external and system triggers to their corresponding external and system events, but this will not be discussed here.

We will use the two-dimensional view of temporal database evolution to give a formal semantics to triggers. *Internal, external* and *system* triggers are seen as *non-persistent predicates.* We have to distinguish between trigger predicates and data predicates. When we defined the update semantics of data predicates in Section 3.4, whatever data was neither inserted nor deleted had its validity persisting into the following database state at the following transaction time.

However, trigger predicates have a 'fixed duration' of a single transaction time unit. The insertion a trigger atomic predicate in the database corresponds to the *firing of a trigger* with the respective set of parameters. A trigger fired (i.e. inserted) at transaction time t will hold at the next transaction time, $t+1$. It will only hold at transaction time $t+2$ if has been re-fired (i.e. re-inserted) at $t+1$; otherwise it is removed from the database.

With this dynamic semantics for trigger predicates, we can construct *trigger expressions* by combining trigger predicates with Boolean operators. A guarded rule of the form

$$\mathit{trigger\_expression} : \mathsf{Condition} \leadsto \mathsf{Action}$$

is then evaluated simply as

$$(\mathit{trigger\_expression} \wedge \mathsf{Condition}) \leadsto \mathsf{Action}.$$

The difference between *trigger_expression* and Condition remains solely in the dynamic behaviour of its components.

7 Conclusion

The main contribution of this paper is the suggestion of an appropriate executable temporal logic for declarative management of temporal databases. By using a two-dimensional temporal language and restricting our attention to a certain simple form of formula, we can express many useful patterns of updates and yet provide them with a direct procedural effect. We have demonstrated the usefulness of this language in a small example: in future work we hope to apply the language and develop the techniques in a much larger example. It will be interesting to demonstrate the technique in action on a two-dimensional temporal database where the use of Imperative History rules in database management will be slightly different.

In other future work we hope to address questions of implementability of various restricted Imperative History languages.

Marcelo Finger
Universidade de São Paulo, Brasil.

Mark Reynolds
Murdoch University, Western Australia.

References

1. James F. Allen. An interval based representation of temporal knowledge. In *Proc. 7th IJCAI*, 1981.
2. Howard Barringer, Michael Fisher, Dov M. Gabbay, Graham Gough, and Richard P. Owens. METATEM: A Framework for Programming in Temporal Logic. In *REX Workshop on Stepwise Refinement of Distributed Systems: Models, Formalism, Correctness*, volume 430 of *LNCS*, pages 94–129, Mook, Netherlands, 1989. Springer-Verlag.

3. Howard Barringer, Michael Fisher, Dov Gabbay, Richard Owens, and Mark A. Reynolds, editors. *The Imperative Future, Vol. 1.* Research Studies Press, 1996.
4. Marcelo Finger and Dov M. Gabbay. Adding a Temporal Dimension to a Logic System. *Journal of Logic Language and Information*, 1:203–233, 1992.
5. Marcelo Finger and Dov Gabbay. Combining temporal logic systems. *Notre Dame Journal of Formal Logic*, 37(2):204–232, 1996.
6. Marcelo Finger. Handling Database Updates in Two-dimensional Temporal Logic. *J. of Applied Non-Classical Logic*, 2(2):201–224, 1992.
7. Marcelo Finger. *Changing the Past: Database Applications of Two-dimensional Temporal Logics.* PhD thesis, Imperial College, Department of Computing, February 1994.
8. Michael Fisher and Richard P. Owens, editors. *Proceedings of IJCAI Workshop on Executable Modal and Temporal Logics, Chambery, France 1993*, volume 897 of *LNAI*. Springer-Verlag, 1995.
9. Dov M. Gabbay, Ian M. Hodkinson, and Mark A. Reynolds. *Temporal Logic – Mathematical Foundations and Computational Aspects, Vol. 1.* Oxford University Press, 1994.
10. Dov M. Gabbay. The Declarative Past and the Imperative Future. In B. Banieqbal, H. Barringer, and A. Pnueli, editors, *Coloquium on Temporal Logic and Specifications – Lecture Notes in Computer Science 389*, Manchester, April 1987. Springer-Verlag.
11. Christopher S. Jensen, James Clifford, Shashi K. Gadia, Arie Segev, and Richard T. Snodgrass. A glossary of temporal database concepts. *SIGMOD RECORD*, 21(3):35–43, September 1992.
12. Hans Kamp. *Tense Logic and the Theory of Linear Order.* PhD thesis, Michigan State University, 1968.
13. Hans Kamp. Formal properties of now. *Theoria*, 35:227–273, 1971.
14. L. Edwin McKenzie, Jr. and Richard T. Snodgrass. Evaluation of relational algebra incorporating the time dimension in databases. *ACM Computing Surveys*, 23(4):501–544, December 1991.
15. Robin Milner. *Communication and Concurrency.* Prentice-Hall, 1989.
16. Amir Pnueli and Roni Rosner. On the synthesis of a reactive module. In *Proceedings of the Sixteenth Symposium of Principles of Programming Languages*, pages 179 – 190, 1989.
17. Mark A. Reynolds. Towards first-order concurrent meta-tem. In *[8]*. 1995.
18. Mark Reynolds. Axiomatising first-order temporal logic: Until and since over linear time. To appear in *Studia Logica*, 1996.
19. Richard T. Snodgrass and Ilso Ahn. A taxonomy of time in databases. In *ACM SIGMOD International Conference on Management of Data*, pages 236–246, Austin, Texas, May 1985.
20. Suri Sripada. A Basis for Historical Deductive Databases. Internal report, Imperial College, Department of Computing, March 1990.
21. Wayne P. Stevens, Glenford Myers, and Larry L. Constantine. Structured design. *IBM Systems Journal*, 13(12):115–139, 1974.
22. Yde Venema. *Many-dimensional Modal Logic.* PhD thesis, University of Amsterdam, 1992.

DIAGRAMMATIC REASONING IN PROJECTIVE GEOMETRY

PHILIPPE BALBIANI AND LUIS FARIÑAS DEL CERRO

1 Introduction

Diagrammatic representation appears in many human activities: mathematics, physics, economics, politics, etc. It constitutes a natural framework for the formalization and the mechanization of reasoning. As an example, Euclid, pioneer of the formal methods at the beginnings of the study of the algorithms, took advantage, proving the theorems of his geometry, of the visual properties of the geometrical figures. It is generally believed that figures are informal objects and that they cannot be used in a formal proof: their use can lead to fallacious arguments, famous examples being the proof that there exists a triangle with two right angles or the proof that every triangle is isoceles [7].

Logical systems are the favoured frameworks for the formalization and the mechanization of reasoning. This marked preference can be explained by the unambiguous character of the well-formed expressions that constitute their languages. Consequently, the logicians have rarely considered that a formal proof could be based on diagrammatic expressions. The truth of the matter is, fallacious arguments come from an improper use of diagrams which are mostly considered as the instances rather than the general form of a concept. It turns out that diagrams can support formal reasoning as soon as they are associated to a precise semantics and that a clear definition of their use can constitute the framework of a synthetic and readable form of representation and reasoning [6].

Shin [12] has recently presented a formal system based on Venn diagrams that is sound and complete for some specific form of syllogistic reasoning. Broadly speaking, a diagram is a general concept and covers different forms of representation. It usually consists in the topological description of the relationship between its constituents. The diagrammatic representation of syllogism can be applied to other forms of reasoning as well. Answering

H.J. Ohlbach and U. Reyle (eds.), Logic, Language and Reasoning, 99–114.

the question of the advantages of a diagrammatic representation, this paper proves that diagrams can be used for mechanical theorem proving in projective geometry.

A projective structure is a relational structure of the form (P, L, in), with P the set of points, L the set of lines and in the binary relation of incidence between points and lines [8]. In projective geometry, a figure is completely described by the set of its geometrical beings together with the incidence relation between these beings. The spectrum of its representations goes from linguistic expressions to diagrammatic objects, from well-formed formulas to well-formed diagrams, from linear and unambiguous sequences of symbols to plane and vague drawings. The mechanization of reasoning on linguistic expressions appeals to logical rules that are not directly attached to our natural comprehension of the geometrical figures that they represent. The mechanization of reasoning on diagrammatic objects appeals to rules that are directly connected to our visual perception of these figures. We have chosen to use matrices of incidence for the diagrammatic representation and the mechanization of reasoning on projective structures.

In the literature, matrices of incidence are a widely-known framework for the representation of finite projective structures [1].

$$\begin{array}{c} \\ A \\ B \\ C \\ D \\ E \\ F \\ G \end{array} \begin{array}{c} \begin{array}{ccccccc} a & b & c & d & e & f & g \end{array} \\ \begin{pmatrix} 0 & 1 & 1 & 1 & 0 & 0 & 0 \\ 1 & 0 & 1 & 0 & 1 & 0 & 0 \\ 1 & 1 & 0 & 0 & 0 & 1 & 0 \\ 1 & 0 & 0 & 1 & 0 & 0 & 1 \\ 0 & 1 & 0 & 0 & 1 & 0 & 1 \\ 0 & 0 & 1 & 0 & 0 & 1 & 1 \\ 0 & 0 & 0 & 1 & 1 & 1 & 0 \end{pmatrix} \end{array}$$

Our matrices are slightly different since they are associated to geometrical statements which do not directly define finite projective structures. For us, a matrix of incidence is an array M whose entries $M(X, x)$ are either equal to 0 or 1, for every point X and for every line x. To every geometrical statement of projective geometry is associated a matrix of incidence M such that, for every point X and for every line x of the statement, X is incident with x in the statement only if $M(X, x) = 1$.

EXAMPLE 1.1 Let:

$$M = \begin{array}{c} \\ A \\ B \\ C \\ D \end{array} \begin{array}{c} \begin{array}{cccc} a & b & c & d \end{array} \\ \begin{pmatrix} 1 & 1 & 0 & 0 \\ 1 & 1 & 1 & 0 \\ 0 & 1 & 1 & 1 \\ 0 & 0 & 1 & 1 \end{pmatrix} \end{array}$$

If any two distinct rows or columns of a matrix M are compared then there might be more than one position in which they have both a 1. If (X, x),

(Y,x), (X,y) and (Y,y) are these positions then the points X and Y are both incident with the lines x and y in the geometrical structure associated to the matrix. In this case, either the row of X and the row of Y should be identified or the column of x and the column of y should be identified.

EXAMPLE 1.2 Let us identify in the matrix M of Example 1.1 the column of a and the column of b:

$$\begin{array}{c} \\ A \\ B \\ C \\ D \end{array} \begin{array}{c} \begin{array}{ccc} a & c & d \end{array} \\ \begin{pmatrix} 1 & 0 & 0 \\ 1 & 1 & 0 \\ 1 & 1 & 1 \\ 0 & 1 & 1 \end{pmatrix} \end{array}$$

Let us identify in the matrix M of Example 1.1 the row of C and the row of D:

$$\begin{array}{c} \\ A \\ B \\ C \end{array} \begin{array}{c} \begin{array}{cccc} a & b & c & d \end{array} \\ \begin{pmatrix} 1 & 1 & 0 & 0 \\ 1 & 1 & 1 & 0 \\ 0 & 1 & 1 & 1 \end{pmatrix} \end{array}$$

The heart of our thesis is that matrices of incidence can be used for mechanical theorem proving in projective geometry. To every geometrical statement of projective geometry is associated a matrix of incidence whose normal form is computed after successive identifications of rows and columns. Our main result is that the geometrical statement implies such or such property (incidence between a point and a line) if and only if the normal form of the associated matrix contains a 1 at such or such entry.

Obviously, matrices of incidence are formal objects that completely describe the relation of incidence between the geometrical beings of the statements of projective geometry. In other respects, they are not linear sequences of symbols but plane representations of relational structures. Furthermore, the duality between rows and columns exactly reflects the duality between points and lines in projective geometry. For all these reasons, matrices of incidence can be considered as diagrammatic expressions.[1] That the mechanization of reasoning on matrices of incidence appeals to rules that are directly connected to our visual perception is another aspect of their diagrammatic nature.

Section 2 briefly discusses the pros and cons of the synthetic approach versus the analytic approach to geometrical reasoning. Section 3 introduces projective geometry as a first-order theory while Section 4 defines the geometrical statements of projective geometry. Section 5 formally introduces the matrices of incidence while Section 6 presents their relations with the

[1] Although there is no criterion that would enable us to distinguish a diagrammatic form of representation from a linguistic one.

geometrical statements. Section 7 defines the mechanization of reasoning on the matrices of incidence and Section 8 proves the theorem of mechanization of projective geometry.

2 Mechanical Theorem Proving in Geometry

It is customary to distinguish between the synthetic approach and the analytic approach to geometrical reasoning [11]. In the synthetic approach, we directly argue about the relations between the geometrical beings of a figure of Euclidean geometry while, in the analytic approach, we first represent these relations by polynomial equations and then use the techniques of algebraic manipulation.

The weakness and the difficulty of a direct reasoning about the figures of Euclidean geometry stem from the wide range of the particular cases that one should consider to carry a proof through to a successful conclusion: inequality between two points or two lines, non-incidence between a point and a line, non-parallelism or non-orthogonality between two lines, etc. The value of algebraic reasoning about polynomials lies in the absence of such particular cases, only the problem of the division by 0 should be taken into account. The success and the efficiency of the algebraic approaches to geometrical reasoning are partly based on this absence of particular cases [7, 13, 14].

Translating, in a first step, a geometrical statement into a polynomial family $f_1, \ldots, f_n, g$ and proving, in a second step, with polynomial rewriting techniques, the belonging of g to the ideal generated by $f_1, \ldots, f_n$, they produce de facto in a few seconds the proofs of the most difficult theorems of elementary geometry: Desargues, Pappus, Pascal, Simson, etc. The main defect of these approaches is that they prevent a synthetic reading of the proofs that they produce. Moreover, the non-belonging of g to the ideal generated by $f_1, \ldots, f_n$ does not constitute a sufficient condition for the falsity of the geometrical statement [7], in other words: the analytic approaches to geometrical reasoning are not complete.

The linguistic basis of the synthetic approach to mechanical theorem proving in projective geometry proposed by Balbiani, Dugat, Fariñas del Cerro and Lopez [2, 4, 10] is the predicate calculus with equality. Translating, in a first step, a geometrical statement into a couple of terms s, t and proving, in a second step, with term rewriting techniques, the belonging of t to the equational class generated by s, this approach produces de facto in a few seconds the synthetic proofs of the theorems of projective geometry. Balbiani and Fariñas del Cerro [3, 5] extend this approach to other geometries. Nevertheless, the mechanization of reasoning on the terms of the predicate calculus with equality appeals to rewriting rules that are very far from being directly connected to our perception of the geometrical figures that they represent.

The reasons for it are as follows. The terms of the predicate calculus with equality are linear sequences of symbols while the geometrical figures of projective geometry that they represent are plane representations of relational structures. The matrices of incidence that we define in Section 5 do offer more similarity with the geometrical figures of projective geometry than the terms of the predicate calculus with equality did. Our purpose is to prove that matrices of incidence can be used for mechanical theorem proving in projective geometry.

3 Projective Geometry

This section is devoted to the definition of the language of projective geometry and to the succinct description of its model theory.

3.1 Language

The language of projective geometry consists of a non-empty set V_P of variables of type P denoted by capital letters, a non-empty set V_L of variables of type L denoted by small letters and a relation symbol *in* of arity $P \times L$. The formulas of projective geometry are defined by induction in the following way:

- for every $X, Y \in V_P$, $X = Y$ is a formula;
- for every $x, y \in V_L$, $x = y$ is a formula;
- for every $X \in V_P$ and for every $x \in V_L$, $X \; in \; x$ is a formula;
- if A is a formula then $\neg A$ is a formula;
- if A and B are formulas then $A \wedge B$ is a formula;
- if A is a formula then, for every $X \in V_P$, $(\forall X)A$ is a formula;
- if A is a formula then, for every $x \in V_L$, $(\forall x)A$ is a formula.

3.2 Models

A *projective frame* is a relational structure of the form $\mathcal{P} = (P_{\mathcal{F}}, L_{\mathcal{F}}, in_{\mathcal{F}})$ where $P_{\mathcal{F}}$ is a non-empty set of *points*, $L_{\mathcal{F}}$ is a non-empty set of *lines* and $in_{\mathcal{F}}$ is a subset of $P_{\mathcal{F}} \times L_{\mathcal{F}}$ such that:

- for every $X, Y \in P_{\mathcal{F}}$, there exists $x \in L_{\mathcal{F}}$ such that $X \; in_{\mathcal{F}} \; x$ and $Y \; in_{\mathcal{F}} \; x$;
- for every $x, y \in L_{\mathcal{F}}$, there exists $X \in P_{\mathcal{F}}$ such that $X \; in_{\mathcal{F}} \; x$ and $X \; in_{\mathcal{F}} \; y$;
- for every $X, Y \in P_{\mathcal{F}}$ and for every $x, y \in L_{\mathcal{F}}$, if $X \; in_{\mathcal{F}} \; x$, $Y \; in_{\mathcal{F}} \; x$, $X \; in_{\mathcal{F}} \; y$ and $Y \; in_{\mathcal{F}} \; y$ then $X = Y$ or $x = y$.

A *projective model* is a structure of the form $\mathcal{G} = (P_\mathcal{F}, L_\mathcal{F}, in_\mathcal{F}, v_{P_\mathcal{F}}, v_{L_\mathcal{F}})$ where $(P_\mathcal{F}, L_\mathcal{F}, in_\mathcal{F})$ is a projective frame, $v_{P_\mathcal{F}}$ is a mapping on V_P to $P_\mathcal{F}$ and $v_{L_\mathcal{F}}$ is a mapping on V_L to $L_\mathcal{F}$.

3.3 Satisfiability

The satisfiability relation between a model $\mathcal{G} = (P_\mathcal{F}, L_\mathcal{F}, in_\mathcal{F}, v_{P_\mathcal{F}}, v_{L_\mathcal{F}})$ and a formula A is defined by induction on A in the following way:

$\mathcal{G} \models X = Y$ iff $v_{P_\mathcal{F}}(X) = v_{P_\mathcal{F}}(Y)$.
$\mathcal{G} \models x = y$ iff $v_{L_\mathcal{F}}(x) = v_{L_\mathcal{F}}(y)$.
$\mathcal{G} \models X \text{ in } x$ iff $v_{P_\mathcal{F}}(X)\ in_\mathcal{F}\ v_{L_\mathcal{F}}(x)$.
$\mathcal{G} \models \neg A$ iff $\mathcal{G} \not\models A$.
$\mathcal{G} \models A \wedge B$ iff $\mathcal{G} \models A$ and $\mathcal{G} \models B$.
$\mathcal{G} \models (\forall X)A$ iff $(P_\mathcal{F}, L_\mathcal{F}, in_\mathcal{F}, v'_{P_\mathcal{F}}, v_{L_\mathcal{F}}) \models A$ for every mapping $v'_{P_\mathcal{F}}$ that possibly differs from $v_{P_\mathcal{F}}$ in the value given to X.
$\mathcal{G} \models (\forall x)A$ iff $(P_\mathcal{F}, L_\mathcal{F}, in_\mathcal{F}, v_{P_\mathcal{F}}, v'_{L_\mathcal{F}}) \models A$ for every mapping $v'_{L_\mathcal{F}}$ that possibly differs from $v_{L_\mathcal{F}}$ in the value given to x.

A formula is *valid* when it is satisfied in every model.

4 Constructive Geometrical Statement

Following the tradition inaugurated by Euclid and formalized by Hilbert [9] in the context of elementary geometry, this section is devoted to the definition of the geometrical statement of constructive type of projective geometry. In the context of elementary geometry, a geometrical statement is of constructive type if the points of the statement can be effectively constructed with a ruler and a pair of compasses. In the context of projective geometry, a geometrical statement is of constructive type if the points of the statement can be effectively constructed with a ruler only.

More precisely, an *elementary construction* is one of the following English sentences:

construction 1 'Take an arbitrary point X'. This sentence constructs X with respect to the empty set of variables of type L.
construction 2 'Draw an arbitrary line x'. This sentence constructs x with respect to the empty set of of variables of type P.
construction 3 'Take an arbitrary point X on a line x'. This sentence constructs X with respect to the set $\{x\}$ of variables of type L.
construction 4 'Draw an arbitrary line x on a point X'. This sentence constructs x with respect to the set $\{X\}$ of variables of type P.
construction 5 'Take the intersection X of two lines x, y'. This sentence constructs X with respect to the set $\{x, y\}$ of variables of type L.

construction 6 'Draw the line x on two points X, Y'. This sentence constructs x with respect to the set $\{X, Y\}$ of variables of type P.

A *geometrical statement of constructive type* is a structure of the form $(\Pi, \Lambda, <, S)$ where Π is a non-empty set of variables of type P, Λ is a non-empty set of variables of type L, $<$ is a total order on $\Pi \cup \Lambda$ and S is a mapping on $\Pi \cup \Lambda$ to the set of the elementary constructions such that:

- for every $X \in \Pi$, $S(X)$ constructs X with respect to a set λ of variables of type L such that, for every $x \in \lambda$, $x < X$;
- for every $x \in \Lambda$, $S(x)$ constructs x with respect to a set π of variables of type P such that, for every $X \in \Pi$, $X < x$.

EXAMPLE 4.1 The following sequence of elementary constructions implicitly defines a geometrical statement of constructive type:

1. draw an arbitrary line a;
2. take an arbitrary point A on a;
3. take an arbitrary point B on a;
4. draw the line b on A and B.

In this case, $\Pi = \{A, B\}$, $\Lambda = \{a, b\}$ and $a < A < B < b$.

EXAMPLE 4.2 The following sequence of elementary constructions implicitly defines a geometrical statement of constructive type:

1. take an arbitrary point A;
2. draw an arbitrary line a on A;
3. draw an arbitrary line b on A;
4. take the intersection B of a and b.

In this case, $\Pi = \{A, B\}$, $\Lambda = \{a, b\}$ and $A < a < b < B$.

EXAMPLE 4.3 The following sequence of elementary constructions implicitly defines a geometrical statement of constructive type:

1. take an arbitrary point A;
2. draw an arbitrary line a on A;
3. take an arbitrary point B on a;
4. draw the line b on A and B;
5. draw an arbitrary line c on B;
6. take the intersection C of b and c;
7. draw an arbitrary line d on C;
8. take the intersection D of c and d.

In this case, $\Pi = \{A, B, C, D\}$, $\Lambda = \{a, b, c, d\}$ and

$$A < a < B < b < c < C < d < D.$$

5 Matrix Representation

This section is devoted to the definition of the matrix of incidence and to the definition of the restriction of a matrix to one of its geometrical beings.

5.1 Matrix of Incidence

A *matrix of incidence* is a structure of the form $(\Pi, \Lambda, <, M)$ where Π is a non-empty set of variables of type P, Λ is a non-empty set of variables of type L, $<$ is a total order on $\Pi \cup \Lambda$ and M is a mapping on $\Pi \times \Lambda$ to $\{0, 1\}$ such that:

- for every $X \in \Pi$, $Card(\lambda(X)) \leq 2$;
- for every $x \in \Lambda$, $Card(\pi(x)) \leq 2$;

where:

- $\lambda(X) = \{x\colon x \in \Lambda,\ x < X \text{ and } M(X, x) = 1\}$;
- $\pi(x) = \{X\colon X \in \Pi,\ X < x \text{ and } M(X, x) = 1\}$.

Let $deg(\mathcal{M}) = Card(\Pi) + Card(\Lambda)$ be the *degree* of $\mathcal{M}$. Let $\underline{\mathcal{M}}$ be the conjunction of all the formulas $X\ in\ x$ such that $X \in \Pi$, $x \in \Lambda$ and $M(X, x) = 1$. Let $\leq$ be the reflexive closure of $<$.

EXAMPLE 5.1 Let $\Pi = \{A, B, C, D\}$, $\Lambda = \{a, b, c, d\}$ and $<$ be the total order on $\Pi \cup \Lambda$ defined by $A < a < B < b < c < C < d < D$. Let:

$$M = \begin{array}{c} \\ A \\ B \\ C \\ D \end{array} \begin{array}{c} \begin{array}{cccc} a & b & c & d \end{array} \\ \begin{pmatrix} 1 & 1 & 0 & 0 \\ 1 & 1 & 1 & 0 \\ 0 & 1 & 1 & 1 \\ 0 & 0 & 1 & 1 \end{pmatrix} \end{array}.$$

Then $\mathcal{M} = (\Pi, \Lambda, <, M)$ is a matrix of incidence. Moreover, $deg(\mathcal{M}) = 8$. Finally, $\underline{\mathcal{M}} = A\ in\ a \wedge A\ in\ b \wedge B\ in\ a \wedge B\ in\ b \wedge B\ in\ c \wedge C\ in\ b \wedge C\ in\ c \wedge C\ in\ d \wedge D\ in\ c \wedge D\ in\ d$.

In some sense, the matrix M of Example 5.1 is a diagrammatic representation of the geometrical statement of Example 4.3. The relation between matrices and geometrical statements will be further developed in Section 6.

5.2 Restriction of a Matrix to one of its Points

Let $X \in \Pi$. The *restriction of* $\mathcal{M}$ *to* X is the matrix of incidence $\mathcal{M}_X = (\Pi_X, \Lambda_X, <_X, M_X)$ defined in the following way:

- $\Pi_X = \{Y\colon Y \in \Pi \text{ and } Y \leq X\}$;
- $\Lambda_X = \{x\colon x \in \Lambda \text{ and } x < X\}$;
- $<_X$ is the restriction of $<$ to $\Pi_X \cup \Lambda_X$;
- M_X is the restriction of M to $\Pi_X \times \Lambda_X$.

EXAMPLE 5.2 The restriction of the matrix $\mathcal{M}$ of Example 5.1 to C is:

$$M_C = \begin{array}{c} \\ A \\ B \\ C \end{array} \begin{array}{c} \begin{array}{ccc} a & b & c \end{array} \\ \begin{pmatrix} 1 & 1 & 0 \\ 1 & 1 & 1 \\ 0 & 1 & 1 \end{pmatrix} \end{array}.$$

5.3 Restriction of a Matrix to one of its Lines

Let $x \in \Lambda$. The *restriction of $\mathcal{M}$ to x* is the matrix of incidence $\mathcal{M}_x = (\Pi_x, \Lambda_x, <_x, M_x)$ defined in the following way:

- $\Pi_x = \{X\colon X \in \Pi \text{ and } X < x\}$;
- $\Lambda_x = \{y\colon y \in \Lambda \text{ and } y \leq x\}$;
- $<_x$ is the restriction of $<$ to $\Pi_x \cup \Lambda_x$;
- M_x is the restriction of M to $\Pi_x \times \Lambda_x$.

EXAMPLE 5.3 The restriction of the matrix $\mathcal{M}$ of Example 5.1 to c is:

$$M_c = \begin{array}{c} \\ A \\ B \end{array} \begin{array}{c} \begin{array}{ccc} a & b & c \end{array} \\ \begin{pmatrix} 1 & 1 & 0 \\ 1 & 1 & 1 \end{pmatrix} \end{array}.$$

6 Matrices and Geometrical Statements

This section is devoted to the proof of the equivalence between the set of the matrices of incidence and the set of the geometrical statements of constructive type.

6.1 From Matrices to Geometrical Statements

Let $\mathcal{M} = (\Pi, \Lambda, <, M)$ be a matrix of incidence. For every $X \in \Pi$, let $S(M)(X)$ be the elementary construction defined in the following way:

- if $\lambda(X) = \emptyset$ then $S(M)(X) =$ 'Take an arbitrary point X';
- if $\lambda(X) = \{x\}$ then $S(M)(X) =$ 'Take an arbitrary point X on x';
- if $\lambda(X) = \{x, y\}$ where x and y are distinct variables of type L then $S(M)(X) =$ 'Take the intersection X of x and y'.

For every $x \in \Lambda$, let $S(M)(x)$ be the elementary construction defined in the following way:

- if $\pi(x) = \emptyset$ then $S(M)(x) =$ 'Draw an arbitrary line x';
- if $\pi(x) = \{X\}$ then $S(M)(x) =$ 'Draw an arbitrary line x on X';
- if $\pi(x) = \{X, Y\}$ where X and Y are distinct variables of type P then $S(M)(x) =$ 'Draw the line x on X and Y'.

Direct calculations would lead to the conclusion that:

LEMMA 6.1 $\mathcal{S}(\mathcal{M}) = (\Pi, \Lambda, <, S(M))$ *is a geometrical statement of constructive type.*

$\mathcal{S}(\mathcal{M})$ is the *geometrical statement over* $\mathcal{M}$.

6.2 From Geometrical Statements to Matrices

Let $\mathcal{S} = (\Pi, \Lambda, <, S)$ be a geometrical statement of constructive type. For every $X \in \Pi$ and for every $x \in \Lambda$, let $M(S)(X, x)$ be the value in $\{0, 1\}$ defined in the following way:

- if $X < x$ then $M(S)(X, x) =$ if $S(x)$ defines x with respect to, among other variables of type P, X then 1 else 0;
- if $x < X$ then $M(S)(X, x) =$ if $S(X)$ defines X with respect to, among other variables of type L, x then 1 else 0.

Direct calculations would lead to the conclusion that:

LEMMA 6.2 $\mathcal{M}(\mathcal{S}) = (\Pi, \Lambda, <, M(S))$ *is a matrix of incidence.*

$\mathcal{M}(\mathcal{S})$ is the *matrix of incidence over* $\mathcal{S}$.

6.3 Equivalence Between Matrices and Geometrical Statements

Direct calculations would lead to the conclusion that:

LEMMA 6.3 *Let* $\mathcal{M}$ *be a matrix of incidence. Then* $\mathcal{M}(\mathcal{S}(\mathcal{M})) = \mathcal{M}$.

LEMMA 6.4 *Let* $\mathcal{S}$ *be a geometrical statement of constructive type. Then* $\mathcal{S}(\mathcal{M}(\mathcal{S})) = \mathcal{S}$.

7 Matrix Reasoning

This section is devoted to the mechanization of reasoning on the set of the matrices of incidence.

7.1 Equational Matrix of Incidence

For every $X, Y \in V_P$, the formula $X = Y$ is an *equation of type* P and the formula $X \neq Y$ is a *disequation of type* P. For every $x, y \in V_L$, the formula $x = y$ is an *equation of type* L and the formula $x \neq y$ is a *disequation of type* L. An *equation* is either an equation of type P or an equation of type L. A *disequation* is either a disequation of type P or a disequation of type L. An *equational matrix of incidence* consists of a matrix of incidence $\mathcal{M} = (\Pi, \Lambda, <, M)$, a set σ of equations and a set q of disequations such that:

- for every $X, Y \in \Pi$, if $X =_\sigma Y$ then $X = Y$;

– for every $x, y \in \Lambda$, if $x =_\sigma y$ then $x = y$

where $=_\sigma$ is the smallest relation of equivalence on $\Pi \cup \Lambda$ containing σ. If $\sigma = \emptyset$ and $q = \emptyset$ then $\mathcal{M}$ is an *initial equational matrix of incidence.*

7.2 Identification of two Rows

Let $\mathcal{M} = (\Pi, \Lambda, <, M, \sigma, q)$ be an equational matrix of incidence. Let $X, Y \in \Pi$ and $x, y \in \Lambda$ be such that $X < Y$, $x < y$, $M(X, x) = 1$, $M(Y, x) = 1$, $M(X, y) = 1$ and $M(Y, y) = 1$. If $y < Y$ then Y is constructed 'after' y in the geometrical statement $S(\mathcal{M})$ over $\mathcal{M}$. In this case, the row of X and the row of Y should be identified. Let $\mathcal{M}_Y^X = (\Pi_Y^X, \Lambda_Y^X, <_Y^X, M_Y^X, \sigma_Y^X, q_Y^X)$ be the structure defined in the following way:

- $\Pi_Y^X = \Pi \setminus \{Y\}$;
- $\Lambda_Y^X = \Lambda$;
- $<_Y^X$ is the restriction of $<$ to $\Pi_Y^X \cup \Lambda_Y^X$;
- M_Y^X is the mapping on $\Pi_Y^X \times \Lambda_Y^X$ to $\{0, 1\}$ defined by $M_Y^X(Z, z) =$ if $Z = X$ then $max(M(X, z), M(Y, z))$ else $M(Z, z)$;
- $\sigma_Y^X = \sigma \cup \{X = Y\}$;
- $q_Y^X = q \cup \{x \neq y\}$.

Direct calculations would lead to the conclusion that:

LEMMA 7.1 *$\mathcal{M}_Y^X$ is an equational matrix.*

In other respects, let us remarked that $deg(\mathcal{M}) > deg(\mathcal{M}_Y^X), deg(\mathcal{M}_y)$.

EXAMPLE 7.2 Let $\mathcal{M} = (\Pi, \Lambda, <, M)$ be the matrix of Example 5.1. Let $X = C$, $Y = D$, $x = c$ and $y = d$. Then:

$$M_D^C = \begin{array}{c} \\ A \\ B \\ C \end{array} \begin{array}{c} \begin{array}{cccc} a & b & c & d \end{array} \\ \begin{pmatrix} 1 & 1 & 0 & 0 \\ 1 & 1 & 1 & 0 \\ 0 & 1 & 1 & 1 \end{pmatrix} \end{array}.$$

7.3 Identification of two Columns

Let $\mathcal{M} = (\Pi, \Lambda, <, M, \sigma, q)$ be an equational matrix of incidence. Let $X, Y \in \Pi$ and $x, y \in \Lambda$ be such that $X < Y$, $x < y$, $M(X, x) = 1$, $M(Y, x) = 1$, $M(X, y) = 1$ and $M(Y, y) = 1$. If $Y < y$ then Y is constructed 'before' y in the geometrical statement $S(\mathcal{M})$ over $\mathcal{M}$. In this case, the column of x and the column of y should be identified. Let $\mathcal{M}_y^x = (\Pi_y^x, \Lambda_y^x, <_y^x, M_y^x, \sigma_y^x, q_y^x)$ be the structure defined in the following way:

- $\Pi_y^x = \Pi$;
- $\Lambda_y^x = \Lambda \setminus \{y\}$;
- $<_y^x$ is the restriction of $<$ to $\Pi_y^x \cup \Lambda_y^x$;

- M_y^x is the mapping on $\Pi_y^x \times \Lambda_y^x$ to $\{0,1\}$ defined by $M_y^x(Z,z) =$ if $z = x$ then $max(M(Z,x), M(Z,y))$ else $M(Z,z)$;
- $\sigma_y^x = \sigma \cup \{x = y\}$;
- $q_y^x = q \cup \{X \neq Y\}$.

Direct calculations would lead to the conclusion that:

LEMMA 7.3 *$\mathcal{M}_y^x$ is an equational matrix.*

In other respects, let us remarked that $deg(\mathcal{M}) > deg(\mathcal{M}_y^x), deg(\mathcal{M}_Y)$.

EXAMPLE 7.4 Let $\mathcal{M} = (\Pi, \Lambda, <, M)$ be the matrix of Example 5.1. Let $X = A$, $Y = B$, $x = a$ and $y = b$. Then:

$$M_b^a = \begin{array}{c} \\ A \\ B \\ C \\ D \end{array} \begin{array}{c} \begin{array}{ccc} a & c & d \end{array} \\ \begin{pmatrix} 1 & 0 & 0 \\ 1 & 1 & 0 \\ 1 & 1 & 1 \\ 0 & 1 & 1 \end{pmatrix} \end{array}.$$

7.4 Reduction Relation

Let $n \geq 2$. Let $\rightarrow_n$ be a decidable and terminating reduction relation on the set of the equational matrices of incidence $\mathcal{M}$ such that $deg(\mathcal{M}) \leq n$. Direct calculations would lead to the conclusion that:

LEMMA 7.5 *There exists a decidable and terminating reduction relation $\rightarrow_{n+1}$ on the set of the equational matrices of incidence $\mathcal{M}$ such that $deg(\mathcal{M}) \leq n+1$ where $\mathcal{M} = (\Pi, \Lambda, <, M, \sigma, q) \rightarrow_{n+1} \mathcal{M}'$ iff there exists $X, Y \in \Pi$ and there exists $x, y \in \Lambda$ such that $X < Y$, $x < y$, $M(X,x) = 1$, $M(Y,x) = 1$, $M(X,y) = 1$, $M(Y,y) = 1$ and either:*

- *$y < Y$;*
- *for every normal form $\mathcal{M}_y^* = (\Pi_y^*, \Lambda_y^*, <_y^*, M_y^*, \sigma_y^*, q_y^*)$ of $\mathcal{M}_y$ with respect to $\rightarrow_n$, $x \neq_{\sigma_y^*} y$;*
- *$\mathcal{M}' = \mathcal{M}_Y^X$,*

or:

- *$Y < y$;*
- *for every normal form $\mathcal{M}_Y^* = (\Pi_Y^*, \Lambda_Y^*, <_Y^*, M_Y^*, \sigma_Y^*, q_Y^*)$ of $\mathcal{M}_Y$ with respect to $\rightarrow_n$, $X \neq_{\sigma_Y^*} Y$;*
- *$\mathcal{M}' = \mathcal{M}_y^x$.*

Let $\rightarrow = \bigcup_{n \geq 2} \rightarrow_n$. Direct calculations would lead to the conclusion that:

THEOREM 7.6 *$\rightarrow$ is a decidable and terminating reduction relation on the set of the equational matrices of incidence where $\mathcal{M} = (\Pi, \Lambda, <, M, \sigma, q) \rightarrow \mathcal{M}'$ iff there exists $X, Y \in \Pi$ and there exists $x, y \in \Lambda$ such that $X < Y$, $x < y$, $M(X,x) = 1$, $M(Y,x) = 1$, $M(X,y) = 1$, $M(Y,y) = 1$ and either:*

- *$y < Y$;*

– *for every normal form* $\mathcal{M}_y^* = (\Pi_y^*, \Lambda_y^*, <_y^*, M_y^*, \sigma_y^*, q_y^*)$ *of* $\mathcal{M}_y$ *with respect to* $\rightarrow$, $x \neq_{\sigma_y^*} y$;
– $\mathcal{M}' = \mathcal{M}_Y^X$;

or:

– $Y < y$;
– *for every normal form* $\mathcal{M}_Y^* = (\Pi_Y^*, \Lambda_Y^*, <_Y^*, M_Y^*, \sigma_Y^*, q_Y^*)$ *of* $\mathcal{M}_Y$ *with respect to* $\rightarrow$, $X \neq_{\sigma_Y^*} Y$;
– $\mathcal{M}' = \mathcal{M}_y^x$.

Moreover:

COROLLARY 7.7 *Let* $\mathcal{M} = (\Pi, \Lambda, <, M, \sigma, q)$ *be an equational matrix of incidence such that* $\sigma \cup q$ *is consistent. If* $\mathcal{M} \rightarrow (\Pi', \Lambda', <', M', \sigma', q')$ *then* $\sigma' \cup q'$ *is consistent.*

Let $\downarrow$ be the binary relation on the set of the equational matrices of incidence defined by $\mathcal{M} \downarrow \mathcal{M}'$ iff there exists an equational matrix $\mathcal{M}''$ such that $\mathcal{M} \rightarrow \mathcal{M}''$ and $\mathcal{M}' \rightarrow \mathcal{M}''$.

7.5 Confluence

This section is devoted to the proof of the confluence of the reduction relation on the set of the equational matrices of incidence. Let $\mathcal{M} = (\Pi, \Lambda, <, M, \sigma, q)$ be an equational matrix, $X, Y, Z, T \in \Pi$, $x, y, z, t \in \Lambda$ such that $X < Y$, $x < y$, $M(X, x) = 1$, $M(Y, x) = 1$, $M(X, y) = 1$, $M(Y, y) = 1$, $Z < T$, $z < t$, $M(Z, z) = 1$, $M(T, z) = 1$, $M(Z, t) = 1$ and $M(T, t) = 1$. Suppose that $Y < y$ and, for every normal form $\mathcal{M}_Y^* = (\Pi_Y^*, \Lambda_Y^*, <_Y^*, M_Y^*, \sigma_Y^*, q_Y^*)$ of $\mathcal{M}_Y$ with respect to $\rightarrow$, $X \neq_{\sigma_Y^*} Y$, so that $\mathcal{M} \rightarrow \mathcal{M}_y^x$.

First case: $T < t$**.** Suppose that for every normal form $\mathcal{M}_T^* = (\Pi_T^*, \Lambda_T^*, <_T^*, M_T^*, \sigma_T^*, q_T^*)$ of $\mathcal{M}_T$ with respect to $\rightarrow$, $Z \neq_{\sigma_T^*} T$, so that $\mathcal{M} \rightarrow \mathcal{M}_t^z$. If $t < Y$ then $\mathcal{M}_Y \rightarrow (\mathcal{M}_Y)_t^z = (\mathcal{M}_t^z)_Y$. Consequently, $\mathcal{M}_t^z \rightarrow (\mathcal{M}_t^z)_y^x$. If $Y < t$ then $(\mathcal{M}_t^z)_Y = \mathcal{M}_Y$. Therefore, $\mathcal{M}_t^z \rightarrow (\mathcal{M}_t^z)_y^x$. A similar argument would lead to the conclusion that $\mathcal{M}_y^x \rightarrow (\mathcal{M}_y^x)_t^z$. Since $(\mathcal{M}_t^z)_y^x = (\mathcal{M}_y^x)_t^z$, then $\mathcal{M}_y^x \downarrow \mathcal{M}_t^z$.

Second case: $t < T$**.** Suppose that for every normal form $\mathcal{M}_t^* = (\Pi_t^*, \Lambda_t^*, <_t^*, M_t^*, \sigma_t^*, q_t^*)$ of $\mathcal{M}_t$ with respect to $\rightarrow$, $z \neq_{\sigma_t^*} t$, so that $\mathcal{M} \rightarrow \mathcal{M}_T^Z$. If $T < Y$ or $T = Y$ then $\mathcal{M}_Y \rightarrow (\mathcal{M}_Y)_T^Z = (\mathcal{M}_T^Z)_Y$. Consequently, $\mathcal{M}_T^Z \rightarrow (\mathcal{M}_T^Z)_y^x$. If $Y < T$ then $(\mathcal{M}_T^Z)_Y = \mathcal{M}_Y$. Therefore, $\mathcal{M}_T^Z \rightarrow (\mathcal{M}_T^Z)_y^x$. A similar argument would lead to the conclusion that $\mathcal{M}_y^x \rightarrow (\mathcal{M}_y^x)_T^Z$. Since $(\mathcal{M}_T^Z)_y^x = (\mathcal{M}_y^x)_T^Z$, then $\mathcal{M}_y^x \downarrow \mathcal{M}_T^Z$.

Consequently:

LEMMA 7.8 *For every equational matrix* $\mathcal{M}, \mathcal{M}', \mathcal{M}''$, *if* $\mathcal{M} \rightarrow \mathcal{M}'$ *and* $\mathcal{M} \rightarrow \mathcal{M}''$ *then* $\mathcal{M}' \downarrow \mathcal{M}''$.

8 Completeness

This section is devoted to the proof of the completeness of the reduction relation on the set of the equational matrices of incidence. Let $\mathcal{M} = (\Pi, \Lambda, <, M, \emptyset, \emptyset)$ be an initial equational matrix of incidence. Let $\mathcal{M}^* = (\Pi^*, \Lambda^*, <^*, M^*, \sigma^*, q^*)$ be the normal form of $\mathcal{M}$ with respect to $\to$. Direct calculations would lead to the conclusion that $\models_{PG} \underline{\mathcal{M}} \wedge q^* \to \underline{\mathcal{M}}^* \wedge \sigma^*$. Consequently, for every $X \in \Pi$ and for every $x \in \Lambda$, if $M^*(\sigma^*(X), \sigma^*(x)) = 1$ then $\models_{PG} \underline{\mathcal{M}} \wedge q^* \to X \textit{ in } x$, where $\sigma^*(X)$ denotes the only $Y \in \Pi^*$ such that $X =_{\sigma^*} Y$ and $\sigma^*(x)$ denotes the only $y \in \Lambda^*$ such that $x =_{\sigma^*} y$.

THEOREM 8.1 *Let $\mathcal{M} = (\Pi, \Lambda, <, M, \emptyset, \emptyset)$ be an initial equational matrix of incidence. Let $\mathcal{M}^* = (\Pi^*, \Lambda^*, <^*, M^*, \sigma^*, q^*)$ be the normal form of $\mathcal{M}$ with respect to $\to$. For every $X \in \Pi$ and for every $x \in \Lambda$, if $M(\sigma^*(X), \sigma^*(x)) = 1$ then $\models_{PG} \underline{\mathcal{M}} \wedge q^* \to X \textit{ in } x$.*

EXAMPLE 8.2 Let $\mathcal{M} = (\Pi, \Lambda, <, M)$ be the matrix of Example 5.1. Direct calculations would lead to the conclusion that the normal form of $(\Pi, \Lambda, <, M, \emptyset, \emptyset)$ with respect to $\to$ is $(\Pi^*, \Lambda^*, <^*, M^*, \sigma^*, q^*)$ where $\Pi^* = \{A, B\}$, $\Lambda^* = \{a, c, d\}$, $A <^* a <^* B <^* c <^* d$, $M^* = ((M_D^C)_C^B)_b^a$, $\sigma^* = \{C = D, B = C, a = b\}$ and $q^* = \{c \neq d, b \neq c, A \neq B\}$. Since $M^*(B, c) = 1$, $\sigma^*(B) = B$ and $\sigma^*(d) = c$, then $\models_{PG} \underline{\mathcal{M}} \wedge A \neq B \wedge b \neq c \wedge c \neq d \to B \textit{ in } d$.

Let $\mathcal{M} = (\Pi, \Lambda, <, M, \emptyset, \emptyset)$ be an initial equational matrix of incidence. Let $\mathcal{M}^* = (\Pi^*, \Lambda^*, <^*, M^*, \sigma^*, q^*)$ be the normal form of $\mathcal{M}$ with respect to $\to$. If there exists $X, Y \in \Pi$ and there exists $x, y \in \Lambda$ such that $X < Y$, $x < y$, $M(X, x) = 1$, $M(Y, x) = 1$, $M(X, y) = 1$, $M(Y, y) = 1$ then either $y < Y$ and there exists a normal form of $\mathcal{M}_y^*$ with respect to $\to$ during the computation of which x and y are identified or $Y < y$ and there exists a normal form of $\mathcal{M}_Y^*$ with respect to $\to$ during the computation of which X and Y are identified. According to Lemma 7.8, in both cases, $\mathcal{M}^*$ is reducible with respect to $\to$, a contradiction. Therefore, there does not exist any $X, Y \in \Pi$ and there does not exist any $x, y \in \Lambda$ such that $X < Y$, $x < y$, $M(X, x) = 1$, $M(Y, x) = 1$, $M(X, y) = 1$, $M(Y, y) = 1$. Consequently, there exists a projective model $\mathcal{G}$ such that:

- $\mathcal{G} \models X = Y$ iff $X =_{\sigma^*} Y$;
- $\mathcal{G} \models x = y$ iff $x =_{\sigma^*} y$;
- $\mathcal{G} \models X \textit{ in } x$ iff $M^*(\sigma^*(X).\sigma^*(x)) = 1$.

Therefore, $\mathcal{G} \models \sigma^* \wedge q^*$ and $\mathcal{G} \models \underline{\mathcal{M}}^*$. Since $\models \underline{\mathcal{M}}^* \wedge \sigma^* \to \underline{\mathcal{M}}$, then $\mathcal{G} \models \underline{\mathcal{M}}$. Consequently, for every $X \in \Pi$ and for every $x \in \Lambda$, if $\models_{PG} \underline{\mathcal{M}} \wedge q^* \to X \textit{ in } x$ then $M^*(\sigma^*(X), \sigma^*(x)) = 1$.

THEOREM 8.3 *Let $\mathcal{M} = (\Pi, \Lambda, <, M, \emptyset, \emptyset)$ be an initial equational matrix of incidence. Let $\mathcal{M}^* = (\Pi^*, \Lambda^*, <^*, M^*, \sigma^*, q^*)$ be the normal form of $\mathcal{M}$ with respect to $\to$. For every $X \in \Pi$ and for every $x \in \Lambda$, if $\models_{PG} \underline{\mathcal{M}} \wedge q^* \to X \textit{ in } x$ then $M^*(\sigma^*(X), \sigma^*(x)) = 1$.*

EXAMPLE 8.4 Let $\mathcal{M} = (\Pi, \Lambda, <, M)$ be the matrix of Example 5.1 and $(\Pi^*, \Lambda^*, <^*, M^*, \sigma^*, q^*)$ be its normal form. Since $M^*(A, c) = 0$, $\sigma^*(A) = A$ and $\sigma^*(d) = c$, then $\not\models_{PG} \underline{\mathcal{M}} \wedge A \neq B \wedge b \neq c \wedge c \neq d \rightarrow A\ in\ d$.

9 Conclusion

The original feature of our synthetic approach to geometrical reasoning consists in the matrix representation of the statements of projective geometry. The mechanization of reasoning on the statements of projective geometry is obtained through a decidable, terminating and confluent reduction relation on the matrices of incidence: a statement implies such or such property if and only if the normal form of the associated matrix contains a 1 at such or such entry. In the course of the computation of the normal form of the matrix associated to a statement of projective geometry, non-degenerate conditions are generated as disequations. Without these disequations, the properties of the considered statement might not be valid.

The decidability, the termination and the confluence of the reduction relation on the matrices of incidence secure some efficiency for our approach. The diagrammatic representation of the statements of projective geometry provides some readability of the proofs that our approach produces. Nevertheless, for the time being, we do not pretend that we can supplant the algebraic approaches to geometrical reasoning which, if they are not readable, produce the proofs of the most difficult theorems of elementary geometry in a few seconds. However, let us remark that these approaches cannot mechanize the projective geometry. Wu [14] even conjectures that the projective geometry of space is not mechanizable.

The synthetic approach that we have developed shows how a matrix representation of the statements of projective geometry furthers the emergence of a theorem of mechanization for projective geometry. Our ambition is now to extend our approach to the projective geometry of space. Its completion would refute the conjecture of Wu.

Philippe Balbiani
Université Paris-Nord, France.

Luis Fariñas del Cerro
Université Paul Sabatier, France.

References

1. Andrian Albert and Reuben Sandler. An Introduction to Finite Projective Planes. Holt, Rinehart and Winston, 1968.
2. Philippe Balbiani. Equation solving in projective planes and planar ternary rings. In G. Levi and M. Rodríguez-Artalejo (editors), *Algebraic and Logic Programming*, 4th

International Conference, ALP '94, Madrid, Spain, September 1994, Proceedings. Lecture Notes in Computer science 850, 95-113, Springer-Verlag, 1994.
3. Philippe Balbiani. Equation solving in geometrical theories. In N. Dershowitz and N. Lindenstrauss (editors), *Conditional and Typed Rewriting Systems*, 4th International Workshop, CTRS-94, Jerusalem, Israel, July 1994, Proceedings. Lecture Notes in Computer Science 968, 31-50, Springer-Verlag, 1995.
4. Philippe Balbiani, V. Dugat, Luis Fariñas del Cerro and A. Lopez. *Eléments de géométrie mécanique*. Hermès, 1994.
5. Philippe Balbiani and Luis Fariñas del Cerro. Affine geometry of collinearity and conditional term rewriting. In H. Comon and J.-P. Jouannaud (editors), *Term Rewriting*, French School on Theoretical Computer Science, Font Romeux, France, May, 1993, Advanced Course. Lecture Notes in Computer science 909, 196-213, Springer-Verlag, 1995.
6. Jon Barwise and Eric Hammer. Diagrams and the concept of logical system. In D. Gabbay (editor), *What is a Logical System?* 73-107, Oxford University Press, 1994.
7. Shang-Ching Chou. Mechanical Geometry Theorem Proving. Reidel, 1988.
8. Harold S. Coxeter. Projective Geometry. Blaisdell, 1964.
9. David Hilbert. Foundations of Geometry. Second english edition, Open Court, 1971.
10. Anne Lopez. Déduction automatique en géométrie par réduction de figures. Thèse de l'université Paul Sabatier, Toulouse, 1995.
11. John Greenlees Semple and G.T. Kneebone. Algebraic Projective Geometry. Oxford University Press, 1952.
12. Sun-Joo Shin. *The Logical Status of Diagrams*. Cambridge University Press, 1994.
13. Dongming Wang. Elimination procedures for mechanical theorem proving in geometry. In H. Hong, D. Wang and F. Winkler (editors), *Algebraic Approaches to Geometrical Reasoning*. Annals of Mathematics and Artificial Intelligence, Volume 13, Numbers 1,2, 1-24, 1995.
14. Wen-Tsün Wu. Basic Principles of Mechanical Theorem Proving in Geometry. Springer-Verlag, 1994.

ON SENTENCES OF THE KIND "SENTENCE '*P*' IS ABOUT TOPIC *T*"

ROBERT DEMOLOMBE AND ANDREW J. I. JONES

1 Introduction

In philosophy, and in the study of non-classical logics, a good deal of interest has been shown in the phenomenon of *aboutness*, and in the formal explication of the idea that a given sentence is about some particular class of topics. More recently, in computer science, it has become clear that *aboutness* is of practical interest for a number of application domains.

These trends, both theoretical and practical, are described and discussed below in Sections 3 and 4, while in Sections 1 and 2 we first present our own steps towards an analysis of *aboutness*.

Four fundamental assumptions guide our approach in this paper:

1. Sentences of the type "sentence 'p' is about topic t" are shorthand for "the proposition expressed by the sentence 'p' is about topic t", where the notion of proposition is interpreted in a way which conforms to our use of a three-valued semantics of the Bochvar kind [2]. According to this interpretation, a proposition can be viewed as a set of partially defined worlds, that is, a set of worlds in relation to which some of the members of the set of all propositional variables may be undefined, but where the given proposition itself is true.
2. We shall assume that any particular proposition concerns a set of topics. And thus we shall say that the sentence 'p' is about some topic t provided t is a member of that set of topics with which the sentence 'p' is concerned.
3. Suppose that two sentences are equivalent in classical propositional calculus, and that they contain just the same atoms. Then we shall suppose that the one is about a given topic t if and only if the other is also about t. But note the proviso concerning "contains the same atoms". For instance, although the following is a truth of classical

H.J. Ohlbach and U. Reyle (eds.), Logic, Language and Reasoning, 115–133.

propositional logic:

$$p \leftrightarrow (p \wedge (q \vee \neg q))$$

we are not prepared to allow that the sentences on the left-hand side and on the right-hand side of the equivalence sign are necessarily about the same topic.

Our reason may be explained as follows: suppose that the sentence 'q' is about some topic t, but that the sentence 'p' is not about t. Given that the sentence 'q' is about t, we would want to accept that '$q \vee \neg q$' might also be about topic t, and thus that t *may be* among the topics that '$p \wedge (q \vee \neg q)$' is about. From which it follows, obviously, that although the sentences 'p' and '$p \wedge (q \vee \neg q)$' are logically equivalent, they nevertheless *need not* be about the same topics.

One objection to this argument would be the contention that tautologies are not about any topic at all. We shall have a little more to say about that issue in our discussion below of related work by Nelson Goodman and David Lewis.

4. We shall assume that the set of topics a *compound* sentence is about need not be independent of its *mode of composition.* Consider, for instance, the following two sentences: 'Maria is married to Jim', 'Maria is married to Jules'. Considered separately, and in the absence of any specific presuppositions to the contrary, neither of these two sentences is about the topic of bigamy, although their conjunction most certainly is. So the conjunctive mode of combining the two sentences may itself alter the class of topics concerned.

2 Syntax and Semantics

In this section we first present formal definitions of the language, of the models, and of the truth-conditions. We then consider several additional model-theoretic conditions which, were they to be adopted, would give a richer structure to a set of sentences that are about some topic.

2.1 Formal Characterisation of the Language L

Let CPC be *the language of* classical propositional calculus. Let σ be a set of names for sentences of CPC. Let τ be a set of names of topics. Let $A(-,-)$ be a sorted binary predicate whose first argument is of sort name of topic (τ), and whose second argument is of sort name of sentence (σ).

The language L is defined by the following rules:

- any atomic sentence of CPC, and any ground atom of the form $A(x, y)$ are sentences of L. (The arguments x, y must of course satisfy the restrictions as to sort mentioned above.)

- if p and q are any sentences of L, then $\neg p$ and $p \vee q$ are sentences of L;
- the logical connectives $\wedge$, $\rightarrow$ and $\leftrightarrow$, are introduced in the usual way, in terms of negation and disjunction;
- all the sentences of L are defined by the above rules.

Notation: 'p' names the sentence occurring between the quotation marks in 'p'. Intuitively, we read sentences of the form $A(t, \text{'}p\text{'})$ as: "the sentence 'p' is about the topic t".

Some illustrative examples of sentences of L:

- $A(t, \text{'}p\text{'}) \wedge (p \rightarrow q)$, "the sentence '$p$' is about t and if p then q";
- $A(t, \text{'}p \wedge q\text{'}) \rightarrow A(t, \text{'}q \wedge p\text{'})$, " if the sentence '$p \wedge q$' is about t, then the sentence '$q \wedge p$' is about t";
- $A(t, \text{'}p\text{'}) \rightarrow A(t', \text{'}p\text{'})$, "if the sentence '$p$' is about t, then the sentence 'p' is about t' ".

2.2 Models

A model M of the logic is a tuple $M = \langle W, I, J, \mathcal{T}, \mathcal{S}, N, T, F \rangle$ where:

- W is a set of worlds;
- I is a function that assigns to each topic name in L an element of $\mathcal{T}$;
- J is a function that assigns to each sentence name in L an element of $\mathcal{S}$;
- $\mathcal{T}$ is a set of topics;
- $\mathcal{S}$ is the set of sentences of CPC;
- N is a function which assigns sets of topics to pairs of sets of worlds. The definition domain of N is: $2^W \times 2^W \rightarrow 2^{\mathcal{T}}$;
- T is a function which assigns to each atom in CPC a set of worlds;
- F is a function which assigns to each atom in CPC a set of worlds.

We impose on M the constraint that the name "p" is assigned, by the function J, the sentence 'p' itself.

We also impose on the models M the constraint $T(p) \cap F(p) = \emptyset$. The intuition is that, in a given world, a sentence 'p' can be true or false, or neither true nor false, but it cannot be both true and false.

The functions T and F are extended to any sentence in CPC by the following rules:

$$\begin{aligned}
T(\neg p) &= F(p) \\
F(\neg p) &= T(p) \\
T(p \vee q) &= (T(p) \cap D(q)) \cup (T(q) \cap D(p)) \\
F(p \vee q) &= F(p) \cap F(q)
\end{aligned}$$

where $D(p)$ is an abbreviation of $T(p) \cup F(p)$.

The functions T and F define a 3-valued logic for sentences in CPC. This logic was initially defined by Bochvar in [2]. Independently of our work the same logic has also been used by Buvač, Mason and McCarthy in [3, 16] to characterize the set of meaningful sentences in a given context.

It has been shown by Demolombe in [9] that for any sentences 'p' and 'q' of CPC we have $T(p) = T(q)$ and $F(p) = F(q)$ for every model M iff we have $\vdash p \leftrightarrow q$, in classical propositional calculus, and 'p' and 'q' are formed with the same atoms. However, if we have only the property $T(p) = T(q)$ for every model, then 'p' and 'q' are not necessarily formed with the same atoms. Consider, for instance, $p = a \wedge \neg a$ and $q = b \wedge \neg b$, where a and b are atoms.[1]

These results indicate that in order to characterize the set of topics of a given sentence 'p' in a way which conforms to our third initial assumption, it is not enough to consider $T(p)$ alone; an alternative which suggests itself, and which we have employed here, is to consider the pair: $\langle T(p), F(p)\rangle$.

If we give up the constraint $T(p) \cap F(p) = \emptyset$ we get a 4-valued logic of the kind suggested by David Lewis in [15]. However, it has been shown in [9] that, even in this 4-valued logic, the fact that $\vdash p \leftrightarrow q$ (where 'p' and 'q' are formed with the same atoms) is not equivalent to the fact that $T(p) = T(q)$ for every model. Consider for instance, $p = (a \wedge \neg a) \wedge b$ and $q = (a \wedge \neg a) \wedge \neg b$. In the 4-valued logic $T(p)$ and $T(q)$ are different whenever there exist a world w_1 where a is true and false and b is only true, and a world w_2 where a is true and false and b is only false.

The truth conditions for sentences in L are:

$$
\begin{array}{lll}
M, w \models p & \text{iff} & w \in T(p), \text{ if } p \text{ is an atom of CPC.} \\
M, w \models \neg p & \text{iff} & M, w \not\models p. \\
M, w \models p \vee q & \text{iff} & M, w \models p \text{ or } M, w \models q. \\
M, w \models A(t, \text{`}p\text{'}) & \text{iff} & I(t) \in N(T(J(\text{`}p\text{'})), F(J(\text{`}p\text{'}))).
\end{array}
$$

(In the following, we shall abbreviate $J(\text{`}p\text{'})$ to p.)

The last of the four truth conditions corresponds to our second assumption, above, and says that '$A(t, \text{`}p\text{'})$' is true if and only if the topic t is one of the topics assigned by the function N to the proposition expressed by the sentence 'p'.

2.3 Some Interesting Schemas

We understand validity of a sentence schema in the usual way, as truth in all worlds in all models. Among the schemas which the semantical conditions given in the previous section do *not* validate is:

$$A(t, \text{`}p \wedge q\text{'}) \rightarrow (A(t, \text{`}p\text{'}) \vee A(t, \text{`}q\text{'})). \qquad (i)$$

[1]The appendix to this paper presents a summary of some of the central results of [9].

This (negative) result is important in relation to our fourth initial assumption, above, where we claimed that the mode of composition of a compound sentence may play a significant role in determining the class of topics the compound is about. Thus '$p \wedge q$' might be about the topic of bigamy, whilst neither 'p' nor 'q' is about bigamy.

By contrast, a good case can perhaps be made for maintaining that where two or more sentences are each about the same topic, then their conjunction is also about this topic, i.e.,

$$(A(t, `p\text{'}) \wedge A(t, `q\text{'})) \rightarrow A(t, `p \wedge q\text{'}). \qquad (ii)$$

The validity of (ii) may be secured by imposing the following constraint on function N:

$$\begin{array}{ll} \text{If} & I(t) \in N(T(p), F(p))) \\ \text{and} & I(t) \in N(T(q), F(q))) \\ \text{then} & I(t) \in N(T(p \wedge q), F(p \wedge q))). \end{array} \qquad (cii)$$

Furthermore, we have found no good reason to deny that if a sentence is about some topic t, then its negation is also about that topic[2]:

$$A(t, `p\text{'}) \rightarrow A(t, `\neg p\text{'}). \qquad (iii)$$

The corresponding constraint on N is:

$$\begin{array}{ll} \text{If} & I(t) \in N(T(p), F(p))) \\ \text{then} & I(t) \in N(T(\neg p), F(\neg p))). \end{array} \qquad (ciii)$$

As was explained above in 2.2, when two sentences containing just the same atoms are classically equivalent, then the one is about a given topic if and only if the other is about that topic. That is:

$$\begin{array}{ll} \text{If} & \models p \leftrightarrow q \text{ and } p \text{ and } q \text{ contain just the same atoms} \\ \text{then} & \models A(t, `p\text{'}) \leftrightarrow A(t, `q\text{'}). \end{array} \qquad (iv)$$

From (iv) and (iii) it follows that:

$$A(t, `p\text{'}) \leftrightarrow A(t, `\neg p\text{'}) \qquad (v)$$

is a valid schema.

Given now that our fourth basic assumption above, dictates that schema (i) is not valid, a consequence is that the following schema is also not valid:

$$A(t, `p \vee q\text{'}) \rightarrow (A(t, `p\text{'}) \vee A(t, `q\text{'})). \qquad (vi)$$

---

[2]In connection with formula (iii), we should like to mention a very interesting comment made by Alice ter Meulen at the AAAI workshop *Formalizing Context* (Boston, 1995), where an earlier draft of this paper was presented. She pointed out that (iii) could be accepted only if one assumes that the presuppositions of sentence 'p' remain unchanged when 'p' is negated. For instance, if we assume that the sentence 'Venus is a star' is about the topic astronomy, then the sentence 'Venus is not a star' is also about astronomy, provided that the presupposition that Venus is an object in the sky remains unchanged. If this presupposition were to be changed, for instance if Venus were presupposed to be a film actress, then 'Venus is not a star' would no longer be about astronomy. The assumption that presuppositions must be kept unchanged applies, presumably, also to (ii), and quite generally in the evaluation of questions about validity/invalidity of schemas containing several occurrences of formulas of the type '$A(t, `p\text{'})$'.

For suppose $A(t, \text{`}p \wedge q\text{'})$; then by (iv) $A(t, \text{`}\neg(\neg p \vee \neg q)\text{'})$ and by (v) and (iv) $A(t, \text{`}\neg p \vee \neg q\text{'})$. Acceptance of (vi) would then yield $(A(t, \text{`}\neg p\text{'}) \vee A(t, \text{`}\neg q\text{'}))$, which by (v) gives $(A(t, \text{`}p\text{'}) \vee A(t, \text{`}q\text{'}))$, which would mean that (i) is valid.

The existing literature (see Section 4) hardly reflects a consensus regarding the logical properties of 'aboutness'; so, no doubt, some will disagree with our claims regarding which schemas are valid, and which are not. Our *firmest* intuitions lead us to accept (iv) and our inclination is to accept (ii) and to reject (i); this means that (iii) and (vi) cannot *both* be accepted. These are just preliminary conclusions. Much remains to be done by way of further investigation of possible constraints on the function N. However, as will become clear from Section 3 below, a final decision about which axiom schemas to adopt can be made only in relation to consideration of a particular domain of application.

2.4 Axiomatics

The axiomatic characterization of the logic is the result of a rather straightforward translation of the schemas which are retained for the semantics.

In the axiomatics, in addition to the axiom schemas of classical propositional calculus and the inference rule Modus Ponens, we have the following weakened rule of equivalence:

$$\text{(REA)}\ \frac{\vdash p \leftrightarrow q \text{ and } p \text{ and } q \text{ contain the same atoms}}{A(t, \text{`}p\text{'}) \leftrightarrow A(t, \text{`}q\text{'})}.$$

If we accept schemas (ii) and (iii) we have the corresponding axiom schemas:

$$(A(t, \text{`}p\text{'}) \wedge A(t, \text{`}q\text{'})) \rightarrow A(t, \text{`}p \wedge q\text{'}) \qquad (aii)$$

$$A(t, \text{`}p\text{'}) \rightarrow A(t, \text{`}\neg p\text{'}). \qquad (aiii)$$

A proof of completeness of the logic similar to the proof of completeness given in [5, 4], and based on results in [9], should be forthcoming without any particular difficulties. The proof of soundness is quite easy.

3 Application Domains for Topics

We present below three application domains where we have to characterize sets of sentences in terms of their meaning, and where topics seem to be appropriate for this purpose.

When an agent puts a query to a database system, the system can help the agent by providing him with additional information relevant to the query. This kind of system behaviour is usually called 'cooperative answering'. In [6, 7] Cuppens and Demolombe have developed a method for cooperative answering based on the use of topics. Roughly speaking, if the query is the sentence 'p', and if 'p' is about the topic t, then t is identified as a topic of interest for the agent, and the system returns, in addition to the answer to '*p*', other sentences that are about the agent's topics of interest, and which are consequences of the database. Here, topics permit the characterization of the set of sentences an agent is interested in. The idea that, if an agent a is interested in a topic t he is interested in all the sentences '*p*' about this topic, could be represented by an axiom schema of the form: (1) $IT_a(t) \wedge A(t, \text{`}p\text{'}) \rightarrow I_a(p)$, where $IT_a(t)$ means that a is interested in all the sentences about t, and $I_a(p)$ means that a is interested in '*p*'.[3]

In this context, if it happens that the database system concerned cannot answer the query '*p*', but it can answer the query '$p \vee q$', it will be useful for the agent to know at least whether '$p \vee q$' is true, provided that '$p \vee q$' is informative in relation to '*p*' in the context of this database.[4] Then, we would like to be able to infer from the fact that agent a is in interested in '*p*' that he is also interested in '$p \vee q$'. This requirement could be satisfied in two different ways. We could add to the axiom schema (1) either the axiom schema: (2) $I_a(p) \rightarrow I_a(p \vee q)$ or the axiom schema: (vii) $A(t, \text{`}p\text{'}) \rightarrow A(t, \text{`}p \vee q\text{'})$. Indeed, from the assumptions that agent a is interested in topic t, and that sentence '*p*' is about t, represented by $IT_a(t)$ and $A(t, \text{`}p\text{'})$, we can infer from (1) and (vii) that $I_a(p \vee q)$. We see that the structure of the set of sentences about the topic $t$ is imposed on the set of sentences agent $a$ is interested in. More formally from (1) and (vii) we can infer: $IT_a(t) \rightarrow ((A(t, \text{`}p\text{'}) \rightarrow A(t, \text{`}p \vee q\text{'})) \rightarrow (I_a(p) \rightarrow I_a(p \vee q)))$.

Another application domain for topics is the characterization of the reliability of agents who insert data in a database. Informally, Demolombe and Jones in [10] define an agent to be reliable for the sentence '*p*' iff it is guaranteed that, if he inserts '*p*' in the database, then '*p*' is true of the world. It might be much more convenient to define reliability of agents in terms of topics rather than in terms of sentences. For instance, in a company, one agent may be known to be reliable for all sentences about the topic 'accounting', another one for all sentences about the topic 'health'. The idea that if an agent a is reliable for a topic t and a sentence '*p*' is about topic t, then he is reliable for the sentence '*p*', could be represented by the axiom schema: (4) $RT_a(t) \wedge A(t, \text{`}p\text{'}) \rightarrow R_a(p)$. Here, if we accept that $A(t, \text{`}p\text{'})$ implies $A(t, \text{`}p \vee q\text{'})$, that is (vii), it follows that if an agent is reliable for t, he is reliable for '*p*' and also for '$p \vee q$', that is: $R_a(p) \rightarrow R_a(p \vee q)$. This consequence in general is not acceptable because the reason why an agent believes p (if does so), and the reason why he believes $p \vee q$ (if he does so), may be completely independent. Then, in this application domain, the axiom schema (vii) should be rejected.

[3]These comments anticipate future planned work. At this stage $IT_a(t)$ and $I_a(p)$ have not been formally defined, they should be considered just as convenient notations.

[4]We shall not go into the question of how to give a formal definition of 'informative' here, but some proposals are to be found in the database literature.

Finally, a third possible application domain of topics is the characterization of the set of sentences an agent is permitted or prohibited to access in a database. For example, an agent may be permitted or prohibited to access all the data which are about the topic 'nuclear energy'. Permissions to know and prohibitions to know, in terms of topics or in terms of sentences, could be represented by the axiom schema: (5) $PKT_a(t) \wedge A(t, `p') \rightarrow PK_a(p)$, where $PKT_a(t)$ means that agent $a$ is permitted to know all the sentences about topic $t$, and $PK_a(p)$ means that agent $a$ is permitted to know $p$; a similar axiom schema could be accepted for prohibition: (6) $FKT_a(t) \wedge A(t, `p') \rightarrow FK_a(p)$. In [8] Cuppens and Demolombe have shown that if an agent is permitted to know p, he is also permitted to know all the consequences of p, and, in particular $p \vee q$, that is: $PK_a(p) \rightarrow PK_a(p \vee q)$. Then the structure induced by the axiom schema (vii) on $PK_a(p)$ is acceptable. However, it has also been shown that if an agent is forbidden to know p he is not necessarily forbidden to know the consequences of p, and in particular $p \vee q$. That means that the structure induced by (vii) is not accepable for prohibitions. Notice also that the axiom schema (ii) $(A(t, `p') \wedge A(t, `q')) \rightarrow A(t, `p \wedge q')$, that we above indicated a willingness to accept, would induce on $PK_a(p)$ the axiom schema $PK_a(p) \wedge PK_a(q) \rightarrow PK_a(p \wedge q)$, and it has been argued in [8] that in general this result would not be acceptable.

4 Related Works

There are many works in the area of relevance logic (see [1, 11] and the collection of papers in [17]) which seem, at first glance, to have connections with our work. In fact there are many relevance logics and there are significant differences in their objectives. Nevertheless, roughly speaking, all of them have the objective of restricting derivations that can be made in classical logic to those derivations where consequences are relevant to antecedents. For example, in [12], Epstein defines several logics (called relatedness logics or dependence logics) in an attempt to formalize the notion of 'topic' or 'subject matter' or 'referential content'. For each logic two possible formalizations are proposed.

The first one is based on a 'relatedness' relation on sentences. Two sentences 'p' and 'q' are related, and this is denoted by $R(p, q)$, if their subject matters have something in common. For instance the subject matter of 'Ralph is a dog' is related to the subject matter of 'Dogs are faithful', or to that of 'George is a duck', but it is not related to the subject matter of the sentence '$2 + 2 = 4$'. It is assumed that one can say, for each pair of sentences, whether $R(p, q)$ holds or not. The structure of the relatedness relation is defined by the following properties (which are equivalent to properties described at p. 67 in [12]): R1: $R(p, q)$ iff $R(p, \neg q)$, R2: $R(p, q \wedge r)$

iff $R(p,q)$ or $R(p,r)$, R3: $R(p, q \to r)$ iff $R(p,q)$ or $R(p,r)$, R4: $R(p,p)$, R5: $R(p,q)$ iff $R(q,p)$. It is shown in [12] that two sentences are related iff they contain a common atom. A very important feature of the relatedness relation is that it is independent of the logical connectives that appear in the sentences and depends only on the relatedness of their atoms. "The subject matter of a proposition is independent of its truth value. As I see it, a virtual consequence of this assumption is that the logical connectives are neutral with respect to what a proposition is related to...", Epstein says. But he gives no justification for this statement.

Then he says: "one question that may strike you is: why don't we take 'relatedness' as a primitive in the language? The answer is because the binary relation of subject matter relatedness, call it R, is not a connective." And then, to reinforce his answer he adds: "It makes no sense to iterate it." However, Epstein does not consider the possibility of defining a language that extends classical propositional calculus with a predicate $R(x,y)$ in a way similar to our introduction of the predicate $A(x,y)$. The relation R is used in the definition of the semantics only to assign to the connective $\to$ a more restrictive meaning. According to his definition $p \to q$ is true iff it is true in the classical sense and $R(p,q)$ holds.

The second formalization is based on a function $s(p)$ which assigns to each sentence 'p' a set of topics. It is assumed that a set of topics is assigned to each atom, and that the set of topics of a compound sentence is just the union of the sets of topics assigned to each component. Here again we see that logical connectives play no role in the determination of the topics a sentence is about. It is possible, from a given relatedness relation, to define the set assignment function s, and vice versa.

Here there are significant differences from our approach. The first is that the notion of topic is not explicitly represented in the language; it is implicitly represented in the form of appropriate axiom schemas, and this observation holds for all the relevance logics we know. Therefore, it is not possible to reason about the consequences of assumptions concerning the fact that such and such a sentence is about such and such a topic. For instance, if we assume p, we can know which consequences of 'p' are relevant to 'p', or related to 'p'. But, what we cannot do in these relevant logics is to assume that a sentence 'p' is about a given topic t, and to infer from this assumption which other sentences are also about t. The second difference is that the role played by logical connectives in the determination of the subject matter of a sentence is ignored in relevant logics.

Buvač *et. al.* in [3] define the notion of meaningful sentence in a given context in the following way. A vocabulary (a set of atoms) is assigned to each context and all the sentences formed with the vocabulary of a given context are meaningful in this context. That means that the fact that a sentence is, or is not, meaningful is independent of the logical connectives that appear in this sentence. In this respect Buvač *et.al* have a different approach to the definition of meaningfulness from the one we have to aboutness.

However, an interesting common feature is that the notion of meaningfulness is formalized with the same 3-valued logic of Bochvar. They have introduced in the language a binary predicate $ist(c,p)$ whose meaning is: 'p' is meaningful in the context c, and 'p' holds in the context c. That means that the predicate $ist(c,p)$ represents in fact two notions: truth and meaningfulness. If some component of the the sentence 'p' is meaningless in the context c, then the overall sentence 'p' is meaningless in this context. Notice that an important difference between $ist(c,p)$ and $A(t,`p\text{'})$ is that $A(t,`p\text{'})$ holds independently of whether 'p' is in fact true.

Another difference is that the logic they have defined validates the properties: $ist(c,p \wedge q) \rightarrow ist(c,p) \vee ist(c,q)$ and $ist(c,p) \rightarrow \neg ist(c,\neg p)$. The first one corresponds to the property (i) we are not inclined to accept, and the second one is incompatible with the property (iii) that we are ready to accept. Also, the rule of substitutivity of equivalent sentences is restricted to sentences which are formed with the vocabulary of the given context. That is, we can infer $ist(c,p) \leftrightarrow ist(c,q)$ from $\vdash p \leftrightarrow q$, provided '$p$' and '$q$' are formed with the vocabulary of c. But that does not mean that 'p' and 'q' are formed with the same atoms. This situation arises, for example, if 'p' is '$a \vee \neg a$' and q is '$b \vee \neg b$', and a and b are two atoms in the vocabulary of c.

One may notice that, as in the dependence logic presented by Epstein, logical connectives play no role in the determination of meaningfulness. But, an important difference from Epstein, and a similarity to the logic we have presented, is that the predicate $ist(c,p)$ allows the indirect representation in the language of the notion of meaningfulness.

If we want to formalize the part of the meaning of $ist(c,p)$ that refers to meaningfulness by using the predicate $A(t,`p\text{'})$, we may take a context to be a topic and assume, for every atom $a$ in the vocabulary of $c$, that we have $A(c,`a\text{'})$. Then, if we accept for the predicate A the axiom schemas: $A(c,`p \vee q\text{'}) \leftrightarrow A(c,`p\text{'}) \vee A(c,`q\text{'})$ and $A(c,`\neg p\text{'}) \leftrightarrow A(c,`p\text{'})$, we have $A(c,`p\text{'}) \leftrightarrow A(c,`q\text{'})$ iff '$p$' and '$q$' are formed with the vocabulary of $c$; and we would expect to be able to prove the inference rule: from $\vdash p \leftrightarrow q$ and $A(c,`p\text{'}) \leftrightarrow A(c,`q\text{'})$ infer $ist(c,p) \leftrightarrow ist(c,q)$.

In [13], Nelson Goodman offers a very interesting analysis of aboutness from a perspective quite different from ours. "Our sole problem" he says (p. 3), "...is to determine what a sentence is about, *given* what its terms designate". So, for instance, he considers that the sentence 'Paris is growing' is about Paris, and that 'The capital city of France is growing' is also about Paris. In his analysis of what he calls 'absolute' aboutness, his preliminary conjecture is that a sentence S is about (say) Paris, if some sentence T that mentions Paris follows logically from S (p. 4). However, Goodman then immediately points out that this preliminary idea must be refined, because it leads to unaccepatble consequences. The sentence 'Paris is growing or London is growing' mentions London and is implied by 'Paris is growing'. But we do not want to conclude that 'Paris is growing' is about London!

Goodman proposes some refinements which avoid this problem. However, as he says, the notion of 'absolute' aboutness he defines is purely extensional (p. 10). In keeping with this, tautologies are not about anything; so, although 'Paris is growing' mentions Paris and is about Paris, and 'Paris is not growing' also mentions Paris and is about Paris, the sentence 'Paris is growing or Paris is not growing' mentions Paris but is not about anything at all, Goodman claims. Likewise, self-contradictory sentences are not about anything, and Goodman criticizes Carnap's account of aboutness because it fails to "...meet the requirement that logically equivalent statements are about the same things." (p. 9, footnote 1).

Thus Goodman's approach to the analysis of aboutness is very different from our account of sentences of the kind 'sentence '*p*' is about topic *t*'. The notion of *selectivity* is at the core of Goodman's understanding of aboutness: "...'about' behaves somewhat as 'choose' does. If I ask Johnny to choose some presents and he replies 'I choose everything', he has not chosen anything. Choosing something involves not choosing something else. That Johnny chooses every x is always false. Likewise, saying so and so about an object involves not saying so and so about some other." (p. 5).

There seem to be very strong similarities to fundamental aspects of information theory here. According to Shannon and Weaver's account of 'amount of information', a signal can carry information only if its occurrence reduces uncertainty, and it reduces uncertainty only if its occurrence eliminates some other possibilities. In other words, if a signal α was bound to happen anyway, in the sense that there was no possibility that any signal other than α could have occurred, then the occurrence of α has no surprisal value and thus, on this view, it carries no information. Compare with Goodman's claim "...that nothing can be said about every object, or about every class of objects, or about every class of classes of objects, etc." (p. 6). Nothing can be said about everything, because 'saying about' requires the possibility of contrast, but there are no other things with which to make the contrast if the subject is 'everything'.

Whether or not information-theoretic ideas influenced Goodman's approach to aboutness, it is clear that our proposals regarding the topics a sentence is about are not based on consideration of what information that sentence carries, at least not 'information' in the sense of information theory. The sentences 'It is raining or it is not raining' and 'It is snowing or it is not snowing' are, of course, indistinguishable from the purely extensional point of view, and neither of them carries any information, in the sense of information theory, because neither of them eliminates any possibilities. But we agree with David Lewis (see [14]) that there is a finer-grained level of analysis (than the extensional) at which it would be correct to say that these tautologies do not have the same meaning. And, we conjecture, it is differences in meaning of that finer-grained kind which will have to be captured, if we are to understand the logic of sentences of the kind 'sentence '*p*' is about topic *t*'. Our main suggestion in this paper has been that the intensionality of such sentences might be articulated in terms of the semantical framework of Bochvar's 3-valued logic.

There is another difference between our proposal and Goodman's, with regard to their respective objectives. Goodman does confront the question: "What is it that determines what a given sentence is about?". And his answer, as we have indicated, takes its point of departure in a consideration of what the terms of the given sentence designate. However, we do not pretend to have supplied an answer to *Goodman's* question. What we have done is to specify truth conditions for sentences of the kind 'sentence '*p*' is about topic *t*', and this specification just requires *t* to be a member of the set of topics '*p*' is about. But we say no more by way of characterization of the latter set than it is the value of a function whose argument is the pair consisting of the truth-set of the sentence '*p*' names, and the falsity-set of the sentence '*p*' names, where the notions of truth and falsity are defined as in the semantics for a 3-valued logic of the Bochvar kind. Beyond this, we have given no further account of what it is for a proposition to be about a set of topics.

So what, then, can we be said to have achieved, as regard a contribution to the understanding of topics and aboutness? The answer is that we have supplied a formal-semantical framework within which it is possible to determine, in a systematic fashion, the consistency or inconsistency of sets of sentences which themselves make claims about which topics some given sentences are about. And thus we may also check which implication relations do, or do not, hold between sentences which make claims of these kinds. In this respect, our approach is comparable to that which has dominated applications of modal logic in the last 40 years. For instance, possible-worlds semantics for alethic modal logic does not supply an answer to the question "What is it that determines whether a given sentence is possible/necessary?" – at least, it does not supply an answer of a non-circular kind, in which no appeal is made to some prior, intuitive understanding of the concepts of possibility and necessity. But a possible world semantics for sentences of alethic modal logic nevertheless *does* provide a tool for the systematic investigation of questions about implication and consistency, in regard to sets of sentences about possibilities and necessities.

Further investigation of the question of what it is that determines what a given proposition is about remains for us as an issue for future work. One point, however, seems clear to us already: that the designations of the terms occurring in the sentence expressing a proposition need not by themselves exhaust the class of topics the proposition may be about. That the proposition expressed by 'Maria is married to Jim and Maria is married to Jules' is about the topic bigamy, is a case in point.

Finally, to close this comparison with other related works, we have to mention some other work by David Lewis. In [15] Lewis defines a 'subject matter' as 'part of the world in intension'. An equivalence relation on worlds is defined from a given subject matter as follows: two worlds are in the same equivalence class if and only if they are exactly alike for that part of the world defined by the subject matter. Moreover, a proposition corresponds to each equivalence class. This equivalence relation suggests an extensional view of subject matters.

Then Lewis informally analyses the structure of subject matters. This analysis is based on the inclusion relation on subject matters. From his point of view a subject matter M1 is included in a subject matter M2, if the part of the world described by M1 is included in the part of the world described by M2. For instance the subject matter '1680' is included in the subject matter '17th century', since the description of the 17th century contains the description of every year in the 17th century, and in particular of the year 1680.

Notice that, as in our approach, Lewis accepts schema (iii) ('a proposition and its negation should be exactly alike with respect to what they are about'), and he does not accept schema (i). In order to avoid assigning the same subject matter to all the non-contingent propositions he suggests considering a four -valued logic, but only some guidelines about this logic are given.

5 Conclusion

We have presented a formal logic for sentences of the kind 'the sentence 'p' is about the topic t'. The main features of this logic are that it is based on a 3-valued logic of the Bochvar type, and that assumptions of the form 'the sentence 'p' is about the topic t' can explicitly be represented in the language by $A(t, `p')$. Several possible additional axiom schemas have been discussed, and we have argued that a decision about their acceptance or rejection must depend on the intended application domains. For some of these, it may be that none of the additional axiom schemas should be accepted, in which case consequences of assumptions of the form $A(t, `p')$ can be drawn only by using the inference rule (REA). Some might maintain that in that case, the logic is so weak that it has no practical interest. However that is not true, because in the absence of any inference rule, for the characterization of the set of sentences about a given topic, one should have to give an extensional definition of *all* the sentences in this set. For instance, if the sentence '$p \wedge \neg(q \vee r)$' is about t, one should have to say that, the following sentences, among others, also are about the topic t: '$p \wedge \neg(r \vee q)$', '$\neg(q \vee r) \wedge p$', '$\neg(r \vee q) \wedge p$', '$p \wedge \neg q \wedge \neg r$', '$p \wedge \neg r \wedge \neg q$', '$\neg q \wedge \neg r \wedge p$', '$\neg r \wedge \neg q \wedge p$', and '$\neg r \wedge p \wedge \neg q$'.

The comparison of our work with other related research has suggested to us some possible directions for future work. In particular, the analysis of Goodman's approach shows that it is interesting to consider the structure of atomic sentences, that is, in formal terms, to move from propositional calculus to predicate calculus. We could then investigate principles which determine how the topics the proposition expressed by a sentence is about are dependent upon the names and predicates occurring in that sentence, and we could try to relate such principles to the 3-valued approach to the analysis of propositions presented here. Another possible direction for future works is to investigate the definition of a structure on sets of topics. For instance, a relation of the form $t > t'$, whose meaning would be: 'the topic t is more specific than the topic t'', would permit representation of a hierarchy, and could be used to represent axiom schemas of the form: $t > t' \rightarrow (A(t, \text{'}p\text{'}) \rightarrow A(t', \text{'}p\text{'}))$, whose intended meaning is that, if the topic t is more specific than the topic t', and the sentence 'p' is about t, then it is also about t'.

Acknowledgements:

We are very grateful to Sylvie Cazalens whose previous joint work with us played a role in the development of this paper. We are also grateful to Sača Buvač for several fruitful discussions we have had about the logic of contexts. This work was partially supported by the ESPRIT BRA project MEDLAR2.

Appendix

DEFINITION 5.1 (Propositional calculus language) *Let* VAR *be a set of propositional variables, the associated propositional calculus language is defined as usual from* VAR *using the logical connectives* $\neg$, *for negation, and* $\vee$ *for disjunction.*

The connectives $\wedge$ *and* $\rightarrow$ *are defined as usual from negation and disjunction.* ■

DEFINITION 5.2 (Structure) *A structure is a tuple* $S = \langle W, T, F \rangle$ *such that* W *is a set of worlds,* T *is a function from* VAR *to* 2^W *and* F *is a function from* VAR *to* 2^W. ■

From an intuitive point of view, if v is a propositional variable, T (resp. F) assigns to v the set of worlds where v is true (resp. false).

The functions T and F are extended to compound sentences by the following rules:

$$\begin{aligned} T(\neg p) &= F(p) \\ F(\neg p) &= T(p) \\ T(p \vee q) &= (T(p) \cap D(q)) \cup (T(q) \cap D(p)) \\ F(p \vee q) &= F(p) \cap F(q). \end{aligned}$$

The truth values 'defined' and 'undefined' are defined from 'true' and 'false'. The set of worlds where a sentence p is defined (resp. undefined) is denoted by $D(p)$ (resp. $U(p)$). These functions are defined by:

$$D(p) \stackrel{\text{def}}{=} T(p) \cup F(p) \quad \text{and} \quad U(p) \stackrel{\text{def}}{=} W \setminus D(p).$$

In the case where several structures are under consideration we adopt the notation $T_S(p)$ to denote the set of worlds where p is true in the structure S. Similar notations are adopted for $F(p)$, $D(p)$ and $U(p)$.

For a given Popositional Calculus Language the set of all the possible structures such that:

$$T_S(v) \cap F_S(v) = \emptyset$$

is denoted by Σ_3.

DEFINITION 5.3 (Two-valued logic associated to a three-valued logic.) *Let $S = \langle W, T, F \rangle$ be a structure in Σ_3. The associated two-valued structure s is the tuple $s = \langle W, t \rangle$ where t is a function from* VAR *to 2^W such that:*

- *for a propositional variable $v : t(v) = T(v)$;*
- *$t(\neg p) = W \setminus t(p)$;*
- *$t(p \vee q) = t(p) \cup t(q)$.* ■

To denote the set of worlds where a proposition p is false we adopt the notation $f(p)$, and we have:

$$f(p) \stackrel{\text{def}}{=} W \setminus t(p)$$

Notation: the fact that a sentence p is a tautology of classical propositional calculus (CPC) is denoted by: $\models_{CPC} p$.

THEOREM 5.4 *Let $S = \langle W, T, F \rangle$ be a given struture in Σ_3, we define the structure S^+ in function of S by:*

$W^+ = W$

For every propositional variable v:

$T_{S^+}(v) = T_S(v)$

$F_{S^+}(v) = W \setminus T_S(v)$.

We have for all sentence p: $T_{S^+}(p) = t_S(p)$ and $F_{S^+}(p) = f_S(p)$.

Proof: The proof is by induction on the complexity of sentences.
If p is an atomic sentence, by definition of T_{S^+} and F_{S^+}, we have $T_{S^+}(v) = T_S(v)$ and $F_{S^+}(v) = W \setminus T_S(v)$. By definition of $t_S(p)$ and $f_S(p)$ we have $t_S(p) = T_S(p)$ and $f_S(p) = W \setminus T_S(p)$. Therefore we have $T_{S^+}(p) = t_S(p)$ and $F_{S^+}(p) = f_S(p)$.
For a sentence of the form $\neg p$ we have by definition of T: $T_{S^+}(\neg p) = F_{S^+}(p)$. By induction hypothesis we have: $F_{S^+}(p) = f_S(p)$, and by definition of f we have: $f_S(p) = t_S(\neg p)$. Therefore we have $T_{S^+}(\neg p) = t_S(\neg p)$. For similar reasons we have $F_{S^+}(\neg p) = T_{S^+}(p) = t_S(p) = f_S(\neg p)$.
For a sentence of the form $p \vee q$ by definition of T we have: $T_{S^+}(p \vee q) = (T_{S^+}(p) \cap (T_{S^+}(q) \cup F_{S^+}(q))) \cup (T_{S^+}(q) \cap (T_{S^+}(p) \cup F_{S^+}(p)))$. By induction hypothesis we have $T_{S^+}(p) = t_S(p)$ and $F_{S^+}(p) = f_S(p)$. Since we have $t_S(p) \cup f_S(p) = W$ we also have $T_{S^+}(p) \cup F_{S^+}(p) = W$. For the same reasons we have $T_{S^+}(q) \cup F_{S^+}(q) = W$. Therefore we have $T_{S^+}(p \vee q) = T_{S^+}(p) \cup T_{S^+}(q)$. From the induction hypothesis we have $T_{S^+}(p) \cup T_{S^+}(q) = t_S(p) \cup t_S(q)$, and by definition of t we have $t_S(p) \cup t_S(q) = t_S(p \vee q)$, and therefore we have $T_{S^+}(p \vee q) = t_S(p \vee q)$. For similar reasons we have: $F_{S^+}(p \vee q) = F_{S^+}(p) \cap F_{S^+}(q) = f_S(p) \cap f_S(q) = f_S(p \vee q)$. ■

THEOREM 5.5 *If for all structure S in Σ_3 we have $T_S(p) \subseteq T_S(q)$ then for all structure S in Σ_3 we have $t_S(p) \subseteq t_S(q)$.*

Proof: By contraposition, Theorem 5.5 is equivalent to: if there exists a structure S in Σ_3 such that $t_S(p) \not\subseteq t_S(q)$ then there exists a structure S in Σ_3 such that $T_S(p) \not\subseteq T_S(q)$.
Let us assume $t_S(p) \not\subseteq t_S(q)$. From Theorem 5.4 we have $T_{S^+}(p) = t_S(p)$ and $F_{S^+}(p) = f_S(p)$, then we have $T_{S^+}(p) \not\subseteq T_{S^+}(q)$. Therefore there exists a structure in Σ_3, namely S^+, such that we have $T_{S^+}(p) \not\subseteq T_{S^+}(q)$. ■

THEOREM 5.6 *For all sentence p and for all structure S in Σ_3 we have: $T_S(p) \subseteq t_S(p)$ and $F_S(p) \subseteq f_S(p)$.*

Proof: The proof is by induction on the complexity of sentences.
If p is an atomic sentence, by definition of t, we have $T_S(p) = t_S(p)$, then we have $T_S(p) \subseteq t_S(p)$. From the definition of the three-valued logic we have: $T_S(p) \cap F_S(p) = \emptyset$. Then, if some world w is in $F_S(p)$, it is not in $t_S(p)$, and therefore it is in $f_S(p)$. Then we also have $F_S(p) \subseteq f_S(p)$.
For a sentence of the form $\neg p$ we have $T_S(\neg p) = F_S(p)$ and, by induction hypothesis, we have $F_S(p) \subseteq f_S(p)$. Since, by definition of f, we have $f_S(p) = t_S(\neg p)$, we finally have: $T_S(\neg p) \subseteq t_S(\neg p)$. For similar reasons we have $F_S(\neg p) = T_S(p) \subseteq t_S(p) = f_S(\neg p)$.
For a sentence of the form $p \vee q$, by definition of T, we have: $T_S(p \vee q) = (T_S(p) \cap (T_S(q) \cup F_S(q))) \cup (T_S(q) \cap (T_S(p) \vee F_S(p)))$. By induction hypothesis we have: $T_S(p) \subseteq t_S(p)$ and $F_S(p) \subseteq f_S(p)$, then we have $T_S(p \vee q) \subseteq (t_S(p) \cup t_S(q))$ and, since we have $t_S(p) \cup t_S(q) = t_S(p \vee q)$, we have $T_S(p \vee q) \subseteq$

$t_S(p \vee q)$. From the definition of F we have $F_S(p \vee q) = F_S(p) \cap F_S(q)$, and by induction hypothesis we have $F_S(p) \subseteq f_S(p)$ and $F_S(q) \subseteq f_S(q)$, therefore we have $F_S(p \vee q) \subseteq f_S(p \vee q)$. ■

THEOREM 5.7 *For all sentence p and for all structure S in Σ_3 we have: $t_S(p) \subseteq T_S(p) \cup U_S(p)$ and $f_S(p) \subseteq F_S(p) \cup U_S(p)$.*

Proof: The proof is by induction on the complexity of sentences.
If p is an atomic sentence, by definition of t, we have $T_S(p) = t_S(p)$, then we have $t_S(p) \subseteq T_S(p) \cup U_S(p)$. By definition of f, we have $f_S(p) = W \setminus T_S(p)$, and in the three-valued logic we have $W \setminus T_S(p) = F_S(p) \cup U_S(p)$, then we have $f_S(p) \subseteq F_S(p) \cup U_S(p)$.
For a sentence of the form $\neg p$, by definition of t, we have $t_S(\neg p) = f_S(p)$, and by induction hypothesis we have $f_S(p) \subseteq F_S(p) \cup U_S(p)$. Since we have $F_S(p) = T_S(\neg p)$ and $U_S(p) = U_S(\neg p)$, we have $t_S(\neg p) \subseteq T_S(\neg p) \cup U_S(\neg p)$. For similar reasons we have: $f_S(\neg p) = t_S(p) \subseteq T_S(p) \cup U_S(p) = F_S(\neg p) \cup U_S(\neg p)$.
For a sentence of the form $p \vee q$, by definition of t, we have $t_S(p \vee q) = t_S(p) \cup t_S(q)$, and by induction hypothesis, we have $t_S(p) \subseteq T_S(p) \cup U_S(p)$ and $t_S(q) \subseteq T_S(q) \cup U_S(q)$. From the definition of T we have: $T_S(p \vee q) = (T_S(p) \cup T_S(q)) \setminus (U_S(p) \cup U_S(q))$, then we have: $t_S(p \vee q) \subseteq T_S(p \vee q) \cup U_S(p) \cup U_S(q)$,that is $t_S(p \vee q) \subseteq T_S(p \vee q) \cup U_S(p \vee q)$.
For similar reasons we have $f_S(p \vee q) = f_S(p) \cap f_S(q)$ and $f_S(p) \subseteq F_S(p) \cup U_S(p)$ and $f_S(q) \subseteq F_S(q) \cup U_S(q)$. Then we have: $f_S(p \vee q) \subseteq (F_S(p) \cup U_S(p)) \cap (F_S(q) \cup U_S(q))$, and we have: $f_S(p \vee q) \subseteq (F_S(p) \cap F_S(q)) \cup U_S(p) \cup U_S(q)$, that is: $f_S(p \vee q) \subseteq F_S(p \vee q) \cup U_S(p \vee q)$. ■

THEOREM 5.8 *The facts $\models_{CPC} p \rightarrow q$ and $Var(q) \subset Var(p)$ imply that for all structure S in Σ_3 we have $T_S(p) \subseteq T_S(q)$.*

Proof: Let S be a structure in Σ_3, from Theorem 5.6 we have: $T_S(p) \subseteq t_S(p)$. Since we have $\models_{CPC} p \rightarrow q$ we have $t_S(p) \subseteq t_S(q)$. From Theorem 5.7 we have $t_S(q) \subseteq T_S(q) \cup U_S(q)$, therefore we have $T_S(p) \subseteq T_S(q) \cup U_S(q)$. Since we have $Var(q) \subseteq Var(p)$, from the definition of U, we have $U_S(q) \subseteq U_S(p)$. Then we have $T_S(p) \subseteq T_S(q) \cup U_S(p)$. From the definition of T and U, we have $T_S(p) \cap U_S(p) = \emptyset$. Therefore we have $T_S(p) \subseteq T_S(q)$. ■

THEOREM 5.9 *The facts $\models_{CPC} p \rightarrow q$ and $Var(p) \subseteq Var(q)$ imply that for all structure S in Σ_3 we have $F_S(q) \subseteq F_S(p)$.*

Proof: Since $\models_{CPC} p \rightarrow q$ implies $f_S(q) \subseteq f_S(p)$, and $Var(p) \subseteq Var(q)$ implies $U_S(p) \subseteq U_S(q)$, a similar proof as for Theorem 5.8 allows to show that we have $F_S(q) \subseteq F_S(p)$. ■

THEOREM 5.10 *The facts $\models_{CPC} p \leftrightarrow q$ and $Var(p) = Var(q)$ imply that for all structure S in Σ_3 we have $T_S(p) = T_S(q)$ and $F_S(p) = F_S(q)$.*

Proof: This theorem is a direct consequence of Theorems 5.8 and 5.9. ∎

THEOREM 5.11 *If for all structure S in Σ_3 we have $T_S(p) = T_S(q)$ and $F_S(p) = F_S(q)$, then we have Var(p)=Var(q).*

Proof: Let us assume that for all sentences p and for all structures S in Σ_3 we have $T_S(p) = T_S(q)$ and $F_S(p) = F_S(q)$.
Let us assume that we have $Var(q) \not\subseteq Var(p)$, then there exists a propositional variable v such that $v \in Var(q)$ and $v \notin Var(p)$.
Let S be a structure in Σ_3 and w be a world in S. We have either $w \in t_S(p)$ or $w \in f_S(p)$. Let us assume first that we have $w \in t_S(p)$.
We define a world w' of a structure S' from w and S in the following way:

If a variable u is in Var(p) then:
if $w \in T_S(u)$ then $w' \in T_{S'}(u)$,
if $w \notin T_S(u)$ then $w' \in F_{S'}(u)$.
If a variable u is not in $Var(p)$ then $w' \in U_{S'}(u)$.

According to this definition we have $w' \in t_{S'}(p)$, because we have $w \in t_S(p)$, and the fact $w' \in t_{S'}(p)$ (resp. $w \in t_S(p)$) only depends on the variables u such that $w' \in T_{S'}(u)$ (resp. $w \in T_S(u)$), and for the variables u in p we have $w \in T_S(u)$ iff $w' \in T_{S'}(u)$.
From Theorem 5.7 we have $t_{S'}(p) \subseteq T_{S'}(p) \cup U_{S'}(p)$, then we have $w' \in T_{S'}(p)$ or $w' \in U_{S'}(p)$. From the definition of w' none of the variables in p is undefined in w' then we do not have $w' \in U_{S'}(p)$, therefore we have $w' \in T_{S'}(p)$. Since we have $T_{S'}(p) = T_{S'}(q)$, we also have $w' \in T_{S'}(q)$.
Since the variable v of q is not in p, by definition of w', we have $w' \in U_{S'}(v)$, and, by definition of U, we have $w' \in U_{S'}(q)$, which contradicts the fact $w' \in T_{S'}(q)$. Therefore we have $Var(q) \subseteq Var(p)$.
If we assume now that we have $w \in f_S(p)$, a similar proof, based on the fact $F_{S'}(p) = F_{S'}(q)$, also allows to infer $Var(q) \subseteq Var(p)$.
Then, in both cases we have $Var(q) \subseteq Var(p)$.
Since p and q play a similar role, we can also prove that $Var(p) \subseteq Var(q)$, and finally we have $Var(p) = Var(q)$. ∎

THEOREM 5.12 *If for all structure S in Σ_3 we have $T_S(p) = T_S(q)$ then we have $\models_{CPC} p \leftrightarrow q$.*

Proof: If for all structure S in Σ_3 we have $T_S(p) = T_S(q)$, then, from Theorem 5.6, we have for all structure S in Σ_3 $t_S(p) = t_S(q)$. Since in Σ_3 we have all the possible assignments for t, if for all structure S in Σ_3 we have $t_S(p) = t_S(q)$, we have $\models_{CPC} p \leftrightarrow q$. Therefore we have $\models_{CPC} p \leftrightarrow q$. ∎

THEOREM 5.13 *For all sentences p, if for all structures S in Σ_3 we have $T_S(p) = T_S(q)$ and $F_S(p) = F_S(q)$, then we have $\models_{CPC} p \leftrightarrow q$ and $Var(p) = Var(q)$.*

Proof: This result collates the results of Theorems 5.11 and 5.12. ■

THEOREM 5.14 *We have* $\models_{CPC} p \leftrightarrow q$ *and* $Var(p)=Var(q)$ *iff for all structure* S *in* Σ_3 *we have* $T_S(p) = T_S(q)$ *and* $F_S(p) = F_S(q)$.

Proof: This theorem is a direct consequence of Theorems 5.10 and 5.13. ■

Robert Demolombe
ONERA-CERT, France.

Andrew J. I. Jones
University of Oslo, Norway.

References

1. Alan R. Anderson and Nuel D. Belnap. *Entailment.* Princeton University Press, 1975.
2. D.A. Bochvar. Two papers on partial predicate calculus. Technical report STAN-CS-280-72, Stanford University, 1972. Translation of two papers from 1938 and 1943.
3. Sača Buvač, Vanja Buvač, and Ian A. Mason. Metamathematics of contexts. *Fundamenta Informaticae*, 23(3), 1995.
4. Sylvie Cazalens. *Formalisation en Logique non standard de certaines méthodes de raisonnement pour fournir des réponses coopératives, dans des systèmes de Bases de Données et de Connaissances.* PhD thesis, Université Paul Sabatier, Toulouse France, 1992.
5. Sylvie Cazalens, Robert Demolombe, and Andrew J.I. Jones. A Logic for Reasoning about Is About. Technical report, ONERA-CERT, 1992.
6. Frédéric Cuppens and Robert Demolombe. Cooperative Answering: a methodology to provide intelligent access to Databases. In *Proc. of 2nd Int. Conf. on Expert Database Systems*, Tysons Corner, Virginia, 1988.
7. Frédéric Cuppens and Robert Demolombe. How to recognize interesting topics to provide cooperative answering. *Information Systems*, 14(2), 1989.
8. Frédéric Cuppens and Robert Demolombe. A deontic logic for reasoning about confidentiality. In *Proc. of 3rd International Workshop on Deontic Logic in Computer Science*, 1996.
9. Robert Demolombe. Multivalued logics and topics. Technical report, CERT-ONERA, 1994.
10. Robert Demolombe and Andrew J.I. Jones. Deriving answers to safety queries. In R. Demolombe and T. Imielinski, editors, *Nonstandard queries and non-standard answers.* Oxford University Press, 1994.
11. Michael J. Dunn. Relevance logic and entailment. In D. Gabbay and F. Guenthner, editor, *Handbook of Philosophical Logic.* D. Reidel Publishing Company, 1984.
12. Richard L. Epstein. *The Semantic Foundations of Logic, Volume1: Propositional Logic.* Kluwer Academic, 1990.
13. Nelson Goodman. About. *Mind*, LXX(277), 1961.
14. David K. Lewis. General semantics. In D. Davidson and G.Harman, editor, *Semantics of Natual Language.* D. Reidel Publishing Company, 1972.
15. David K. Lewis. Relevant implication. *Theoria*, LIV(3), 1988.
16. John McCarthy and Saša Buvač. Formalizing contexts. Technical Report STAN-CS-TN-94-13, Stanford University, 1994.
17. Jean Norman and Richard Sylvan, editors. *Directions in Relevant Logics.* Kluwer Academic, 1989.

TWO TRADITIONS IN THE LOGIC OF BELIEF: BRINGING THEM TOGETHER

KRISTER SEGERBERG

1 The Two Traditions

The terms 'epistemic logic' and 'doxastic logic' were introduced by Georg Henrik von Wright in his 1951 *An essay in modal logic* [8], but it was not until Jaakko Hintikka published his seminal book *Knowledge and Belief* [4] that the discipline named by them took off. During the following ten to fifteen years epistemic/doxastic logic received a good deal of attention in the philosophical community, but towards the end of the 1970s the philosophers seem to have considered that the theme had been played out.[1] As often happens, however, interest in the subject was rekindled in a different quarter: computer scientists discovered or re-invented epistemic logic, as they prefer to call the subject; at the present time it is flourishing.[2]

This development, named by von Wright but started by Hintikka, is one of the two traditions of doxastic logic alluded to in the title of the paper. The other is the logic of theory change, brought to prominence by a trio of scholars often referred to by the initials AGM of their surnames, Carlos Alchourrón and David Makinson on the one hand and Peter Gärdenfors on the other; their paper [1] from 1985 is already classic. It may sound as if there is nothing doxastic about theory change; however, the kind of theory change that has been most thoroughly investigated is belief revision, and it is this fact that motivates our view that the logic of theory change can – and perhaps should – be seen as a kind of doxastic logic.

What distinguishes Hintikka type doxastic logic from AGM-type belief revision is that the former is static, the latter dynamic. Hintikka studies

[1] For a good account of this period, see [5].

[2] According to von Wright's terminology, epistemic logic is the logic of knowledge *(episteme)* while doxastic logic is the logic of belief *(doxa)*. The distinction between knowledge and belief is difficult to draw, and more often than not today's modal logicians, especially in the computer science camp, seem uninterested in trying to draw one. Here we prefer the term doxastic logic as the more inclusive term.

H.J. Ohlbach and U. Reyle (eds.), Logic, Language and Reasoning, 135–147.

the unchanging beliefs of a certain rational agent, AGM how rational belief might change in the face of new information.[3] At the same time, the two traditions have a great deal in common. It is the thesis of this paper that it will be fruitful to cast the logic of belief revision as a generalization of ordinary Hintikka type doxastic logic. Van Benthem may have been the first to have this insight, which was then taken further by his former student de Rijke [2, 3]. Other ways of trying to use dynamic logic in the study of belief revision have been explored by numerous authors.

2 Trying to Find a Suitable Object Language

AGM is not really logic; it is a theory about theories. A typical statement will be of the kind $\varphi \in T$, where φ is a formula and T is a theory, intuitively a belief-set of some fixed agent. In doxastic logic the same fact is expressed by the formula $\mathbf{B}\varphi$, where $\mathbf{B}$ is Hintikka's familiar operator. Thus we have two families of formal expressions with the following reading lists:

$\varphi \in T$	'φ is in the agent's belief-set T';
$\varphi \notin T$	'φ is not in the agent's belief-set T';
$\neg\varphi \in T$	'$\neg\varphi$ is in the agent's belief-set T';
$\neg\varphi \notin T$	'$\neg\varphi$ is not in the agent's belief-set T';
$\mathbf{B}\varphi$	'the agent believes that φ';
$\neg\mathbf{B}\varphi$	'the agent does not believe that φ';
$\mathbf{B}\neg\varphi$	'the agent believes that not φ';
$\neg\mathbf{B}\neg\varphi$	'the agent does not believe that not φ'.

One striking difference here is that the language of belief revision theory is what can perhaps be called absolute, while the language of doxastic logic is indexical. Given φ and T it is either true or false, absolutely, that $\varphi \in T$. But, given φ, it depends on the context whether it is true or false that $\mathbf{B}\varphi$. We will return to this point.

Three kinds of change are prominent in current theories about belief revision: expansion and revision on the one hand, and contraction on the other. *Expansion* of a theory consists in adding something to the theory, *revision* in adding something and, if necessary, to take appropriate measures

[3] Nowadays there is general agreement on the importance of distinguishing between collecting information about an unchanging world and collecting information about a world in change. It is belief revision of the former kind that AGM theory studies. The latter kind of information collection is often called *updating* and requires a special, more complicated analysis. In this paper the focus is on belief change in a world that does not change; nevertheless we have gone to great length (putting up with a good deal of cumbersome complication) in trying to work out a theory within whose framework update theory can be developed.

to ensure that the result will be consistent (assuming that the new piece added is consistent in its own right). *Contraction* is giving up belief in something – not coming to believe something new, just ceasing to believe it. In AGM, theories were required to be closed under logical consequence, and the three operations just mentioned were taken to be functional. Thus if T is a given theory and φ is a formula, the following theories were thought to be well-defined:

$T + \varphi$	the result of expanding T by φ;
$T * \varphi$	the result of revising T by φ;
$T - \varphi$	the result of contracting T by φ.

Now to expand or revise or contract is to do something. Thus it is possible to think of expansion, revision and contraction as actions of a certain kind – epistemic or doxastic actions. This means that we can look to dynamic logic for a language that lets us reason about belief change. In general, if α is a term and φ is a formula, then $[\alpha]\varphi$ and $\langle\alpha\rangle\varphi$ are well-formed formulae of dynamic logic, meaning 'after every way of performing α, it is the case that φ' and 'after some way of performing α, it is the case that φ', respectively. For each φ, then, let us think of the following three doxastic actions:

$+\varphi$	expanding by φ;
$*\varphi$	revision by φ;
$-\varphi$	contraction by φ.

Thus we have the following way of rendering belief revision talk in dynamic doxastic logic:

$\chi \in T + \varphi$	$[+\varphi]\mathbf{B}\chi$;
$\chi \in T * \varphi$	$[*\varphi]\mathbf{B}\chi$;
$\chi \in T - \varphi$	$[-\varphi]\mathbf{B}\chi$.

Even though AGM did not say much about it, there is nothing in syntax to preclude iteration. For example, $\chi \in ((T + \varphi) * \psi) - \theta$ is a well-formed expression and is rendered by the formula $[+\varphi][*\psi][-\theta]\mathbf{B}\chi$.

Without defining a particular object language we list the conditions that it is assumed to satisfy. There are two categories of well-formed expressions, formulae and terms. The terms are divided into three subgroups: purely doxastic terms, purely real terms and mixed terms. The primitive formulae are the propositional letters, either finitely or denumerably many. The propositional operators include a truth-functionally complete set of Boolean connectives as well as the belief operator $\mathbf{B}$. A purely Boolean formula is one built exclusively from propositional letters and Boolean connectives. $\mathbf{B}\varphi$ is a formula if φ is a purely Boolean formula but not otherwise. There are doxastic operators $+$, $*$, and $-$ taking purely Boolean formulae to terms; more specifically, $+\varphi$ and $*\varphi$ and $-\varphi$ are purely doxastic terms if φ is a purely Boolean formula but not defined otherwise. If the language does not contain **b**, the dual of **B**, as a primitive operator, then we will consider it as defined by the convention

$$\mathbf{b}\chi \overset{\text{def}}{=} \neg\mathbf{B}\neg\chi;$$

note that **B** is a *box* operator and **b** is a *diamond* operator. Similarly, we assume that for each box operator $[\alpha]$ there is a corresponding diamond operator $\langle\alpha\rangle$, either primitive or defined.

In principle, belief can be about anything, even about one's own beliefs. However, in this paper we restrict our analysis to the agent's beliefs about the world; and we do not regard his beliefs as part of the world. This decision is reflected in the requirement that **B** and the doxastic operators operate only on purely Boolean formulae. This is a technically important point, and to avoid every possibility of misunderstanding we stress it: *Throughout this paper, in expressions* $\mathbf{B}\varphi, +\varphi, *\varphi, -\varphi$ *it is assumed that* φ *is a purely Boolean formula.*

3 Relational Semantics

Let U be some given *universe*, the elements of which we call *points* or *possible world states*. An *action* in U is a binary relation in $\mathcal{P}U \times U$. A *frame* is a structure (U, R) where R, the *repertoire*, is a set of actions. A *valuation* in a frame (U, R) is a function mapping each propositional letter to a subset of U and each term to an action in R; in particular, if α is a purely doxastic term, then, for all $X, X' \subseteq U$ and $u, u' \in U$, $((X,u),(X',u')) \in V(\alpha)$ implies that $u = u'$, and if α is a purely real term, then, for all $X, X' \subseteq U$ and $u, u' \in U$, $((X,u),(X',u')) \in V(\alpha)$ implies that $X = X'$. A *model* is a structure (U, R, V), where (U, R) is a frame and V a valuation in it.

Let $\mathcal{M} = (U, R, V)$ be a model. We will give an inductive definition of *truth* of a formula φ in $\mathcal{M}$ with respect to X and u, in symbols $(X, u) \models^{\mathcal{M}} \varphi$ or more simply $X, u \models^{\mathcal{M}} \varphi$, where $X \subseteq U$ and $u \in U$. (For simplicity, from now on we suppress explicit reference to $\mathcal{M}$ in the notation.) A formula is false with respect to X and u if and only if not true. We adopt the following convention: if φ is a purely Boolean formula then $\|\varphi\|$ denotes the *truth-set* of φ; given V, the definition is the standard extension of V to the set of all purely Boolean formulae. This convention makes it easier to express the inductive definition:

$$
\begin{array}{lll}
X, u \models \varphi & \text{iff} & u \in \|\varphi\|, \text{if } \varphi \text{ is a purely Boolean formula;} \\
X, u \models \varphi \wedge \psi & \text{iff} & X, u \models \varphi \text{ and } X, u \models \psi; \\
X, u \models \neg\varphi & \text{iff} & \text{not } X, u \models \varphi,
\end{array}
$$

and similarly for other Boolean connectives. Furthermore:

$$X, u \models \mathbf{B}\varphi \text{ iff } X \subseteq \|\varphi\|;$$
$$X, u \models \mathbf{b}\varphi \text{ iff } X \cap \|\varphi\| \neq \emptyset$$

(whether **b** is primitive or defined). Finally, if α is any term,

$$X, u \models [\alpha]\chi \text{ iff } \forall X' \, \forall u' \, (((X, u), (X', u')) \in V(\alpha) \Rightarrow X', u' \models \chi).$$

A formula is *valid* in a certain frame if it is true in all models on the frame with respect to all subsets of the universe and all points in the universe.

4 Normal Logics

Consider the following postulates, where **o** is **B** or $[+\varphi]$ or $[*\varphi]$ or $[-\varphi]$:

(#0) $\vdash \theta$, if θ is a tautology;
(#1) if $\vdash \varphi$ and $\vdash \varphi \supset \psi$ then $\vdash \psi$;
(#2o) $\vdash \mathbf{o}(\varphi \supset \psi) \supset (\mathbf{o}\varphi \supset \mathbf{o}\psi)$;
(#3o) if $\vdash \varphi$ then $\vdash \mathbf{o}\varphi$;
(#4) if $\vdash \varphi \equiv \psi$ then $\vdash [+\varphi]\chi \equiv [+\psi]\chi$;
(#5) if $\vdash \varphi \equiv \psi$ then $\vdash [*\varphi]\chi \equiv [*\psi]\chi$;
(#6) if $\vdash \varphi \equiv \psi$ then $\vdash [-\varphi]\chi \equiv [-\psi]\chi$.

If (##0–1) hold, our logic contains all truth-functional tautologies and arguments. If (##2o–3o) hold, **o** is a normal operator in the sense of modal logic. If in addition (##4–6) hold, then the resulting logic is closed under replacement of provably equivalent formulae. We will say that a logic that satisfies all these postulates is *normal*.

Let us now go further and ask the following question: what conditions would be sufficient for $+$ and $*$ and $-$ to deserve to be regarded as an expansion operator, a revision operator and a contraction operator, respectively? The question is of course normative. Nevertheless, in working out an answer we will at least articulate some of our own preformal intuitions. The conditions listed below exemplify those that an analyst might reasonably consider when it comes to providing an answer; it is not suggested that the list is complete. There are several groups of model theoretic and corresponding syntactical conditions. (It is not claimed that they are independent.)

Noninterference conditions

There is already such a condition implicit in our model theory: the requirement on valuations V that $((X, u), (X', u')) \in V(\alpha)$ implies $u = u'$, for all purely doxastic terms α. There are the following corresponding syntactic postulates:

(#N+) $\langle +\varphi\rangle\top \supset (\chi \equiv [+\varphi]\chi)$, where χ is purely Boolean.

(#N*) $\langle *\varphi\rangle\top \supset (\chi \equiv [*\varphi]\chi)$, where χ is purely Boolean.

(#N–) $\langle -\varphi\rangle\top \supset (\chi \equiv [-\varphi]\chi)$, where χ is purely Boolean.

The generality of our definition of action becomes unnecessarily cumbersome when the discussion is limited to belief change in a static world.[4] To ease the exposition, at the cost of some generality, we assume that $((X,u),(X',u)) \in V(\alpha)$ if and only if $((X,v),(X',v)) \in V(\alpha)$ for all purely doxastic terms α. Then we introduce the definition $V^\circ(\alpha) \stackrel{\text{def}}{=} \{(X,X') : \exists u((X,u),(X',u)) \in V(\alpha)\}$. For simplicity we omit the symbol $^\circ$ from now on; no confusion need result.

Vacuity conditions

If $X \subseteq \|\varphi\|$, then $(X,X') \in V(+\varphi)$ if and only if $X = X'$.
If $X \subseteq \|\varphi\|$, then $(X,X') \in V(*\varphi)$ if and only if $X = X'$.
If $X - \|\varphi\| \neq \emptyset$, then $(X,X') \in V(-\varphi)$ if and only if $X = X'$.

(#V+) $\mathbf{B}\varphi \supset (\chi \equiv [+\varphi]\chi)$.

(#V*) $\mathbf{B}\varphi \supset (\chi \equiv [*\varphi]\chi)$.

(#V–) $\neg\mathbf{B}\varphi \supset (\chi \equiv [-\varphi]\chi)$.

Triviality conditions

$V(+\top) = \{(X,X') : X = X'\}$; and $(X,X') \in V(+\bot)$ only if $X' = \emptyset$.
If $X \neq \emptyset$ then $V(*\top) = \{(X,X') : X = X'\}$; and $V(*\bot) = \emptyset$.
$V(-\top) = \emptyset$; and if $X \neq \emptyset$ then $V(-\bot) = \{(X,X') : X = X'\}$.

(#T+) $\chi \equiv [+\top]\chi$ $\qquad [+\bot]\mathbf{B}\bot$.

(#T*) $\mathbf{b}\top \supset (\chi \equiv [*\top]\chi)$; $\qquad [*\bot]\bot$.

(#T–) $[-\top]\bot$; $\qquad \mathbf{b}\top \supset (\chi \equiv [-\bot]\chi)$.

Success conditions

$(X,X') \in V(+\varphi)$ only if $X' \subseteq \|\varphi\|$.
$(X,X') \in V(*\varphi)$ only if $X' \subseteq \|\varphi\|$.
$(X,X') \in V(-\varphi)$ only if $X' - \|\varphi\| \neq \emptyset$.

(#S+) $[+\varphi]\mathbf{B}\varphi$.

(#S*) $[*\varphi]\mathbf{B}\varphi$.

(#S–) $[-\varphi]\neg\mathbf{B}\varphi$.

---

[4] Cf. footnote 2, page 135.

Independence conditions

$(X, X') \in V(+\varphi)$ only if $X' \subseteq X$.
$(X, X') \in V(-\varphi)$ only if $X \subseteq X'$.

(#I+) $\mathbf{B}\chi \supset [+\varphi]\mathbf{B}\chi$.

(#I−) $\langle -\varphi \rangle \mathbf{B}\chi \supset \mathbf{B}\chi$ or, equivalently, $\neg\mathbf{B}\chi \supset [-\varphi]\neg\mathbf{B}\chi$.

Consistency conditions

If $X \cap \|\varphi\| \neq \emptyset$, then $(X, X') \in V(+\varphi)$ only if $X' \neq \emptyset$.
$(X, X') \in V(-\varphi)$ only if $X' \neq \emptyset$.

(#C+) $\mathbf{b}\varphi \supset [+\varphi]\mathbf{b}\top$.

(#C−) $[-\varphi]\mathbf{b}\top$.

A *reasonable* consistency condition for $V(*\varphi)$ would perhaps be

If $\|\varphi\| \neq \emptyset$, then $(X, X') \in V(*\varphi)$ only if $X' \neq \emptyset$.

However, this condition can be expressed only if the object language has certain resources. For example, if we had a modal operator $\Box$ for universal necessity with the dual $\Diamond$, then the given condition would be rendered by

$$\Diamond\varphi \supset [*\varphi]\mathbf{b}\top.$$

In order to find an example of a consistency condition for $V(*\varphi)$ that is generally expressible within our present family of languages, define the set

$$preX = \bigcup\{Y : \exists\varphi((X, Y) \in V(-\varphi)\}$$

as the *presupposition* or the *prejudice* at X. Then there is the following condition and the corresponding schema:

If $preX \cap \|\varphi\| \neq \emptyset$, then $(X, X') \in V(*\varphi)$ only if $X' \neq \emptyset$.

(#C∗) $\langle -\psi \rangle \mathbf{b}\varphi \supset [*\varphi]\mathbf{b}\top$.

Feasibility conditions

There is some X' such that $(X, X') \in V(+\varphi)$.

(#F+) $\langle +\varphi \rangle \top$.

As before, the two following conditions are not generally expressible:

if $\|\varphi\| \neq \emptyset$, then there is some X' such that $(X, X') \in V(*\varphi)$;
if $X - \|\varphi\| \neq \emptyset$, then there is some X' such that $(X, X') \in V(-\varphi)$.

However, there are the following surrogates:

if $preX \cap \|\varphi\| \neq \emptyset$, then there is some X' such that $(X, X') \in V(*\varphi)$;
if $preX - \|\varphi\| \neq \emptyset$, then there is some X' such that $(X, X') \in V(-\varphi)$.

(#F*) $\langle -\psi \rangle \mathbf{b}\varphi \supset \langle * \varphi \rangle \top$ or, equivalently, $[*\varphi]\bot \supset [-\psi]\mathbf{B}\neg\varphi$.

(#F−) $\langle -\psi \rangle \mathbf{b}\neg\varphi \supset \langle -\varphi \rangle \top$ or, equivalently, $[-\varphi]\bot \supset [-\psi]\mathbf{B}\varphi$.

The great advantage of recasting belief revision theory as dynamic doxastic logic is that it puts at our disposal the rich meta-theory developed in the study of modal and dynamic logic. For example, it is not difficult to prove the following result:

THEOREM 4.1 *Let L be the smallest normal logic providing any (one or several) of the above schemata. Then L is determined by the corresponding semantic condition or conditions. Furthermore, L has the finite model property and is decidable.*

It is arguable that, with one exception, the conditions listed are implicit in AGM. The exception is triviality. According to AGM, $T - \top = T$, for every theory T. On the other hand, our postulate $\vdash [-\top]\bot$ may be said to express the view that it is impossible to contract by a tautology. Of these alternatives, ours would seem to be closer to logical tradition.

5 From Relations to Hypertheories

The point of this paper is to suggest the indexical language of dynamic doxastic logic, not as a replacement of, but as an alternative to the absolute language of belief revision theory. In the preceding sections we have sketched a relational semantics of the kind that is traditional in modal and dynamic logic. But however natural this analysis is from a technical point of view, and however fruitful it might be for deriving technical results, it does not do very much for intuition. The situation might be compared with that in conditional logic: Chellas's relational semantics is general and technically useful, but philosophically it is modellings like those of Stalnaker and Lewis that make semantic sense of conditional logic. By the same token, one would wish for something more intuitive for dynamic doxastic logic that the mere relations of the relational semantics.

Let us review the situation. We are dealing with an agent in a certain environment, called the world. Actions are modelled as sets of pairs of total states; the total state is made up of the state of the agent and the state of the world. The only kind of agent state we are dealing with in this paper is belief state. Thus the technical problem is to represent, within a given universe U of possible states of the world, the belief state of the agent and the state of the world.

While the world state is no problem, the state of the agent is more complicated. The agent has a certain belief-set; that is, the total set of propositional beliefs (about the world) that he holds. Each belief can be represented in U by a subset of U: the set of points u such that if u is the real world state, then the belief is true. Thus also the agent's belief-set can be represented as a certain subset of U, namely, as the intersection of the representations of his beliefs. But this *belief-set* cannot be the same as his *belief state*, for it seems obvious that two agents may hold identical beliefs about the world (as to what is the actual state of the world) and yet react very differently to new information. In other words, it is not enough to represent the agent's current set of beliefs about the world, we must also represent what might be called his *doxastic dispositions* – the way or ways he is disposed to act in the face of new information; in fact the way or ways he is disposed to act in the face of any sequence of new pieces of information.

This may seem like excessive idealization. Nonetheless, if this idealization is not accepted there will be no general theory of doxastic behaviour. The surprising assumption made by AGM was that it makes sense to try to give rules that characterize the doxastic behaviour of a rational believer. The same assumption is made by anyone who wants to construct an automated, *reasonable* system for handling changing information.[5] It is this assumption, however unrealistic, that we shall have to accept.

The next question is, how to represent the agent's doxastic dispositions in terms of a given universe. Confronted with a new piece of information, the agent will perhaps move from his current belief-set-cum-dispositions to a new belief-set-cum-dispositions. If it is a question of weakening beliefs, we may say that he will *fall back* on a new belief-set (the term is due to Lindström and Rabinowicz). If it is a question of strengthening belief, we may coin an analogous new term and say that he will *push on* to a new belief-set. There are certainly more complex cases when the agent will go to a new belief-set that is neither weaker nor stronger than the current one; but those can perhaps be seen as derivative, as achievable by a combination of weakenings and strengthenings.

So there are fall-backs and push-ons; and if we knew all the possible fall-backs and push-ons of the agent as seen from his present belief-set, we would know a great deal about his doxastic dispositions. And if in addition we knew all the possible fall-backs and push-ons of the agent as seen from those new fall-backs and push-ons, and all the possible fall-backs and push-ons as seen from there, and so on, then perhaps we would know enough for our purposes about his doxastic dispositions.

The preceding remarks are meant to paint the informal background for the technical definitions that follow. In particular there are three desiderata on the theory we are about to develop. One is that we want the belief state of the agent to be represented by what we will call a *hypertheory*, which

[5] One qualification is important: we offer no theory for when information should be accepted. Our theory gives a partial description of what happens if the agent performs a certain doxastic action but extends no advice on whether to perform it.

will be a set of subsets of the universe (*theories*). In addition there will be the belief-set of the agent (a particular *theory*). The elements of the hypertheory that include the belief-set are called fall-backs, those that are included in the belief-set push-ons. We assume that every element of the hypertheory is either a fall-back or a push-on.

Another desideratum is that there be two basic kinds of doxastic action, *basic expansion* and *basic contraction*. As one develops particular modellings within the general framework we envisage, one will have to decide which among the indefinitely many possible actions deserve the name of doxastic actions. An interesting further question will then be whether every doxastic action thus defined can be constructed in set-theoretical terms from a small number of basic actions; in particular, in terms of basic expansion and basic contraction. (Cf. Levi's Commensuration Requirement [7], p. 65.)

A third desideratum is that the treatment of basic expansion and basic contraction should be symmetric. Philosophical consideration will perhaps later tell against symmetry, and in the end, as the theory is developed, symmetry may have to be given up. At the outset, however, we wish to treat the two kinds of belief change in an even-handed manner.

In order to express these informal ideas formally it is necessary to go over part of the same ground as in Section 3. As before we assume a set U (the *universe*), the elements of which are *points* or *possible states* of the world. There is also a designated set A of subsets of U closed under the Boolean operations; the elements of A are the *propositions* of the frame. A *theory* in A is any intersection, even infinite, of elements of A. Any theory in A is a possible *belief-set*. A *hypertheory* in A is a non-empty set of theories (hence a non-empty set of possible belief-sets). A *doxology* in U is a non-empty set of hypertheories in U.[6] If D is a doxology in U, then a binary relation on $D \times \mathcal{P}U \times U$ is called an *action* in U and D. A *repertoire* in U and D is a set of actions in U and D. A repertoire will always contain two particular basic actions, *(basic) expansion* and *(basic) contraction*, denoted by $\oplus$ and $\ominus$, respectively. Thus, if P is any proposition, then $\oplus P$ denotes expansion by P and $\ominus P$ denotes contraction by P.

A *frame* is a structure (U, A, D, R) where U is a universe, A is a set of propositions, D is a doxology in U and A and R is a repertoire in U and D. A *valuation* suitable for a frame (U, A, D, R) is a function V that assigns to each propositional letter a value in A and to each term a value in R, subject to the following conditions (as before, if φ is a purely Boolean formula we write $\|\varphi\|$ for the *truth-set* of φ): in particular, $V(+\varphi) = \oplus\|\varphi\|$ and $V(-\varphi) = \ominus\|\varphi\|$; in general, if α is a doxastic term

[6] The word 'doxology' is formed in analogy with the word 'epistemology'. There is no connection with the religious homonym.

then $((H,X,u),(H',X',u')) \in V(\alpha)$ implies $u = u'$, if α is a real term then $((H,u),(H',u')) \in V(\alpha)$ implies $H = H'$ and $X = X'$.

A *model* is a structure (U, A, D, R, V) where (U, A, D, R) is a frame and V is a suitable valuation for the frame. We define the *truth-value* of any formula φ in the model with respect to a hypertheory $H \in D$ a belief-set X and a point $u \in U$, in symbols $(H,X,u) \models \varphi$ or $H,X,u \models \varphi$ for simplicity – as follows:

$$\begin{array}{lll} H,X,u \models \varphi & \text{iff} & u \in \|\varphi\|, \text{if } \varphi \text{ is a purely Boolean formula;} \\ H,X,u \models \varphi \wedge \psi & \text{iff} & H,X,u \models \varphi \text{ and } H,X,u \models \psi; \\ H,X,u \models \neg\varphi & \text{iff} & \text{not } H,X,u \models \varphi; \end{array}$$

and similarly for other Boolean connectives. Furthermore:

$$\begin{array}{lll} H,X,u \models \mathbf{B}\varphi & \text{iff} & X \subseteq \|\varphi\|; \\ H,X,u \models \mathbf{b}\varphi & \text{iff} & X \cap \|\varphi\| \neq \emptyset \end{array}$$

(whether **b** is primitive or defined). Finally, if α is any term,

$$\begin{array}{l} H,X,u \models [\alpha]\chi \text{ iff} \\ \qquad \forall H' \forall X' \forall u' (((H,X,u),(H',X',u')) \in V(\alpha) \Rightarrow H',X',u' \models \chi). \end{array}$$

Note that in this definition, H plays no role; the importance of H is in connection with what we called the two basic doxastic actions. For any $Z \subseteq U$ define

$$\begin{array}{lll} H \uparrow Z & = & \{X \in H : Z \subseteq X\}; \\ H \downarrow Z & = & \{X \in H : X \subseteq Z\}. \end{array}$$

Assuming that P is a proposition, define

$$\begin{array}{lll} H \Cap P & = & \{Y \in H : Y \cap \|P\| \neq \emptyset\}; \\ H \dot{-} P & = & \{Y \in H : Y - \|P\| \neq \emptyset\}. \end{array}$$

We say that a set X is *maximal* in a set S of sets if $X \in S$ and, for all $Y \in S$, if $X \subseteq Y$ then $X = Y$; similarly, that X is *minimal* in S if $X \in S$ and, for all $Y \in S$, if $Y \subseteq X$ then $X = Y$.

A *frame for belief revision* (a *BR frame*) is a frame such that

$$\begin{array}{rcl} (H,X,u) \in field(\mathbf{a}) & \Rightarrow & X \in H, \text{for all actions } \mathbf{a} \in R, \\ ((H,X,u),(H',X',u')) \in \oplus P & \Rightarrow & X \supseteq X' \ \&\ u = u' \ \& \\ & & X' \text{ is maximal in } (H \downarrow X) \Cap P, \\ ((H,X,u),(H',X',u')) \in \ominus P & \Rightarrow & X \subseteq X' \ \&\ u = u' \ \& \\ & & X' \text{ is minimal in } (H \uparrow X) \dot{-} P. \end{array}$$

A *BR model* is a model on a BR frame. A formula is *valid (BR valid)* if true in all models (all BR models) with respect to all hypertheories in the doxology, all belief-sets and all points in the universe. The set of BR valid formulae is denoted by BR.[7]

6 Discussion

It is clear that the hypertheoretical semantics is more general than the relational one. In fact, the relational semantics can be seen as a special case of hypertheoretical semantics, namely, when the doxology contains only a single, fixed hypertheory. The philosophical question here is whether such restriction is reasonable. Are the doxastic dispositions of a rational agent fixed once and for all? In other words, is belief change just change in belief-set, or does it at least occasionally involve a change in one's disposition to change?

This point is worth dwelling on. Suppose that $\alpha_0, \ldots, \alpha_{m-1}$ and $\beta_0, \ldots, \beta_{n-1}$ are sequences of terms, each of which is an expansion or a contraction by a formula. For greater ease, let us write $[\alpha]$ and $[\beta]$ for $[\alpha_0] \ldots [\alpha_{m-1}]$ and $[\beta_0] \ldots [\beta_{n-1}]$, respectively. Then the following statement is true for the relational semantics but not for the hypertheoretical one:

$$\text{if } [\alpha]\mathbf{B}\chi \equiv [\beta]\mathbf{B}\chi \text{ is valid, for all } \chi, \text{ then } [\alpha]\chi \equiv [\beta]\chi \text{ is valid, for all } \chi.$$

If Levesque's operator **E** is available [6], the same point can be brought out by saying that the schema

$$(\langle\alpha\rangle\mathbf{E}\gamma \wedge \langle\beta\rangle\mathbf{E}\gamma) \supset ([\alpha](\mathbf{E}\gamma \supset \chi) \equiv [\beta](\mathbf{E}\gamma \supset \chi))$$

is valid in the relational but not in the hypertheoretical semantics. Thus it is possible to make distinctions in the hypertheoretical semantics which cannot be made in the relational one. We give two examples, functionality and recovery; in each case there is a pair of conditions where the second condition is stronger than the first.

Belief-set functionality

If α is any purely doxastic action, then $((H, X, u), (H', X', u')) \in V(\alpha)$ and $((H, X, u), (H'', X'', u'')) \in V(\alpha)$ implies $X' = X''$.

(#BF) $\langle\alpha\rangle\mathbf{B}\chi \supset [\alpha]\mathbf{B}\chi$.

Belief state functionality

If α is any purely doxastic action, then $((H, X, u), (H', X', u')) \in V(\alpha)$ and $((H, X, u), (H'', X'', u'')) \in V(\alpha)$ implies $H' = H''$ and $X' = X''$.

[7]It is clear that some schemata are BR valid but not valid, for example, $\mathbf{B}\chi \supset [+\varphi]\mathbf{B}\chi$ and $\langle-\varphi\rangle\mathbf{B}\chi \supset \mathbf{B}\chi$.

(#HF) $\langle\alpha\rangle\chi \supset [\alpha]\chi$.

Belief-set recovery

If $((H,X,u),(H',X',u')) \in V(-\varphi)$ and $((H',X',u'),(H'',X'',u'')) \in V(+\varphi)$, then $X = X'$.

(#BR) $[-\varphi][+\varphi]\mathbf{B}\chi \equiv \mathbf{B}\chi$.

Belief state recovery

If $((H,X,u),(H',X',u')) \in V(-\varphi)$ and $((H',X',u'),(H'',X'',u'')) \in V(+\varphi)$, then $H = H'$ and $X = X'$.

(#HR) $[-\varphi][+\varphi]\chi \equiv \chi$.

Acknowledgements

As usual, the author is indebted to Sten Lindström and Wlodek Rabinowicz for countless discussions on belief revision.

Uppsala University, Sweden.

References

1. Carlos Alchourrón, Peter Gärdenfors and David Makinson. On the logic of theory change: partial meet contraction and revision functions. *The Journal of Symbolic Logic*, 50:510–530, 1985.
2. Maarten de Rijke. *Extending Modal Logic*. PhD thesis, University of Amsterdam, 1993.
3. Maarten de Rijke. Meeting some neighbours: a dynamic modal logic meets theories of change and knowledge representation. In Jan van Eijck and Albert Visser, editors, *Logic and Information Flow*, pages 170–196. MIT Press, Cambridge MA, 1994.
4. Jaakko Hintikka. *Knowledge and Belief: an Introduction to the Logic of the two Notions*. Cornell University Press, Ithaca, N. Y, 1962.
5. Wolfgang Lenzen. *Recent Work in Epistemic Logic*. Acta Philosophica Fennica, fasc. 30, North-Holland, Amsterdam, 1978.
6. Hector Levesque. All I know: a study in autoepistemic logic. *Artificial intelligence*, 42:263–309, 1990.
7. Isaac Levi. *The Fixation of Belief and its Undoing*. Cambridge University Press, Cambridge, 1991.
8. Georg Henrik von Wright. *An Essay in Modal Logic*. North-Holland, Amsterdam, 1951.

ELIMINATION OF PREDICATE QUANTIFIERS

ANDREAS NONNENGART, HANS JÜRGEN OHLBACH AND
ANDRZEJ SZAŁAS

1 Introduction

Formulae of higher-order predicate logic are difficult to handle with automated inference systems. Some of these formulae, however, are equivalent to formulae of first-order predicate logic or even propositional logic. For example the formula of second-order predicate logic $\exists P\ (P \wedge \neg P)$ is trivially equivalent to the propositional constant *false*. In applications where formulae of higher-order predicate logic occur naturally it is very useful to determine whether the given formula is in fact equivalent to a simpler formula of first-order or propositional logic. Typical applications where this occurs are predicate minimization by circumscription, correspondence theory in non-classical logic, and simple versions of set theory. In these areas we are faced with formulae of second-order predicate logic with existentially or universally quantified predicate variables and we want to simplify them by computing equivalent first-order formulae.

In general this problem is not even semi-decidable, so no complete quantifier elimination algorithms for predicate quantifiers can exist. To illustrate the complexity of the problem, consider the following formula:

$$\exists C \left(\begin{array}{l} \forall x\ C(x,R) \vee C(x,Y) \vee C(x,G) \wedge \\ \forall x \forall y \forall z\ (C(x,y) \wedge C(x,z)) \Rightarrow y = z \wedge \\ \forall x \forall y\ E(x,y) \Rightarrow \neg \exists z\ (C(x,z) \wedge C(y,z)) \end{array} \right) \tag{1}$$

where R, Y, G are constant symbols. If $C(x,y)$ is interpreted as 'node x in a graph has colour y' and $E(x,y)$ as 'y is adjacent to x in a graph' then this formula expressess the graph 3-colourability problem. The first conjunct states that each node is coloured with one of the three colours. The second conjunct says that each node has at most one colour and finally the last conjunct requires adjacent nodes to have different colours. If there existed a sound algorithm which in this case could eliminate the predicate C and reduce the formula to something first-order we had a proof that

H.J. Ohlbach and U. Reyle (eds.), Logic, Language and Reasoning, 149–171.

$P = NP$. Namely, if the algorithm would eliminate the C and compute an equivalent first-order formula ϕ in terms of the predicates E and $=$, we had a polynomial algorithm[1] for deciding whether a graph can be coloured with three colours or not, and thus proved $P = NP$.

Although this shows that one's expectations should not be too high, some of the proposed algorithms are quite powerful and can solve hard problems in the areas mentioned in the beginning.

Wilhelm Ackermann has investigated the quantifier elimination problem in the following form: given a formula $\exists P\ \Phi[P]$ with a predicate variable P and a first-order formula Φ containing P in predicate position, find a first-order formula Φ' not containing P any more such that $(\exists P\ \Phi[P]) \equiv \Phi'$. The corresponding problem with a *universally* quantified predicate variable can be reduced to the problem with existentially quantified predicate variable by negating the formula, eliminating P and then negating the result. This way, formulae with arbitrary prefixes of either universally or existentially quantified predicates, but not mixed ones, can be treated.

Ackermann [1, 2] developed two different methods for finding such Φ', the first one is essentially a generalization of the Modus Ponens inference rule, and the second one can be seen as a simplified version of the resolution principle. This approach exploits the fact that a predicate P occurring in a subset Γ' of some set Γ of formulae is not needed any more if Γ contains already all the consequences of the formulae with P [10]. This principle itself is not yet very useful because usually there are infinitely many consequences. It turns out, however, that a subset of the set of consequences of the formulae with P which allows us to derive the full set of consequences is sufficient. The set of resolvents with P is such a sufficient subset, and this is the kernel of Ackermann's second approach.

In recent years Ackermann's two approaches have been rediscovered. Dov Gabbay and various other authors have extended and refined the method and turned it into implementable and in fact implemented algorithms. In this chapter we give an overview on the different approaches and discuss some applications.

2 Ackermann's Quantifier Elimination Methods

We briefly sketch Ackermann's two quantifier elimination methods.

[1] ϕ would specify a property of the graph which is necessary and sufficient for the graph to be colourable. Thus in order to decide this one would have to check whether the graph is a model for ϕ. An algorithm for doing this is: compute the clause normal form for ϕ and test for each clause whether it is satisfied by the graph. If the clause has, say n variable symbols and the graph has m nodes then m^n instances of the clause need to be tested. Since the number of clauses is fixed and the maximal number of variables in the clauses is bounded, this is a polynomial algorithm. It is in fact linear in terms of the number m of nodes.

2.1 Direct Elimination of Quantifiers

We say that a formula Φ is *positive* w.r.t. a predicate P iff there is no occurrence of $\neg P$ in the negation normal form of Φ. Dually, we say that Φ is *negative* w.r.t. P iff every of its occurrences in the negation normal form of Φ is of the form $\neg P$.
The following lemma was proved by Ackermann in [1] and can also be found in Szałas [21].

LEMMA 2.1 *Let P be a predicate variable and let Φ and $\Psi(P)$ be first-order formulae such that $\Psi(P)$ is positive w.r.t. P and Φ contains no occurrences of P at all. Then*

$$\exists P\, \forall \overline{x}\, (\neg P(\overline{x}) \vee \Phi) \wedge \Psi(P) \quad \equiv \quad \Psi\left[P(\overline{\alpha}) \leftarrow [\Phi]^{\overline{x}}_{\overline{\alpha}}\right]$$

and similarly if the sign of P is switched and Ψ is negative w.r.t. P. The right-hand formula is to be read as: Every occurrence of P in Ψ is to be replaced by Φ, where the actual argument of P, say $\overline{\alpha}$, replaces the variables of $\overline{x}$ in Φ (and the bound variables are renamed if necessary). ■

Hence, if the second-order formula under consideration has the syntactic form of the left-hand side of the equivalence given in Lemma 2.1 then this lemma can immediately be applied for the elimination of P.

Let us illustrate the application of the method with a small example from modal logic. The standard translation of the modal logic T-axiom $\Box p \Rightarrow p$ is

$$\forall P\, \forall w\, ((\forall x\, R(w,x) \Rightarrow P(x)) \Rightarrow P(w)). \tag{2}$$

The negation of this formula is equivalent to

$$\exists w\, \exists P\, \forall x\, (P(x) \vee \neg R(w,x)) \wedge \neg P(w). \tag{3}$$

The $\exists P \ldots$ part of the formula has the right syntactic structure for lemma 2.1 with exchanged signs and with $\Phi \equiv \neg R(w,x)$ and $\Psi \equiv P(w)$. Replacing $P(w)$ with Φ^x_w yields $\neg R(w,w)$. Thus, (3) is equivalent to $\exists w\, \neg R(w,w)$ and therefore (2) is equivalent to $\forall w\, R(w,w)$.

As we have seen, the syntactic form required in Lemma 2.1 is usually not given initially and we are therefore forced to apply some well-known equivalence preserving transformations of classical logic in order to obtain this form. The following one was found particularly useful in [1]:

$$P(\overline{x}) \;\equiv \forall \overline{y}\, (P(\overline{y}) \vee \overline{x} \neq \overline{y}).$$

This technique was substantially strengthened in Doherty, Łukaszewicz, Szałas [3] by applying the following equivalence:

$$P(\overline{x_1}) \vee \cdots \vee P(\overline{x_n}) \;\equiv \exists \overline{z}\, [(\overline{z} = \overline{x_1} \vee \cdots \vee \overline{z} = \overline{x_n}) \wedge P(\overline{z})] \tag{4}$$

where $\overline{z} = \overline{x_n}$ denotes a componentwise conjunction. The following extended purity deletion rule, formulated in Szałas [21], is also useful:

> if there is a predicate Q among the list of predicates to be eliminated such that Q occurs with mixed sign in some clauses and either only with positive or only with negative sign in the other clauses then all clauses containing Q are deleted. For example in the two clauses $\neg Q(x), Q(f(x))$ and $Q(a)$ there is no clause containing Q only with negative sign. If these are the only clauses with Q, they can be deleted. (Since Q is existentially quantified, a model making Q true everywhere satisfies the clauses.)

Other quite well-known transformation techniques turn out to be useful as well as there are, e.g., the transformation into conjunctive normal form, renaming, and the following second-order Skolemization that preserves equivalence of formulae and allows us to eliminate existential quantifiers (similar formulation can be found, e.g., in van Benthem [23]):

$$\forall \overline{x}\, \exists y\, \Phi \quad \equiv \quad \exists f\, \forall \overline{x}\, \Phi^{y}_{f(\overline{x})}.$$

Second-order Skolemization may make Lemma 2.1 applicable, thus eliminating P, but this way we end up again with a second-order formula. Sometimes, however, it is possible to do the transformation and then turn the Skolem-function again into an existential quantifier ('un-Skolemization').

An algorithm for eliminating predicate quantifiers based on Lemma 2.1 was defined in Szałas [21]. This algorithm was later developed and strengthened in [3]. The strengthened version of the algorithm, called DLS, is sketched in Section 2.3.

One of the major restrictions in Lemma 2.1 is that it does not allow disjunctions which contain both positive and negative occurrences of the predicate symbol to be eliminated. Ackermann found a way to overcome this problem.

2.2 The Resolution-based Approach

On page 401 of Ackermann [1] there is the definition of a *contraction operation*

$$\mathcal{A}^{x_1,\ldots,x_n,z}_{y_1,\ldots,y_m} \wedge \mathcal{B}^{p_1,\ldots,p_n}_{q_1,\ldots,q_l,z} \rightarrow \mathcal{A}^{x_1,\ldots,x_n}_{y_1,\ldots,y_m} \vee \mathcal{B}^{p_1,\ldots,p_n}_{q_1,\ldots,q_l},$$

where the subscripts y_i stand for $P(y_i)$ and the superscripts x_i stand for $\neg P(x_i)$ in a clause containing also literals $\mathcal{A}$ or $\mathcal{B}$ respectively. Thus, contraction on z actually means resolution between $P(z)$ and $\neg P(z)$. The variables can and must of course be renamed appropriately for making a contraction step possible.

Ackermann proved that, given a formula $\exists P\, \Phi$, where P is a one-place predicate variable and Φ is a set of clauses written in the above form, the P-free subset of the conjunction of all contractions is equivalent to $\exists P\, \Phi$.

This holds even if this conjunction is infinite. As an example, consider the formula

$$\exists P\ P(a) \wedge \neg P(b) \wedge \forall x \forall y\ \neg P(x) \vee P(y) \vee N(x,y).$$

In Ackermann's clause notation this is

$$(x \neq a)_x \wedge (x \neq b)^x \wedge N(x,y)^x_y.$$

We obtain an infinite set of contractions which simplifies to

$$\begin{array}{c} a \neq b \wedge N(a,b) \wedge \\ (\forall v\ N(a,v) \vee N(v,b)) \wedge \\ (\forall v \forall w\ N(a,v) \vee N(v,w) \vee N(w,b)) \wedge \ldots \end{array}$$

Ackermann defined his resolution method for one-place predicate variables only. For n-place predicate variables he proposed a transformation which turns the n-place predicate variables into one-place predicate variables. This transformation, however, is complex and unnecessary. If one gives up the tensor notation Ackermann used and thinks in terms of resolution steps, a much more natural generalization to n-place predicate variables is possible (cf. Section 4).

2.3 The DLS-algorithm

The DLS algorithm was defined in Doherty, Łukaszewicz, Szałas [3]. It is a strengthened version of an algorithm given in Szałas [21]. The algorithm tries to transform the input formula into the form suitable for application of Lemma 2.1.

More precisely, the algorithm takes a formula of the form $\exists P\ \Phi$, where Φ is a first-order formula, as an input and returns its first-order equivalent or reports failure.[2] Of course, the algorithm can also be used for formulae of the form $\forall P\ \Phi$, since the latter formula is equivalent to $\neg \exists P\ \neg \Phi$. Thus, by repeating the algorithm one can deal with formulae containing many arbitrary second-order quantifiers.

The elimination algorithm consists of three phases: (1) preprocessing; (2) preparation for Lemma 2.1; (3) application of Lemma 2.1. These phases are described below. It is always assumed that (1) whenever the goal specific for a current phase is reached, then the remaining steps of the phase are skipped, (2) every time the extended purity deletion rule (see Section 2.1) is applicable, it should be applied.

[2]The failure of the algorithm does not mean that the second-order formula at hand cannot be reduced to its first-order equivalent. The problem we are dealing with is not even partially decidable, for first-order definability of the formulae we consider is not an arithmetical notion (see, for instance van Benthem [23]).

1. *Preprocessing.* The purpose of this phase is to transform the formula $\exists P\ \Phi$ into a form that separates positive and negative occurrences of the quantified predicate variable P. The form we want to obtain is

$$\exists\bar{x}\exists P[(\Phi_1(P) \wedge \Psi_1(P)) \vee \cdots \vee (\Phi_n(P) \wedge \Psi_n(P))],$$

where, for each $1 \leq i \leq n$, $\Phi_i(P)$ is positive w.r.t. P and $\Psi_i(P)$ is negative w.r.t. P. The steps of this phase are the following. (i) Eliminate the connectives $\Rightarrow$ and $\equiv$ using the usual definitions. Remove redundant quantifiers. Rename individual variables until all quantified variables are different and no variable is both bound and free. Using the usual equivalences, move the negation connective to the right until all its occurrences immediately precede atomic formulae. (ii) Move universal quantifiers to the right and existential quantifiers to the left, applying as long as possible the usual quantifier rules. (iii) In the matrix of the formula obtained so far, distribute all top-level conjunctions over the disjunctions that occur among their conjuncts. (iv) If the resulting formula is not in the required form, then report the failure of the algorithm. Otherwise replace the input formula by its equivalent given by

$$\exists\bar{x}(\exists P(\Phi_1(P) \wedge \Psi_1(P)) \vee \cdots \vee \exists P(\Phi_n(P) \wedge \Psi_n(P))).$$

Try to find first-order equivalent of the above formula by applying the next phases in the algorithm to each its disjunct separately. If the first-order equivalents of each disjunct are successfully obtained then return their disjunction, preceded by the prefix $\exists\bar{x}$, as the output of the algorithm.
2. *Preparation for the Ackermann lemma.* The goal of this phase is to transform a formula of the form $\exists P(\Phi(P) \wedge \Psi(P))$, where $\Phi(P)$ (respectively, $\Psi(P)$) is positive (respectively, negative) w.r.t. P, into one of the forms required in Lemma 2.1. Both forms can always be obtained by using equivalences given in Section 2.1 and both transformations should be performed because none, one or both forms may require Skolemization. Un-Skolemization, which occurs in the next phase, could fail in one form, but not the other. In addition, one form may be substantially smaller than the other.
3. *Application of the Ackermann Lemma.* The goal of this phase is to eliminate the second-order quantification over P, by applying Lemma 2.1, and then to un-Skolemize the function variables possibly introduced. This latter step employs the second-order Skolemization equivalence.
4. *Simplification.* Generally, application of Lemma 2.1 in step (3) often involves the use of equivalences mentioned in Section 2.1 in the left to right direction. If so, the same equivalences may often be used after

application in the right to left direction, substantially shortening the resulting formula.

Observe that the above algorithm can be used in connection with both Lemma 2.1 and Lemma 3.1. Lemma 3.1 should be applied in case DLS reports failure in step 1(iv).

The DLS algorithm was implemented in the University of Linköping, IDA, by Joakim Gustafsson [7]. It is accessible via World Wide Web (URL: http://www.ida. liu.se/labs/kplab/projects/dls/). The program can be executed remotely by filling out a html form in a WWW browser and clicking the 'submit' button. The contents of the html form is then sent to a central server, which activates the program and returns the answer to the user. The system provides the user with a form for eliminating predicate quantifiers and a separate form for computing circumscription.

3 A Fixpoint Approach to Quantifier Elimination

Recently (see Nonnengart, Szałas [16]), the idea that lead to Lemma 2.1 has been generalized along the lines of Ackermann's observations as described in Section 2.2, e.g., infinite formulae are allowed and in order to be able to finitely represent these infinite formulae the original syntax got extended by fixpoint operators.

As an example consider again the second-order formula

$$\exists P\ P(a) \land \neg P(b) \land \forall x \forall y\ \neg P(x) \lor P(y) \lor N(x,y).$$

The problem we have if we try to apply Lemma 2.1 is that we are not able to separate the positive from the negative occurrences of P such that the requirements for the lemma are fulfilled. This is certainly not too surprising for otherwise we would be able to find an equivalent first-order formula which is impossible as Ackermann's result shows. The idea is therefore to describe these many conjunctive elements in a finite language and that with the help of fixpoint operators as follows:

$$\Big[\nu P(x).\ x \neq a \land \forall y\ P(y) \lor N(y,x)\Big]_b^x$$

where $[\Phi]_b^x$ is meant to express that every occurrence of x in Φ is to be replaced by b. This fixpoint formula indeed represents the infinite result obtained by Ackermann and why this is so is described below.

3.1 Fixpoint Calculus

Let $\mathcal{L}_\mathcal{I}$ be the classical first-order logic. In order to define the *fixpoint calculus* $\mathcal{L}_\mathcal{F}$ we extend $\mathcal{L}_\mathcal{I}$ by allowing the least fixpoint operator $\mu P.\Phi(P)$, where

Φ is positive w.r.t. P. We abbreviate a formula of the form $\neg(\mu P.\neg\Phi(P))$ by $\nu P.\Phi(P)$. Similarly, if Φ is negative w.r.t. P we consider the formulae $\mu\overline{P}.\Phi(\overline{P})$ and $\nu\overline{P}.\Phi(\overline{P})$ respectively. It is sometimes convenient to indicate the individual variables which are bound by the fixpoint operators μ and ν. We write $\mu P(\bar{x})$ and $\nu P(\bar{x})$ to indicate that the tuple $\bar{x}$ of variables is bound by a fixpoint operator.

Let us now recall some useful well-known facts. The formulation we give is adapted to the particular problems we deal with. In particular, the partial order we consider is the following:

- the carrier is the set of formulae of $\mathcal{L}_{\mathcal{F}}/_{\equiv}$, where we do not distinguish between logically equivalent formulae (formally the carrier is the quotient set $\mathcal{L}_{\mathcal{F}}/_{\equiv}$; in order to simplify the considerations the equivalence classes and formulae are identified);
- the formulae are ordered by implication, i.e. Φ is less or equal to Ψ iff $\Phi \Rightarrow \Psi$ is a tautology.

Note that fixpoint operator formulae are also formulae. Thus the partial order we consider is complete in the sense that for every formula $\Phi(P)$ which is positive w.r.t. P the set $\{\Phi^i(\bot) \mid i \in \omega\}$ has a least upper bound. Every such formula is monotone and therefore we have by the Knaster & Tarski fixpoint theorem that the fixpoints we consider are well defined. Moreover, the fixpoints have the following nice characterization[3]:

$$\mu P.\Phi(P) \equiv \bigvee_{\beta\in\alpha} \Phi^{\beta}(\bot)$$

for some ordinal α (the least such ordinal is called the *closure ordinal for* $\Phi(P)$). In the case of fixpoint formulae we deal with the closure ordinal is always ω. In fact, we always have the following equivalences (see also Lemma 3.1):

$$\begin{aligned} \mu P.\Phi(P) &\equiv \textstyle\bigvee_{\beta\in\omega} \Phi^{\beta}(\bot) \\ \nu P.\Phi(P) &\equiv \textstyle\bigwedge_{\beta\in\omega} \Phi^{\beta}(\top). \end{aligned}$$

Note that $\mu P(\bar{x}).\Phi(P)$ is the least (w.r.t. the partial order defined above) formula $\Psi(\bar{x})$ such that

$$\Psi(\bar{x}) \equiv \Phi(P \leftarrow \Psi(\bar{x})).$$

Now let us come back to the formula

$$\Big[\nu P(x).\ x \neq a \wedge \forall y\ P(y) \vee N(y,x)\Big]_b^x.$$

[3]Given $\Phi(P)$ we write $\Phi(A)$ to indicate that we want to consider Φ with each occurrence of P replaced by A. Thus $\Phi(\bot)$ is Φ with P replaced by $\bot$ (false) and $\Phi^{\beta}(\bot) \equiv \Phi(\Phi^{\beta-1}(\bot))$.

In this case we have that $\Phi(P) \equiv x \neq a \wedge \forall y\, P(y) \vee N(y,x)$ and therefore

$$\begin{array}{lcl}
\Phi^0(\top) & \equiv & \top \\
\Phi^1(\top) & \equiv & x \neq a \\
\Phi^2(\top) & \equiv & x \neq a \wedge \forall y\, y \neq a \vee N(y,x) \\
 & \equiv & x \neq a \wedge N(a,x) \\
\Phi^2(\top) & \equiv & x \neq a \wedge \forall y\, (y \neq a \wedge N(a,y) \vee N(y,x)) \\
 & \equiv & x \neq a \wedge N(a,x) \wedge \forall y\, (N(a,y) \vee N(y,x)) \\
\dots & & \dots
\end{array}$$

such that the above fixpoint formula – with x replaced by b – becomes

$$a \neq b \wedge N(a,b) \wedge \forall y\, (N(a,y) \vee N(y,b)) \wedge \dots$$

just as desired.

3.2 The Fixpoint Lemma

Let us now show how a fixpoint formula can be obtained from a given second-order formula. The following lemma, proved in Nonnengart, Szałas [16], is a generalization of Lemma 2.1 and introduces fixpoint formulae.

LEMMA 3.1 *If Φ and Ψ are positive w.r.t. P then the closure ordinal for $\Phi(P)$ is less than or equal to ω and*

$$\exists P\, \forall \overline{x}\, (\neg P(\overline{x}) \vee \Phi(P)) \wedge \Psi(P) \;\equiv\; \Psi\left[P(\overline{\alpha}) \leftarrow \left[\nu P(\overline{x}).\Phi(P)\right]^{\overline{x}}_{\overline{\alpha}}\right]$$

and similarly for the case where the sign of P is switched and Φ and Ψ are negative w.r.t. P. ∎

Note the strong similarities between Lemma 2.1 and Lemma 3.1. In fact, it can quite easily be observed that this fixpoint result subsumes the former result as described in Lemma 2.1 for in case that Φ does not contain any P at all we have that $\nu P(\overline{y}).\Phi$ is equivalent to Φ. Hence Lemma 3.1 is a proper generalization of Lemma 2.1.

Again it is usually necessary to apply some equivalence preserving transformations in order to obtain a formula in the form required for applying Lemma 3.1. This can be done by the initial phases of the DLS algorithm (see Section 2.3). Recall that the syntactic form required in Lemma 2.1 cannot always be obtained. This is not the case any more for Lemma 3.1 for any formula can be transformed into the form required provided second-order Skolemization is allowed. This Skolemization evidently cannot be avoided in general for otherwise every second-order formula could be transformed into a (possibly infinite) first-order formula. Nevertheless, the lemma can always be applied and returns some result which is usually a fixpoint formula and sometimes another second-order formula. Such fixpoints can be tried to be simplified then and in particular in case where the fixpoint is bounded a final first-order result can be found (see [4]).

A prototypic implementation of the fixpoint algorithm developed by M. J. Gabbay has been finished, but it is not yet available.

4 Quantifier Elimination by the SCAN Algorithm

4.1 The SCAN Algorithm

The SCAN algorithm[4] was proposed by Gabbay and Ohlbach [6] as a refinement of Ackermann's resolution method.[5] SCAN takes as input second-order formulae of the form

$$\alpha \equiv \exists P_1 \ldots \exists P_k \; \Phi$$

with existentially quantified predicate variables P_i and a first-order formula Φ. SCAN eliminates all predicate variables at once. The following three steps are performed by SCAN:

1. Φ is transformed into clause form.
2. All C-resolvents and C-factors with the predicate variables $P_1, \ldots, P_k$ are generated. *C-resolution* ('C' is short for constraint) is defined as follows:

   $$\frac{\begin{array}{ll} P(s_1, \ldots s_n) \vee C & P(\ldots) \text{ and } \neg P(\ldots) \\ \neg P(t_1, \ldots, t_n) \vee D & \text{are the } \textit{resolution literals} \end{array}}{C \vee D \vee s_1 \neq t_1 \vee \ldots \vee s_n \neq t_n}$$

   and the *C-factorization* rule is defined analogously:

   $$\frac{P(s_1, \ldots, s_n) \vee P(t_1, \ldots, t_n) \vee C}{P(s_1, \ldots, s_n) \vee C \vee s_1 \neq t_1 \vee \ldots \vee s_n \neq t_n}.$$

   When *all* resolvents and factors between a particular literal and the rest of the clause set have been generated (the literal is then said to be 'resolved away'), the clause containing this literal is deleted (this is called 'purity deletion'). If all clauses have been deleted this way, we know α is a tautology. If an empty clause is generated, we know α is inconsistent.
3. If step 2 terminates and the set of clauses is non-empty then the quantifiers for the Skolem functions are reconstructed.

[4]SCAN means 'Synthesizing Correspondence Axioms for Normal Logics'. The name was chosen before the general nature of the procedure was recognized.

[5]The SCAN algorithm was discovered independently. Only afterwards Andrzej Szałas found Ackermann's paper.

The next example illustrates the various steps of the SCAN algorithm in detail. The input is:

$$\exists P\ \forall x \forall y\ \exists z\ (\neg P(a) \vee Q(x)) \wedge (P(y) \vee Q(a)) \wedge P(z).$$

In the first step the clause form is computed:

$$\begin{array}{ll} C_1 & \neg P(a), Q(x) \\ C_2 & \ \ P(y), Q(a) \\ C_3 & \ \ P(f(x,y)) \end{array}$$

f is a Skolem function.

In the second step of SCAN we begin by choosing $\neg P(a)$ to be resolved away. The resolvent between C_1 and C_2 is $C_4 \equiv Q(x), Q(a)$ which is equivalent to $Q(a)$ (this is one of the equivalence preserving simplifications). The C-resolvent between C_1 and C_3 is $C_5 \equiv (a \neq f(x,y), Q(x))$. There are no more resolvents with $\neg P(a)$, so C_1 is deleted. We are left with the clauses

$$\begin{array}{ll} C_2 & P(y), Q(a) \\ C_3 & P(f(x,y)) \\ C_4 & Q(a) \\ C_5 & a \neq f(x,y), Q(x). \end{array}$$

Selecting the next two P-literals to be resolved away yields no new resolvents, so C_2 and C_3 can be deleted as well. All P-literals have now been eliminated. Restoring the quantifiers, we then get

$$\forall x\ \exists z\ Q(a) \wedge (a \neq z \vee Q(x))$$

as the final result (y is no longer needed).

There are two critical steps in the algorithm. First of all the C-resolution loop need not always terminate. This may but need not indicate that there is no first-order equivalent for the input formula. If the resolution eventually terminates the next critical step is the un-Skolemization. Since this is essentially a quantifier elimination problem for existentially quantified function variables, there is also no complete solution. The algorithm we usually use is heuristics based.

Preventing C-resolution from looping is a difficult control issue. Some equivalence preserving transformations on clause sets turned out to be quite useful. In the algorithm we have implemented each new resolvent can be tested whether it is implied by the non-parent clauses. In the affirmative case it is deleted even if more resolutions are possible. Another technique is the application of (4) for collapsing multiple occurrences of the predicate P in the same clause. In the graph colouring axiomatization (1) the latter method can successfully be applied. The clause normal form of (1) is

$$\begin{array}{l} C(x,R), C(x,Y), C(x,G) \\ \neg C(x,y), \neg C(x,z), x=z \\ \neg C(x,z), \neg C(y,z), \neg E(x,y). \end{array}$$

Given this to SCAN and asking it to eliminate C the resolution loops. If we however replace the first clause by the equivalent formula $\forall x\ \exists c\ (c = R \vee c = Y \vee c = G) \wedge C(x,c)$, whose clause normal form is

$$\begin{array}{l} c(x) = R, c(x) = Y, c(x) = G \\ C(x, c(x)), \end{array}$$

there is no problem any more. The two successive resolutions with the second clause yields a tautology. The two successive resolutions with the third clause yields $c(x) \neq c(y), \neg E(x,y)$. The result is now $\exists c\ (\forall x\ c(x) = R \vee c(x) = Y \vee c(x) = G) \wedge \forall x, y\ E(x,y) \Rightarrow c(x) \neq c(y)$ which is again second-order. It is just a reformulation of the original formula in terms of the colouring function c. As we have explained in the introduction, it would be quite surprising to get a better result.

4.2 The SCAN Program

The SCAN program has been implemented as a modified version of the OTTER theorem-prover devolved by William McCune at Argonne National Laboratory [15]. The modifications were implemented by Thorsten Engel at the Max-Planck Institute in Saarbrücken [5].

SCAN maintains two lists of formulae, the SOS list and the USABLE list. The SOS list contains the formulae with the predicates to be eliminated. The USABLE list may contain extra information which may hold in the given domain and which may be used to simplify formulae. That means if α is in the usable list, β is in the SOS list and $P_1, \ldots, P_n$ are the predicates to be eliminated, SCAN tries to compute a formula γ such that

$$\alpha \Rightarrow ((\exists P_1 \ldots P_n\ \beta) \equiv \gamma)$$

holds.

SCAN performs the following steps (if not deactivated by corresponding options).

1. All formulae are converted into conjunctive normal form (clause form).
2. Certain simplifications are performed, in particular elimination of tautologies and subsumed clauses. If possible, unit-deletion is applied. Unit-deletion is a resolution step followed by a subsumption step that deletes one parent clause. The net effect is the deletion of a literal from a clause. As an example, resolution between the unit clause $p(x)$ and

the non-unit clause $\neg p(a), q$ yields the resolvent q, which subsumes the non-unit parent clause. The same effect is achieved by simply deleting $\neg p(a)$ from the parent clause. (Sometimes this brings about surprising effects, but it is a very useful equivalence preserving transformation.)

3. For each predicate symbol P in the list of predicates to be eliminated:
   - a clause C containing P is chosen from the SOS list;
   - all C-factors are generated;
   - a literal L in C containing P is chosen and all C-resolvents with this literal and other clauses in the SOS list are created.

     Each resolvent (i) is simplified by means of the unit-deletion rule, (ii) is deleted if it is a tautology, (iii) is deleted if it is subsumed by other clauses, (iv) is deleted if it is implied by the other non-parent clauses and the USABLE list (in order to check this, the SCAN process is forked and operates for some time in the OTTER theorem proving mode), (v) causes the deletion of other subsumed clauses.
   - Once the literal L is 'worked off', the clause C is deleted;
   - apply the extended purity deletion rule (see Section 2.1), if possible.
4. If the previous steps terminate (which cannot be guaranteed) then the resulting clause set is checked for redundancy. To this end, an attempt is made to prove this clause from the other clauses (again by forking the process and running it in a theorem proving mode). Clauses which can be proved from other clauses in a certain time are deleted.
5. Finally the remaining clause set is un-Skolemized, if possible. Reconstructing existential quantifiers for Skolem functions is in fact a quantifier elimination operation for second-order formulae with existentially quantified function variables, so the un-Skolemization algorithm can be used as a quantifier elimination algorithm for function variables. There are three possible outcomes of the un-Skolemization routine:
   (i) it may generate a normal first-order formula,
   (ii) it may simply tell us that un-Skolemization is not possible, or
   (iii) it may generate an (again second-order) parallel Henkin quantifier. For example the un-Skolemized version of the clause $P(x, y, f(x)), Q(x, y, g(y))$ is

$$\begin{bmatrix} \forall x & \exists u \\ \forall y & \exists v \end{bmatrix} P(x, y, u) \vee Q(x, y, v).$$

As already mentioned, SCAN may not terminate and go on generating C-resolvents forever. If this happens, two measures may help. Sometimes changing the order of the list of predicates to be eliminated may cause the system to terminate. As an example, consider the clauses

$$\begin{array}{ll} C_1 & \neg P(x), P(f(x)), Q(x) \\ C_2 & P(a) \\ C_3 & Q(b) \\ C_4 & -P(b). \end{array}$$

Starting by resolving P triggers an infinite loop due to the self-resolving clause C_1. If we instead start with Q then C_1 is replaced by its resolvent $\neg P(b), P(f(b))$, which is no longer self-resolving. Eliminating P is now possible in a finite number of steps. (In principle an algorithm could recognize such situations and control the resolution process more intelligently, but this has not been implemented.)

The second measure for terminating the resolution loop is application dependent. Sometimes extra information is available which make resolvents redundant. Checking this and deleting redundant resolvents may terminate a loop. As a simple example for the application of this method consider the clauses

$$\begin{array}{ll} C_1 & \neg P(x), P(f(x)), Q(x) \\ C_2 & P(a) \\ C_3 & Q(a), \end{array}$$

where P is to be eliminated. The self resolving clause C_1 triggers an infinite loop. The first resolvent is $P(f(a)), Q(a)$, and the second resolvent is $P(f(f(a))), Q(a), Q(f(a))$. Further resolutions with P are possible. However already the first resolvent is subsumed by C_3 and can therefore be deleted. The loop is stopped if C_3 is available.

There is no exact characterization of the formulae for which SCAN terminates. It very much depends on the selection heuristics for the resolution steps.

The system is accessible via World Wide Web (URL: http://www.mpi-sb.mpg.de/guide/staff/ohlbach/scan/scan.html). The program can be executed remotely by filling out a html form in a WWW browser and clicking the "submit" button. The contents of the html form are then sent to a central server, which activates the program and returns the answer to the user. Besides the basic functionality for eliminating predicate quantifiers it contains two preprocessors which allow you to compute circumscription and to automate certain aspects of correspondence theory as explained below. So far, access to the program is not restricted.

5 Applications

5.1 Circumscription

Circumscription was proposed by John McCarthy as a logically simple and clear means of doing default reasoning. As an example consider the database consisting of the single entry $flies(Tweety)$. From this database you can of course prove that Tweety flies, but if you ask $flies(Woodstock)$? the database either replies with 'don't know' or responds brutally with 'no'. If you have evidence that your database is complete then the answer 'no' is justified, but in this case you conclude $\neg flies(Woodstock)$ from the fact that $flies(Woodstock)$ is not provable from the database. Since predicate logic is only semi-decidable, this is not a complete procedure. Moreover, there is no clear semantics which allows one to justify this step.

McCarthy's circumscription idea solves this problem on the semantic level. The idea is to axiomatize in a certain sense the information that "this is all I know about a particular predicate P", i.e. I want to consider only those interpretations for P in which $P(x)$ is true only for the absolutely minimum number of x necessary to satisfy the database. This *minimization* of the extension of predicate symbols is called circumscription. Unfortunately the formula which axiomatizes the minimized predicate is second-order. In the simplest case it is as follows:

$$circ(\Phi[P], P) \equiv \Phi[P] \wedge \forall P^*(\Phi[P^*] \wedge (P^* \rightarrow P)) \Rightarrow (P \rightarrow P^*)$$

where $\Phi[P]$ is an arbitrary first-order formula containing the predicate P which is to be minimized. $\Phi[P^*]$ is like Φ, but all occurrences of P are replaced by P^*. $P^* \rightarrow P$ is short for $\forall x_1 \ldots \forall x_n\ (P^*(x_1, \ldots, x_n) \Rightarrow P(x_1, \ldots, x_n))$. You can also have a list of predicates to be minimized simultaneously, in which case $P^* \rightarrow P$ stands for the conjunction of all these implications. As an example consider our little database above with the entry $flies(Tweety)$.

EXAMPLE 5.1 According to the definition of circumscription,

$$\begin{aligned}circ(flies(Tweety), flies) \equiv\ & flies(Tweety) \wedge \\ & \forall flies^*(flies^*(Tweety) \wedge (\forall x\ flies^*(x) \Rightarrow flies(x))) \\ & \quad \Rightarrow (\forall x\ flies(x) \Rightarrow flies^*(x)).\end{aligned}$$

This calls for a quantifier elimination procedure to eliminate the predicate $flies^*$. If we do this, we find as a result

$$\begin{aligned}& circ(flies(Tweety), flies) \\ & \quad \equiv flies(Tweety) \wedge (\forall x\ flies(x) \Rightarrow x = Tweety),\end{aligned}$$

i.e. $Tweety$ is the only thing that flies. ∎

In an extended version of circumscription one can minimize certain predicates at the cost of certain other predicates which are allowed to vary. That is, if P are the predicates to be minimized and Z are the predicates allowed to vary then $circ(\Phi[P,Z],P,Z)$ is a formula from which one might be able to prove additional positive facts about Z which are not provable from $\Phi[P,Z]$. The circumscription formula for this version is

$$circ(\Phi[P,Z],P,Z) \equiv \Phi[P,Z] \wedge \forall P^*,Z^* \ (\Phi[P^*,Z^*] \wedge (P^* \rightarrow P)) \rightarrow (P \rightarrow P^*)$$

The current implementation of SCAN contains a module for realizing this general version of circumscription by generating the circumscription formula according to the above schema and then applying SCAN to the second-order part.

Similarly, the DLS algorithm, sketched in Section 2.3, appeared quite powerful when applied to elimination of predicate quantifiers from circumscription formulae. In particular, it subsumes most known results about reducing circumscription (see [12, 13, 14, 17]). Moreover, in [3] it is proved to be substantially stronger than most of those results. Let us show an example of application of the DLS algorithm in reducing circumscription (see also [3]).

EXAMPLE 5.2 This is an example considered by Doherty, Łukaszewicz and Szałas in [3]. It is a variant of the Vancouver example of Reiter. Rather than using the function *city* as Reiter does, we will use a relation $C(x,y)$ with suitable axioms. The intention is that $C(x,y)$ holds iff y is the home town of x. Let $\Phi(Ab,C)$ be the theory

$$[\forall x \forall y \forall z (\neg Ab(x) \wedge C(x,y) \wedge C(wife(x),z)) \Rightarrow y = z] \wedge \\ [\forall x \forall y \forall z \ (C(x,y) \wedge C(x,z)) \Rightarrow y = z].$$

The circumscription of $\Phi(Ab,C)$ with Ab minimized and C varied is

$$circ(\Phi(Ab,C); Ab; C) \equiv \Phi(Ab,C) \wedge \forall P^* \forall Z^* [\Phi(P^*,Z^*) \wedge [P^* \leq Ab] \Rightarrow [Ab \leq P^*]],$$

where

$$\Phi(P^*,Z^*) \equiv [\forall x \forall y \forall z \neg P^*(x) \wedge Z^*(x,y) \wedge Z^*(wife(x),z) \Rightarrow y = z] \wedge \\ [\forall x \forall y \forall z \ Z^*(x,y) \wedge Z^*(x,z) \Rightarrow y = z].$$

The DLS algorithm reduces the second-order part of circumscription:

$$\forall P^* \forall Z^* [\Phi(P^*,Z^*) \wedge [P^* \leq Ab] \Rightarrow [Ab \leq P^*]].$$

After two iterations (the first for reducing P^* and the second for reducing Z^*) one obtains the result $\forall t \neg Ab(t)$. Consequently,

$$circ(\Phi(Ab,C); Ab; C) \equiv \Phi(Ab,C) \wedge \forall t \neg Ab(t).$$

■

For more information about quantifier elimination and circumscription see Doherty, Łukaszewicz and Szałas [3], Lifschitz, 1994 [14] and Kartha, Lifschitz [11].

5.2 Correspondence Theory in Non-classical Logics

The correspondence problem comes in non-classical logics, in particular in modal logics, as well as in certain algebras. In modal logics it is the problem of finding for a given Hilbert axiom a corresponding characteristic property of the underlying possible worlds structure (*frame properties*). For example the modal axiom $\Box p \Rightarrow p$ corresponds to reflexivity of the accessibility relation $\forall x\ R(x, x)$. As another example, $\Box p \Rightarrow \Box\Box p$ corresponds to the transitivity of the accessibility relation.

An algebraic version of the same problem turns up when considering Boolean algebras with operators. Jónsson and Tarski [9] have shown that under certain conditions the operators with binary relations in the same way as modal operators can be represented with accessibility relations (at a certain level of abstraction there is no longer any difference). The 'correspondence problem' here is to find the correspondences between additional axioms in terms of the operators on the one side and the underlying relation on the other.

We explain briefly the general construction. We start with an axiomatic presentation of an extension of propositional logic or just a Boolean algebra with extra operators. In the first case the Boolean algebra is obtained as the Lindenbaum–Tarski algebra whose elements are equivalence classes of provably (from the axioms and rules) equivalent formulae. Stone's famous representation theorem for Boolean algebras maps a Boolean algebra isomorphically to a field of sets. That means for an element x of the Boolean algebra there is an isomorphic image U_x consisting of all ultrafilters (or maximally consistent set of formulae in the logic case) containing x. The Boolean connectives $\wedge, \vee, \neg$ are mapped to the set functions $\cap, \cup, '$ defined for the complete and atomic set algebra consisting of the full powerset 2^U of the set U of all ultrafilters in the Boolean algebra.

The extra functions, for example a unary function f (think of it as the algebraic version of the modal $\Diamond$-operator) are so far only defined for those sets of ultrafilters U_x which are images of some element x in the Boolean algebra. In this case $f'(U_x) = U_{f(x)}$. How f' operates on arbitrary sets of ultrafilters is not defined. Jónsson and Tarski could show that in case f is *normal*, i.e. $f(0) = 0$ and f is *additive*, i.e. $f(x \vee y) = f(x) \vee f(y)$, which correspond to the necessitation rule and the K-axiom in modal logic, there is a proper extension f'' of f' to the full powerset. The definition is

$$w \in f''(U) \equiv \exists u\ u \in U \wedge \forall U_x\ u \in U_x \Rightarrow w \in U_f(x),$$

or, since $u \in U_x$ iff $x \in u$:

$$w \in f''(U) \equiv \exists u\ u \in U \wedge \forall x\ x \in u \Rightarrow f(x) \in w. \tag{5}$$

If we abbreviate $\forall x\ x \in u \Rightarrow f(x) \in w$ by $R(w,u)$ we get a shorter notation

$$w \in f''(U) \equiv \exists u\ R(w,u) \wedge u \in U$$

which is familiar from the Kripke semantics of the modal $\Diamond$-operator.

Although f'' is a proper extension of f' which means $f''(U_x) = f'(U_x)$ for the 'representable' sets U_x, this does not guarantee that f'' inherits all properties of f'. As a positive example, consider the property $\forall x\ x \leq f(x)$ which is an algebraic version of the modal T-axiom $P \Rightarrow \Diamond P$. In terms of the set representation this means $\forall U_x\ U_x \subseteq U_{f(x)}$, or alternatively

$$\forall x\ \forall w\ x \in w \Rightarrow f(x) \in w. \tag{6}$$

The question is now: does this imply $\forall U\ U \subseteq f''(U)$? In fact it does. By the definition of f'' (5) we get $\forall U\ \forall w\ w \in U \Rightarrow \exists u\ u \in U \wedge \forall x\ x \in u \Rightarrow f(x) \in w$ which is implied by (6) (choose $u = w$).

Thus, $\forall x\ x \leq f(x)$ in the Boolean algebra implies $\forall U\ U \subseteq f''(U)$ in its set representation. So called *preservation theorems* give syntactic characterizations of properties which transfer to the full powerset algebra (see Sahlqvist [19] and Jónsson [8].)

Quantifier elimination comes into the play if we want to express the given property of f'' in terms of the accessibility relation R introduced above. For example $\forall U\ U \subseteq f''(U)$ can be written as $\forall U\ \forall w\ w \in U \Rightarrow \exists u\ R(w,u) \wedge u \in U$. Since $\forall U$ quantifies over the whole powerset of some basic set, this is equivalent to quantifying over a predicate variable: $\forall P\ \forall w\ P(w) \Rightarrow \exists u\ R(w,u) \wedge P(u)$ and this is equivalent to $\forall w\ R(w,w)$.

As we have seen, correspondence theory rises essentially two problems, showing that the given axiom continues to hold in the full powerset structure of the set representation (which in the logic case is nothing other than the canonical model) and then finding for the formulation in terms of the second-order variables a hopefully but not necessarily first-order formulation in terms of the accessibility relation only. Our quantifier elimination algorithms automate to a certain extend this second step.

The following example shows an application of Lemma 2.1 to a correspondence theory problem (see also Szałas [21]).

EXAMPLE 5.3 Consider the Hilbert axiom $\Box p \Rightarrow \Box\Box p$. Similarly to the previous example we get the following corresponding second-order formula:

$$\forall P\ \forall u\ \neg\forall v(\neg R(u,v) \vee P(v)) \vee \forall v(\neg R(u,v) \vee \forall w(\neg R(v,w) \vee P(w)))$$

After negation and transformation to a form required in Lemma 2.1 (e.g., using the DLS algorithm) we get:

$$\exists u \exists P \forall v[P(v) \vee \neg R(u,v)] \wedge [\exists v\ R(u,v) \wedge \exists w(R(v,w) \wedge \neg P(w))].$$

The application of Lemma 2.1 results in:

$$\exists u \exists v\ R(u,v) \wedge \exists w(R(v,w) \wedge \neg R(u,w)).$$

This formula has to be unnegated, i.e. we get:

$$\forall u \forall v \neg R(u,v) \vee \forall w(\neg R(v,w) \vee R(u,w)),$$

which is equivalent to

$$\forall u \forall v \forall w(R(u,v) \Rightarrow (R(v,w) \Rightarrow R(u,w)),$$

i.e. to the transitivity of R, which is the desired frame property. ∎

Consider the following example of Nonnengart and Szałas [16], where SCAN loops and DLS fails, but Lemma 3.1 can successfully be applied.

EXAMPLE 5.4 Consider the temporal logic formula

$$\Box(p \Rightarrow \circ p) \quad \Rightarrow \quad (p \Rightarrow \Box p).$$

where $\Box$ should be interpreted as *always* or *henceforth* and $\circ$ as *at the next moment of time*. This formula corresponds to the following second-order formula, where $R_\Box$ and $R_\circ$ are accessibility relations for modalities $\Box$ and $\circ$, respectively:

$$\forall P\, \forall u\, [\forall v\, (R_\Box(u,v) \Rightarrow (P(v) \Rightarrow \forall w\, (R_\circ(v,w) \Rightarrow P(w))))] \Rightarrow \\ [P(u) \Rightarrow \forall x\, (R_\Box(u,x) \Rightarrow P(x))].$$

After negating and transforming this formula into a form required in Lemma 3.1 we obtain:[6]

$$\exists u \exists x\, \exists P \left[\begin{array}{c} \forall w\, (P(w) \vee \\ (u \neq w \wedge \forall v\, (\neg R_\Box(u,v) \vee \neg R_\circ(v,w) \vee \neg P(v)))) \\ \wedge \\ R_\Box(u,x) \wedge \neg P(x) \end{array} \right]$$

After application of Lemma 3.1 we get:

$$\exists u \exists x\ R_\Box(u,x) \wedge \nu \neg P(x).[u \neq x \wedge \forall v\, (\neg R_\Box(u,v) \vee \neg R_\circ(v,x) \vee \neg P(v))].$$

Unnegating the formula results in:

$$\forall u \forall x\ R_\Box(u,x) \Rightarrow \mu P(x).[u = x \vee \exists v\, (R_\Box(u,v) \wedge R_\circ(v,x) \wedge P(v))].$$

Thus the initial formula is equivalent to:

[6] Observe that the positive and negative occurrences of P are not separated, thus Lemma 2.1 cannot be applied

$$\forall u \forall x R_{\Box}(u,x) \Rightarrow \{R_{\Box}(u,u) \wedge [u = x \vee R_{\circ}(u,x) \vee \bigvee_{i \in \omega} \exists v_0 \ldots \exists v_i (R_{\Box}(u,v_0) \wedge \ldots \wedge R_{\Box}(u,v_i) \wedge R_{\circ}(u,v_0) \wedge R_{\circ}(v_0,v_1) \wedge \ldots \wedge R_{\circ}(v_{i-1},v_i) \wedge R_{\circ}(v_i,x))]\}.$$

I.e. this formula states that $R_{\Box}$ is the reflexive and transitive closure of $R_{\circ}$, a property which is not expressible by means of classical logic but expressible by means of fixpoint logic. ■

Other applications of Lemma 2.1 to correspondence theory are described in [21, 22]. For more application examples of Lemma 3.1 see [16].

6 Discussion of Other Approaches

6.1 Lifschitz Results

In the last ten years V. Lifschitz published a number of results on second-order quantifier elimination techniques in the context of circumscription (see Lifschitz [14]). Most of these results are subsumed by the DLS algorithm. The only exception is formulated in the following theorem of [14].

THEOREM 6.1 *Let $\Phi_1(P)$, $\Phi_2(P)$ be any first-order formulae such that $\Phi_1(P)$ is positive w.r.t. P and $\Phi_2(P)$ is negative w.r.t. P.*
Then $circ(\Phi_1(P) \wedge \Phi_2(P), P)$ is equivalent to a first-order sentence. ■

Similarly, some formulae that are reducible by Theorem 6.1 are not reducible by SCAN. This indicates the necessity of combining general quantifier elimination algorithms with particular, specialized solutions, like the one formulated above.

6.2 The Sahlqvist–van Benthem Algorithm

The Sahlqvist–van Benthem algorithm was motivated by the modal correspondence theory (see [19, 23]). It is based on the idea of finding "minimal" substitutions for the eliminated predicates. The key role is played here by second-order Sahlqvist formulae that reflect a particular class of modal axioms (for a general definition see de Rijke [18]). The Sahlqvist-van Benthem algorithm is based on the following theorem:

THEOREM 6.2 *Let Φ be a Sahlqvist formula. Then Φ reduces to a first-order formula via suitable substitutions. Moreover, these substitutions can be effectively obtained from Φ.* ■

It can now be observed that negated Sahlqvist formulae are of the form suitable for the DLS algorithm. Moreover, the substitutions mentioned in Theorem 6.2 are obtained by the DLS algorithm (some of them during applications of the Ackermann lemma and some of them during applications of the extended purity deletion rule).

Thus the Sahlqvist-van Benthem algorithm is subsumed by the DLS algorithm. Moreover, the subsumption is strict. Also the SCAN algorithm extends the Sahlqvist-van Benthem algorithm (see de Rijke [18]).

6.3 The Simmons Algorithm

An algorithm for eliminating second-order quantifiers in the context of modal correspondence theory is also given in Simmons [20]. The main idea of this algorithm is similar to that of the Sahlqvist–van Benthem algorithm. It depends on looking for first-order equivalents by finding suitable substitutions of the eliminated predicates. However, in addition to the substitution technique, Simmons applies second-order Skolemization (see Section 2.1), which strengthens the Sahlqvist–van Benthem algorithm.

7 Summary

The development of algorithms for eliminating predicate variables has become a small but quite active area of research. For particular applications like circumscription and correspondence theory a number of methods and results had been known, but for the general case not much happened after Ackermann's early papers. Only after Gabbay and Ohlbach's first paper in the KR92 conference a few people became interested in this problem and began exploring different alternatives. Since the problem is not even semidecidable there is much room for special methods and heuristics. An ideal implementation of a quantifier elimination procedure seems to be a kind of expert system which analyses the formula first and then applies the most appropriate method. Since new ideas and methods are coming up quite frequently it might still be too early to start developing such a complicated system.

There is some indication that such a system would be quite useful. In areas where quantifier elimination plays a role, for example in correspondence theory people so far have only investigated cases with quite small formulae (which nevertheless may be tricky). The method employed was more or less nothing else than guessing and verifying. A program which can deal with really big and complex formulae can open the door to the investigation of systems which are currently out of reach.

Andreas Nonnengart
Max-Planck-Institut für Informatik, Germany.

Hans Jürgen Ohlbach
King's College, London.

Andrzej Szałas
University of Warsaw, Poland.

References

1. Wilhlem Ackermann. Untersuchung über das Eliminationsproblem der mathematischen Logik. *Mathematische Annalen*, 110:390–413, 1935.
2. Wilhlem Ackermann. Zum Eliminationsproblem der Mathematischen Logik. *Mathematische Annalen*, 111:61–63, 1935.
3. Patrick Doherty, Witold Łukaszewicz, and Andrzej Szałas. Computing circumscription revisited: a reduction algorithm. Technical Report LiTH-IDA-R-94-42, Institutionen för Datavetenskap, University of Linköping, 1994. A preliminary report published in *Proceedings* 4th *IJCAI*, Morgan Kaufmann Pub. Inc., pp 1502–1508, 1995. To appear in *Journal of Automated Reasoning.*
4. Patrick Doherty, Witold Łukaszewicz, and Andrzej Szałas. A characterization result for circumscribed normal logic programs. Technical Report LiTH-IDA-R-95-20, Institutionen för Datavetenskap, University of Linköping, 1995. To appear in *Fundamenta Informaticae.*
5. Thorsten Engel. *Elimination of Predicate and Function Quantifiers.* Diploma Thesis. Max-Planck-Institut für Informatik, Saarbrücken, 1996.
6. Dov M. Gabbay and Hans Jürgen Ohlbach. Quantifier elimination in second-order predicate logic. In Bernhard Nebel, Charles Rich, and William Swartout, editors, *Principles of Knowledge Representation and Reasoning (KR92)*, 425–435. Morgan Kaufmann, 1992. Also published in the *South African Computer Journal*, 7:35–43, 1992.
7. Joakim Gustafsson. An implementation and optimization of an algorithm for reducing formulae in second-order logic. Technical Report LiTH-MAT-R-96-04, Dept. of Mathematics, Linköping University, Sweden, 1996.
8. Bjarni Jónsson. A survey of Boolean algebras with operators. In Rosenberg and Sabidussi, editors *Algebra and Orders*, pp. 239–286, 1994.
9. Bjarni Jónsson and Alfred Tarski. Boolean algebras with operators, part I. American Journal of Mathematics, Vol. 73, 891–939, 1951.
10. Georg Kreisel and Jean-Louis Krivine. *Éléments de Logique Mathématique. Théorie des modèles.* Société Mathématique de France, 1966.
11. G. Neelakantan Kartha and Vladimir Lifschitz. A simple formalization of actions using circumscription. In *Proceedings of IJCAI 95*, 1995.
12. Phokion G. Kolaitis and Christos H. Papadimitriou. Some computational aspects of circumscription. In *AAAI-88: Proceedings of the 7th National Conference on Artificial Intelligence*, 465–469, 1988.
13. Vladimir Lifschitz. Computing circumscription. In *Proceedings of the 9th Int'l Joint Conference on Artificial Intelligence*, volume 1, pages 121–127, 1985.
14. Vladimir Lifschitz. Circumscription. In D.M. Gabbay, C.J. Hogger, J.A. Robinson editors, *Handbook of Logic in Artificial Intelligence and Logic Programming*, vol. 3, Clarendon Press, Oxford, 297–352, 1994.
15. William McCune. OTTER 2.0. In Mark Stickel, editor, *Proc. of* 10th *International Conference on Automated Deduction, LNAI 449*, 663–664. Springer Verlag, 1990.
16. Andreas Nonnengart and Andrzej Szałas. A fixpoint approach to second-order quantifier elimination with applications to correspondence theory. Technical Report MPI-I-95-2-007, Max-Planck-Institut für Informatik, Saarbrücken, 1995. To appear in E. Orłowska (ed.), *Logic at Work. Essays Dedicated to the Memory of Helena Rasiowa*, Kluwer.
17. A. Rabinov. A generalization of collapsible cases of circumscription. *Artificial Intelligence*, 38:111–117, 1989.
18. Maarten de Rijke. *Extending Modal Logic.* Ph.D. Thesis, Institute for Logic, Language and Computation, University of Amsterdam, 1993.
19. Henrik Sahlqvist. Completeness and correspondence in the first and second-order semantics for modal logic. In S. Kanger, editor, *Proc. 3rd Scandinavian Logic Symposium*, North Holland, 110–143, 1975.

20. Harold Simmons. The monotonous elimination of predicate variables. *Journal of Logic and Computation*, 4:23–68, 1994.
21. Andrzej Szałas. On the correspondence between modal and classical Logic: an automated approach. Technical Report MPI-I-92-209, Max-Planck-Institut für Informatik, Saarbrücken, 1992. Also published in *Journal of Logic and Computation*, 3:605–620, 1993.
22. Andrzej Szałas. On an automated translation of modal proof rules into formulas of the classical logic. *Journal of Applied Non-Classical Logics*, 4:119–127, 1994.
23. Johan van Benthem. *Modal Logic and Classical Logic*. Bibliopolis, Naples, 1983.

LABELLED NATURAL DEDUCTION

RUY J. G. B. DE QUEIROZ AND DOV M. GABBAY

1 Overview

The functional interpretation of logical connectives is concerned with a certain harmony between, on the one hand, a indexfunctional!calculus functional calculus on the expressions built up from the recording of the deduction steps (the *labels*), and, on the other hand, a logical calculus on the formulae. It has been associated with Curry's early discovery of the correspondence between the axioms of intuitionistic implicational logic and the type schemes of the so-called 'combinators' of Combinatory Logic [12], and has been referred to as the *formulae-as-types* interpretation. Howard's [80] extension of the formulae-as-types paradigm to full intuitionistic first-order predicate logic meant that the interpretation has since been referred to as the 'Curry–Howard' functional interpretation. Although Heyting's [75, 76] intuitionistic logic did fit well into the formulae-as-types paradigm, it seems fair to say that, since Tait's [117, 118] intensional interpretations of Gödel's [69] *Dialectica* system of functionals of finite type, there has been enough indication that the framework would also be applicable to logics beyond the realm of intuitionism. Ultimately, the foundations of a *functional* approach to formal logic are to be found in Frege's [47, 50, 51] system of 'concept writing', not in Curry, or Howard or, indeed, Heyting.

In an attempt to account for some of the less declarative aspects of certain non-classical logics, in a way that those aspects could be handled directly in the object language, D. Gabbay has recently set up a novel research programme in his book on *Labelled Deductive Systems* [54]. The idea, which may be seen as the seeds of a more general alternative to the type-theoretic interpretation of two-dimensional logical systems (i.e. 'terms alongside formulae'), is that the declarative unit of a logical system is to be seen as a *labelled* formula 't : **A**' (read 't labels **A**'). From this perspective, a logical system is taken to be not simply a calculus of logical deductions on formulae, but a suitably harmonious combination of a functional calculus

H.J. Ohlbach and U. Reyle (eds.), Logic, Language and Reasoning, 173–250.

on the labels and a logical calculus on the formulae. A logic will then be defined according to the meta-level features of the conceptual norm that the logic is supposed to formalise: the allowable logical moves will then be 'controlled' by appropriate constraints on 'what has been done so far' (has the assumption been used at all; have the assumptions been used in a certain order; has the assumption been used more than once; etc.).

Here we wish to present a framework for studying the mathematical foundations of *Labelled Deductive Systems.* We could also regard it (quite pretentiously) as an attempt at a reinterpretation of Frege's logical calculus where abstractors and functional operators work harmoniously alongside logical connectives and quantifiers. In other words, the functional interpretation (sometimes referred to as the Curry–Howard!–Tait interpretation) can be viewed in a wider perspective of a labelled deductive system which can be used to study a whole range of logics, including some which may not abide by the tenets of the intuitionistic interpretation (e.g. classical implicational logic, many-valued logics, etc.). The result is a labelled natural deduction system which we would like to see as a reinterpretation of Frege's 'functional' account of logic: it is as if the theory of functions of *Grundgesetze* is put together with the theory of predicates of *Begriffsschrift*, in such a way that a formula is true (valid) if and only if a deduction of it can be constructed where the label contains no free variable (i.e. its proof-construction is a 'complete' object, which means that the truth of the formula relies on no assumptions). The weaker the logic, the stricter are the ways by which assumptions may be withdrawn. Classical implicational logic, for example, will have a procedure for withdrawing implicational assumptions depending on the history of the deduction, which its intuitionistic counterpart will not have. So, we need to look for a paradigm for two-dimensional logical systems (terms alongside formulae) which can account for a general perspective on the harmony between withdrawal of assumptions in the logic and abstraction of variables in the functional calculus. We are beginning to find what seems to be a reasonable architecture for such a methodology underlying logical systems based on term calculi: *Grundgesetze* alongside *Begriffsschrift* [22].

1.1 Labels and Gentzen's programme

In order to prove the *Hauptsatz*, which could not be done in the natural deduction calculi NJ and NK because of lack of symmetry in NJ and lack of elegance in NK, Gentzen went on to develop the 'logistic' calculi.

> "In order enunciate and prove the *Hauptsatz* in a convenient form, I had to provide a logical calculus especially suited to the purpose. For this the natural calculus proved unsuitable. For, although it already contains the properties essential to the validity of the *Hauptsatz*, it does so only

with respect to its intuitionist form, in view of the fact that the law of excluded middle, as pointed out earlier, occupies a special position in relation to these properties." [59, Opening Section, §2]

A major improvement on Gentzen's original programme of analysis of deduction via analysis of connectives was put forward by D. Prawitz in his monograph on natural deduction [108]. The main features of Prawitz' framework can be summarized as follows:

- definition of normalization (i.e. the so-called 'reduction' rules) for NJ, therefore 'pushing' the *cut* principle down to the level of connectives, rather than the level of consequence relation; e.g.:

$$\dfrac{\begin{matrix}\Sigma_1\\ \mathbf{A}\end{matrix} \qquad \dfrac{\begin{matrix}[\mathbf{A}]\\ \Sigma_2\\ \mathbf{B}\end{matrix}}{\mathbf{A}\rightarrow\mathbf{B}}\rightarrow\textit{-intr}}{\mathbf{B}}\rightarrow\textit{-elim} \qquad \twoheadrightarrow \qquad \begin{matrix}\Sigma_1\\ [\mathbf{A}]\\ \Sigma_2\\ \mathbf{B}\end{matrix}$$

  where the Σs (i.e. Σ_1, Σ_2) stand for whole deduction trees.
- definition of (classical) *reductio ad absurdum*, i.e.:

$$\dfrac{\begin{matrix}[\sim \mathbf{A}]\\ \Lambda\end{matrix}}{\mathbf{A}}$$

  where **A** is atomic and different from Λ, '$\sim$' stands for negation, and 'Λ' is the distinguished propositional constant for absurdity. With the addition of this rule to the intuitionistic system, Prawitz provided an inferential counterpart to Gentzen's special place for the axiom of the excluded middle;
- proof theory is based on the *subformula principle*, which compromised the credibility of natural deduction systems (especially the full fragment, i.e. with $\vee$, $\exists$), on what concerned decision procedures;
- little emphasis on the formulation of a proof theory for classical logic, perhaps due to the philosophical underpinnings of his [110, 111] programme (joint with M. Dummett [33, 35]) on a language-based philosophical account of intuitionism.

1.1.1. *Adding an Extra Dimension*

The main features of a system of natural deduction where there is an additional dimension of labels alongside formulae, can be summarized as follows:

- it is 'semantics driven': by bringing meta-level information back into the object-level, it is bringing a little of the semantics (i.e. names of individuals and dependency functions, names of possible worlds, etc.) into the proof-calculus;

- it retakes Gentzen's programme of analysis of logical deduction via an analysis of connectives (via introduction/elimination rules, and a distinction between assumptions, premisses and conclusions), by introducing the extra dimension (the *label*) which will take care of deduction in a more direct fashion than the sequent calculus. That is to say, the extra dimension will take care of eventual dependencies among referents. In other words, the shift to the sequent calculus was motivated by the need for recovering symmetry (i.e. hypotheses and conclusions stand in opposition to each other in terms of polarity) and local control (for each inference rule, no need for side conditions or global control), but in fact the calculus can only do 'bookkeeping' on formulae, but not on individuals, dependency functions or possible worlds. The handling of *inclusive* logics, i.e. logics which also deal with empty domains, is much improved by the explicit introduction of individuals as variables of the functional calculus on the labels. Thus, a formula like '$\forall x.\mathbf{P}(x) \to \exists x.\mathbf{P}(x)$' is not a theorem in a labelled system, but its counterpart with the explicit domain of quantification '$\mathbf{D} \to (\forall x^{\mathbf{D}}.\mathbf{P}(x) \to \exists x^{\mathbf{D}}.\mathbf{P}(x))$', to be interpreted as 'if the domain of quantification $\mathbf{D}$ is non-empty, then if for all x in $\mathbf{D}$, $\mathbf{P}$ is true of x, then there exists an x in the domain such that $\mathbf{P}$ is true of it'.
- it recovers symmetry (non existent in either Gentzen's [59] NJ or Prawitz' [108] **I**) by allowing a richer analysis of the properties of '$\to$', whose labels may play the role of *function* or *argument*;
- it recovers a 'connective-based' account of deduction, for reasons already mentioned above;
- it replaces the *subformula principle* by the *subdeduction principle.* With this proviso we shall avoid the complications introduced when the straightforward notion of *branch* is replaced either by more complex notions such as *path* or *track.* Cf. [109], where the notion of *path* is replacing the notion of *branch*, and [122], where the complicated notion of *track* has to be defined in order to account for the proof of the subformula property. As we shall see, the use of *subdeduction*, instead of *subformula* is especially useful for the so-called 'Skolem'-type connectives such as disjunction, existential quantifier and propositional equality, just because their *elimination* rules may violate the *subformula* property of a deduction, while they will always respect the *subdeduction* property;
- from the properties of implication, it is easy to obtain a *generalized reductio ad absurdum* for classical positive implicational logic:

$$\frac{\dfrac{[x : \mathbf{A} \to \mathbf{B}]}{b(x, \ldots, x) : \mathbf{B}}}{\lambda x.b(x, \ldots, x) : \mathbf{A}} \quad \boxed{\begin{array}{c}\text{`}\mathbf{A} \to \mathbf{B}\text{'} \\ \text{as minor \&} \\ \text{as ticket}\end{array}}$$

- it incorporates the handling of first-order variables into the calculus, therefore dispensing with special conditions on *Eigenvariables*;
- with the new dimension introduced it is easier to check the connections between the proof procedures and the model-theoretic-based resolutions (e.g. Skolem's, ,Herbrand's) because variables for dependency functions (the Skolem functions) and for 'justification of equalities' (substitution) (the Herbrand functions) are introduced and manipulated accordingly in the functional calculus on the labels, yet without appearing in the logical calculus on the formulae;
- the *Hauptsatz* is recast in a more 'realizability'-like presentation, as in the Tait's [117] method: cut elimination is replaced by normalization, i.e. main measure of redundancy is in the label;
- it recovers the 'continuation' aspect of classical logic, which, unlike Gentzen's [59] NJ or Prawitz' [108] **C**, the sequent calculus did manage to keep by allowing more than one formula to the right of the turnstile. Via a more careful analysis of what is at stake when 'new' branches have to be open in a proof tree whenever a Skolem-type connective is being eliminated, one observes the importance of reasoning about 'our proof so far' which we can do with a labelled system due to the added dimension. For example, the replacement of *subformula* by *subdeduction* would seem to facilitate the proof of decidability results via Gentzen-type techniques. It also makes it possible to define *validity* on the basis of *elimination* rules, something which is not easily done with plain systems. (Cf. [109, page 290], on the definition of validity through elimination rules breaking down the induction for the cases of disjunction and existential quantification.)
- recovering duality by establishing that any label will either play the role of a *function* or that of an *argument*;
- the additional dimension is crucial in the establishment of a proof theory for equality, given that referents and dependency functions are handled directly by the two-dimensional calculus;
- the definition of normal derivations becomes easier than the one given in [109, II.3, page 248], because it is to be based on the normality of the expression in the label, the latter containing the encoding of the steps taken, even if they involved Skolem-type connectives (those which may violate the subformula property, but which in a labelled system will not violate the subdeduction property).

1.2 Labels and Computer Programming

There are a number of features of *labelled* systems that can have significant benefits on the logical treatment of computer programming issues. Some of these features were already pointed out in P. Martin-Löf's [100] seminal paper, such as the connections between constructive mathematics and computer programming. It has also been pointed out that the essence of the correspondence between proof theory (i.e. logic as a deduction system) and computer programming are to be found in the so-called *conversion* rules such as those of the Curry–Howard isomorphism.[1]

Furthermore, developments on computational interpretations of logics have demonstrated that there is more to the connections between *labelled* proof theory and computer science, such as, e.g. the establishment of a logical perspective on computer programming issues like:

- iteration vs recursion
- potential infinity and lazy evaluation
- implementation of a type into another
- use of resources
- flow of control
- order of evaluation.

While the first three topics were addressed in [28], [29], the remaining ones are dealt with in the functional interpretation (sequent-style) of linear logic given by S. Abramsky [2].

1.3 Labels and Information Flow

There has been a significant amount of research into the characterization of the concept of *information flow* as a general paradigm for the semantics of logical systems. It has been said, for example, that , Girard's [62] *linear logic* is the *right* logic to deal with information flow.[2] It is also claimed that Barwise's [7] *situation theory* is the most general mathematical formulation of the notion of information flow.

The approach generally known as the *construction-theoretic* interpretation of logic (cf. e.g. [119]) which underlies Gabbay's *Labelled Deductive Systems*, is more adequate for the formalization of the concept of *information flow* and its applications to computer science. Some of the reasons for this claim could be enumerated as follows:

---

[1]Cf.:

> "These equations [arising out of the Curry–Howard term-rewriting rules] (and the similar ones we shall have occasion to write down) are the essence of the correspondence between logic and computer science." [67, Section **Computational significance**, subsection **Interpretation of the rules**]

and:

> "The idea that a reduction (normalization) rule can be looked at as a semantic instrument should prove a useful conceptual view that could allow the unification of techniques from theories of abstract data type specification with techniques from proof theory, constructive mathematics and λ-calculus." [28, page 408]

[2]V. Pratt's contribution to 'linear-logic' e-mail list, Feb 1992.

- it is neither a specific logic nor a specific semantical approach, but a general (unifying) framework where the integration of techniques from both proof theory and model theory is the driving force;
- it accounts for 'putting together' meta-language and object-language in a disciplined fashion;
- it can be viewed as an attempt to benefit from the devices defined in *Begriffsschrift* (i.e. connectives and quantifiers) on the one hand, and *Grundgesetze* (i.e. functional operations, abstractors, etc.) on the other hand, by having a functional calculus on the labels harmonised with a logical calculus on the formulae. In fact, by developing the basis of formal logic in terms of function and argument, Frege is to be credited as the real pioneer of the functional interpretation of logic, not Curry, Howard, or indeed Heyting;
- it is closer to the realizability interpretation than the (intuitionistic) Curry–Howard interpretation, thus giving a more general account of the paradigm 'formulae and the processes which realize them'. A formula is a theorem if it can be proved with a 'complete object' (no free variable) as its label. The label can be thought of as the 'evidence' (the 'reason') for the validity of the formula. Thus, by appropriately extending the means by which one can 'close' the term labelling a formula one extends the stock of valid formulae; cf. [93, page 47]:

  > "Logicians should note that a deductive system is concerned not just with unlabelled entailments or sequents $\mathbf{A} \to \mathbf{B}$ (as in Gentzen's proof theory), but with deductions or proofs of such entailments. In writing $f : \mathbf{A} \to \mathbf{B}$ we think of f as the 'reason' why $\mathbf{A}$ entails $\mathbf{B}$."

  In a paper to appear in the *JSL* [58] (abstract in [57]), we demonstrate how to extend the interpretation to various logics, including classical positive implication, with a generalized form of *reductio ad absurdum* involving some form of self-application in the labels.

  For a philosophical account of the generality of the construction-theoretic interpretation see, e.g. Tait's 'Against Intuitionism: Constructive Mathematics is part of Classical Mathematics' [119, page 182]:

  > "I believe that, with certain modifications, this idea [propositions as types of their proofs] provides an account of the meaning of mathematical propositions which is adequate, not only for constructive mathematics, but for classical mathematics as well. In particular, the pseudo Platonism implicit in the truth functional account of classical mathematics is, on this view, eliminated. The distinction between constructive and classical rests solely on what principles are admitted for constructing an object of a given type."
- it is *resource aware*: disciplines of abstraction on label-variables reflect the disciplines of assumption withdrawing peculiar to the logic being

considered;

- it is not limited to logics with/without Gentzen's *structural rules*, such as *contraction*, *exchange* and *weakening*, though these are indeed reflected naturally in the disciplines of abstraction. (Here we could think of structures (*constellations of labels*) other than sets, multisets.)
- it is *natural-language-friendly* in the sense that it provides a convenient way of modelling natural language phenomena (such as anaphora, 'universal' indefinites, etc. Cf. Gabbay and Kempson's [65, 66] work on relevance, labelled abduction and *wh*-construal) by the underlying device of *keeping track of proof steps*, thus accounting for dependencies. (Here it may be worth mentioning the potential connections with H. Kamp's *Discourse Representation Theory* [83] and with K. Fine's [40] account of *Reasoning with Arbitrary Objects.*)
- it provides a 'natural' environment whereby the connections between model-theoretic (Skolem, Herbrand) and proof-theoretic (Gentzen, Prawitz) accounts of the theory of provability are more 'visible'. The division of tasks into two distinct (yet harmonious) dimensions, namely label-formula (i.e. functional-logical), allows the handling of 'second-order' objects such as function-names to be done via the functional calculus with abstractors, thus keeping the 'logical' calculus first-order. Cf. Gabbay and de Queiroz' [26] 'The functional interpretation of the existential quantifier', presented at *Logic Colloquium '91*, Uppsala;
- it opens the way to a closer connection between Lambek and Scott's [93] equational interpretation of deductive systems as categories, and proof-theoretic accounts of proof equivalences. (Here we would draw attention to the potential for spelling out the connections between the unicity conditions of mappings in pullbacks, pushouts and equalisers, and the inductive role of η-equality for '$\wedge$', '$\vee$' and '$\rightarrow$', respectively.)
- it offers a *deductive* (as opposed to *model-theoretic*) account of the connections between modal logics and its propositional counterparts when world-variables are introduced in the functional calculus on the labels (i.e. when a little of the semantics is *brought to the syntax*, so to speak).

  E.g.:

$$\Box\textit{-introduction} \qquad \frac{\begin{array}{c}[\mathbb{W} : \mathcal{U}]\\ F(\mathbb{W}) : \mathbf{A}(\mathbb{W})\end{array}}{\Lambda\mathbb{W}.F(\mathbb{W}) : \Box\mathbf{A}} \qquad\qquad \Box\textit{-elimination} \qquad \frac{\mathbb{T} : \mathcal{U} \qquad l : \Box\mathbf{A}}{\mathcal{EXTR}(l, \mathbb{T}) : \mathbf{A}(\mathbb{T})}$$

  where '$\mathcal{U}$' would be a collection of 'worlds' (where a world can be taken to be, e.g. structured collections (lists, bags, trees, etc.) of labelled formulae) and '$F(\mathbb{W})$' is an expression which may depend on the world-variable '$\mathbb{W}$'. The conditions on $\Lambda\mathbb{W}$-abstraction will distinguish different $\Box$s, in a way which is parallel to the distinction of various implications by conditions on λx-abstraction in:

$$\begin{array}{cc}
\rightarrow\textit{-introduction} & \rightarrow\textit{-elimination} \\
\begin{array}{c}[x:\mathbf{A}] \\ b(x):\mathbf{B} \\ \hline \lambda x.b(x):\mathbf{A}\rightarrow\mathbf{B}\end{array} & \begin{array}{c} \\ a:\mathbf{A} \qquad f:\mathbf{A}\rightarrow\mathbf{B} \\ \hline \mathtt{APP}(f,a):\mathbf{B}\end{array}
\end{array}$$

- it offers a convenient framework where various notions of equality can be studied (including the λ-calculus-like β-, η- and ξ- equalities), and whose applications to the formalization of propositional equality and the definite article, as well as a proof theory for descriptions, are of general interest;
- by incorporating means of manipulating referents and dependency functions (the objects from the 'functional' side) it provides an adequate (and logic-free) middle ground between procedural and declarative approaches to logic, where it makes sense to ask both 'what is the proof theory of model theory?' and 'what is the model theory of proof theory?'. (A forthcoming paper entitled 'Situating Labelled Entailments' [8] investigates the possibilities of combining Situation Theory with *Labelled Deductive Systems*.)
- it offers a natural deduction based explanation for the disjunction-conjunction ambiguity which may appear in some ordinary language interpretations of logic. The most illustrious example is Girard's [63] defence of a 'disjunctive conjunction' as finding its counterpart in ordinary language when 'proofs are interpreted as actions' (see example later in the section on 'resource handling').

1.4 Labels and 'Constructivity as Explicitation'

In a recent paper on a sequent calculus for classical linear logic Girard rightly points out the intimate connections between constructivity and explicitation:

> "Constructivity should not be confused with its ideological variant 'constructiv*ism*' which tries to build a kind of *countermathematics* by an *a priori* limitation of the methods of proofs; it should not either be characterized by a list of technical properties: e.g. disjunction and existence properties. Constructivity is the possibility of extracting the information *implicit* in proofs, i.e. constructivity is about *explicitation*." [64, page 255]

Now, one of the aims of inserting a label alongside formulae (accounting for the steps made to arrive at each particular point in the deduction) is exactly that of making *explicit* the use of formulae (and instances of formulae and individuals) throughout a deduction. At this stage it may be relevant to ask how one can be more explicit than this: the functional aspect (related to names of individuals, instances of formulae, names of contexts, etc.) is handled by devices which are of a different nature and origin from the ones which handle the logical aspect, namely, connectives and quantifiers. By using labels/terms alongside formulae, we can:

1. keep track of proof steps (giving local control)
2. handle 'arbitrary' names (via variable abstraction operators)

and our labelled natural deduction system gives us at least three advantages over the usual plain natural deduction systems:

1. it benefits from the harmony between
   - the functional calculus on terms and
   - the logical calculus on formulae;
2. it takes care of 'contexts' and 'scopes' in a more explicit fashion;
3. normalization theorems may be proved via techniques from term rewriting.

As an example of how explicitation is indeed at the heart of a labelled system, let us look at how the inference rules for quantifiers are formulated:

$\forall$-*introduction*

$$\frac{\begin{array}{c}[x : \mathbf{D}]\\ f(x) : \mathbf{F}(x)\end{array}}{\Lambda x.f(x) : \forall x^{\mathbf{D}}.\mathbf{F}(x)}$$

$\forall$-*elimination*

$$\frac{a : \mathbf{D} \qquad c : \forall x^{\mathbf{D}}.\mathbf{F}(x)}{\mathtt{EXTR}(c, a) : \mathbf{F}(a)}$$

$\exists$-*introduction*

$$\frac{a : \mathbf{D} \qquad f(a) : \mathbf{F}(a)}{\varepsilon x.(f(x), a) : \exists x^{\mathbf{D}}.\mathbf{F}(x)}$$

$\exists$-*elimination*

$$\frac{e : \exists x^{\mathbf{D}}.\mathbf{F}(x) \qquad \begin{array}{c}[t : \mathbf{D}, g(t) : \mathbf{F}(t)]\\ d(g, t) : \mathbf{C}\end{array}}{\mathtt{INST}(e, \sigma g.\sigma t.d(g, t)) : \mathbf{C}}$$

Note that the individuals are explicitly introduced as labels (new variables) alongside the domain of quantification, the latter being explicitly introduced as a formula: e.g. '$a : \mathbf{D}$', 'a' being an individual from domain '$\mathbf{D}$'.

Some of the difficulties of other systems of natural deduction can be easily overcome. For example, the handling of *inclusive* logics, cf. [40, Chapter 21, page 205]:

> "An *inclusive* logic is one that is meant to be correct for both empty and non-empty domains. There are certain standard difficulties in formulating a system of inclusive logic. If, for example, we have the usual rules of UI [$\forall$-*elim*], EG [$\exists$-*intr*] and conditional proof [$\rightarrow$-*intr*], then the following derivation of the theorem $\forall xFx \supset \exists xFx$ goes through (...) But the formula $\forall xFx \supset \exists xFx$ is not valid in the empty domain; the antecedent is true, while the consequent is false."

Here the difficulty of formulating a system of inclusive logic does not exist simply because the individuals are taken to be part of the calculus: recall that the labelled natural deduction presentation system is made of a functional calculus on the terms, and a logical calculus of deductions on the formulae. It requires that the names of individuals be introduced in the functional part in order for the quantifiers to be introduced and eliminated. This is not the case for plain natural deduction systems: there is no direct way to handle either terms or function symbols in a deduction without the labels. E.g. in:

$$\dfrac{\dfrac{\dfrac{[\forall x.\mathbf{F}(x)]}{\mathbf{F}(t)}}{\exists x.\mathbf{F}(x)}}{\forall x.\mathbf{F}(x) \to \exists x.\mathbf{F}(x)}$$

the term t is not explicitly introduced as an extra assumption, as it would be the case in the informal reading of the above deduction ('let t be an arbitrary element from the domain').

Using the functional interpretation, where the presence of terms and of the domains of quantification make the framework a much richer instrument for deduction calculi, we have:

$$\dfrac{\dfrac{\dfrac{[t : \mathbf{D}] \qquad [z : \forall x^{\mathbf{D}}.\mathbf{F}(x)]}{\mathtt{EXTR}(z,t) : \mathbf{F}(t)}}{\varepsilon x.(\mathtt{EXTR}(z,x),t) : \exists x^{\mathbf{D}}.\mathbf{F}(x)}}{\lambda z.\varepsilon x.(\mathtt{EXTR}(z,x),\boxed{t}) : \forall x^{\mathbf{D}}.\mathbf{F}(x) \to \exists x^{\mathbf{D}}.\mathbf{F}(x)}$$

Here the presence of a free variable (namely 't') indicates that the assumption '$[t : \mathbf{D}]$' remains to be discharged. By making the domain of quantification explicit one does not have the antecedent (vacuously) true and the consequent trivially false in the case of empty domain: the proof of the proposition is still depending on the assumption 'let t be an element from **D**', i.e. that the type '**D**' is presumably non-empty. To be categorical the above proof would still have to proceed one step, as in:

$$\dfrac{\dfrac{\dfrac{\dfrac{[t : \mathbf{D}] \qquad [z : \forall x^{\mathbf{D}}.\mathbf{F}(x)]}{\mathtt{EXTR}(z,t) : \mathbf{F}(t)}}{\varepsilon x.(\mathtt{EXTR}(z,x),t) : \exists x^{\mathbf{D}}.\mathbf{F}(x)}}{\lambda z.\varepsilon x.(\mathtt{EXTR}(z,x),\boxed{t}) : \forall x^{\mathbf{D}}.\mathbf{F}(x) \to \exists x^{\mathbf{D}}.\mathbf{F}(x)}}{\underbrace{\lambda t.\lambda z.\varepsilon x.(\mathtt{EXTR}(z,x),t)}_{\text{no free variable}} : \mathbf{D} \to (\forall x^{\mathbf{D}}.\mathbf{F}(x) \to \exists x^{\mathbf{D}}.\mathbf{F}(x))}$$

Now we look at the proof-construction ('$\lambda t.\lambda z.\varepsilon x.(\mathtt{EXTR}(z,x),t)$') we can see no free variables, thus the corresponding proof is categorical, i.e. does not rely on any assumption.

An alternative to the explicitation of the first-order variables and their domains via labels and formulae is given in Lambek and Scott's [93] definition of an intuitionistic type theory with equality. The idea is to define the consequence relation with the set of variables to be used in a derivation made explicit as a subscript to the '$\vdash$':

> "We write $\vdash$ for $\vdash_\varnothing$, that is, for $\vdash_X$ when X is the empty set. The reason for the subscript X on the entailment symbol becomes apparent when we look at the following 'proof tree'
>
> $$\dfrac{\dfrac{\dfrac{\forall_{x\in A}\varphi(x) \vdash \forall_{x\in A}\varphi(x)}{\forall_{x\in A}\varphi(x) \vdash_x \varphi(x)} \qquad \dfrac{\exists_{x\in A}\varphi(x) \vdash \exists_{x\in A}\varphi(x)}{\varphi(x) \vdash_x \exists_{x\in A}\varphi(x)}}{\forall_{x\in A}\varphi(x) \vdash_x \exists_{x\in A}\varphi(x)}}{\forall_{x\in A}\varphi(x) \vdash \exists_{x\in A}\varphi(x)}$$
>
> where the last step is justified by replacing every free occurrence of the variable x (there are none) by the closed term a of type A, *provided* there is such a closed term. Had we not insisted on the subscripts, we could have deduced this in any case, even when A is an empty type, that is, when there are no closed terms of type A." (p. 131)

Note that the introduction of a subscript to the symbol of consequence relation is rather *ad hoc* device, whereas in our case the pattern will fit within the general framework of labels and formulae, being also applicable to identify the phenomenon of *inclusiveness* in modal logics, i.e. the case of *serial* modal logics where there is always an accessible world from the current world (axiom schema '$\Box\mathbf{A} \to \Diamond\mathbf{A}$').

1.5 Labels, Connectives, Consequence Relation, and Structures

Theorems of a certain logic are well-formed formulae which can be demonstrated to be true regardless of other formulae being true. That is, a complete proof of a theorem relies on no assumptions. So, whenever we construct proofs in natural deduction, we need to look at the rules which discharge assumptions. This is only natural, because when starting from hypotheses and arriving at a certain thesis we need to say that the hypotheses imply the thesis. So, we need to look at the rules of inference which allow us to 'discharge' assumptions (hypotheses) without introducing further assumptions. It so happens that the *introduction* rules for the conditionals (namely, implication, universal quantifier, necessity) do possess this useful feature. They allow us to 'get rid of hypotheses' by making a step from 'given the hypotheses, and arriving at premise' to 'hypotheses imply thesis'. Let us look at the introduction rules for implication in the plain natural deduction style:

→ -introduction

$$\frac{\begin{array}{c}[\mathbf{A}]\\ \mathbf{B}\end{array}}{\mathbf{A}\rightarrow\mathbf{B}}$$

Note that the hypothesis '**A**' was *discharged*, and by the introduction of the implication, the conclusion '**A** → **B**' (hypothesis '**A**' implies thesis '**B**') was reached.

Now, if we introduce labels alongside formulae this 'discharge' of hypotheses will be reflected on the label of the intended conclusion by a device which makes the arbitrary name introduced as the label of the corresponding assumption 'loose its identity', so to speak. It is the device of 'abstracting' a variable from a term containing one or more 'free' occurrences of that variable. So, let us look at how the rule given above looks like when augmented by inserting labels alongside formulae:

$$\frac{\begin{array}{c}[x:\mathbf{A}]\\ b(x):\mathbf{B}\end{array}}{\lambda x.b(x):\mathbf{A}\rightarrow\mathbf{B}}$$

Notice that when we reach the conclusion the arbitrary name 'x' looses its identity simply because the abstractor 'λ' binds its free occurrences in the term '$b(x)$' (which in its turn may have none, one or many free occurrence(s) of 'x').[3] Just think of the more usual variable binding mechanism on the formulae: being simply a place-marker, the 'x' has no identity whatsoever in x-quantified formulae such as $\forall x.\mathbf{P}(x)$ and $\exists x.\mathbf{P}(x)$.

As we can see from the rule *→-introduction* (and generally from the *introduction* rule of any conditional) the so-called 'improper' inference rules, to use a terminology from Prawitz' [108] *Natural Deduction*, leave *room for manoeuvre* as to how a particular logic can be handled just by adding conditions on the discharge of assumptions that would correspond to the particular logical discipline one is adopting (linear, relevant, ticket entailment, intuitionistic, classical, etc.). The side conditions can be 'naturally' imposed, given that a degree of 'vagueness' is introduced by the form of those improper inference rules, such as the rule of *→-introduction*:

$$\frac{\begin{array}{c}[x:\mathbf{A}]\\ b(x):\mathbf{B}\end{array}}{\lambda x.b(x):\mathbf{A}\rightarrow\mathbf{B}}$$

[3]N.B. The notation '$b(x)$' indicates that '$b(x)$' is a functional term which depends on 'x', and not the application of 'b' to 'x'.

Note that one might (as some authors do) insert an explicit sign between the assumption '$[x : \mathbf{A}]$' and the premise of the rule, namely '$b(x) : \mathbf{B}$', such as, e.g. the three vertical dots, making the rule look like:

$$\frac{\begin{array}{c}[x : \mathbf{A}]\\ \vdots \\ b(x) : \mathbf{B}\end{array}}{\lambda x.b(x) : \mathbf{A} \to \mathbf{B}}$$

to indicate the element of vagueness. There is no place, however, for the introduction of side conditions on those rules which do not allow for such a 'room for manoeuvre', namely those rules which are not improper inference rules. In his account of linear logic via a (plain) natural deduction system, Avron [6] introduces what we feel rather 'unnatural' side conditions in association with an inference rule which is not improper, namely the rule of $\wedge$-*introduction.*[4]

Now, while in our labelled natural deduction we explore the nuances in the interpretation of conditionals simply by a careful analysis of the properties of the connective itself, there may be cases where the discipline of assumption withdrawal is dictated from the 'outside', so to speak. In the logics which deal with exceptions, priorities, defaults, revisions, etc. there is the need to impose a structure on the collection of assumptions

[4]See p. 163:

$$\frac{A \quad B}{A \wedge B}(*)$$

and condition (3) on p. 164: "For $\wedge$-Int we have side condition that A and B should depend on exactly the same multiset of assumptions (condition $(*)$). Moreover, the shared hypotheses are considered as appearing *once*, although they seem to occur *twice*".

Nevertheless, in the framework of our labelled natural deduction we are still able to handle different notions of conjunction and disjunction in case we need to handle explicitly the *contexts* (structured collections of formulae). This is done by introducing names (variables) for the contexts as an extra parameter in all our *introduction* and *elimination* rules for the logical connectives, and considering those names as identifiers for structured collections (sets, multisets, lists, trees, etc.) of labelled formulae: e.g. '$\mathbf{A}(\mathbb{S})$' would read 'an occurrence of the formula $\mathbf{A}$ is stored in the structure $\mathbb{S}$'. So, in the case of conjunction, for example, we would have:

$\wedge$-*introduction*

$$\frac{a_1 : \mathbf{A}_1(\mathbb{S}) \qquad a_2 : \mathbf{A}_2(\mathbb{T})}{\langle a_1, a_2\rangle : (\mathbf{A}_1 \wedge \mathbf{A}_2)(\mathbb{S} \circledcirc \mathbb{T})} \qquad \left(\text{in sequent calculus: } \frac{\mathbb{S} \vdash a_1 : \mathbf{A}_1 \qquad \mathbb{T} \vdash a_2 : \mathbf{A}_2}{\mathbb{S} \circledcirc \mathbb{T} \vdash \langle a_1, a_2\rangle : \mathbf{A}_1 \wedge \mathbf{A}_2}\right)$$

where the '$\circledcirc$' operator would be compatible with the data structures represented by $\mathbb{S}$ and $\mathbb{T}$. For example, if $\mathbb{S}$ and $\mathbb{T}$ are both taken to be sets, and '$\circledcirc$' is taken to be set union, then we have a situation which is similar to the rule of classical Gentzen's sequent calculus (augmented with labels alongside formulae). If, on the other hand, we take the structures to be multisets, and '$\circledcirc$' multiset union, we would have Girard's linear logic and the corresponding conjunctions ($\otimes/\&$) depending on whether '$\mathbb{S}$' is distinct or identical to '$\mathbb{T}$'.

(hypotheses) in such a way that it disturbs as little as possible the basic properties of the connectives. It is for these logics that we need to define structured *constellations* of labelled formulae, and an appropriate proof theory for them. It is as if one needs to study the (meta-)logical sign used for consequence relation, i.e. the turnstile '⊢', *as a connective.* Thus, we shall need to define the notion (and its formal counterpart) of *structural cut*, via the use of explicit data type operations (*à la* Guttag [71]) over the structured collection of formulae.

The proof theory for the (meta-)logical connective '⊢' which relates *structured constellation of labelled formulae* will be pursued in detail in a paper on a 'labelled sequent calculus' [24].

1.6 Labels and Non-normal Modal Logics

The use of labels alongside formulae allows for the use of analogies to be made which may be useful in understanding certain concepts in logic where there is less declarative content, such as *relevance* and *non-normality*. Having accounted for relevance in the proof theory of relevant implication, we shall use the analogy between implication and necessity made clear when both are characterized by labelled proof calculus, in order to carry over the reasoning to obtain a proof theory of non-normal necessity.

It is common to define *non-normal* modal logics as systems of modal logics where the *necessitation* rule, i.e.:

$$\frac{\vdash \mathbf{A}}{\vdash \Box\mathbf{A}}$$

is not a valid inference rule. The class of *regular* modal logics is defined as non-normal modal logics where the rule of *regularity*, i.e.:

$$\frac{\vdash \mathbf{A} \to \mathbf{B}}{\vdash \Box\mathbf{A} \to \Box\mathbf{B}}$$

replaces the necessitation rule, somehow making the *necessity* weaker. In order to use modal connectives to formalise concepts like belief and knowledge, one would like to avoid pathologies arising out of the way the inference rule of necessitation, namely: that all tautologies are believed (resp. known); that from 'believed **A**' and 'believed **A** → **B**' one infers 'believed **B**' (i.e. *omniscience*).

Now, notice that the expression 'provable **A**' (i.e. '⊢ **A**') is *vague*, but if we know *how* **A** was proved, we might wish to be more careful in inferring that **A** is *necessarily true.* One way of knowing how **A** was proved is by looking at the whole deduction of **A**. Another way is by *keeping track of proof steps* via a labelling mechanism. After all, one of the main motivations for developing a *labelled* proof theory is to bring some objects from the meta-level (i.e. names of individuals, function symbols, 'resource'-related information, names of collections of formulae, etc.) back into the proof calculus itself. It is as if we wish to make proof theory more *semantics-driven*, yet retaining the good features of Gentzen's [59] analysis of deduction.

2 Preamble

The functional interpretation of logical connectives, the so-called Curry–Howard!interpretation [12, 80], provides an adequate framework for the establishment of various calculi of logical inference. Being an 'enriched' system of natural deduction, in the sense that terms representing proof-constructions (the *labels*) are carried alongside formulae, it constitutes an apparatus of great usefulness in the formulation of logical calculi in an operational manner. By uncovering a certain harmony between a functional calculus on the labels and a logical calculus on the formulae, it proves to be instrumental in giving mathematical foundations for systems of logic presentation designed to handle meta-level features at the object-level via a labelling mechanism, such as, e.g. D. Gabbay's [54] *Labelled Deductive Systems*.

Here we demonstrate that the introduction of 'labels' is of great usefulness not only for the understanding of the proof-calculus itself, but also for the clarification of its connections with model-theoretic interpretations. For example, the Skolem-type procedures of introducing 'new' names (in order to eliminate the existential quantifier, for example) and 'discharging' them at the end of a deduction (i.e. concluding that the deduction could have been made without those new names previously introduced) are directly reflected in the *elimination* rules for connectives like $\vee$ and $\exists$. From this perspective one realizes that for plain systems it may be true that "natural deduction is only satisfactory for $\rightarrow$, $\forall$, $\wedge$; the connectives $\vee$ and $\exists$ receive a very *ad hoc* treatment: the elimination rule for '$\vee$' [and '$\exists$'] is with the presence of an extraneous formula **C**," [63, page 34] although the situation becomes very different in the presence of labels alongside formulae. In a natural deduction system with labels alongside formulae we realize that there is nothing 'extraneous' about the arbitrary formula '**C**'. For example, the rule of $\exists$-*elimination* (shown below) states that:

(1) if an arbitrary formula **C** can be obtained from a deduction which uses both the individual's name newly introduced ('let it be t') and the assumption that it does have the required property ('let $g(t)$ be the evidence for $\mathbf{P}(t)$', g being a newly introduced name for the way in which the evidence for $\mathbf{P}(t)$ depends on t), then
(2) this formula **C** can be obtained *without* using the new names at all.

Condition (1) is reflected in the label expression alongside the formula **C** in the premise being required to show a free occurrence of the new names (i.e. t and g), and requirement (2) is shown to be met when the new names become bound (by the σ-abstractor) in the label expression alongside the same formula **C** in the conclusion.

$\vee$-*elimination*

$$\frac{p : \mathbf{A}_1 \vee \mathbf{A}_2 \quad \begin{array}{c}[s_1 : \mathbf{A}_1]\\ d(s_1) : \mathbf{C}\end{array} \quad \begin{array}{c}[s_2 : \mathbf{A}_2]\\ e(s_2) : \mathbf{C}\end{array}}{\texttt{CASE}(p, \upsilon s_1.d(s_1), \upsilon s_2.e(s_2)) : \mathbf{C}}$$

$\exists$-*elimination*

$$\frac{e : \exists x^{\mathbf{D}}.\mathbf{P}(x) \quad \begin{array}{c}[t : \mathbf{D}, g(t) : \mathbf{P}(t)]\\ d(g,t) : \mathbf{C}\end{array}}{\texttt{INST}(e, \sigma g.\sigma t.d(g,t)) : \mathbf{C}}$$

Here 'υ' (nu) and 'σ' (sigma) are abstractors which bind free variables in a similar way to the more usual λ-abstractor. In our labelled natural deduction, the λ-abstractor shall be reserved for the constructor of label expressions associated with the connective of implication, i.e. '$\rightarrow$'. Although all abstractors (λ, υ, ε, σ, etc.) originate in Frege's value-range technique, there is an important difference: whereas the λ-abstractor accompanies an introduction of an implication in the logical calculus, here our υ- and σ-abstractors reflect the discharge of assumptions by binding the corresponding variables, but are not accompanied by the introduction of a conditional connective on the logical side. In other words, we shall need a more general view of the device of abstraction, which may not necessarily be connected with the implication connective, but rather with the discharge of assumptions in the logical calculus.

In passing, note that the abstractors are all reminiscent of Frege's device of transforming functions into objects by forming value-range terms with the help of the notation '$\grave{\varepsilon}f(\varepsilon)$', where the sign (' ̓ ', which is the smooth breathing for Greek vowels, according to Dummett [36]) plays the role of an abstractor [49].[5] (We shall come back to this point later on when we describe the role of the labels in our framework.)

By means of a suitable harmony between, on the one hand, a functional calculus with abstractors, and, on the other hand, a logical calculus with connectives and quantifiers, we want to show that the labelling device is more than just an extra syntactical ornament. It is a useful device to 'put back into the object language' the (lost) capability of handling names (and eventual dependencies), e.g. names of instances of formulae, names of individuals, names of collections of formulae, etc. which is kept at the meta-level in plain logic presentation systems. The lack of this capability makes it difficult (or at least 'unnatural') for the latter systems to handle a class of logics called 'resource' logics where non-declarative features such as 'how many times a formula was used to obtain another formula', 'which

[5]See, e.g. the partial English translation of *Grundgesetze I* [53]. See also Peano's [106] original device for functional notation.

order certain formulae were used in order to obtain another formula as a conclusion', etc. do have a crucial role to play. The lack of a naming device such as the labelling mechanism, we shall contend, also obscures the connections with some fundamental theorems given by the model-theoretic interpretations.

Moreover, by using labels one can keep track of all proof steps. This feature of labelled natural deduction systems helps to recover the 'local' control virtually lost in plain natural deduction systems. The 'global' character of the so-called *improper* inference rules (i.e. those rules who do not only manipulate premisses and conclusions, but also involve assumptions, such as, e.g. $\rightarrow$-*introduction*, $\vee$-*elimination*, $\exists$-*elimination*, etc.) is made 'local' by turning the 'discharge functions' into appropriate disciplines for variable-binding via the device of 'abstractors'. As a matter of fact, the use of abstractors to make arbitrary names loose their identity is not a new device: it was already used by Frege in his *Grundgesetze I* [50]. (Think of the step from 'assuming $x : \mathbf{A}$ and arriving at $b(x) : \mathbf{B}$' to 'assert $\lambda x.b(x) : \mathbf{A} \rightarrow \mathbf{B}$' as making the arbitrary name 'x' loose the identity it had in $b(x)$ as a name. The binding transforms the name into a mere place-marker for arguments in substitutions.) From this perspective, critical remarks of the sort "Natural Deduction uses global rules (...) which apply to whole deductions, in contrast to rules like elimination rule for $\rightarrow$, which apply to formulae" [63, page 35] will have much less impact than they have w.r.t. 'plain' natural deduction systems.

In what follows we demonstrate how labelled natural deduction presentation systems based on our 'Fregean' functional interpretation (resulting from a generalization of the Curry–Howard interpretation) can help us provide a more intuitive and more general alternative to most logic presentation systems in the Gentzen tradition. We shall discuss how some 'problematic' aspects of plain natural deduction can be overcome by the use of labels alongside formulae.

In particular, we shall be dealing with the following aspects of natural deduction:

- the loss of 'local' control caused by the structure of the so-called *improper* inference rules (i.e. rules which discharge assumptions);
- the lack of an appropriate treatment of η-type normalization (an *elimination* followed by an *introduction*) for connectives other than implication; this shall point the way to providing an answer to the problem related to non-confluence of Natural Deduction systems with disjunction and η-normalization[6]; furthermore, we shall be looking at the connections between the role of η-equality for $\wedge$, $\vee$ and $\rightarrow$ in guaranteeing unicity in the forms of the proofs, and the unicity condition involved

[6]Recently answered in [15].

in the category-theoretic devices of *pullbacks*, *pushouts* and *equalisers*, respectively;

- the absence of a formal account of the role of *permutative* reductions with respect to the functional calculus on the labels, as well as, to some extent, with respect to the logical calculus on the formulae; we shall define these reductions as ζ-reductions;
- the lack of a generalized form of the (classical) *reductio ad absurdum*; we shall demonstrate what form the rule of *reductio ad absurdum* should take if one does not have a distinguished propositional constant (namely '$\mathcal{F}$', the *falsum*);
- the absence of a clear account of first-order quantification and the role of labels in predicate formulae;
- the lack of an appropriate link with classical results in proof theory obtained via model-theoretic means, such as Skolem's and Herbrand's resolution theorems;[7]
- the absence of appropriate devices for directly handling referents, dependency functions, and equality, thus making it difficult to see how natural deduction techniques can be used to handle descriptions, or even offer a reasonable account of Herbrand's decision procedure for predicate logic, as well as of unification;
- the lack of a 'natural' substitute for the semantic notion of accessibility relation, thus making it difficult to provide a generalized proof theory for modal logics.

3 The Role of the Labels

Unlike axiomatic systems, natural deduction proofs need not start from axioms. More interested in the *structure* of proofs, Gentzen conceived his natural deduction as a logical system which would make explicit the *structural properties of connectives*. Inference rules would be made of not simply premisses and conclusions (as inference rules in Hilbert-style axiomatic systems happened to be, e.g. *Modus Ponens*), but also of *assumptions*. They would also be framed into a pattern of *introduction* and *elimination* rules, with a logical *principle of inversion* underlying the harmony between those two kinds of rules. Connectives which behaved like conditionals (most notably, implication, but also the universal quantifier) would have *introduction* rules

[7] e.g. to the best of our knowledge, there is no *deduction*-based analysis of Leisenring's [87] studies of the connections between ε-like calculi, choice principles and the resolution theorems of Skolem and Herbrand. The two 'worlds' (proof theory and model theory) are usually seen as so disjoint as to disencourage a more integrating approach. With an analysis of deduction via our labelled system, we would wish to bridge some of the gaps between these two worlds.

doing the job of *withdrawing assumptions* in favour of the corresponding conditional statement.

As defined in the logic literature, a *valid* statement would be one which would rely on no assumptions. Now, in order to test for the validity of statements using Gentzen's method of natural deduction, one would need to make sure that, by the time a deduction of the statement was achieved, *all* assumptions had been withdrawn. Thus, whenever we were to construct proofs in natural deduction, we would need to look at the rules which could withdraw assumptions. This would only be too natural, because when starting from hypotheses and arriving at a certain thesis we need to say that the hypotheses imply the thesis. So, we need to look at the rules of inference which allow us to 'discharge' assumptions (hypotheses) without introducing further assumptions. As we mentioned above, it so happens that the *introduction* rules for the conditionals (namely, implication, universal quantifier, necessity) do indeed perform the task for us. They allow us to 'get rid of the hypotheses' by making a step to 'hypotheses imply thesis'. Let us look at the introduction rules for implication in the plain natural deduction style:

$\rightarrow$ *-introduction*

$$\frac{\begin{array}{c}[\mathbf{A}]\\ \mathbf{B}\end{array}}{\mathbf{A}\rightarrow\mathbf{B}}$$

Note that the hypothesis '$\mathbf{A}$' is discharged, and by the introduction of the implication, the conclusion '$\mathbf{A} \rightarrow \mathbf{B}$' (i.e. hypothesis '$\mathbf{A}$' implies thesis '$\mathbf{B}$') is reached.

Now, if we introduce labels alongside formulae this 'discharge' of hypotheses will be reflected on the label of the intended conclusion by a device which makes the arbitrary name introduced as the label of the corresponding assumption 'loose its identity', so to speak. It is the device of 'abstracting' a variable from a term containing one or more 'free' occurrences of that variable. So, let us look at how the rule given above looks like when augmented by putting labels alongside formulae:

$$\frac{\begin{array}{c}[x:\mathbf{A}]\\ b(x):\mathbf{B}\end{array}}{\lambda x.b(x):\mathbf{A}\rightarrow\mathbf{B}}$$

Notice that now by the time we reach the conclusion the arbitrary name 'x' looses its identity because the abstractor 'λ' binds its free occurrences in the term 'b', which in its turn may have one or more free occurrences of 'x' (the notation '$b(x)$' indicates that 'b' is a functional term which depends on 'x').

The moral of the story here is that the last inference rule of any complete proof must be the introduction rule of a conditional, simply because those are the rules which do the job we want: discharging assumptions already made, without introducing any further assumptions. We are now speaking in more general terms ('conditional', rather than 'implication') because the introduction rules for the universal quantifier and the necessitation connectives, namely:

$$\begin{array}{cc} \forall\text{-}introduction & \Box\text{-}introduction \\ \begin{array}{c}[x : \mathbf{D}] \\ f(x) : \mathbf{P}(x) \\ \hline \Lambda x.f(x) : \forall x^{\mathbf{D}}.\mathbf{P}(x)\end{array} & \begin{array}{c}[\mathbb{W} : \mathcal{U}] \\ F(\mathbb{W}) : \mathbf{A}(\mathbb{W}) \\ \hline \Lambda \mathbb{W}.F(\mathbb{W}) : \Box\mathbf{A}\end{array} \end{array}$$

also discharge old assumptions without introducing new ones. (The treatment of 'necessity' – '□' – by means of the functional interpretation will be sketched in a later section, and shall be given in more detail in [56].)

Our motto here is that all labelled assumptions must be discharged by the end of a deduction, and we should be able to check this very easily just by looking at the label of the intended conclusion and check if all 'arbitrary' names (labels) of hypotheses are bound by any of the available abstractors. (As we will see later on, abstractors other than 'λ' and 'Λ' will be used. We have already mentioned 'υ', 'ε' and 'σ'.) So, in a sense our proofs will be *categorical* proofs, to use a terminology from [1].[8] The connection with the realizability interpretation will be made in the following sense: e realizes **P** iff e is a *complete* object (no free-variables). For stronger logics (e.g. classical positive logic) there will be additional ways of binding free variables of the label expression which may establish an extended harmony between the functional calculus on the labels and the logical calculus on the formulae. (Here we have in mind the *generalized reductio ad absurdum* defined in [58], where a λ-abstraction binds a variable occurring as function and as argument in the label expression, and there is no introduction of an implication.)

The device of variable-binding, and the idea of having terms representing incomplete 'objects' whenever they contain free variables, were both introduced in a systematic way by Frege in his *Grundgesetze*. As early as 1893 Frege developed in his *Grundgesetze I* what can be seen as the early origins of the notions of *abstraction* and *application*, when showing techniques for transforming functions (expressions with free variables) into value-range terms (expressions with no free variables) by means of an 'introductory' operator of abstraction producing the *Werthverlauf* expression,[9]

[8] "A proof is *categorical* if all hypotheses in the proof have been discharged by use of →-I, otherwise *hypothetical*; and A is a *theorem* if A is the last step of a categorical proof." [1, page 9].

[9] [50, §3, page 7], translated as *course-of-values* in [53, page 36], and *value-range* in most other translations of Frege's writings, including the translation of [49] (published in [95]) where the term first appeared.

e.g. '$\grave{\varepsilon}f(\varepsilon)$', and the effect of its corresponding 'eliminatory' operator '$\frown$' on a *value-range* expression.[10]

The idea of forming *value-range* function-terms by abstracting from the corresponding free variable is in fact very useful in representing the handling of assumptions within a natural deduction style calculus. In particular, when the natural deduction presentation system is based on a 'labelling' mechanism the binding of free variables in the labels corresponds to the discharge of respective assumptions. In the sequel we shall be using 'abstractors' (such as 'λ' in '$\lambda x.f(x)$') to bind free-variables and discharge the assumption labelled by the corresponding variable.

We consider a formula to be a theorem of the logical system if it can be derived in a way that its corresponding 'label' contains no free variable, i.e. the deduction rests on no assumption. In other words, a formula is a theorem if admits a categorical proof to be constructed.

3.1 Dividing the Tasks: a Functional Calculus on the Labels, a Logical Calculus on the Formulae

We have seen that the origins of variable binding mechanisms, both on the formulae of logic (the propositions) and on the expressions of the functional calculus (the terms), go back at least as far as Frege's early investigations on a 'language of concept writing'. Although the investigations concerned essentially the establishment of the basic laws of logic, for Frege the functional calculus would have the important role of demonstrating that arithmetic could be formalised simply by defining its most basic laws in terms of rules of the 'calculus of concept writing'. Obviously, the calculus defined in *Begriffsschrift* [47], in spite of its functional style, was primarily concerned with the 'logical' side, so to speak. The novel device of binding free variables, namely the universal quantifier, was applicable to propositional functions. Thus, *Grundgesetze* [50, 51] was written with the intention of fulfilling the ambitious project of designing a language of concept-writing

[10]Cf. [50, §34, pages 52ff], (translated in [53, page 92]):

"(...) it is a matter only of designating the value of the function $\Phi(\xi)$ for the argument Δ, i.e., $\Phi(\Delta)$, by means of 'Δ' and '$\grave{\varepsilon}\Phi(\varepsilon)$'. I do so in this way:

$$\text{“}\Delta \frown \grave{\varepsilon}\Phi(\varepsilon)\text{”}$$

which is to mean the same as '$\Phi(\Delta)$'."

(Note the similarity to the rule of functional *application*, where 'Δ' is the argument, '$\grave{\varepsilon}\Phi(\varepsilon)$' is the function, and '$\frown$' is the application operator 'APP'.)

Expressing how important he considered the introduction of a variable-binding device for the functional calculus (recall that the variable-binding device for the logical calculus had been introduced earlier in *Begriffsschrift*), Frege says:

"The introduction of a notation for courses-of-values [value-ranges] seems to me to be one of the most important supplementations that I have made of my *Begriffsschrift* since my first publication on this subject." (*Grundgesetze I*, §9, p. 15f.)

which could be useful to formalise mathematics. Additional mechanisms to handle the functional aspects of arithmetic (e.g. equality between number-expressions, functions over number-expressions, etc.) had to be incorporated. The outcome of Frege's second stage of investigations also brought pioneering techniques of formal logic, this time with respect to the handling of functions, singular terms, definite descriptions, etc. An additional mechanism of variable binding was introduced, this time to bind variables of functional expressions, i.e. expressions denoting individuals, not truth-values.

Summarising, we can see the pioneering work of Frege in its full significance if we look at the two sides of formal logic he managed to formulate a calculus for:

1. the 'logical' calculus on formulae (*Begriffsschrift*)
2. the 'functional' calculus on terms (*Grundgesetze*)

As a pioneer in any scientific activity one is prone to leave gaps and loopholes to be later filled by others. It happened with Frege that a big loophole was discovered earlier than he would himself have expected: Russell's discovery of the antinomies of his logical notion of set was a serious challenge. There may be several ways of explaining why the resulting calculus was so much susceptible to that sort of challenge. We feel particularly inclined to think that the use of devices which were designed to handle the so-called 'objects', i.e. expressions of the functional calculus, ought to have been kept apart from, and yet harmonised with, the logical calculus on the formulae. Thus, here we may start wondering what might have been the outcome had Frege kept the two sides separate and yet harmonious.

Let us for a moment think of a connection to another system of language analysis which would seem to have some similarity in the underlying ontological assumption, with respect to the idea of dividing the logical calculus into two dimensions, i.e. functional vs. logical. The semantical framework defined in Montague's [104] intensional logic makes use of a distinction among the semantic *types* of the objects handled by the framework, namely e, t and s, in words: *entities*, *truth-values*, and *senses*. The idea was that logic (language) was supposed to deal with objects of three kinds: names of entities, formulae denoting truth-values, and possible-worlds/contexts of use. Now, here when we say that we wish to have the bi-dimensional calculus, we are saying that the entities which are *namable* (i.e. individuals, possible-worlds, etc.) will be dealt with separately from (yet harmoniously with) the logical calculus on the formulae, by a calculus of functional expressions. Whereas the variables for individuals are handled 'naturally' in the interpretation of first-order logic with our labelled natural deduction, the introduction of variables to denote *contexts*, or *possible-worlds* (structured collection of labelled formulae), as in [56], is meant to account for Montague's *senses*.

3.2 Reassessing Frege's Two-dimensional Calculus

In our attempt to reassess the benefits of having those two sides working together, we would like to insist on the two sides being treated separately. Thus, instead of binding variables on the formulae with the device of forming 'value-range' expressions as Frege does,[11] we shall have a clear separation of functional vs. logical devices. We still want to have the device of forming propositional functions, so we still need to have the names of variables of the functional calculus being carried over to take part in the formulae of the logical side. That will be dealt with accordingly when we describe what it means to have predicate formulae in a labelled system. Nevertheless, abstractors shall only bind variables occurring in expressions of the functional calculus, and quantifiers shall bind variables occurring in formulae of the logical calculus. For example, in:

$$\forall\textit{-introduction}\quad \frac{\begin{array}{c}[x:\mathbf{D}]\\ f(x):\mathbf{P}(x)\end{array}}{\Lambda x.f(x):\forall x^{\mathbf{D}}.\mathbf{P}(x)} \qquad \exists\textit{-introduction}\quad \frac{a:\mathbf{D}\qquad f(a):\mathbf{P}(a)}{\varepsilon x.(f(x),a):\exists x^{\mathbf{D}}.\mathbf{P}(x)}$$

whilst the abstractors 'Λ' and 'ε' bind variables of the functional calculus, the quantifiers '$\forall$' and '$\exists$' bind variables of the logical calculus, even if the same variable name happens to be occurring in the functional expression as well as in the logical formula.

Notice that although we are dealing with the two sides independently, the harmony seems to be maintained: to each discharge of assumption in the logical calculus there will correspond an abstraction in the functional calculus. In the case of our quantifier rules, we observe that the *introduction* of the universal quantifier is made with the arbitrary name x being bound in both sides (functional and logical) at the same time. In the existential case the 'witness' a is kept unbound in the functional calculus, whilst in the formula the binding is performed.

Here is not really the place to discuss the paradoxes of Frege's formalised set theory, but it might be helpful to single out one particularly relevant

---

[11] Cf. the following opening lines of *Grundgesetze I*, §10:

"Although we laid it down that the combination of signs '$\grave{\varepsilon}\Phi(\varepsilon) = \grave{\alpha}\Psi(\alpha)$' has the same denotation as '$\overset{\mathfrak{a}}{\smile}\ \Phi(\mathfrak{a}) = \Psi(\mathfrak{a})$', this by no means fixes completely the denotation of a name like '$\grave{\varepsilon}\Phi(\varepsilon)$'."

Note that both the abstractor ' ' ' and the universal quantifier '$\smile$' are used for binding free variables of formulae of the logical calculus such as 'Φ' and 'Ψ'. In our labelled natural deduction we shall take the separation 'functional vs. logical' more strictly than Frege himself did. While the abstractors will be used to bind variables in the functional calculus, the quantifiers will be used to bind variables in the logical calculus. Obviously, variables may be occurring in both 'sides', but in each side the appropriate mechanism will be used accordingly.

facet of his 'mistake'. First, let us recall that the development of mechanisms for handling both sides of a calculus of concept writing, namely the logical and the functional, would perhaps recommend special care in the harmonising these two sides. We all know today (thanks to the intervention of the likes of M. Furth [53], P. Aczel [4], M. Dummett [32, 36], H. Sluga [115], and others) that one of the fundamental flaws of Frege's attempt to put the two sides together was the so-called 'Law V' of *Grundgesetze*, which did exactly what we shall avoid here in our 'functional interpretation' of logics: using functions where one should be using propositions, and vice versa. The 'Law V' was stated as follows:

$$\vdash (\grave{\varepsilon}f(\varepsilon) = \grave{\alpha}g(\alpha)) = (\overset{\mathfrak{a}}{\smile} f(\mathfrak{a}) = g(\mathfrak{a}))$$

Here we have equality between terms – i.e. $\grave{\varepsilon}f(\varepsilon) = \grave{\alpha}g(\alpha)$ and $f(\mathfrak{a}) = g(\mathfrak{a})$ – on a par with equality between truth-values – i.e. the middle equality sign.

In his thorough analysis of Frege's system, Aczel [4] makes the necessary distinction by introducing the sign for propositional equality:

$$(\lambda x.f(x) \doteq \lambda x.g(x)) \leftrightarrow \forall x.(f(x) \doteq g(x)) \text{ is true}$$

where '$\doteq$' stands for propositional equality, and '$\leftrightarrow$' is to mean logical equivalence (i.e. 'if and only if').[12]

Despite the challenges to his theories of formal logic, Frege's tradition has remained very strong in mathematical logic. Indeed, there is a tendency among the formalisms of mathematical logic to take the same step of 'blurring' the distinction between the functional and the logical side of formal logic. As we have already mentioned, Frege introduced in the *Grundgesetze* the device of binding variables in the functional calculus,[13] in addition to the variable-binding device presented in *Begriffsschrift*, but allowed variables occurring in the formulae to be bound not only by the quantifier(s), but also by a device of the functional calculus, namely the 'abstractors'. One testimony to the strength of Frege's legacy which is particularly relevant to our purposes here is the formalism described in Hilbert and Bernays'

[12]Later in this monograph, (and in more detail in a forthcoming report [27]) we shall be dealing with the problem of handling equality on the 'logical side', so to speak: we demonstrate how to provide an analysis of deduction (Gentzen style, i.e. via rules of *introduction* and *elimination* with appropriate labelling discipline) for a proposition saying that two expressions of the functional calculus denote the same object. In order to explain the properties of this new propositional connective we will be discussing the issues of 'extensional vs. intensional' approaches to equality. An analysis of propositional equality via our labelled natural deduction may serve as the basis for a proof theory for descriptions.

[13]In fact, Frege had already introduced the device which he called *Werthverlauf* in his article on 'Function and Concept' [49], which, in its turn, may have been inspired by Peano's [106] functional notation.

[73, 74] book where various calculi of singular terms are established. One of these calculi was called the ε-calculus, and consisted of an extension of first-order logic by adding the following axiom schema:

$$(\varepsilon_1) \qquad \mathbf{A}(a) \to \mathbf{A}(\varepsilon_x \mathbf{A}(x)) \qquad \text{(for any term `}a\text{')}$$
$$(\varepsilon_2) \qquad \forall x.(\mathbf{A}(x) \leftrightarrow \mathbf{B}(x)) \to (\varepsilon_x \mathbf{A}(x) = \varepsilon_x \mathbf{B}(x))$$

where any term of the form '$\varepsilon_x \mathbf{A}(x)$' is supposed to denote a term 't' with the property that '$\mathbf{A}(t)$' is true, if there is one such term.

Now, observe that the addition of these new axioms has to be proven 'harmless' to the previous calculus, namely the first-order calculus with bound variables, in the sense that no formulae involving only the symbols of the language of the old calculus which was not previously a theorem, is a theorem of the new calculus. For that one has to prove the fundamental theorems stating that the new calculus is only a 'conservative extension' the old calculus (First and Second ε-Theorems).

The picture becomes slightly different when we follow somewhat more strictly the idea of dividing, as sharply as we can, the two tasks: let all that has to do with entities to be handled by the functional calculus on the labels, and leave only what is 'strictly logical' to the logical calculus on the formulae. So, in the case of ε-terms, we shall not simply replace an existentially quantified variable in a formula (e.g. 'x' in '$\exists x.\mathbf{A}(x)$') by an ε-term involving a formula (e.g. '$\mathbf{A}(\varepsilon_x \mathbf{A}(x))$'). Instead, we shall use 'ε' as an abstractor binding variables of the functional calculus, as we have seen from our rule of $\exists$-*introduction* shown previously. In other words, we do not have the axioms (or rules of inference) for the existential quantifier plus other axiom(s) for the ε-symbol. We shall be presenting the existential quantifier with its usual logical calculus on the formulae, alongside our ε-terms taking care of the 'functional' side. And so be it:

$\exists$-*introduction* $\qquad\qquad$ $\exists$-*elimination*

$$\frac{a : \mathbf{D} \qquad f(a) : \mathbf{P}(a)}{\varepsilon x.(f(x), a) : \exists x^{\mathbf{D}}.\mathbf{P}(x)} \qquad\qquad \frac{e : \exists x^{\mathbf{D}}.\mathbf{P}(x) \qquad \begin{array}{c}[t : \mathbf{D}, g(t) : \mathbf{P}(t)] \\ d(g,t) : \mathbf{C}\end{array}}{\mathtt{INST}(e, \sigma g.\sigma t.d(g,t)) : \mathbf{C}}$$

Notice that here our concern with the 'conservative extension' shall be significantly different from the one Hilbert and Bernays [73, 74] had. We have the ε-symbol appearing on the label (the functional side, so to speak), and it is only introduced alongside the corresponding existential formula. (More details of our treatment of the peculiarities of the existential quantifier are given in [26].)

4 Canonical Proofs and Normalization

Since Heyting's [75] definition of each (intuitionistic) logical connective in terms of proof conditions (as opposed to the then usual truth-valuation technique), there emerged a whole tradition within mathematical logic of replacing the declarative concept of truth-functions by its procedural counterpart proof-conditions. By providing a 'language-based' (as opposed to Brouwer's language*less*) explanation of intuitionistic mathematics, Heyting put forward a serious alternative approach to the usual truth-table based definitions of logical connectives which was adequate for a certain tradition in the philosophy of language and philosophy of mathematics, namely the so-called 'anti-realist' tradition.

With the advent of Gentzen's [59] 'mathematical theory of proofs', its corresponding classification of 'natural deduction' inference rules into *introduction* and *elimination*, and the principle (advocated by Gentzen himself) saying that the conditions under which one can assert a logical proposition (formalised by the *introduction* rules) define the meaning of its major connective,[14] an intuitionistic proof-theoretic approach to semantics was given a (meta-)mathematical status. Later, the philosophical basis of this particular approach to intuitionism via proof theory found in Dummett [31, 33] and Prawitz [108, 110] its main advocates. In his book on the foundations of (language-based) intuitionistic mathematics Dummett advocates that "the meaning of each [logical] constant is to be given by specifying, for any sentence in which that constant is the main operator, what is to count as a proof of that sentence, it being assumed that we already know what is to count as a proof of any of the constituents." [34, page 12]

A further refinement of the notion of meaning as being determined by the proof-conditions is given by P. Martin-Löf when he elegantly points out the crucial and often neglected distinction between *propositions* and *judgements*, and introduces the notion of *canonical* (or *direct*) proof. By making an attempt to formalise the basic principles of a particular strand of intuitionism called *constructive* mathematics, as practiced by, e.g. E. Bishop in [9] *Foundations of Constructive Analysis*, his explanations of meaning in terms of *canonical* proofs further advocates the replacement of truth-valuation-based accounts of meaning by a canonical-proof-based one.[15]

[14]When commenting on the role of the natural deduction rules of *introduction* and *elimination*, Gentzen says:

"The introductions represent, as it were, the 'definitions' of the symbols concerned, and the eliminations are no more, in the final analysis, than the consequences of these definitions." ([59], p. 80 of the English translation.)

[15]In a series of lectures entitled 'On the meanings of the logical constants and the justifications of the logical laws', Martin-Löf [102] presents philosophical explanations concerning the distinction between propositions and judgements, as well as the connec-

It does not seem unreasonable, however, to say that the use of natural deduction and the semantics of assertability conditions advocated by the intuitionists is not the only way to provide a language-based account of the meaning of logical symbols. If for nothing else, the guiding principle that assertability conditions constitute the main semantical device is

---

tions between, on the one hand, Heyting's explanation of propositions in terms of proofs rather than truth-values, and, on the other hand, the principle that "the meaning of a proposition is determined by what it is to verify it, or what counts as a verification of it" [102, page 43]. Although in those lectures the emphasis appears to be on a sort of phenomenological interpretation of the notions of proposition and judgement, in what concern the formalization of the concepts, the explanations suggest the important role of the definition of direct/canonical proofs for the proof-based account of meaning. In a subsequent paper the interpretation is carried a step further, and the connections with Gentzen's claim is spelled out in a more explicit fashion:

"The intuitionists explain the notion of proposition, not by saying that a proposition is the expression of its truth conditions, but rather by saying, in Heyting's words, that a proposition expresses an expectation or an intention, and you may ask, An expectation or an intention of what? The answer is that it is an expectation or an intention of a proof of that proposition. And Kolmogorov phrased essentially the same explanation by saying that a proposition expresses a problem or task (Ger. *Aufgabe*). Soon afterwards, there appeared yet another explanation, namely, the one given by Gentzen, who suggested that the introduction rules for the logical constants ought to be considered as so to say the definitions of the constants in question, that is, as what gives the constants in question their meaning. What I would like to make clear is that these four seemingly different explanations actually all amount to the same, that is, they are not only compatible with each other but they are just different ways of phrasing one and the same explanation." (...) "If you interpret truth conditions in this way, you see that they are identical with the introduction rules for the logical constants as formulated by Gentzen. So I have now explained why, suitably interpreted, the explanation of a proposition as the expression of its truth conditions is no different from Gentzen's explanation to the effect that the meaning of a proposition is determined by its introduction rules."

[103, pages 410 and 411]

Cf. also: "The introduction rules say what are the canonical elements (and equal canonical elements) of the set, thus giving its meaning."

[101, page 24]

Observe that the principle of 'meaning is determined by the assertability (proof) conditions' is unequivocally advocated. The introduction rules of Gentzen-type natural deduction are said to constitute definitions (as Gentzen himself had advocated earlier). Here one might start to wonder what class of definition (abbreviation, presentation, etc.) these introduction rules would fall into. And indeed, as it stands, the principle of 'introduction rules as definitions' does not seem to find a place in virtually any classification of definitions used in mathematical logic, even less so in, e.g. Frege's [52] classification of definitions into *constructive* and *analytic* [16].

Perhaps it should be remarked here that in spite of the 'unequivocal' position expressed by P. Martin-Löf, there still seems to be room for interpretation. Cf., e.g. the following observation in M. Dummett's recently published *The Logical Basis of Metaphysics*:

"Intuitively, Gentzen's suggestion that the introduction rules be viewed as fixing the meanings of the logical constants has no more force than the converse suggestion, that they are fixed by the elimination rules; intuitive plausibility oscillates between these opposing suggestions as we move from one logical constant to another. Per Martin-Löf has, indeed, constructed an entire meaning-theory for the language of mathematics on the basis of the assumption that it is the elimination rules that determine meaning." [37, page 280]

clearly not in tune with the proclaimed source of inspiration for the critique of the truth-valuation approach and its replacement by explanations of proof conditions.[16] So, there appears to be room for the adoption of a different one, such as, e.g. instead of advocating the replacement of the truth-valuation systems by explanations of proof-conditions, one can propose to have the explanation of how the *elimination* inferences act on the result of *introduction* steps, i.e. the β-normalization procedure, as the main semantical device. The normalization procedure would be looked at, not merely as a meta-mathematical device introduced to prove consistency (as it is usually seen by proof-theorists), but it would be seen as the formal explanation of the 'functionality' of the corresponding logical sign. Although some other principle might be found to be more appropriate, this is, in fact, what we adopt here.

4.1 Canonical Proofs

Instead of Heyting's explanation of the logical constants solely in terms of proofs (or, 'canonical' proofs as in [101]), the explanations given by the approach to the functional interpretation taken here involve both the notion of *canonical proofs* and that of *normalization of non-canonical proofs.*

The *canonical* proofs are explained as:

[16]Cf., e.g.:

"*We no longer explain the sense of a statement by stipulating its truth-value in terms of the truth-values of its constituents, but by stipulating when it may be asserted in terms of the conditions under which its constituents may be asserted.* The justification for this change is how we in fact learn to use these statements: furthermore, the notions of truth and falsity cannot be satisfactorily explained so as to form a basis for an account of meaning once we leave the realm of effectively decidable statements." [31, page 161] (The underlining is ours.)

Cf. also:

"As pointed out by Dummett, this whole way of arguing with its stress on communication and the role of the language of mathematics is inspired by ideas of Wittgenstein and is very different from Brouwer's rather solipsistic view of mathematics as a languageless activity. Nevertheless, as it seems, it constitutes the best possible argument for some of Brouwer's conclusions. (...)

I have furthermore argued that the rejection of the platonistic theory of meaning depends, in order to be conclusive, on the development of an adequate theory of meaning along the lines suggested in the above discussion of the principles concerning meaning and use. Even if such a Wittgensteinian theory did not lead to the rejection of classical logic, it would be of great interest in itself." [110, page 18]

We have previously endeavoured to demonstrate that the so-called 'semantics of use' advocated by Wittgenstein did not involve simply assertability conditions, but it also accounted for the explanation of (immediate) consequences [18]. So, as it seems, it would be unreasonable to call a theory of meaning based on *assertability conditions* a 'Wittgensteinian theory'.

a proof of the proposition:	has the canonical form of:
$\mathbf{A}_1 \wedge \mathbf{A}_2$	$\langle a_1, a_2\rangle$, where a_1 is a proof of $\mathbf{A}_1$ and a_2 is a proof of $\mathbf{A}_2$
$\mathbf{A}_1 \vee \mathbf{A}_2$	$\mathtt{inl}(a_1)$, where a_1 is a proof of $\mathbf{A}_1$ or $\mathtt{inr}(a_2)$, where a_2 is a proof of $\mathbf{A}_2$ ('inl' and 'inr' abbreviate 'into the left disjunct' and 'into the right disjunct', respectively)
$\mathbf{A} \to \mathbf{B}$	$\lambda x.b(x)$, where $b(a)$ is a proof of $\mathbf{B}$ provided a is a proof of $\mathbf{A}$
$\forall x^{\mathbf{D}}.\mathbf{P}(x)$	$\Lambda x.f(x)$, where $f(a)$ is a proof of $\mathbf{P}(a)$ provided a is an arbitrary individual chosen from the domain $\mathbf{D}$
$\exists x^{\mathbf{D}}.\mathbf{P}(x)$	$\varepsilon x.(f(x), a)$, where a is an individual (witness) from the domain $\mathbf{D}$, and $f(a)$ is a proof of $\mathbf{P}(a)$

As the reader can easily notice, the explanation of the logical connectives in terms of canonical proofs only cover the rules of *introduction*.[17] They constitute an explanation of the conditions under which one can form a canonical (direct) proof of the corresponding proposition. Its counterpart in Gentzen's natural deduction is the *introduction* rule, now enriched with the 'witnessing' construction which shall be handled by the functional calculus on the labels which we have mentioned above.

Thus, the corresponding formal presentations *à la* natural deduction are, e.g.:

$\wedge$-*introduction*

$$\frac{a_1 : \mathbf{A}_1 \qquad a_2 : \mathbf{A}_2}{\langle a_1, a_2\rangle : \mathbf{A}_1 \wedge \mathbf{A}_2}$$

$\vee$-*introduction*

$$\frac{a_1 : \mathbf{A}_1}{\mathtt{inl}(a_1) : \mathbf{A}_1 \vee \mathbf{A}_2} \qquad \frac{a_2 : \mathbf{A}_2}{\mathtt{inr}(a_2) : \mathbf{A}_1 \vee \mathbf{A}_2}$$

[17] When looking at some of those informal explanations of intuitionistic connectives, such as the one via canonical proofs (or, indeed, the one via a realizability predicate given in [82]), one is often tempted to see the explanations given for '$\wedge$', '$\to$' and '$\forall$' as covering the procedures corresponding to *elimination* rules. For the sake of the argument, however, let us stick to the usual intuitionistic account of Heyting's semantics.

→-introduction

$$\frac{\begin{array}{c}[x : \mathbf{A}]\\ b(x) : \mathbf{B}\end{array}}{\lambda x.b(x) : \mathbf{A} \to \mathbf{B}}$$

∀-introduction[18]

$$\frac{\begin{array}{c}[x : \mathbf{D}]\\ f(x) : \mathbf{P}(x)\end{array}}{\Lambda x.f(x) : \forall x^{\mathbf{D}}.\mathbf{P}(x)}$$

∃-introduction

$$\frac{a : \mathbf{D} \qquad f(a) : \mathbf{P}(a)}{\varepsilon x.(f(x), a) : \exists x^{\mathbf{D}}.\mathbf{P}(x)}$$

When constructing an 'ε'-term, we make an 'inverse' substitution: we replace all the occurrences of 'a' in '$f(a)$' by a new variable 'x' which is bound by the '$\varepsilon x.$'-`constructor`.

4.2 Normalization

Within the functional interpretation the operators forming the canonical proofs are usually referred to as the '`constructors`' (e.g. 'λ', '$\langle , \rangle$', '`inl/inr`', 'Λ', 'ε'), whereas the eliminatory operators which form the non-canonical proofs are referred to as the '`DESTRUCTORS`' [28].

To recall our previous discussion, we know that, according to the intuitionistic semantics of canonical proofs, a proposition is characterized by the explanation of the conditions under which one can prove it. The procedure of exhibiting the canonical elements of a type (the canonical proofs of a proposition), which gets formalised by the *introduction* rules, is the key semantical device for the intuitionistic account of meaning. The functional interpretation, however, which accounts for the match between the functional calculus on the labels and the logical calculus on the formulae, does not have to abide by the Heyting-like account of meaning. One can, for

[18]Note that in our formulation, where the domain over which one is quantifying is explicitly stated, the *introduction* of the universal quantifier does require the discharge of an assumption, namely the assumption which indicates a choice of an arbitrary individual from the domain. This account of the Universal Generalization does not run into the difficulties related to the classification of *∀-introduction* either as proper or as improper inference rule, because, similarly to the case of *→-introduction*, this rule requires the discharge of an assumption and would therefore have to be classified as an improper inference rule if one were to use Prawitz' [108] terminology. Cf. Fine's remark on such difficulties concerning the classification of *∀-introduction*:

"Some of the rules require the discharge of suppositions and so have to be classified as improper. Others are so obviously proper that it seems absurd to classify them in any other way. The only real choice concerns universal generalization (∀I); this requires no discharge of suppositions and might, intuitively, be classified as either proper or improper. In considering any proposed account of validity therefore, it must be decided what status this rule is to have." [40, page 72].

example, take the explanation of the convertibility (normalization) relation as the key semantical device, such as it is done in Tait's [117, 118] intensional interpretation of Gödel's [69] T, and the result is an account of the meaning of logical signs which does not rely on intuitionistic principles.[19]

An important step in the characterization of the logical connectives which is not covered by the (language-based) intuitionistic account of meaning based on proofs is the explanation of the functionality of the logical sign in the calculus. In the formal apparatus, this means that the explanation as to how the DESTRUCTORS operate on terms built up by the constructors, i.e. the explanation of what (immediate) consequences one can draw from the corresponding proposition, does not play a major semantical role for the intuitionists.[20] Within the general (not just intuitionistic) functional interpretation, which we believe might have been implicit in Tait's [117, 118] semantics of convertibility, this aspect is given by the so-called β-normalization rules. They have the role of spelling out the effect of an *elimination* inference on the result of *introduction* steps.

4.2.1. *β-Type Reductions*

The explanation of the normalization of non-canonical proofs, i.e. those which contain 'redundant' steps identified by an *introduction* inference immediately followed by an *elimination* inference, are framed in the following way (where '$\twoheadrightarrow_\beta$' represents 'β-converts/normalises to'):

∧-reduction

$$\dfrac{\dfrac{a_1 : \mathbf{A}_1 \qquad a_2 : \mathbf{A}_2}{\langle a_1, a_2\rangle : \mathbf{A}_1 \wedge \mathbf{A}_2}\ \wedge\textit{-intr}}{\mathtt{FST}(\langle a_1, a_2\rangle) : \mathbf{A}_1}\ \wedge\textit{-elim} \qquad \twoheadrightarrow_\beta \qquad a_1 : \mathbf{A}_1$$

$$\dfrac{\dfrac{a_1 : \mathbf{A}_1 \qquad a_2 : \mathbf{A}_2}{\langle a_1, a_2\rangle : \mathbf{A}_1 \wedge \mathbf{A}_2}\ \wedge\textit{-intr}}{\mathtt{SND}(\langle a_1, a_2\rangle) : \mathbf{A}_2}\ \wedge\textit{-elim} \qquad \twoheadrightarrow_\beta \qquad a_2 : \mathbf{A}_2$$

---

[19] When explaining the doctrine that a proposition is the type of its proofs, and suggesting that this might have been implicit in Brouwer's writings, Tait says:

"Now, although I shall not have time to explain my views, I believe that, with certain modifications, this idea [propositions are types of their proofs] provides an account of the meaning of mathematical propositions which is adequate, not only for constructive mathematics, but for classical mathematics as well." [119, page 182]

This is clearly a departure from Heyting's strictly intuitionistic principles, and, as it seems, it comes as no surprise, given that Tait's semantical instrument is *convertibility* (*normalization*), rather than canonical proofs.

[20] In an analysis of the relevant aspects of proof-theoretic semantics [21], we have suggested that the *introduction* rules only cover one aspect, namely the grammatical (formational) aspect: they only say how to construct a proof, leaving untouched the aspect as to how to de-construct (challenge) this same proof.

∨-reduction

$$\dfrac{\dfrac{a_1 : \mathbf{A}_1}{\mathtt{inl}(a_1) : \mathbf{A}_1 \vee \mathbf{A}_2}\vee\textit{-intr} \quad \begin{matrix}[s_1 : \mathbf{A}_1] & [s_2 : \mathbf{A}_2] \\ d(s_1) : \mathbf{C} & e(s_2) : \mathbf{C}\end{matrix}}{\mathtt{CASE}(\mathtt{inl}(a_1), vs_1.d(s_1), vs_2.e(s_2)) : \mathbf{C}}\vee\textit{-elim} \quad \twoheadrightarrow_\beta \quad \begin{matrix}[a_1 : \mathbf{A}_1] \\ d(a_1/s_1) : \mathbf{C}\end{matrix}$$

$$\dfrac{\dfrac{a_2 : \mathbf{A}_2}{\mathtt{inr}(a_2) : \mathbf{A}_1 \vee \mathbf{A}_2}\vee\textit{-intr} \quad \begin{matrix}[s_1 : \mathbf{A}_1] & [s_2 : \mathbf{A}_2] \\ d(s_1) : \mathbf{C} & e(s_2) : \mathbf{C}\end{matrix}}{\mathtt{CASE}(\mathtt{inr}(a_2), vs_1.d(s_1), vs_2.e(s_2)) : \mathbf{C}}\vee\textit{-elim} \quad \twoheadrightarrow_\beta \quad \begin{matrix}[a_2 : \mathbf{A}_2] \\ e(a_2/s_2) : \mathbf{C}\end{matrix}$$

where 'v' is an abstractor which forms value-range terms such as '$vx.d(x)$' where 'x' is bound, discharging the corresponding assumption labelled by x.

→-reduction

$$\dfrac{a : \mathbf{A} \quad \dfrac{\begin{matrix}[x : \mathbf{A}] \\ b(x) : \mathbf{B}\end{matrix}}{\lambda x.b(x) : \mathbf{A} \to \mathbf{B}}\to\textit{-intr}}{\mathtt{APP}(\lambda x.b(x), a) : \mathbf{B}}\to\textit{-elim} \quad \twoheadrightarrow_\beta \quad \begin{matrix}[a : \mathbf{A}] \\ b(a/x) : \mathbf{B}\end{matrix}$$

∀-reduction

$$\dfrac{a : \mathbf{D} \quad \dfrac{\begin{matrix}[x : \mathbf{D}] \\ f(x) : \mathbf{P}(x)\end{matrix}}{\Lambda x.f(x) : \forall x^{\mathbf{D}}.\mathbf{P}(x)}\forall\textit{-intr}}{\mathtt{EXTR}(\Lambda x.f(x), a) : \mathbf{P}(a)}\forall\textit{-elim} \quad \twoheadrightarrow_\beta \quad \begin{matrix}[a : \mathbf{D}] \\ f(a/x) : \mathbf{P}(a)\end{matrix}$$

∃-reduction

$$\dfrac{\dfrac{a : \mathbf{D} \quad f(a) : \mathbf{P}(a)}{\varepsilon x.(f(x), a) : \exists x^{\mathbf{D}}.\mathbf{P}(x)}\exists\textit{-intr} \quad \begin{matrix}[t : \mathbf{D}, g(t) : \mathbf{P}(t)] \\ d(g, t) : \mathbf{C}\end{matrix}}{\mathtt{INST}(\varepsilon x.(f(x), a), \sigma g.\sigma t.d(g, t)) : \mathbf{C}}\exists\textit{-elim} \quad \twoheadrightarrow_\beta$$

$$\begin{matrix}[a : \mathbf{D}, f(a) : \mathbf{P}(a)] \\ d(f/g, a/t) : \mathbf{C}\end{matrix}$$

where 'σ' is an abstractor which binds the free variables of the label, discharging the corresponding assumptions made in eliminating the existential quantifier, namely the 'Skolem'-type assumptions '$[t : \mathbf{D}]$' and '$[g(t) : \mathbf{P}(t)]$', forming the value-range term '$\sigma g.\sigma t.d(g, t)$' where both the Skolem-constant 't', and the Skolem-function 'g', are bound. In the ∃-*elimination* the variables 't' and 'g' must occur free at least once in the term alongside the formula '$\mathbf{C}$' in the premise, and will be bound alongside the same formula in the conclusion of the rule.

It is useful to compare our definition of the β-normalization rules with the original definitions given by Prawitz [108, 109] for plain natural deduction systems. Whereas in the latter there was the need to refer to whole branches of deductions (which in Prawitz' terminology were referred to as $\Pi_1(a)$, $\Sigma(t)$, $\Sigma_2(a)$, etc.), here we only need to refer to assumptions, premisses and conclusions. The relevant information on the dependency of premisses from names (variables, constants, etc.) occurring in the assumptions are to be recorded in the label alongside the formula in the respective premise by whatever proof step(s) eventually made from assumptions to premisses. It would seem fair to say that this constitutes an improvement on the formal presentation of proof reductions, reflecting the (re-)gain of local control by the use of labels.[21]

4.2.2. *β-Equality*

By using equality to represent the β-convertibility ('$\twoheadrightarrow_\beta$') relation between terms we can present the *reductions* in the following way:[22]

∧-β-equality

$$\texttt{FST}(\langle a_1, a_2\rangle) =_\beta a_1 \qquad\qquad \texttt{SND}(\langle a_1, a_2\rangle) =_\beta a_2$$

∨-β-equality

$$\begin{array}{l}\texttt{CASE}(\texttt{inl}(a_1), vs_1.d(s_1), vs_2.e(s_2)) =_\beta d(a_1/s_1)\\ \texttt{CASE}(\texttt{inr}(a_2), vs_1.d(s_1), vs_2.e(s_2)) =_\beta e(a_2/s_2)\end{array}$$

[21] Obviously, the first steps towards such improvement was already made by Martin-Löf [97, 99] in the definition of an intuitionistic theory of types, but here we want to see it applicable to a wide range of logics.

[22] The reader may find it unusual that we are here indexing the (definitional) equality with its kind (β, η, ξ, ζ, etc.). But we shall demonstrate that it makes sense in the context of the functional interpretation to classify (and name) the equalities: one has distinct equalities according to the distinct logical equivalences on the deductions. For example, in the presentation of a set of proof rules for a certain logical connective, the second *introduction* rule is meant to show when two canonical proofs are to be taken as equal, so it is concerned with ξ-equality. The *reduction* rule shows how non-canonical expressions can be brought to normal form, so it is concerned with β-equality. Finally, the *induction* rule shows that by performing an introduction step right after an elimination inference, one gets back to the original proof (and corresponding term), thus it concerns η-equality.

As it will be pointed out later on, it is important to identify the kind of definitional equality, as well as to have a logical connective of 'propositional equality' in order to be able to reason about the functional objects (those to the left-hand side of the ':' sign). The connective will have an 'existential' flavour: two referents are verified to be equal if there exists a reason (composition of rewrites) for asserting it. For example, one might wish to prove that for any two functional objects of $\rightarrow$-type, if they are equal then their application to all objects of the domain type must result in equal objects of the co-domain type.

→-β-equality

$$\mathtt{APP}(\lambda x.b(x), a) =_\beta b(a/x)$$

∀-β-equality

$$\mathtt{EXTR}(\Lambda x.f(x), a) =_\beta f(a/x)$$

∃-β-equality

$$\mathtt{INST}(\varepsilon x.(f(x), a), \sigma g.\sigma t.d(g,t)) =_\beta d(f/g, a/t)$$

Remark. Here it is useful to think in terms of '`DESTRUCTORS` acting on `constructors`', especially in connection with the fact that a proof containing an *introduction* inference followed by an *elimination* step is only β-normalizable at that point if the *elimination* has as *major* premise the formula produced by the previous *introduction* step. For example, as remarked by Girard *et al.* [67],[23] despite involving an *→-introduction* immediately followed by an *→-elimination*, the following proof fragment is not β-normalizable:

$$\dfrac{\dfrac{\dfrac{[x : \mathbf{A}]}{b(x) : \mathbf{B}}}{\lambda x.b(x) : \mathbf{A} \to \mathbf{B}} \to\textit{-intr} \qquad c : (\mathbf{A} \to \mathbf{B}) \to \mathbf{D}}{\mathtt{APP}(c, \lambda x.b(x)) : \mathbf{D}} \to\textit{-elim}$$

Here the major premise of the *elimination* step is not the same formula as the one produced by the *introduction* inference. Moreover, it is clear that the `DESTRUCTOR` '`APP`' is not acting on the term built up with the `constructor` '$\lambda x.b(x)$' in the previous step, but it is operating on an unanalysed term 'c'.

4.2.3. *η-Type Reductions*

In a natural deduction proof system there is another way of making 'redundant' steps that one can make, apart from the above '*introduction* followed by *elimination*'.[24] It is the exact inverse of this previous way of introducing redundancies: an *elimination* step is followed by an *introduction* step. As it turns out, the convertibility relation will be revealing another aspect of the 'propositions-are-types' paradigm, namely that there are redundant steps which from the point of view of the definition/presentation of propositions/types are saying that given any arbitrary proof/element from the proposition/type, it must be of the form given by the *introduction* rules. In

[23] Chapter **Sums in Natural Deduction**, Section **Standard Conversions**.

[24] Some standard texts in proof theory, such as Prawitz' [109] classic survey, have referred to this proof transformation as 'expansions'. Here we are referring to those proof transformations as *reductions*, given that our main measuring instrument is the label, and indeed the label is reduced.

other words, it must satisfy the '*introduction* followed by an *elimination*' convertibility relation. In the typed λ-calculus literature, this 'inductive' convertibility relation has been referred to as 'η'-convertibility.[25]

The '$\twoheadrightarrow_\eta$'-convertibility relation then defines the *induction* rules:

$\wedge$-*induction*

$$\dfrac{\dfrac{c : \mathbf{A}_1 \wedge \mathbf{A}_2}{\mathtt{FST}(c) : \mathbf{A}_1}\wedge\textit{-elim} \qquad \dfrac{c : \mathbf{A}_1 \wedge \mathbf{A}_2}{\mathtt{SND}(c) : \mathbf{A}_2}\wedge\textit{-elim}}{\langle \mathtt{FST}(c), \mathtt{SND}(c)\rangle : \mathbf{A}_1 \wedge \mathbf{A}_2}\wedge\textit{-intr} \twoheadrightarrow_\eta \qquad c : \mathbf{A}_1 \wedge \mathbf{A}_2$$

$\vee$-*induction*

$$\dfrac{c : \mathbf{A}_1 \vee \mathbf{A}_2 \quad \dfrac{[a_1 : \mathbf{A}_1]}{\mathtt{inl}(a_1) : \mathbf{A}_1 \vee \mathbf{A}_2}\vee\textit{-intr} \quad \dfrac{[a_2 : \mathbf{A}_2]}{\mathtt{inr}(a_2) : \mathbf{A}_1 \vee \mathbf{A}_2}\vee\textit{-intr}}{\mathtt{CASE}(c, \upsilon a_1.\mathtt{inl}(a_1), \upsilon a_2.\mathtt{inr}(a_2)) : \mathbf{A}_1 \vee \mathbf{A}_2}\vee\textit{-elim}$$

$$\twoheadrightarrow_\eta \qquad c : \mathbf{A}_1 \vee \mathbf{A}_2$$

---

[25]The classification of those η-conversion rules as inductive rules was introduced by the methodology of defining types used in our reformulated Type Theory described in [28], and first presented publicly in [30]. It seems to have helped to give a 'logical' status which they were given previously in the literature.

In a discussion about the Curry–Howard isomorphism and its denotational significance, Girard *et al.* say:

"Denotationally, we have the following (*primary*) equations

$$\pi^1\langle u, v\rangle = u \qquad \pi^2\langle u, v\rangle = v \qquad (\lambda x^U.v)u = v[u/x]$$

together with the *secondary* equations

$$\langle \pi^1 t, \pi^2 t\rangle = t \qquad \lambda x^U.tx = t \quad (x \text{ not free in } t)$$

which have never been given adequate status." [67, page 16]

Cf. also:

"Let us note for the record the analogues of $\langle \pi^1 t, \pi^2 t\rangle \rightsquigarrow t$ and $\lambda x.tx \rightsquigarrow t$:

$$\varepsilon_{\mathsf{Emp}} t \rightsquigarrow t \qquad \delta x.(\iota^1 x)y.(\iota^2 y)t \rightsquigarrow t$$

Clearly the terms on both sides of the '$\rightsquigarrow$' are denotationally equal." (Ibid., p. 81.)

Here 'δ' is used instead of 'CASE', and 'ι^1'/'ι^2' are used instead of 'inl'/'inr' resp.

Later on, when discussing the coherence semantics of the lifted sum, a reference is made to a rule which we here interpret as the induction rule for $\vee$-types, no mention being made of the role such 'equation' is to play in the proof calculus:

"Even if we are unsure how to use it, the equation

$$\delta x.(\iota^1 x)y.(\iota^2 y)t = t$$

plays a part in the implicit symmetries of the disjunction." (Ibid., p. 97.)

By demonstrating that these kind of conversion rules have the role of guaranteeing minimality for the non-inductive types such as the logical connectives (not just $\rightarrow$, $\wedge$, $\vee$, but also $\forall$, $\exists$) characterized by types, we believe we have given them adequate status. (That is to say: the rules of η-reduction state that any proof of $\mathbf{A} \rightarrow \mathbf{B}$, $\mathbf{A} \wedge \mathbf{B}$, $\mathbf{A} \vee \mathbf{B}$, will have in each case a unique form, namely $\lambda x.y$, $\langle a, b\rangle$, $\mathtt{inl}(a)/\mathtt{inr}(b)$, resp.)

→-induction

$$\dfrac{\dfrac{[x:\mathbf{A}] \qquad c:\mathbf{A}\to\mathbf{B}}{\texttt{APP}(c,x):\mathbf{B}}\to\textit{-elim}}{\lambda x.\texttt{APP}(c,x):\mathbf{A}\to\mathbf{B}}\to\textit{-intr} \twoheadrightarrow_{\eta} \qquad c:\mathbf{A}\to\mathbf{B}$$

where c does not depend on x.

∀-induction

$$\dfrac{\dfrac{[t:\mathbf{D}] \qquad c:\forall x^{\mathbf{D}}.\mathbf{P}(x)}{\texttt{EXTR}(c,t):\mathbf{P}(t)}\forall\textit{-elim}}{\Lambda t.\texttt{EXTR}(c,t):\forall t^{\mathbf{D}}.\mathbf{P}(t)}\forall\textit{-intr} \twoheadrightarrow_{\eta} \qquad c:\forall x^{\mathbf{D}}.\mathbf{P}(x)$$

where x does not occur free in c.

∃-induction

$$\dfrac{c:\exists x^{\mathbf{D}}.\mathbf{P}(x) \qquad \dfrac{[t:\mathbf{D}] \qquad [g(t):\mathbf{P}(t)]}{\varepsilon y.(g(y),t):\exists y^{\mathbf{D}}.\mathbf{P}(y)}\exists\textit{-intr}}{\texttt{INST}(c,\sigma g.\sigma t.\varepsilon y.(g(y),t)):\exists y^{\mathbf{D}}.\mathbf{P}(y)}\exists\textit{-elim} \quad \twoheadrightarrow_{\eta} \quad c:\exists x^{\mathbf{D}}.\mathbf{P}(x)$$

In the terminology of Prawitz' [109], these rules (with the conversion going from right to left) are called *immediate expansions*. Notice, however, that whilst in the latter the purpose was to bring a derivation in full normal form to expanded normal form where all the minima formulae are atomic, here we are still speaking in terms of *reductions*: the large terms alongside the formulae resulting from the derivation on the left are reduced to the smaller terms alongside the formula on the right. Moreover, the benefit of this change of emphasis is worth pointing out here: whereas in Prawitz' plain natural deduction the principal measure is the degree of formulae (i.e. minimal formulae, etc.) here the labels (or proof constructions) take over the main role of measuring instrument. The immediate consequence of this shift of emphasis is the replacement of the notion of *subformula* by that of *subdeduction*, which not only avoids the complications of proving the subformula property for logics with 'Skolem-type' connectives (i.e. those connectives whose *elimination* rules may violate the subformula property of a deduction, such as $\vee$, $\exists$, $\doteq$), but it also seems to retake Gentzen's analysis of deduction in its more general sense. That is to say, the emphasis is put back into the deductive properties of the logical connectives, rather than on the truth of the constituent formulae.

Remark. Notice that the mere condition of '*elimination* followed by *introduction*' is not sufficient to allow us to perform an η-conversion. We still need to take into consideration what subdeductions we are dealing with. For example, in:

$$\cfrac{c : \mathbf{A} \vee \mathbf{A} \quad \cfrac{[x : \mathbf{A}]}{\mathtt{inl}(x) : \mathbf{A} \vee \mathbf{B}} \vee\text{-}intr \quad \cfrac{[y : \mathbf{A}]}{\mathtt{inl}(y) : \mathbf{A} \vee \mathbf{B}} \vee\text{-}intr}{\mathtt{CASE}(c, \upsilon x.\mathtt{inl}(x), \upsilon y.\mathtt{inl}(y)) : \mathbf{A} \vee \mathbf{B}} \vee\text{-}elim \quad \not\rhd_\eta$$

$$c : \mathbf{A} \vee \mathbf{A}$$

we have a case where an $\vee$-*elimination* is immediately followed by an $\vee$-*introduction*, and yet we are not prepared to accept the proof transformation under η-conversion. Now, if we analyse the subdeductions (via the labels), we observe that

$$\mathtt{CASE}(c, \upsilon x.\mathtt{inl}(x), \upsilon y.\mathtt{inl}(y)) \neq_\eta c$$

therefore, if the harmony between the functional calculus on the labels and the logical calculus on the formulae is to be maintained, we have good enough reasons to reject the unwanted proof transformation.

4.2.4. *η-Equality*

In terms of rewriting systems where '=' is used to represent the 'reduces to' relation, indexed by its kind, i.e. β-, η-, ξ-, ζ-, etc. conversion, the above *induction* rules become:

$\wedge$-*η-equality*

$$\langle \mathtt{FST}(c), \mathtt{SND}(c) \rangle =_\eta c$$

$\vee$-*η-equality*

$$\mathtt{CASE}(c, \upsilon a_1.\mathtt{inl}(a_1), \upsilon a_2.\mathtt{inr}(a_2)) =_\eta c$$

$\rightarrow$-*η-equality*

$$\lambda x.\mathtt{APP}(c, x) =_\eta c$$

provided x does not occur free in c.

$\forall$-*η-equality*

$$\Lambda t.\mathtt{EXTR}(c, t) =_\eta c$$

provided c has no free occurrences of x.

$\exists$-*η-equality*

$$\mathtt{INST}(c, \sigma g.\sigma t.\varepsilon y.(g(y), t)) =_\eta c$$

The presentation taken by each of the rules above does indeed reveal an 'inductive' character: they all seem to be saying that if any arbitrary element 'c' is in the type then it must be reducible to itself via an *elimination* step with the DESTRUCTOR(s) followed by an *introduction* step with the constructor(s).

4.2.5. *ζ-Type Reductions: the permutative Reductions*
For the connectives that make use of 'Skolem'-type procedures of opening local branches with new assumptions, locally introducing new names and making them 'disappear' (or loose their identity via an abstraction) just before coming out of the local context or scope, there is another way of transforming proofs, which goes hand-in-hand with the properties of 'value-range' terms resulting from abstractions.

In natural deduction terminology, these proof transformations are called 'permutative' reductions.

∨-*(permutative) reduction*

$$\dfrac{\dfrac{p:\mathbf{A}_1\vee\mathbf{A}_2 \quad \begin{matrix}[s_1:\mathbf{A}_1]\\ d(s_1):\mathbf{C}\end{matrix} \quad \begin{matrix}[s_2:\mathbf{A}_2]\\ e(s_2):\mathbf{C}\end{matrix}}{\mathtt{CASE}(p, vs_1.d(s_1), vs_2.e(s_2)):\mathbf{C}}}{w(\mathtt{CASE}(p, vs_1.d(s_1), vs_2.e(s_2))):\mathbf{W}} \twoheadrightarrow_\zeta$$

$$\dfrac{p:\mathbf{A}_1\vee\mathbf{A}_2 \quad \dfrac{\begin{matrix}[s_1:\mathbf{A}_1]\\ d(s_1):\mathbf{C}\end{matrix}}{w(d(s_1)):\mathbf{W}} \quad \dfrac{\begin{matrix}[s_2:\mathbf{A}_2]\\ e(s_2):\mathbf{C}\end{matrix}}{w(e(s_2)):\mathbf{W}}}{\mathtt{CASE}(p, vs_1.w(d(s_1)), vs_2.w(e(s_2))):\mathbf{W}}$$

∃-*(permutative) reduction*

$$\dfrac{\dfrac{e:\exists x^{\mathbf{D}}.\mathbf{P}(x) \quad \begin{matrix}[t:\mathbf{D}, g(t):\mathbf{P}(t)]\\ d(g,t):\mathbf{C}\end{matrix}}{\mathtt{INST}(e, \sigma g.\sigma t.d(g,t)):\mathbf{C}}}{w(\mathtt{INST}(e, \sigma g.\sigma t.d(g,t))):\mathbf{W}} \twoheadrightarrow_\zeta \dfrac{e:\exists x^{\mathbf{D}}.\mathbf{P}(x) \quad \dfrac{[t:\mathbf{D}, g(t):\mathbf{P}(t)]}{\begin{matrix}d(g,t):\mathbf{C}\\ w(d(g,t)):\mathbf{W}\end{matrix}}}{\mathtt{INST}(e, \sigma g.\sigma t.w(d(g,t))):\mathbf{W}}$$

4.2.6. *ζ-Equality*
Now, if the functional calculus on the labels is to match the logical calculus on the formulae, we must have the following ζ-equality (read 'zeta'-equality) between terms:

$$w(\mathtt{CASE}(p, vs_1.d(s_1), vs_2.e(s_2)), u) =_\zeta \mathtt{CASE}(p, vs_1.w(d(s_1), u), vs_2.w(e(s_2), u))$$ [26]

[26]When defining 'linearized sum', Girard *et al.* [67] give the following equation as the term-equality counterpart to the permutative reduction:
"Finally, the commuting conversions are of the form

$$\mathsf{E}(\delta\ x.u\ y.v\ t) \rightsquigarrow \delta\ x.(\mathsf{E}u)\ y.(\mathsf{E}v)\ t$$

where E is an elimination." [67, page 103]
Note the restriction on the step corresponding to the operator 'E' (which corresponds to our '*w*'): it has to be an elimination.
In our ζ-equality the operator '*w*' does not have to be an eliminatory operator, but it only needs to be such that it preserves the dependencies of the term coming from the

for disjunction, and

$$w(\texttt{INST}(e, \sigma g.\sigma t.d(g,t)), u) =_\zeta \texttt{INST}(e, \sigma g.\sigma t.w(d(g,t), u))$$

for the existential quantifier.

Note that both in the case of '$\vee$' and '$\exists$' the operator 'w' could be 'pushed inside' the value-range abstraction terms. In the case of disjunction, the operator could be pushed inside the 'υ'-abstraction terms, and in the $\exists$-case, the 'w' could be pushed inside the 'σ'-abstraction term.

In terms of the proof theory, these reductions are stating that the newly opened branches must be independent from the main branch. And, indeed, notice that in the proof-trees above, the step coming after the *elimination* of the connective concerned ($\vee$-, $\exists$-*elimination*) is taken to be as general as possible, provided that it does not affect the dependencies on the main branch (i.e. '$p : \mathbf{A}_1 \vee \mathbf{A}_2$', '$e : \exists x^{\mathbf{D}}.\mathbf{P}(x)$', respectively). (E.g. any deduction step involving discharge of assumptions may disturb the dependencies.) Those reductions will then uncover β-type redundancies which may be hidden by an $\vee$-, $\exists$- *elimination* rule. Perhaps for this reason, in the literature it is common to restrict that particular step to a deduction to an *elimination* rule where the formula '$\mathbf{C}$' is to be its major premise.[27]

The restriction to the case when the step is an *elimination* rule seems to be connected with the idea that the *permutative* conversions are brought

main branch, namely the step must preserve the free variables on which our 'p' depends. In other words, w cannot be an abstraction over free variables of p.

Our generalized ζ-equality also finds parallels in the recent literature on equational counterparts to commutative diagrams of category theory. For example, in the definition of *binary sums* given by A. Poigné [107], the counterpart of our ζ-equality for disjunction appears as:

$$h \circ case(f, g) = case(h \circ f, h \circ g)$$

where '$\circ$' is the basic operation of composition. Note that the function 'h' can be pushed inside the '*case*'-term, similarly to our ζ-equality where the 'w' can be pushed inside the υ-abstraction terms of our CASE-expression.

[27]When commenting on the requirements of permutative reductions, Prawitz remarks:

"It has been remarked by Martin-Löf that it is only necessary to require in the $\vee$E- and $\exists$E-reductions that the lowest occurrence of C is the major premiss of an elimination. A reduction of this kind can then always be carried out and we can sharpen the requirements as to the normal form accordingly." [109, pages 253ff]

And, indeed, for his proof of the Strong Validity Lemma (p. 295) Prawitz needs the condition on the permutative reductions that the step after the $\vee$-*elimination* (resp. $\exists$-*elimination*) be also an *elimination* inference.

No restriction to an *elimination* step is mentioned in [98]. Rather, it is required that the dependencies be preserved:

"(...) the *permutative* rules for $\vee$ and $\exists$, (...) provided the inference from C to D neither binds any free variable nor discharges any assumption in the derivation of $A \vee B$ and $(\exists x)B[x]$, respectively." [98, pages 100f]

Cf. also other standard texts in the literature where the restriction is unnecessarily imposed: Troelstra and van Dalen's [122, pages 534ff] and Girard *et al.*'s [67] definitions of *permutative conversions* have the requirement that the step following the $\vee$-($\exists$-)*elimination* be an 'E-rule' (*Elimination* rule).

in to help recover the so-called *subformula property*.[28] We would prefer to see the role of those rules of proof transformation as that of guaranteeing a 'pact of non-interference' between the main branch and those new branches created by the elimination rules of 'Skolem-type' connectives ($\vee$, $\exists$, $\doteq$).

Thus, in the more general case, it seems as though the restriction (to the case where the formula **C** is a major premise of an *elimination* inference) is unnecessary. And this is because we can have the following conversion using an *introduction* inference instead:

$$\dfrac{\dfrac{p : \mathbf{A}_1 \vee \mathbf{A}_2 \qquad \begin{matrix}[s_1 : \mathbf{A}_1]\\ d(s_1) : \mathbf{C}\end{matrix} \qquad \begin{matrix}[s_2 : \mathbf{A}_2]\\ e(s_2) : \mathbf{C}\end{matrix}}{\texttt{CASE}(p, \upsilon s_1.d(s_1), \upsilon s_2.e(s_2)) : \mathbf{C}}}{\texttt{inl}(\texttt{CASE}(p, \upsilon s_1.d(s_1), \upsilon s_2.e(s_2))) : \mathbf{C} \vee \mathbf{U}}(*) \qquad \twoheadrightarrow_\zeta$$

$$\dfrac{p : \mathbf{A}_1 \vee \mathbf{A}_2 \qquad \dfrac{\begin{matrix}[s_1 : \mathbf{A}_1]\\ d(s_1) : \mathbf{C}\end{matrix}}{\texttt{inl}(d(s_1)) : \mathbf{C} \vee \mathbf{U}} \qquad \dfrac{\begin{matrix}[s_2 : \mathbf{A}_2]\\ e(s_2) : \mathbf{C}\end{matrix}}{\texttt{inl}(e(s_2)) : \mathbf{C} \vee \mathbf{U}}}{\texttt{CASE}(p, \upsilon s_1.\texttt{inl}(d(s_1)), \upsilon s_2.\texttt{inl}(e(s_2))) : \mathbf{C} \vee \mathbf{U}}$$

One can readily notice that the $\vee$-*introduction* step marked '($*$)' does not affect the dependencies (i.e. does not involve any assumption discharge), so the constructor '`inl`' can be pushed inside the υ-abstraction terms. The same holds if, instead of $\vee$-*introduction*, one performs an $\wedge$-*introduction* as in:

$$\dfrac{\dfrac{p : \mathbf{A}_1 \vee \mathbf{A}_2 \qquad \begin{matrix}[s_1 : \mathbf{A}_1]\\ d(s_1) : \mathbf{C}\end{matrix} \qquad \begin{matrix}[s_2 : \mathbf{A}_2]\\ e(s_2) : \mathbf{C}\end{matrix}}{\texttt{CASE}(p, \upsilon s_1.d(s_1), \upsilon s_2.e(s_2)) : \mathbf{C}} \qquad u : \mathbf{U}}{\langle \texttt{CASE}(p, \upsilon s_1.d(s_1), \upsilon s_2.e(s_2)), u\rangle : \mathbf{C} \wedge \mathbf{U}} \qquad \twoheadrightarrow_\zeta$$

$$\dfrac{p : \mathbf{A}_1 \vee \mathbf{A}_2 \qquad \dfrac{\begin{matrix}[s_1 : \mathbf{A}_1]\\ d(s_1) : \mathbf{C}\end{matrix} \qquad u : \mathbf{U}}{\langle d(s_1), u\rangle : \mathbf{C} \wedge \mathbf{U}} \qquad \dfrac{\begin{matrix}[s_2 : \mathbf{A}_2]\\ e(s_2) : \mathbf{C}\end{matrix} \qquad u : \mathbf{U}}{\langle e(s_2), u\rangle : \mathbf{C} \wedge \mathbf{U}}}{\texttt{CASE}(p, \upsilon s_1.\langle d(s_1), u\rangle, \upsilon s_2.\langle e(s_2), u\rangle) : \mathbf{C} \wedge \mathbf{U}}$$

and, clearly:

$$\langle \texttt{CASE}(p, \upsilon s_1.d(s_1), \upsilon s_2.e(s_2)), u\rangle =_\zeta \texttt{CASE}(p, \upsilon s_1.\langle d(s_1), u\rangle, \upsilon s_2.\langle e(s_2), u\rangle)$$

given that the pairing operation can be pushed inside the υ-abstraction terms without disturbing the dependencies. One can readily see that the $\wedge$-*introduction* is harmless with respect to the dependencies. (Note that the same observation applies to ζ-reduction of $\exists$, and, as we shall see later on, to the permutative reduction of $\doteq$.)

[28] Cf. [67, 122].

THEOREM 4.1 ([14]) *In the* LND *system every proof has a unique normal form.*[29]

5 Propositional Equality

One aspect of the functional interpretation (*à la* Curry–Howard) which stirred much discussion in the literature of the past ten years or so, especially in connection with *Intuitionistic Type Theory* [101], was that of whether the logical connective of propositional equality ought to be dealt with 'extensionally' or 'intensionally'. Here we attempt to formulate what appears to be a middle ground solution, in the sense that the intensional aspect is dealt with in the functional calculus on the labels, whereas the extensionality is kept to the logical calculus on the formulae. We also intend to demonstrate that the connective of propositional equality (cf. Aczel's [4] '$\doteq$') needs to be dealt with in a similar manner to 'Skolem-type' connectives (such as disjunction and existential quantification), where the notion of *hiding* plays a crucial role.

5.1 A Proof Theory for Equality in Perspective

The intention here is to show how the framework of labelled natural deduction can help us formulate a proof theory for the logical connective of propositional equality. The connective is meant to be used in reasoning about equality between referents (i.e. the objects of the functional calculus), as well as with a general notion of substitution which is needed for the characterization of the so-called *term declaration logics* [5]. The characterization of propositional equality shall be useful for the establishment of a proof theory for 'descriptions'.

In order to account for the distinction between the equalities that are:

- *definitional*, i.e. those equalities that are given as rewrite rules (equations), or else originate from general functional principles (e.g. β, η, etc.),

and those that are:

- *propositional*, i.e. the equalities that are supported (or otherwise) by an evidence (a composition of rewrites),

we need to provide for an equality sign as a symbol for *rewrite* (i.e. as part of the functional calculus on the labels), and an equality sign as a symbol for a *relation* between referents (i.e. as part of the logical calculus on the formulae).

---

[29] In fact, when we have disjunction and η-rules, uniqueness is not guaranteed. It is necessary to add new transformation rules baptized in [15, 14] as 'ι' ('iota') rules.

Within the framework of the functional interpretation (*à la* Curry–Howard), the definitional equality is often considered by reference to a judgement of the form:

$$a = b : \mathbf{D}$$

which says that a and b are equal elements from domain $\mathbf{D}$. Notice that the 'reason' why they are equal does not play any part in the judgement. This aspect of 'forgetting contextual information' is, one might say, the first step towards 'extensionality' of equality, for whenever one wants to introduce intensionality into a logical system one invariably needs to introduce information of a 'contextual' nature, such as, where the identification of two terms (i.e. equation) comes from.

We feel that a first step towards finding an alternative formulation of the proof theory for propositional equality which takes care of the intensional aspect is to allow the 'reason' for the equality to play a more significant part in the form of judgement. We also believe that from the point of view of the logical calculus, if there is a 'reason' for two expressions to be considered equal, the proposition asserting their equality will be true, regardless of what particular composition of rewrites (definitional equalities) amounts to the evidence in support of the proposition concerned. Given these general guidelines, we shall provide what may be seen as a middle ground solution between the intensional [99] and the extensional [100] accounts of Martin-Löf's propositional equality. The intensionality is taken care by the functional calculus on the labels, while the extensionality is catered by the logical calculus on the formulae. In order to account for the intensionality in the labels, we shall make the composition of rewrites (definitional equalities) appear as indexes of the equality sign in the judgement with a variable denoting a sequence of equality constants (we have seen that in the Curry–Howard functional interpretation there are at least three 'natural' equality constants: β, η and ξ). So, instead of the form above, we shall have the following pattern for the equality judgement:

$$a =_s b : \mathbf{D}$$

where 's' is meant to be a sequence of equality constants. As an example, if a is from domain $\mathbf{D}$, and $\lambda x.f(x)$ is from domain $\mathbf{D} \to \mathbf{C}$, then the rule of β-conversion of the λ-calculus will be written as:

$$\mathtt{APP}(\lambda x.f(x), a) =_\beta f(a/x) : \mathbf{C}$$

and the rule of η-conversion will look like:

$$\lambda y.\mathtt{APP}(c, y) =_\eta c : \mathbf{D} \to \mathbf{C}$$

where 'c' is of type $\mathbf{D} \to \mathbf{C}$ and 'y' is from domain $\mathbf{D}$.

The reader may find it unusual that we are here indexing the equality with its kind (β, η, ξ, ζ, etc.). But we shall demonstrate that it makes sense in the context of the functional interpretation to do this: one has distinct equalities according to the distinct logical equivalences on the deductions. For example, when defining the proof theory for a particular logical connective we need an *introduction* rule which shows when two canonical proofs are meant to be equal, which will be concerned with ξ-equality. The *reduction* rule shows how non-canonical expressions can be brought to normal form, so it is concerned with β-equality. Finally, the *induction* rule shows that by performing an introduction step right after an elimination one gets back to the original proof (and corresponding term), thus it concerns η-equality. With the indexes for equality those rules for implication would look like:

$$\frac{\begin{array}{c}[x:\mathbf{A}]\\ b(x) =_s d(x) : \mathbf{B}\end{array}}{\lambda x.b(x) =_{\xi(s)} \lambda x.d(x) : \mathbf{A} \to \mathbf{B}} \qquad \frac{a:\mathbf{A} \qquad \begin{array}{c}[x:\mathbf{A}]\\ b(x):\mathbf{B}\end{array}}{\mathsf{APP}(\lambda x.b(x), a) =_\beta b(a/x) : \mathbf{B}}$$

$$\frac{c : \mathbf{A} \to \mathbf{B}}{\lambda x.\mathsf{APP}(c, x) =_\eta c : \mathbf{A} \to \mathbf{B}}$$

where in the first case 's' could be any (sequence of) equality and the '( )' stands for the concatenation of 's' with 'ξ' (and each equality step is recorded, in line with the 'keeping track of proof steps' motto of the functional interpretation adopted here).

In the sequel we shall be discussing in some detail the need to identify the kind of definitional equality, as well as the need to have a logical connective of 'propositional equality' in order to be able to reason about the functional objects (those to the left-hand side of the ':' sign). For example, one might wish to prove that for any two functional objects of $\to$-type, if they are equal then their application to all objects of the domain type must result in equal objects of the co-domain type.

5.2 Definitional Equality

Equality in mathematics is more complex to deal with formally than it might first appear. Prior to any attempt at a formalization, one needs to consider the conceptual framework which is more appropriate. We have found Frank Ramsey's [112] 'The Foundations of Mathematics' particularly attractive, especially in what concerns the defence of the idea that many equalities in mathematics are simply definitional (yet not necessarily abbreviatory) equalities. So, in the functional interpretation we have β-equality (*reduction* rules), η-equality (*induction* rules) and ξ-equality (second *introduction* rule, stating when two canonical elements are equal), ζ-equality (*permutative conversions* for Skolem-type connectives) and all are 'definitional' equalities, even if not abbreviatory equalities.[30]

Clearly, there is a need for a careful analysis of equality in its various facets, especially in connection with the formalization of mathematics in the lines of Frege's logic. Recall, for example, that ξ-equality is the formal counterpart to Bishop's [9] constructive principle of definition of a set that says that not only the canonical elements must be shown, but also the condition under which two canonical elements are equal.

The methodological point about defining sets by not only showing how to build its elements, but also by saying when two elements are to be taken as equal, would seem partly to find a counterpart in Frege's [48] idea that for an expression to be a *logical object* it has to have a criterion for identity. Only, there seems to be more to the establishment of a minimum criterion for identity, as there are other kinds of equality than ξ-equality, which is to say that equality among non-canonical and canonical elements (the conversions) also plays an important role in the definition of a *logical* object. Here we are not taking a type to be identical to a set, as it is done in some accounts of constructive mathematics [9, 101], but we are trying to follow the general principles set out by Frege in the sense that the establishment of criteria for identity among the so-called *logical* objects ought to be of prime concern in formal logic.

5.2.1. *Reckonable (Inspectable) Terms*

The notion of '*definitionally equal* reckonable terms' was introduced in Gödel's [69] *Dialectica* interpretation of intuitionistic arithmetic via the system T of of functionals of finite type. Although the notion of *reckonable* (*inspectable*) terms is thoroughly discussed, there is little clarification concerning on what should be understood by equality among two *inspectable* terms. In his intensional interpretation of Gödel's system, Tait [118] defines the notion of definitional equality *relative* to a definite collection of

[30]In his paper on the notion of definitional equality Martin-Löf [98] considers only β-type definitional equalities. But most importantly, he fails to distinguish the two kinds of definitional equality, namely:

1. equality as presentation (which is the one we are dealing with here),
2. equality as abbreviation.

The paper [98] advocates the use of β-type equality as synonymous with abbreviation, so that, for example, the β-equality of λ-calculus is to be understood as an abbreviation:

$$\texttt{APP}(\lambda x.f(x), a) \equiv f(a/x).$$

In the approach we take here, despite considering the various convertibility-based equalities (β, η, ξ, ζ) still as *definitional* equalities, we do not take them to be abbreviatory equalities.

operations with their conversion rules.[31]

Starting from a clear separation of our logical system into a functional calculus on the labels and a logical calculus on the formulae, here we shall keep the term *definitional equality* for those equalities given by the conversion rules (i.e. those *immediate* conversions, such as β, η, ξ, etc.), whereas the term *propositional equality* will be left for the counterpart of Tait's definitional equality: a sequence of conversions leading up to an equality of terms (in the functional calculus on the labels) constitutes a support (evidence) for them being considered propositionally equal (in the logical calculus).

Existential force. Observe that in the present formulation the connective of propositional equality embodies an existential force: the truth conditions will be such that, given that there is a sequence of rewrites to support the assertion, the propositional equality among two referents is taken to be true. The implications for the analysis of deduction (proof theory) will be such that the pattern of introduction rules will involve, as in the case of '$\vee$' and '$\exists$', witness hiding, and those of elimination rules will be concerned with the opening of local branches with the introduction of new local variables.

5.2.2. *Equality in Labelled Natural Deduction*

Following our methodology of 'keeping track of proof steps' via a mechanism of labelling formulae with relevant information about the deduction, we take it that the reason for equality (e.g. β, η, ξ, etc.) must be part of

[31] Commenting on the difficulty in taking equality between terms of the same arbitrary 'finite type' (of Gödel's [69] theory T of functionals of finite type) as *extensional* equality, Tait says:

> "The prime formulae of T_i are equations $s = t$ between terms of the same (arbitrary) f.t. The difficulty in interpreting T_i in U_i [intuitionistic number theory with quantification over functionals of f.t. of order $i - 1$] arises from the fact that $s = t \vee \neg s = t$ is an axiom of T_i, even for equations of non-numerical type, so that $=$ cannot be interpreted simply as extensional equality. Gödel's own interpretation of $s = t$ is this: Terms are to denote *reckonable* (*berechenbare*) functionals, where the reckonable functionals of type 0 are the natural numbers, and the reckonable functionals of type (σ, τ) are operations for which we can constructively prove that, when applied to reckonable functionals of type σ, they uniquely yield ones of type τ. $s = t$ means that s and t denote *definitionally equal* reckonable terms. Lacking a general conception of the kinds of definitions by which an operation may be introduced, the notion of definitional equality is not very clear to me. But if, as in the case of T_i, we can regard the operations ϕ as being introduced by conversion rules
>
> $$\phi t_1 \cdots t_n \Rightarrow s(t_1, \cdots, t_n),$$
>
> then definitional equality has a clear meaning: s and t are *definitionally equal* if they reduce to a common term by means of a sequence of applications of the conversion rules. Of course, this notion makes sense only when we have fixed a definite collection of operations with their conversion rules." [118, opening paragraph]

the information carried by the label. As the separation functional-logical is taken seriously in labelled systems, there will be no place for formulae in the label, so the equality sign used in the label will not be indexed by the domain (as in some accounts of intensional equality given as '$a =_{\mathbf{D}} b$').[32]

Obviously, there will be a place for the formula expressing the domain of both elements said to be equal, such as $\mathbf{D}$ in $a =_s b : \mathbf{D}$, in the logical side of calculus. That is to say, the logical connective of propositional equality will be relative to the given domain in much the same way the quantifiers are defined with respect to an explicitly stated domain of quantification. Thus, we would like to go from:

$$a =_s b : \mathbf{D}$$

which is the (functional) rewriting judgement where s stands for a composition of definitional equalities, to:

$$s(a, b) : \dot{=}_{\mathbf{D}} (a, b)$$

which is the (logical) assertion that a and b are propositionally equal due to the evidence '$s(a, b)$', i.e. 'a' and 'b' are identifiable via s.

Note that while the equality belonging to the functional calculus (i.e. the equalities which are to be seen as *rewrites*) is always relative to a particular (sequence of) definitional equalities, the propositional equality (i.e. the equality belonging to the logical calculus) is always relative to a domain of individuals (similarly to the quantifiers).

5.3 Martin-Löf's Equality Type

There have been essentially two approaches to the problem of characterising a proof theory for propositional equality, both of which originate in P. Martin-Löf's work on *Intuitionistic Type Theory*: the intensional [99] and the extensional [100] formulations.

5.3.1. *The Extensional Version*

In his [100] and [101] presentations of *Intuitionistic Type Theory* P. Martin-Löf defines the type of *extensional* propositional equality '**I**' (here called '$\mathbf{I}_{ext}$') as:

$\mathbf{I}_{ext}$-*formation*

$$\frac{\mathbf{D}\ \mathit{type} \qquad a : \mathbf{D} \qquad b : \mathbf{D}}{\mathbf{I}_{ext}(\mathbf{D}, a, b)\ \mathit{type}}$$

[32] Cf. [122, page 593]: "Let $\vdash \Gamma \Rightarrow A$, $A \equiv I(B, s, t)$ and abbreviate $I(N, t, s)$ as $t =_N s$." and [105, page 62]: "Instead of $\mathsf{Id}(A, a, b)$ we will often write $a =_A b$."

$\mathbf{I}_{ext}$-*introduction*

$$\frac{a = b : \mathbf{D}}{\mathbf{r} : \mathbf{I}_{ext}(\mathbf{D}, a, b)}$$

$\mathbf{I}_{ext}$-*elimination*[33]

$$\frac{c : \mathbf{I}_{ext}(\mathbf{D}, a, b)}{a = b : \mathbf{D}}$$

$\mathbf{I}_{ext}$-*equality*

$$\frac{c : \mathbf{I}_{ext}(\mathbf{D}, a, b)}{c = \mathbf{r} : \mathbf{I}_{ext}(\mathbf{D}, a, b)}$$

Note that the above account of propositional equality does not 'keep track of all proof steps': both in the $\mathbf{I}_{ext}$-*introduction* and in the $\mathbf{I}_{ext}$-*elimination* rules there is a considerable loss of information concerning the deduction steps. While in the $\mathbf{I}_{ext}$-*introduction* rule the 'a' and the 'b' do not appear in the 'trace' (the label/term alongside the logical formula), the latter containing only the canonical element '$\mathbf{r}$', in the rule of $\mathbf{I}_{ext}$-*elimination* all the trace that might be recorded in the label 'c' simply disappears from label of the conclusion. If by 'intensionality' we understand a feature of a logical system which identifies as paramount the concern with issues of *context* and *provability*, then it is quite clear that any logical system containing $\mathbf{I}_{ext}$-type can hardly be said to be 'intensional': as we have said above, neither its *introduction* rule nor its *elimination* rule carry the necessary *contextual* information from the premise to the conclusion.

And, indeed, the well-known statement of the extensionality of functions can be proved as a theorem of a logical system containing the $\mathbf{I}_{ext}$-type such as Martin-Löf's [101] *Intuitionistic Type Theory*. The statement says that if two functions return the same value in their common co-domain when applied to each argument of their common domain (i.e. if they are equal pointwise), then they are said to be (extensionally) equal. Now, we can construct a derivation of the statement written in the formal language as:

$$\forall f, g^{\mathbf{A} \to \mathbf{B}}.(\forall x^{\mathbf{A}}.\mathbf{I}_{ext}(\mathbf{B}, \mathtt{APP}(f, x), \mathtt{APP}(g, x)) \to \mathbf{I}_{ext}(\mathbf{A} \to \mathbf{B}, f, g))$$

by using the rules of proof given for the $\mathbf{I}_{ext}$, assuming we have the rules of proof given for the implication and the universal quantifier.

5.3.2. *The Intensional Version*

Another version of the propositional equality, which has its origins in Martin-Löf's [98, 99] early accounts of *Intuitionistic Type Theory*, and is apparently

[33]The set of rules given in [100] contained the additional *elimination* rule:

$$\frac{c : \mathbf{I}(\mathbf{D}, a, b) \qquad d : \mathbf{C}(\mathbf{r}/z)}{\mathtt{J}(c, d) : \mathbf{C}(c/z)}$$

which may be seen as reminiscent of the previous [99] *intensional* account of propositional equality.

in the most recent, as yet unpublished, versions of type theory, is defined in [122] and [105]. In a section dedicated to the *intensional vs. extensional* debate, Troelstra and van Dalen [122, page 633] state that:

> "Martin-Löf has returned to an intensional point of view, as in Martin-Löf (1975), that is to say, $t = t' \in A$ is understood as "t and t' are definitionally equal". As a consequence the rules for identity types have to be adapted."

If we try to combine the existing accounts of the *intensional* equality type '**I**' [99, 122, 105] (here called '$\mathbf{I}_{int}$'), the rules will look like:

$\mathbf{I}_{int}$*-formation*

$$\frac{\mathbf{D}\ \mathit{type} \qquad a : \mathbf{D} \qquad b : \mathbf{D}}{\mathbf{I}_{int}(\mathbf{D}, a, b)\ \mathit{type}}$$

$\mathbf{I}_{int}$*-introduction*

$$\frac{a : \mathbf{D}}{\mathsf{e}(a) : \mathbf{I}_{int}(\mathbf{D}, a, a)} \qquad \frac{a = b : \mathbf{D}}{\mathsf{e}(a) : \mathbf{I}_{int}(\mathbf{D}, a, b)}$$

$\mathbf{I}_{int}$*-elimination*

$$\frac{a : \mathbf{D}\ \ b : \mathbf{D}\ \ c : \mathbf{I}_{int}(\mathbf{D}, a, b) \quad \begin{matrix}[x : \mathbf{D}] \\ d(x) : \mathbf{C}(x, x, \mathsf{e}(x))\end{matrix} \quad \begin{matrix}[x, y : \mathbf{D}, z : \mathbf{I}_{int}(\mathbf{D}, x, y)] \\ \mathbf{C}(x, y, z)\ \mathit{type}\end{matrix}}{\mathsf{J}(c, d) : \mathbf{C}(a, b, c)}$$

$\mathbf{I}_{int}$*-equality*

$$\frac{a : \mathbf{D} \quad \begin{matrix}[x : \mathbf{D}] \\ d(x) : \mathbf{C}(x, x, \mathsf{e}(x))\end{matrix} \quad \begin{matrix}[x : \mathbf{D}, y : \mathbf{D}, z : \mathbf{I}_{int}(\mathbf{D}, x, y)] \\ \mathbf{C}(x, y, z)\ \mathit{type}\end{matrix}}{\mathsf{J}(\mathsf{e}(a), d(x)) = d(a/x) : \mathbf{C}(a, a, \mathsf{e}(a))}$$

With slight differences in notation, the 'adapted' rules for identity type given in [122] and [105] resembles the one given in [99]. It is called *intensional* equality because there remains no direct connection between judgements like $a = b : \mathbf{D}$ and $s : \mathbf{I}_{int}(\mathbf{D}, a, b)$.

5.4 A Labelled Proof Theory for Propositional Equality

Now, it seems that an alternative formulation of propositional equality within the functional interpretation, which will be a little more elaborate than the extensional $\mathbf{I}_{ext}$-type, and simpler than the intensional $\mathbf{I}_{int}$-type, could prove more convenient from the point of view of the 'logical interpretation'. It seems that whereas in the former we have a considerable loss of information in the $\mathbf{I}_{ext}$*-elimination*, in the latter we have an $\mathbf{I}_{int}$*-elimination* too heavily loaded with (perhaps unnecessary) information.

5.4.1. *Identifiers for (Compositions of) Equalities*
In the functional interpretation, where a functional calculus on the labels go hand in hand with a logical calculus on the formulae, we have a classification of equalities, whose identifications are carried along as part of the deduction: either β-, η-, ξ-, μ[34] or ζ- equality will have been part of an expression labelling a formula containing '$\doteq$'. There one finds the key to the idea of 'hiding' in the *introduction* rule, and opening local (Skolem-type) assumptions in the *elimination* rule. (Recall that in the case of disjunction we also have alternatives: either into the left disjunct, or into the right disjunct.) So, we believe that it is not unreasonable to start off the formalization of our propositional equality with the parallel to the disjunction and existential cases in mind. Only, the witness of the type of propositional equality are not the 'as and 'bs of '$a = b : \mathbf{D}$', but the actual (sequence of) equalities (β-, η-, ξ-, ζ-) that might have been used to arrive at the judgement '$a =_s b : \mathbf{D}$' (meaning '$a = b$' *because* of 's'), 's' being a sequence made up of β-, η-, ξ-, μ, and/or ζ-equalities, perhaps with some of the general equality rules of reflexivity, symmetry and transitivity. So, in the *introduction* rule of the type we need to form the canonical proof as if we were *hiding* the actual sequence. Also, in the rule of *elimination* we need to open a new local assumption introducing a new variable denoting a possible sequence as a (Skolem-type) new constant. That is, in order to eliminate the connective '$\doteq$' (i.e. to deduce something from a proposition like '$\doteq_{\mathbf{D}} (a, b)$'), we start by choosing a new variable to denote the reason why the two terms are equal: 'let t be an expression (sequence of equalities) justifying the equality between the terms'. If we then arrive at an arbitrary formula '$\mathbf{C}$' labelled with an expression where the t still occurs free, then we can conclude that the same $\mathbf{C}$ can be obtained from the $\doteq$-formula regardless of the identity of the chosen t, meaning that the label alongside $\mathbf{C}$ in the conclusion will have been abstracted from the free occurrences of t.

Observe that now we are still able to 'keep track' of all proof steps (which does not happen with Martin-Löf's [100, 101] $\mathbf{I}_{ext}$-type), and we have an easier formulation (as compared with Martin-Löf's [99] $\mathbf{I}_{int}$-type) of how to perform the *elimination* step. Moreover, this will hopefully be coherent with the chosen conceptual framework, namely, Ramsey's [112] idea that mathematical equalities are definitional (though not always abbreviatory) equalities.

[34]Though not usually mentioned explicitly in the literature, the μ-equality concerns the equalities associated to *elimination* rules in a similar way that ξ equalities are associated to *introduction* rules, e.g.:

$$\frac{[a : \mathbf{A}] \qquad [f =_t g : \mathbf{A} \to \mathbf{B}]}{\mathrm{APP}(f, a) =_{\mu(t)} \mathrm{APP}(g, a) : \mathbf{B}}$$

5.4.2. *The Proof Rules*

So, in formulating our propositional equality connective which we shall call '$\doteq$' we shall keep the pattern of inference rules essentially the same as the one used for the other logical connectives, and we shall provide an alternative presentation of propositional equality as follows:

$\doteq$-introduction

$$\frac{a =_s b : \mathbf{D}}{s(a,b) :\doteq_{\mathbf{D}} (a,b)} \qquad \frac{a =_s b : \mathbf{D} \qquad a =_t b : \mathbf{D}}{s(a,b) =_\xi t(a,b) :\doteq_{\mathbf{D}} (a,b)}$$

$\doteq$-reduction

$$\frac{\dfrac{a =_s b : \mathbf{D}}{s(a,b) :\doteq_{\mathbf{D}} (a,b)} \doteq\text{-}intr \qquad \begin{matrix}[a =_t b : \mathbf{D}] \\ d(t) : \mathbf{C}\end{matrix}}{\mathtt{REWR}(s(a,b), \theta t.d(t)) : \mathbf{C}} \doteq\text{-}elim \quad \twoheadrightarrow_\beta \quad \begin{matrix}[a =_s b : \mathbf{D}] \\ d(s/t) : \mathbf{C}\end{matrix}$$

$\doteq$-induction

$$\frac{e :\doteq_{\mathbf{D}} (a,b) \qquad \dfrac{[a =_t b : \mathbf{D}]}{t(a,b) :\doteq_{\mathbf{D}} (a,b)} \doteq\text{-}intr}{\mathtt{REWR}(e, \theta t.t(a,b)) :\doteq_{\mathbf{D}} (a,b)} \doteq\text{-}elim \quad \twoheadrightarrow_\eta \quad e :\doteq_{\mathbf{D}} (a,b)$$

where 'θ' is an abstractor which binds the occurrences of the (new) variable 't' introduced with the local assumption '$[a =_t b : \mathbf{D}]$' as a kind of 'Skolem'-type constant denoting the (presumed) 'reason' why 'a' was assumed to be equal to 'b'. (Recall the Skolem-type procedures of introducing new local assumptions in order to allow for the elimination of logical connectives where the notion of 'hiding' is crucial, e.g. disjunction and existential quantifier – in [26].)

Now, having been defined as a 'Skolem'-type connective, '$\doteq$' needs to have a conversion stating the non-interference of the newly opened branch (the local assumption in the $\doteq$-*elimination* rule) with the main branch. Thus, we have:

$\doteq$-(permutative) reduction

$$\frac{\dfrac{e :\doteq_{\mathbf{D}} (a,b) \qquad \begin{matrix}[a =_t b : \mathbf{D}] \\ d(t) : \mathbf{C}\end{matrix}}{\mathtt{REWR}(e, \theta t.d(t)) : \mathbf{C}}}{w(\mathtt{REWR}(e, \theta t.d(t))) : \mathbf{W}} \quad \twoheadrightarrow_\zeta \quad \frac{e :\doteq_{\mathbf{D}} (a,b) \qquad \dfrac{\begin{matrix}[a =_t b : \mathbf{D}] \\ d(t) : \mathbf{C}\end{matrix}}{w(d(t)) : \mathbf{W}}}{\mathtt{REWR}(e, \theta t.w(d(t))) : \mathbf{W}}$$

provided w does not disturb the existing dependencies in the term e (the main branch). The corresponding ζ-equality is:

$$w(\mathtt{REWR}(e, \theta t.d(t))) =_\zeta \mathtt{REWR}(e, \theta t.w(d(t)))$$

The equality indicates that the operation w can be pushed inside the θ-abstraction term, provided that it does not affect the dependencies of the term e.

As we are defining the logical connective '$\doteq$' as a connective which deals with singular terms, where the 'witness' is supposed to be hidden, we shall not be using direct *elimination* like Martin-Löf's $\mathbf{I}_{ext}$-*elimination*. Instead, we shall be using the following $\doteq$-*elimination*:

$$\frac{e : \doteq_{\mathbf{D}} (a, b) \qquad \begin{matrix} [a =_t b : \mathbf{D}] \\ d(t) : \mathbf{C} \end{matrix}}{\mathtt{REWR}(e, \theta t.d(t)) : \mathbf{C}}$$

The *elimination* involves the introduction of a new local assumption (and corresponding variable in the functional calculus), namely '$[a =_t b : \mathbf{D}]$' (where 't' is the new variable) which is only discharged (and 't' bound) in the conclusion of the rule. The intuitive explanation would be given in the following lines. In order to eliminate the equality $\doteq$-connective, where one does not have access to the 'reason' (i.e. a sequence of 'β', 'η', 'ξ' or 'ζ' equalities) why the equality holds because '$\doteq$' is supposed to be a connective dealing with singular terms (as are '$\vee$' and '$\exists$'), in the first step one has to open a new local assumption supposing the equality holds because of, say 't' (a new variable). The new assumption then stands for 'let t be the unknown equality'. If a third (arbitrary) statement can be obtained from this new local assumption via an unspecified number of steps which does not involve any binding of the new variable 't', then one discharges the newly introduced local assumption binding the free occurrences of the new variable in the label alongside the statement obtained, and concludes that that statement is to be labelled by the term '$\mathtt{REWR}(e, \theta t.d(t))$' where the new variable (i.e. t) is bound by the 'θ'-abstractor.

Another feature of the $\doteq$-connective which is worth noticing at this stage is the equality under 'ξ' of all its elements (see second *introduction* rule). This does not mean that the labels serving as evidences for the $\doteq$-statement are all identical to a constant (cf. constant '$\mathtt{r}$' in Martin-Löf's $\mathbf{I}_{ext}$-type), but simply that if two (sequences of) equality are obtained as witnesses of the equality between, say 'a' and 'b' of domain $\mathbf{D}$, then they are taken to be equal under ξ-equality. It would not seem unreasonable to think of the $\doteq$-connective of propositional equality as expressing the proposition which, whenever true, indicates that the two elements of the domain concerned are equal under some (unspecified, *hidden*) composition of definitional equalities. It is as if the proposition points to the existence of a term (witness) which depends on both elements and on the kind of equality judgements used to arrive at its proof. So, in the logical side, one forgets about what was the actual witness. Cf. the existential generalization:

$$\frac{\mathbf{F}(a)}{\exists x.\mathbf{F}(x)}$$

where the actual witness is in fact 'abandoned'. Obviously, as we are interested in keeping track of relevant information introduced by each proof step, in our labelled natural deduction system the witness is not abandoned, but is carried over as an unbounded name in the label of the corresponding conclusion formula.

$$\frac{a : \mathbf{D} \qquad f(a) : \mathbf{F}(a)}{\varepsilon x.(f(x), a) : \exists x^{\mathbf{D}}.\mathbf{F}(x)}$$

Note, however, that it is carried along *only* in the functional side, the logical side not keeping any trace of it at all.

Now, notice that if the functional calculus on the labels is to match the logical calculus on the formulae, than we must have the resulting label on the left of the '$\twoheadrightarrow_{\beta}$' as β-convertible to the concluding label on the right. So, we must have the convertibility equality:

$$\mathtt{REWR}(s(a,b), \theta t.d(t)) =_{\beta} d(s/t) : \mathbf{C}$$

The same holds for the η-equality:

$$\mathtt{REWR}(e, \theta t.t(a,b)) =_{\eta} e : \doteq_{\mathbf{D}} (a,b)$$

Parallel to the case of disjunction, where two different constructors distinguish the two alternatives, namely '`inl`' and '`inr`', we here have any (sequence of) equality constants 'β', 'η' and 'ξ', etc. as constructors of the $\doteq$-connective denoting the alternatives available.

5.4.3. *General Rules of Equality*

Apart from the already mentioned 'constants' which compose the reasons for equality (i.e. the indices to the equality on the functional calculus), it is reasonable to expect that the following rules are taken for granted:

reflexivity	*symmetry*	*transitivity*
$\dfrac{x : \mathbf{D}}{x =_{\mathtt{refl}} x : \mathbf{D}}$	$\dfrac{x =_t y : \mathbf{D}}{y =_{\mathtt{symm}(t)} x : \mathbf{D}}$	$\dfrac{x =_t y : \mathbf{D} \qquad y =_u z : \mathbf{D}}{x =_{\mathtt{trans}(t,u)} z : \mathbf{D}}$

5.4.4. *Substitution Without Involving Quantifiers*

We know from logic programming, i.e. from the theory of unification, that substitution can take place even when no quantifier is involved. This is justified when, for some reason a certain referent can replace another under some condition for identifying the one with the other.

Now, what would be counterpart to such a 'quantifier-less' notion of substitution in a labelled natural deduction system. Without the appropriate means of handling equality (definitional and propositional) we would hardly be capable of finding such a counterpart. Having said all that, let us think of what we ought to do at a certain stage in a proof (deduction) where the following two premisses would be at hand:

$$a =_g y : \mathbf{D} \qquad \text{and} \qquad f(a) : \mathbf{P}(a)$$

We have that a and y are equal ('identifiable') under some arbitrary sequence of equalities (rewrites) which we name g. We also have that the predicate formula $\mathbf{P}(a)$ is labelled by a certain functional expression f which depends on a. Clearly, if a and y are 'identifiable', we would like to infer that $\mathbf{P}$, being true of a, will also be true of y. So, we shall be happy in inferring (on the logical calculus) the formula $\mathbf{P}(y)$. Now, given that we ought to compose the label of the conclusion out of a composition of the labels of the premisses, what label should we insert alongside $\mathbf{P}(y)$? Perhaps various good answers could be given here, but we shall choose one which is in line with our 'keeping record of what (relevant) data was used in a deduction'. We have already stated how much importance we attach to names of individuals, names of formula instances, and of course, what kind of deduction was performed (i.e. what kind of connective was introduced or eliminated). In this section we have also insisted on the importance of, not only 'classifying' the equalities, but also having variables for the kinds of equalities that may be used in a deduction. Let us then formulate our rule of 'quantifier-less' substitution as:

$$\frac{a =_g y : \mathbf{D} \qquad f(a) : \mathbf{P}(a)}{g(a,y) \cdot f(a) : \mathbf{P}(y)}$$

which could be explained in words as follows: if a and y are 'identifiable' due to a certain g, and $f(a)$ is the evidence for $\mathbf{P}(a)$, then let the composition of $g(a,y)$ (the label for the propositional equality between a and y) with $f(a)$ (the evidence for $\mathbf{P}(a)$) be the evidence for $\mathbf{P}(y)$.

By having this extra rule of substitution added to the system of rules of inference, we are able to validate one half of the so-called 'Leibniz's law', namely:

$$\forall x, y^{\mathbf{D}}.(\dot{=}_{\mathbf{D}}\ (x,y) \to (\mathbf{P}(x) \to \mathbf{P}(y)))$$

5.4.5. *Examples of Deduction*

EXAMPLE 5.1 [Reflexivity] Now that we have introduced the *reflexivity* constant, whose inferential counterpart says that starting from any 'x' from domain '**D**' one has '$x =_{\mathtt{refl}} x : \mathbf{D}$'. We can now prove the proposition which says that for any x, x is (propositionally) equal to itself. The proof that for all elements from domain '**D**' the propositional equality between the element chosen and itself holds, is constructed as:

$$\dfrac{\dfrac{\dfrac{[x:\mathbf{D}]}{x =_{\mathtt{refl}} x : \mathbf{D}}}{\mathtt{refl}(x,x) : \doteq_{\mathbf{D}} (x,x)}}{\Lambda x.\mathtt{refl}(x,x) : \forall x^{\mathbf{D}}. \doteq_{\mathbf{D}} (x,x)}$$

■

EXAMPLE 5.2 [Extensionality on λ-terms] As a more 'concrete' example, let us construct a proof of what is usually called 'the extensionality axiom', namely, the axiom which says that if two functions are equal then they produce the same value for all arguments. Formally:

$$\forall f, g^{\mathbf{A}\to\mathbf{B}}.(\doteq_{\mathbf{A}\to\mathbf{B}} (f,g) \to \forall a^{\mathbf{A}}. \doteq_{\mathbf{B}} (\mathtt{APP}(f,a), \mathtt{APP}(g,a)))$$

The proof goes as follows:

$$\dfrac{\dfrac{\dfrac{\dfrac{\dfrac{[f:\mathbf{A}\to\mathbf{B}] \qquad [g:\mathbf{A}\to\mathbf{B}]}{[e:\doteq_{\mathbf{A}\to\mathbf{B}}(f,g)]} \qquad \dfrac{\dfrac{\dfrac{[a:\mathbf{A}] \qquad [f =_t g:\mathbf{A}\to\mathbf{B}]}{\mathtt{APP}(f,a) =_{\mu(t)} \mathtt{APP}(g,a):\mathbf{B}}}{\mu(t)(\mathtt{APP}(f,a),\mathtt{APP}(g,a)):\doteq_{\mathbf{B}}(\mathtt{APP}(f,a),\mathtt{APP}(g,a))}}{\Lambda a.\mu(t)(\mathtt{APP}(f,a),\mathtt{APP}(g,a)):\forall a^{\mathbf{A}}.\doteq_{\mathbf{B}}(\mathtt{APP}(f,a),\mathtt{APP}(g,a))}}{\mathtt{REWR}(e,\theta t.\Lambda a.\mu(t)(\mathtt{APP}(f,a),\mathtt{APP}(g,a))):\forall a^{\mathbf{A}}.\doteq_{\mathbf{B}}(\mathtt{APP}(f,a),\mathtt{APP}(g,a))}}{\lambda e.\mathtt{REWR}(e,\theta t.\Lambda a.\mu(t)(\mathtt{APP}(f,a),\mathtt{APP}(g,a))):\doteq_{\mathbf{A}\to\mathbf{B}}(f,g)\to\forall a^{\mathbf{A}}.\doteq_{\mathbf{B}}(\mathtt{APP}(f,a),\mathtt{APP}(g,a))}}{\Lambda g.\lambda e.\mathtt{REWR}(e,\theta t.\Lambda a.\mu(t)(\mathtt{APP}(f,a),\mathtt{APP}(g,a))):\forall g^{\mathbf{A}\to\mathbf{B}}.(\doteq_{\mathbf{A}\to\mathbf{B}}(f,g)\to\forall a^{\mathbf{A}}.\doteq_{\mathbf{B}}(\mathtt{APP}(f,a),\mathtt{APP}(g,a)))}}{\Lambda f.\Lambda g.\lambda e.\mathtt{REWR}(e,\theta t.\Lambda a.\mu(t)(\mathtt{APP}(f,a),\mathtt{APP}(g,a))):\forall f,g^{\mathbf{A}\to\mathbf{B}}.(\doteq_{\mathbf{A}\to\mathbf{B}}(f,g)\to\forall a^{\mathbf{A}}.\doteq_{\mathbf{B}}(\mathtt{APP}(f,a),\mathtt{APP}(g,a)))}$$

Unlike Martin-Löf's [100, 101] system with $\mathbf{I}_{ext}$-type described previously, the logical system with our $\doteq$-connective does not allow a closed proof of the converse of the above statement. That is, the proof of:

$$\forall f, g^{\mathbf{A}\to\mathbf{B}}.(\forall a^{\mathbf{A}}. \doteq_{\mathbf{B}} (\mathtt{APP}(f,a), \mathtt{APP}(g,a)) \to \doteq_{\mathbf{A}\to\mathbf{B}} (f,g))$$

will not be categorical, i.e. it will have free variables. ■

EXAMPLE 5.3 [The generality of Herbrand base] Let us take the example which Leisenring [87] uses to demonstrate the application of Herbrand's decision procedure to check the validity of the formula:

$$\exists x^{\mathbf{D}}.\forall y^{\mathbf{D}}.(\mathbf{P}(x) \to \mathbf{P}(y))$$

Herbrand's 'original' procedure. The first step is to find the Herbrand resolution ($\exists$-prenex normal form), which can be done by introducing a new function symbol g, and obtaining:

$$\exists x^{\mathbf{D}}.(\mathbf{P}(x) \to \mathbf{P}(g(x)))$$

As this would be equivalent to a disjunction of substitution instances like:

$$\mathbf{P}(a_1) \to \mathbf{P}(g(a_1)) \quad \vee \quad \mathbf{P}(a_2) \to \mathbf{P}(g(a_2)) \quad \vee \quad \mathbf{P}(a_3) \to \mathbf{P}(g(a_3)) \quad \vee \cdots$$

the second step is to find a p-substitution instance (p finite) which is a tautology. For that, we take the 'Herbrand base' to be $\{a, g\}$, where a is an arbitrary individual from the domain, and g is an arbitrary function symbol which can construct, out of a, further elements of the domain. Thus, taking $a_1 = a$, the 1-substitution instance is:

$$\mathbf{P}(a) \to \mathbf{P}(g(a))$$

which is clearly not a tautology. Now, we can iterate the process, and find the 2-reduction as a disjunction of the 1-reduction and the formula made up with a 2-substitution (taking $a_2 = g(a)$), that is:

$$\mathbf{P}(a) \to \mathbf{P}(g(a)) \quad \vee \quad \mathbf{P}(g(a)) \to \mathbf{P}(g(g(a)))$$

which is a tautology.

Refutation. If, instead, we try to validate the formula by a method made more familiar by the so-called resolution procedure, namely by applying a refutation procedure to the negation of the formula, we have:

1. $\neg\exists x^{\mathbf{D}}.\forall y^{\mathbf{D}}.(\mathbf{P}(x) \to \mathbf{P}(y))$
2. $\forall x^{\mathbf{D}}.\neg\forall y^{\mathbf{D}}.(\mathbf{P}(x) \to \mathbf{P}(y))$
3. $\forall x^{\mathbf{D}}.\exists y^{\mathbf{D}}.\neg(\mathbf{P}(x) \to \mathbf{P}(y))$
4. $\forall x^{\mathbf{D}}.\exists y^{\mathbf{D}}.(\mathbf{P}(x) \wedge \neg\mathbf{P}(y))$
5. $\forall x^{\mathbf{D}}.(\mathbf{P}(x) \wedge \neg\mathbf{P}(g(x)))$ (by Skolemization)
6. $\mathbf{P}(a_1) \wedge \neg\mathbf{P}(g(a_1)) \quad \wedge \quad \mathbf{P}(a_2) \wedge \neg\mathbf{P}(g(a_2)) \quad \wedge \quad \cdots$
7. take $a_1 = a$
8. $\mathbf{P}(a) \wedge \neg\mathbf{P}(g(a))$
9. take $a_2 = g(a)$
10. $\mathbf{P}(a) \wedge \neg\mathbf{P}(g(a)) \quad \wedge \quad \mathbf{P}(g(a)) \wedge \neg\mathbf{P}(g(g(a)))$ (contradiction).

In checking the validity of $\exists x^{\mathbf{D}}.\forall y^{\mathbf{D}}.(\mathbf{P}(x) \to \mathbf{P}(y))$ we needed the following extra assumptions:

1. the domain is non-empty (step 6);
2. there is a way of identifying an arbitrary term with another one (step 7).

As we shall see below, the labelled deduction method will have helped us 'to bring up to the surface' those two (hidden) assumptions.

Now, how can we justify the *generality* of the 'base' $\{a, g\}$? Why is it that it does not matter which a and g we choose, the procedure always works? In other words, why is it that *for any* element a of the domain and *for any* 'function symbol' g, the procedure always works?

In a previous report [26] we have already demonstrated the *universal force* which is given to Skolem functions by the device of *abstraction* in the elimination of the existential quantifier. The point was that although there was no quantification over function symbols being made in the logic (the logical calculus on the formulae, that is), an abstraction on the name for the Skolem function was performed in the functional calculus on the labels. The observation suggested that, as in the statement of Skolem's theorem, *for any* (new) function symbol f we choose when Skolemising $\forall x^{\mathbf{D}}.\exists y^{\mathbf{D}}.\mathbf{P}(x,y)$ to $\forall x^{\mathbf{D}}.\mathbf{P}(x,f(x))$, if an arbitrary statement can be deduced from the latter then it can also be deduced from the former, regardless of the choice of f.

Now, if we were to construct a deduction of the formula used to demonstrate Herbrand's procedure, we would proceed by using the *introduction* rules backward. First of all, assuming the formula is valid, and knowing that its outermost connective is an existential quantifier, its labels must be of the form:

$$\varepsilon x.(f(x),a) : \exists x^{\mathbf{D}}.\forall y^{\mathbf{D}}.(\mathbf{P}(x) \to \mathbf{P}(y))$$

for some functional expression f, and witness a. Now, to have arrived at that conclusion, we must have come from two premisses of the form:

$$a : \mathbf{D} \qquad \text{and} \qquad f(a) : \forall y^{\mathbf{D}}.(\mathbf{P}(a) \to \mathbf{P}(y))$$

The first premisse is already reduced to an atomic formula so we cannot go any further, but the second one has an universal quantifier as its outermost connective. Thus, our $f(a)$ must be of the form:

$$\Lambda y.h(y,a) : \forall y^{\mathbf{D}}.(\mathbf{P}(a) \to \mathbf{P}(y))$$

for some h which depends on both y and a. In its turn, this h will have come from a deduction where we ought to have had:

$[y : \mathbf{D}]$ as an assumption, and $h(y,a) : \mathbf{P}(a) \to \mathbf{P}(y)$ as a premisse

Now, the latter, having implication as its major connective, must be of the form:

$$\lambda u.m(u,y,a) : \mathbf{P}(a) \to \mathbf{P}(y)$$

for some expression m which depends on u, y and a, and must also have come from a deduction where:

$[u(a) : \mathbf{P}(a)]$ is an assumption, and $m(u,y,a) : \mathbf{P}(y)$ as a premisse

(Recall that labels of predicate formulae will be such that they are expressions which depend on the object of predication.)

Now we have reached the atoms, and we now need to move backwards 'solving' the unknowns: we need to find the form of m as an expression depending on u, y and a, in order to find what h is, and so on, until we really obtain the whole label expression of our original formula. Thus, we have the assumptions:

$$a : \mathbf{D}, \quad y : \mathbf{D}, \quad u(a) : \mathbf{P}(a)$$

and we need to arrive at

$$m(u, y, a) : \mathbf{P}(y)$$

As we need to get $\mathbf{P}(y)$ from $\mathbf{P}(a)$, let us then recall our 'quantifier-less' substitution procedure. What extra assumption do we need to make in order to be able to apply our rule of quantifier-less substitution? It is not too difficult to guess:

$$a =_g y : \mathbf{D}$$

for some arbitrary sequence of equalities (rewrites) g. If we have that extra assumption we can apply the rule and get $\mathbf{P}(y)$ as follows:

$$\frac{a =_g y : \mathbf{D} \qquad u(a) : \mathbf{P}(a)}{g(a, y) \cdot u(a) : \mathbf{P}(y)}$$

Having done that, the fact that the validity of the formula must be independent of the choice of g, there must be a step where the g is 'abstracted' away from the label expression at some point down the deduction. This will happen if the assumption $a =_g y : \mathbf{D}$ will have been introduced in the context of an $\doteq$-*elimination* inference.

Prima facie, the degree of *explicitation* may look a little too excessive, but if for nothing else, we have certainly uncovered two extra assumptions on which the validity of the formula $\exists x^{\mathbf{D}}.\forall y^{\mathbf{D}}.(\mathbf{P}(x) \to \mathbf{P}(y))$ must depend. First, the domain must be non-empty, therefore if no specific element is mentioned we simply take an arbitrary one ('let a be an arbitrary element'). Secondly, there must be a way of performing substitutions even after we strip the formula of its quantifiers, which means that if there is no 'function symbol' already specified so that other elements can be built out of a, then take an arbitrary one ('let g be the key to the identifications/substitutions').

If we now reconstruct the whole deduction as a tree, we get the following:

$$\dfrac{\dfrac{[a{:}\mathbf{D}] \qquad \dfrac{\dfrac{\dfrac{\dfrac{[a{:}\mathbf{D}] \qquad [y{:}\mathbf{D}]}{[s(a,y){:}\doteq(a,y)]} \qquad \dfrac{\dfrac{[a{=_g}y{:}\mathbf{D}] \qquad [u(a){:}\mathbf{P}(a)]}{g(a,y)\cdot u(a){:}\mathbf{P}(y)}}{\lambda u.g(a,y)\cdot u(a){:}\mathbf{P}(a)\to\mathbf{P}(y)}}{\mathtt{REWR}(s(a,y),\theta g.\lambda u.g(a,y)\cdot u(a)){:}\mathbf{P}(a)\to\mathbf{P}(y)}}{\lambda s.\mathtt{REWR}(s(a,y),\theta g.\lambda u.g(a,y)\cdot u(a)){:}\doteq(a,y)\to(\mathbf{P}(a)\to\mathbf{P}(y))}}{\Lambda y.\lambda s.\mathtt{REWR}(s(a,y),\theta g.\lambda u.g(a,y)\cdot u(a)){:}\forall y^{\mathbf{D}}.(\doteq(a,y)\to(\mathbf{P}(a)\to\mathbf{P}(y)))}}{\varepsilon x.(\Lambda y.\lambda s.\mathtt{REWR}(s(x,y),\theta g.\lambda u.g(x,y)\cdot u(x)),a){:}\exists x^{\mathbf{D}}.\forall y^{\mathbf{D}}.(\doteq(x,y)\to(\mathbf{P}(x)\to\mathbf{P}(y)))}}{\lambda a.\varepsilon x.(\Lambda y.\lambda s.\mathtt{REWR}(s(x,y),\theta g.\lambda u.g(x,y)\cdot u(x)),a){:}\mathbf{D}\to(\exists x^{\mathbf{D}}.\forall y^{\mathbf{D}}.(\doteq(x,y)\to(\mathbf{P}(x)\to\mathbf{P}(y))))}$$

If we look at abstraction as a device which brings *universal* force to the reading of a functional expression, it is not difficult to see why our 'base', i.e. '$\{a, g\}$', has a generality aspect to it: in the final label:

$$\boxed{\lambda a.}\varepsilon x.(\Lambda y.\lambda s.\mathtt{REWR}(s(x,y), \boxed{\theta g.}\lambda u.\boxed{g}(x,y)\cdot u(x)), \boxed{a})$$

both a and g are bound by the respective abstractors λ and θ. ■

6 Handling Assumptions

As we have seen from the previous sections, the use of labels alongside formulae can be of great help in terms of *controlling* deduction. But 'why should we be concerned with controlled deduction at all?', one may ask. 'Logic is concerned with the validity (or otherwise) of arguments, so we can make do with valuation systems', it could be added. Now, it is quite true that valuation systems can be very powerful in accounting for different logics in a substantially general way. It is also true, however, that whenever a framework is general enough (take for example Kripke-style semantics) there is invariably an element of 'impurity' being introduced in order to account for the generality (i.e. the provision for parameters to be changed according to each particular case): in the case of possible-worlds semantics the relation of accessibility may be seen as the 'extraneous' element, for it stands outside the actual valuation system. (There are valuation systems for modal logics which do not use accessibility relation but functions whose properties are expressed in equational form. But here again the accessibility function is an element standing outside the system of valuation.) In seeking a deductive perspective on the difference between various logics, we are inevitably faced with the question of where the 'extraneous' element should stand, whether in the object language or in the meta-language. The perspective offered by Gentzen's natural deduction, i.e. analysis of deduction in terms of the properties of connectives together with a distinction between assumptions, premisses and conclusions of deduction rules, points us to a 'middle ground': there is a need to account for the global aspects of certain inference rules, thus reaching out to the meta-language, but some of the control will be done in the object-language if we care to 'keep track of proof steps'.

The fact that in some logics the validity of an argument depends not only on the truth of the individual assumptions used, but also on the way these assumptions stand to one another (and to the premisse(s) of an inference rule, for that matter), suggests that a general framework to study logics must be capable of accounting for the way in which assumptions are handled. There are different ways of doing this, and one of them is by providing rules which manipulate directly the structures holding the hypotheses and/or those holding the conclusions, such as in the so-called structural rules of Gentzen's sequent calculus.

6.1 Side Conditions

The so-called 'improper' inference rules, to use a terminology from Prawitz' [108] *Natural Deduction*, leave 'room for manoeuvre' as to how a particular logic could be obtained just by imposing conditions on the discharge of assumptions that would correspond to the logical discipline one is adopting (linear, relevant, intuitionistic, classical, etc.). The side conditions can be 'naturally' imposed, given that a degree of 'vagueness' is introduced by the presentation of those improper inference rules, such as the rule of $\rightarrow$-*introduction*:

$$\frac{\begin{matrix}[x : \mathbf{A}] \\ b(x) : \mathbf{B}\end{matrix}}{\lambda x.b(x) : \mathbf{A} \rightarrow \mathbf{B}}$$

Note that one might (as some authors do) insert an explicit sign between the assumption '$[x : \mathbf{A}]$' and the premisse of the rule, namely '$b(x) : \mathbf{B}$', such as, e.g. the three vertical dots, making the rule look like

$$\frac{\begin{matrix}[x : \mathbf{A}] \\ \vdots \\ b(x) : \mathbf{B}\end{matrix}}{\lambda x.b(x) : \mathbf{A} \rightarrow \mathbf{B}}$$

to indicate the element of vagueness. There is no place, however, for the introduction of side conditions on those rules which do not allow for such a 'room for manoeuvre', namely those rules which are not improper inference rules. In his account of linear logic via a (plain) natural deduction system, Avron [6] introduces what we feel rather 'unnatural' side conditions in association with an inference rule which is not improper, namely the rule of $\wedge$-*introduction*.[35]

[35]See p. 163:

$$\frac{A \quad B}{A \wedge B}(*)$$

and condition (3) on p. 164: "For $\wedge$-Int we have side condition that A and B should depend on exactly the same multiset of assumptions (condition $(*)$). Moreover, the shared hypotheses are considered as appearing *once*, although they seem to occur *twice*".

Nevertheless, in the framework of our labelled natural deduction we are still able to handle different notions of conjunction and disjunction in case we need to handle explicitly the *contexts* (structured collections of formulae). This is done by introducing names (variables) for the contexts as an extra parameter in all our *introduction* and *elimination* rules for the logical connectives, and considering those names as identifiers for structured collections (sets, multisets, lists, trees, etc.) of labelled formulae: e.g. '$\mathbf{A}(\mathbb{S})$' would read 'an occurrence of the formula $\mathbf{A}$ is stored in the structure $\mathbb{S}$'. So, in the case of conjunction, for example, we would have:

6.2 Resource Control

Given that the improper inference rules do leave 'room for manoeuvre', so to speak, one can think of the discipline of handling assumptions characterising a certain logic as based on a prescription as to under what conditions are we to allow formulae to be proved with no assumptions. Now, if we are carrying, through the labels, the record of the all the 'resource' being used in a deduction (resource being seen here as relative to the quantity and/or quality of the stock of assumptions being at our disposal), we should be able to keep control over that resource as we like: the device of abstraction will give us a hand on removing the assumptions we want to get rid of, taking into account whatever factor is more appropriate (the number of occurrences, the order of appearances, etc.).

6.2.1. *Resource Counting*

An interpretation of the conditional in terms of *actions* is proposed in Girard's [63] defence of linear logic. One of the rather unconventional features of linear logic is that two conjunctions and two disjunctions co-exist, due to the absence of the structural rules of weakening and contraction in the logic. An intuitive justification for the difference between the two kinds of conjunction is attempted by reference to the paradigm of 'proofs-as-actions'. An example is given where the 'actions'-interpretation of logic might seem closer to ordinary language. Essentially, the example is concerned with showing why one should accept that indeed there are two kinds of conjunction. The argument is constructed from the situation where one should not accept that 'with one dollar one can get a pack of Camel and a pack of Marlboro' follows from the two assumptions such as 'with one dollar one can get a pack of Camel' and 'with one dollar one can get a pack of Marlboro'. (With the usual notion one has '$\mathbf{A} \to (\mathbf{B} \wedge \mathbf{C})$' as a logical consequence of '$\mathbf{A} \to \mathbf{B}$' and '$\mathbf{A} \to \mathbf{C}$'.) Arguing that 'the real implication

---

$\wedge$-*introduction*

$$\frac{a_1 : \mathbf{A}_1(\mathbb{S}) \qquad a_2 : \mathbf{A}_2(\mathbb{T})}{\langle a_1, a_2\rangle : (\mathbf{A}_1 \wedge \mathbf{A}_2)(\mathbb{S} \circledcirc \mathbb{T})} \quad \text{(in sequent calculus: } \frac{\mathbb{S} \vdash a_1 : \mathbf{A}_1 \qquad \mathbb{T} \vdash a_2 : \mathbf{A}_2}{\mathbb{S} \circledcirc \mathbb{T} \vdash \langle a_1, a_2\rangle : \mathbf{A}_1 \wedge \mathbf{A}_2}\text{)}$$

where the '$\circledcirc$' operator would be compatible with the data structures represented by $\mathbb{S}$ and $\mathbb{T}$. For example, if $\mathbb{S}$ and $\mathbb{T}$ are both taken to be sets, and '$\circledcirc$' is taken to be set union, then we have a situation which is similar to the rule of classical Gentzen's sequent calculus (augmented with labels alongside formulae). If, on the other hand, we take the structures to be multisets, and '$\circledcirc$' multiset union, we would have Girard's linear logic and the corresponding conjunctions ($\otimes$/&) depending on whether '$\mathbb{S}$' is distinct or identical to '$\mathbb{T}$'.

More details of our labelled natural deduction with an extra parameter shall be given in two forthcoming reports: one deals with an attempt at a functional interpretation of the modal necessity [56], and the other one deals with the functional interpretation of the sequent calculus via the use of explicit data type operations (*à la* Guttag [71]) over the structured collection of formulae [24].

is *causal*', the idea is to read the linear conditional '$\mathbf{A} \multimap \mathbf{B}$' as 'an action **A** causes an action **B**'. The answer to the question whether one should accept the conclusion above from the two assumptions, Girard says, is clearly <u>no</u>, for an action of type 'to spend US$1 ⊸ to get a pack of Camel' is a way of replacing any specific US$ by a specific pack of Camel. So, given an action of type 'to spend US$1 ⊸ to get a pack of Camel' and an action of type 'to spend US$1 ⊸ to get a pack of Marlboro', there will be no way of forming an action of type 'to spend US$1 ⊸ (to get a pack of Camel ⊗ to get a pack of Marlboro)' (where '⊗' should be read as one kind of 'and'), since for US$1 one will never get what costs US$2. The line of reasoning followed by Girard is rather ingenious, but it does not seem to be the only way to explain what goes on in that particular situation. One can, for example, demonstrate that a refined notion of the conditional is needed to reject the deduction as invalid, without having to change the notion of conjunction. Moreover, given the properties of disjunction (in particular, the permutative ζ-conversion), the disjunctive reading (i.e. 'does $\mathbf{A} \to (\mathbf{B} \vee \mathbf{C})$ follow from $(\mathbf{A} \to \mathbf{B}) \vee (\mathbf{A} \to \mathbf{C})$?') leaves room for accepting that disjunction may acquire a 'conjunctive' flavouring in some cases.

For the sake of comparison, let us take the usual connectives of implication and conjunction and see how one would construct a proof of

$$((\text{US\$1} \to \text{Camel}) \wedge (\text{US\$1} \to \text{Marlboro})) \to (\text{US\$1} \to (\text{Camel} \wedge \text{Marlboro}))$$

First let us abbreviate as follows (similarly to [63]):

A: to spend US$1
B: to get a pack of Camel
C: to get a pack of Marlboro

Moreover, let us use 'labelled' natural deduction to build the corresponding proof trees.

$$\dfrac{\dfrac{\dfrac{\boxed{[x:\mathbf{A}]} \quad [y:\mathbf{A}\to\mathbf{B}]}{\mathtt{APP}(y,x):\mathbf{B}} \qquad \dfrac{\boxed{[x:\mathbf{A}]} \quad [z:\mathbf{A}\to\mathbf{C}]}{\mathtt{APP}(z,x):\mathbf{C}}}{\langle \mathtt{APP}(y,x), \mathtt{APP}(z,x)\rangle : \mathbf{B}\wedge\mathbf{C}}}{\boxed{\lambda x.}\langle \mathtt{APP}(y,\boxed{x}), \mathtt{APP}(z,\boxed{x})\rangle : \mathbf{A}\to(\mathbf{B}\wedge\mathbf{C})}\,(*)$$

Now, if one is not quite sure about what kind of conditional one is dealing with for this particular instance of logical reasoning, then in order to accept the above proof as a 'good' one, we have to look at the step(s) where the conditionals where introduced and corresponding assumptions discharged. Note that at the step marked with '$(*)$' two occurrences of the assumption 'x : **A**', literally 'x witnesses the existence of US$1', were discharged to conclude that 'US$1→(Camel ∧ Marlboro)'.

Now, if we take the view that in some ordinary uses of 'and' disguise a disjunctive statement, and that is what seems to be happening in this case.

When saying that 'to spend US$1 implies that one gets a pack of Camel and to spend US$1 implies that one gets a pack of Marlboro' one seems to be saying that 'either to spend US$1 implies that one gets a pack of Camel or to spend US$1 implies that one gets a pack of Marlboro'. It is a phenomenon that linguists might be able to explain better. Now, it is easy to see that from the disjunctive assumption one can conclude that 'to spend US$1 implies that one gets either a pack of Camel or a pack of Marlboro'. (Below I shall abbreviate in the same way as before.)

$$\dfrac{[y:(\mathbf{A}\to\mathbf{B})\vee(\mathbf{A}\to\mathbf{C})] \quad \dfrac{\dfrac{\dfrac{\boxed{[x:\mathbf{A}]}\quad [z:\mathbf{A}\to\mathbf{B}]}{\texttt{APP}(z,x):\mathbf{B}}}{\texttt{inl}(\texttt{APP}(z,x)):\mathbf{B}\vee\mathbf{C}}}{\boxed{\lambda x.}\texttt{inl}(\texttt{APP}(z,\boxed{x})):\mathbf{A}\to(\mathbf{B}\vee\mathbf{C})}(*) \quad \dfrac{\dfrac{\dfrac{\boxed{[x:\mathbf{A}]}\quad [w:\mathbf{A}\to\mathbf{C}]}{\texttt{APP}(w,x):\mathbf{C}}}{\texttt{inr}(\texttt{APP}(w,x)):\mathbf{B}\vee\mathbf{C}}}{\boxed{\lambda x.}\texttt{inr}(\texttt{APP}(w,\boxed{x})):\mathbf{A}\to(\mathbf{B}\vee\mathbf{C})}(*)}{\texttt{CASE}(y,vz.\lambda x.\texttt{inl}(\texttt{APP}(z,x)),vw.\lambda x.\texttt{inr}(\texttt{APP}(w,x))):\mathbf{A}\to(\mathbf{B}\vee\mathbf{C})}$$

Observe that in the steps marked with '(*)' *exactly one* occurrence of the free variable is bound by the λ-abstraction, *exactly one* instance of the corresponding assumption is discharged.

Nonetheless, it could still be argued that Girard has a case here. And this is because it is possible to show that even with the disjunction interpretation one has to use non-linear discharge of assumptions. Recall that our permutative ζ-reduction for the disjunction allows us to say that the above proof is equivalent to the following proof:

$$\dfrac{\dfrac{[y:(\mathbf{A}\to\mathbf{B})\vee(\mathbf{A}\to\mathbf{C})] \quad \dfrac{\dfrac{\boxed{[x:\mathbf{A}]}\quad [z:\mathbf{A}\to\mathbf{B}]}{\texttt{APP}(z,x):\mathbf{B}}}{\texttt{inl}(\texttt{APP}(z,x)):\mathbf{B}\vee\mathbf{C}} \quad \dfrac{\dfrac{\boxed{[x:\mathbf{A}]}\quad [w:\mathbf{A}\to\mathbf{C}]}{\texttt{APP}(w,x):\mathbf{C}}}{\texttt{inr}(\texttt{APP}(w,x)):\mathbf{B}\vee\mathbf{C}}}{\texttt{CASE}(y,vz.\texttt{inl}(\texttt{APP}(z,x)),vw.\texttt{inr}(\texttt{APP}(w,x))):\mathbf{B}\vee\mathbf{C}}}{\boxed{\lambda x.}\texttt{CASE}(y,vz.\texttt{inl}(\texttt{APP}(z,\boxed{x})),vw.\texttt{inr}(\texttt{APP}(w,\boxed{x}))):\mathbf{A}\to(\mathbf{B}\vee\mathbf{C})}(*)$$

because, given that y does not depend on x, the following equality holds:

$$\lambda x.\texttt{CASE}(y, vz.\texttt{inl}(\texttt{APP}(z,x)), vw.\texttt{inr}(\texttt{APP}(w,x))) =_{\zeta}$$
$$\texttt{CASE}(y, vz.\lambda x.\texttt{inl}(\texttt{APP}(z,x)), vw.\lambda x.\texttt{inr}(\texttt{APP}(w,x)))$$

It says that the $\lambda x.$-abstraction can be pushed inside the v-abstractions because the first argument of the CASE-term, namely y, does not depend on x.

Now, notice that our disjunction acquires some *conjunctive* flavour when we have the ζ-conversion rules. It is as if we are still discharging two occurrences of the same assumption in one single introduction of the implication, as in step marked '(*)'.

Indeed, in terms of the deduction, notice that in the last step of the equivalent proof-tree (marked '(*)') two free occurrences of the variable x are being bound by the same $\lambda x.$-abstraction. So, strictly speaking, one may say that the disjunction interpretation also violates the discipline of linear discharge of assumptions.[36]

6.2.2. *Resource Awareness*

Let us become devil's advocate and think of a situation where we do need to accept the double use of resource to prove our conditional statement true. The point here is, again, that there are various kinds of conditionals that we use in ordinary language, and that for some conditionals the amount of resource 'consumed' by introducing the conditional is not relevant, provided we do indeed consume at least some of the resource available. In other words, some conditionals need resource counting, others resource availability, etc. However, not all conditionals ought to take care of resource counting.

Keeping the interpretation of *proofs as actions* we want to analyse the situation in which our 'resource' is not US$1, but a doctoral thesis. So, we abbreviate:

A:	to submit a doctoral thesis
B:	to get a PhD degree
C:	to get a Diploma of the College.

Now, if we admit that '**A** $\rightarrow$ **B**' and '**A** $\rightarrow$ **C**', clearly, we must accept that '**A** $\rightarrow$ (**B** $\wedge$ **C**)' just because one does not have to submit 2 (two) doctoral theses to get a PhD degree and a Diploma of the College: one thesis will do. And here our conjunction does not have to be modified, only our conditional. Our '$\wedge$' is not Girard's '&' because in order to perform the action '**A** $\rightarrow$ (**B** $\wedge$ **C**)' we do not have "to choose which among the two possible actions we want to perform, and then to do the one selected." [63, page 6] We can do both at the same time, namely, we can get a PhD degree and a Diploma of the College at the same time, after performing the action of submitting a doctoral thesis. (Of course, if we change 'to submit a doctoral thesis' into 'to hand in a bounded copy of a doctoral thesis' then we need to take care of resource counting (at least at Imperial, anyway): one copy for the PhD and one for the Diploma board are required.)

[36] It should be remarked here that the analysis above does not refer to a specific natural deduction system for linear logic, but rather to an analysis of resource awareness with labelled natural deduction where the discipline of discharging assumptions is reflected directly in the discipline of abstraction on the variables occurring in the labels. In a system of natural deduction where (arguably 'unnatural') restrictions are placed on some rules of inference in order to account for the connectives of Girard's linear logic, it may be that the non-linear discharge of assumptions referred to in the example above would still be taken to be linear because "the shared hypotheses [in $\vee$-*elimination*] are considered as appearing once" [6, page 164].

6.3 Classical Implication, Involution and Self-application

What is it that makes classical implication different from intuitionistic implication? Perhaps there is no unique answer to this question, but it might help us to look at it from the aspect of the classical symmetry. It is from this perspective that there may be an argument in favour of the Sequent Calculus due to its 'natural' symmetry: left and right of the turnstile. According to this same argument, natural deduction is not quite as good, given that the expected symmetry does not spring up as clearly as it does in the Sequent Calculus. While this is true for *plain* natural deduction, it is not exactly true of labelled natural deduction: the labels of implicational formulae live in a kind of 'symmetrical' domain, i.e. a label may appear as an argument and it may also appear as a function (recall that '$\mathtt{APP}(x, y)$' shows x appearing as function and y appearing as argument).

Let us recall that in the propositions-as-types interpretation, by restricting λ-abstractions we *restrict* the supply of definable terms thus getting *less* types $\mathbf{A}$ demonstrably non-empty. Less non-empty types means less theorems which by its turn means a weaker logic. How about stronger logics, intermediate logics between intuitionistic and classical logic? What mechanism do we have for characterising them? Obviously we need to increase the stock of definable (existing) terms so that more types can be shown to be non-empty, i.e. more theorems are available. We do not want to just throw in (stipulate) existence of functionals, in an *ad hoc* manner just to obtain the intermediate logic we want. We should put forward some reasonable principles, the kind natural to λ-calculus functional environment, and show that adopting them yields corresponding logics.

In [58] we chose one simple principle, which is that of completing a functional diagram. If 'φ' is a *monotonic* increasing or a *monotonic* decreasing functional on types, then one can ask the question whether we can complete the following diagram:

$$\begin{array}{ccc} \varphi(\mathbf{A}) & \xrightarrow{\;c\;} & \varphi(\mathbf{B}) \\ a\,\Big\uparrow & & \Big\uparrow\, b \\ \mathbf{A} & \xrightarrow{\;d\;} & \mathbf{B} \end{array}$$

If we assume that we have enough functions to complete the diagram, then we get a logic $L\varphi$ of all formulae (as types) which are non-empty types. If we start with a weaker logic than the intuitionistic (obtained by restricting λ-abstraction) and we add a φ-diagram to be closed, we can get logics which are incomparable with the intuitionistic and yet have a well defined functional interpretation. The lattice of such logics is yet to be studied. To take an example of a φ which can yield classical logic, let '$\varphi(\mathbf{X}) \equiv (\mathbf{A} \rightarrow$

$\mathbf{B}) \to \mathbf{X}$'. The '$a$' and '$b$' arrows, which make, respectively, '$\mathbf{A} \to ((\mathbf{A} \to \mathbf{B}) \to \mathbf{A})$' and '$\mathbf{B} \to ((\mathbf{A} \to \mathbf{B}) \to \mathbf{B})$', are provable with intuitionistic implication. The arrow 'c', which corresponds to '$((\mathbf{A} \to \mathbf{B}) \to \mathbf{A}) \to ((\mathbf{A} \to \mathbf{B}) \to \mathbf{A})$', is also provable with intuitionistic implication if we allow '$\mathbf{A} \to \mathbf{B}$' be used as both ticket and minor premisse of the Modus Ponens (*→-elimination*). So, instead of full classical logic where for any '$\mathbf{A}$' there is a '$\mathbf{B}$' such that either '$\mathbf{A}$' is true or '$\mathbf{A} \to \mathbf{B}$' is true ('$\mathbf{B}$' could be the false, i.e. '$\mathcal{F}$'), thus filling the 'd' arrow, we have a weakened excluded middle: either '$(\mathbf{A} \to \mathbf{B}) \to \mathbf{A}$' or '$\mathbf{A} \to \mathbf{B}$'. As a parallel to Curry's [13] proof that

$$(\mathbf{A} \vee (\mathbf{A} \to \mathbf{B})) \to (((\mathbf{A} \to \mathbf{B}) \to \mathbf{A}) \to \mathbf{A})$$

we show that under a certain extended λ-abstraction discipline, namely that a single abstraction cancels free occurrences of the variable which appear both as higher and as lower type, one can prove that:

$$(((\mathbf{A} \to \mathbf{B}) \to \mathbf{A}) \vee (\mathbf{A} \to \mathbf{B})) \to (((\mathbf{A} \to \mathbf{B}) \to \mathbf{A}) \to ((\mathbf{A} \to \mathbf{B}) \to \mathbf{B})).$$

Taking the consequent of the above proposition as the type-scheme of a combinator we call 'P''', and showing that it can itself be proved under the extended abstraction discipline, we develop a type-theoretic counterpart to *Peirce's law*, which allows us to add as an extra condition to the presentation of the type/proposition '$\mathbf{A} \to \mathbf{B}$' giving us classical implication. The idea is to introduce an extra condition to the λ-abstraction rule which discharges an assumption in the form of '$y : \mathbf{A} \to \mathbf{B}$' forming a λ-abstraction term as a member of the antecedent (lower) type '$\mathbf{A}$', under the condition that '$\mathbf{B}$' is obtained from the assumption '$\mathbf{A} \to \mathbf{B}$', the latter being used as both minor assumption and major assumption of a Modus Ponens (*→-elimination*).

We then formulate a general form of *reductio ad absurdum* which seems to fit more easily within the functional interpretation, and is framed as follows:

$$\frac{\begin{array}{c}[\mathbf{A} \to \mathbf{B}]\\ \mathbf{B}\end{array}}{\mathbf{A}} \quad \text{(provided '}\mathbf{A} \to \mathbf{B}\text{' is used as both minor and ticket)}$$

Note that it almost looks like the *→-introduction* 'upside down', perhaps giving a sign that we are getting closer to the symmetries of classical logic (implication).

In the framework of the functional interpretation it is framed as:

$$\frac{\begin{array}{c}[x : \mathbf{A} \to \mathbf{B}]\\ b(x, \ldots, x) : \mathbf{B}\end{array}}{\lambda x.b(x, \ldots, x) : \mathbf{A}} \boxed{\begin{array}{c}\text{'}\mathbf{A} \to \mathbf{B}\text{'}\\ \text{as minor \&}\\ \text{as ticket}\end{array}}$$

meaning that if from the assumption that a term 'x' belongs to a type of the form '$\mathbf{A} \to \mathbf{B}$' one obtains a term '$b(x)$' belonging to the consequent '$\mathbf{B}$' where 'x' appears both as a 'higher' subterm (i.e. as function) and a 'lower' subterm (i.e. as argument) of '$b(x)$', then we can apply a $\lambda x.$-abstraction over the '$b(x, \ldots, x)$' term and obtain a term of the form '$\lambda x.b(x, \ldots, x)$' belonging to the antecedent '$\mathbf{A}$', discharging the assumption '$x : \mathbf{A} \to \mathbf{B}$'. With that special proviso we can prove *Peirce's axiom* in the following way:

$$\cfrac{\cfrac{\cfrac{\cfrac{[y:\mathbf{A}\to\mathbf{B}] \quad [x:(\mathbf{A}\to\mathbf{B})\to\mathbf{A}]}{\mathtt{APP}(x,y):\mathbf{A}} \qquad [y:\mathbf{A}\to\mathbf{B}]}{\mathtt{APP}(y,\mathtt{APP}(x,y)):\mathbf{B}}}{\lambda y.\mathtt{APP}(y,\mathtt{APP}(x,y)):\mathbf{A}}\,(*)}{\lambda x.\lambda y.\mathtt{APP}(y,\mathtt{APP}(x,y)):((\mathbf{A}\to\mathbf{B})\to\mathbf{A})\to\mathbf{A}}$$

where in the step marked with '$(*)$' we have applied our alternative to Curry's generalized *reductio ad absurdum*. The resulting term '$\lambda x.\lambda y.\mathtt{APP}(y, \mathtt{APP}(x, y))$' which intuitionistically would have a type-scheme of the form '$((\mathbf{A} \to \mathbf{B}) \to \mathbf{A}) \to ((\mathbf{A} \to \mathbf{B}) \to \mathbf{B})$', we would call '$\mathsf{P}'$'. In fact, P' follows from a weakened version of the excluded middle, namely

$$((\mathbf{A} \to \mathbf{B}) \to \mathbf{A}) \vee (\mathbf{A} \to \mathbf{B})$$

in the system of intuitionistic implication extended by allowing two assumptions of the form '$\mathbf{A} \to \mathbf{B}$' (one used as minor premisse and the other used as major premisse of a Modus Ponens) to be discharged by one single *$\to$-introduction.* Unlike the intuitionistic case, the classical implication has 'involution', which means that by switching a formula to the negative (antecedent) side of the implication, and then switching it back again to the positive (consequent) side of the implication, one returns to the original formula. So, in our case, we took '$\mathbf{A}$', switched it to the negative side by making '$\mathbf{A} \to \mathbf{B}$', and then switched it back to the positive side by making '$(\mathbf{A} \to \mathbf{B}) \to \mathbf{A}$'. So, we postulate that this final formula must be equivalent to '$\mathbf{A}$' itself. This is reflected in the terms-side when we see that in the term before the marked inference above 'y' is being applied to a term which represents the application of 'x' to 'y' itself. The 'restricted' self-application comes up to the surface when we use 'labels'.

On the logical side, the resulting alternative presentation of the inferential counterpart to *Peirce's axiom* finds a special case in another one of the standard presentations of the proof-theoretic *reductio ad absurdum*, namely:

$$\cfrac{\begin{array}{c}[\neg\mathbf{A}]\\ \mathcal{F}\end{array}}{\mathbf{A}} \qquad \text{which can also be seen as} \qquad \cfrac{\begin{array}{c}[\mathbf{A}\to\mathcal{F}]\\ \mathcal{F}\end{array}}{\mathbf{A}}$$

with the difference that our '**B**' does not have to be the special *falsum* constant '$\mathcal{F}$', which means that we can have a positive classical logic, i.e. classical implication with no distinguished '$\mathcal{F}$'.

6.3.1. *Can a λ-term Inhabit a Type Which may not be an →-type?*
We have seen above that the η-convertibility rule, namely:

→-induction

$$\frac{c : \mathbf{A} \rightarrow \mathbf{B}}{\lambda x.\mathtt{APP}(c, x) =_\eta c : \mathbf{A} \rightarrow \mathbf{B}}$$

(where c does not have any free occurrence of x) states that for any arbitrary term 'c' which inhabits an →-type, it must be susceptible to an application to an argument, which if it is an arbitrary one, then by abstracting the latter from the result, one should obtain a term definitionally equal to the original λ-abstraction term. Therefore, it must be a λ-abstraction term. It guarantees that no other term inhabits →-types.

In fact, the rules of labelled natural deduction given above say:

1. what a type contains (*introduction*) by showing how to construct a canonical element, as well as when two canonical elements are equal;
2. how an element of the type is de-constructed (*reduction*); and
3. that any term other than those constructed by the *introduction* rules can inhabit the type.

None of the rules (*introduction, reduction, induction*) says anything, however, as to whether a λ-abstraction term can inhabit a type which is not necessarily an →-type. If we remind ourselves that the labelled natural deduction rules given for '→' had the type as the focus of attention, we have no difficulties accepting that the rule of *reductio ad absurdum*, namely:

$$\frac{\begin{array}{c}[x : \mathbf{A} \rightarrow \mathbf{B}] \\ b(x, \ldots, x) : \mathbf{B}\end{array}}{\lambda x.b(x, \ldots, x) : \mathbf{A}} \boxed{\begin{array}{c}\text{'}\mathbf{A} \rightarrow \mathbf{B}\text{'} \\ \text{as minor \&} \\ \underline{\text{as ticket}}\end{array}}.$$

The rule does no harm to the methodology: it only says that an (open) λ-abstraction term can also inhabit a type which is not necessarily an →-type. The rule allows us to conclude that the λ-abstraction term '$\lambda x.b(x, \ldots, x)$' inhabits a type of the form '**A**', which may or may not be an →-type. Notice, however, that the λ-term will not be a closed term, further steps being necessary to turn it into a closed term by discharging the remaining free variables appearing between the two occurrences of x as function and as argument.

7 Finale

In order to summarize our present discussion on our system of labelled natural deduction, we would like to draw attention to two main points. Firstly, with the definition of our labelled system, and an explanation of the resulting bi-dimensional calculus in terms of the devices defined by Frege, we intend to demonstrate that there may be more to the functional interpretation than the somewhat restrictive (intuitionistic) Curry–Howard interpretation. Instead of *terms* and *types*, we can see it as handling *labels* and *formulae*, together with some kind of *label-and-formula* notion of construction. Secondly, given the generality of a framework where the enrichment of the proof calculus is such that entities (which may denote either labels of formulae or proper individuals to be predicated about) are seen as 'arbitrary objects' carrying the computational content of proofs, the potential for the use of such systems in computer science is not to be dismissed.

7.1 Beyond Curry–Howard: the *Label-and-Formula* Notion of Construction

The functional interpretation of logical connectives is concerned with a certain harmony between, on the one hand, a functional calculus on the expressions built up from the recording of the deduction steps (the *labels*), and, on the other hand, a logical calculus on the formulae. It has been associated with Curry's early discovery of the correspondence between the axioms of intuitionistic implicational logic and the type schemes of the so-called 'combinators' of Combinatory Logic [12], and has been referred to as the 'formulae-as-types' interpretation. Howard's [80] extension of the formulae-as-types paradigm to full intuitionistic first-order predicate logic meant that the interpretation has since been referred to as the 'Curry–Howard' functional interpretation. Although Heyting's intuitionistic logic did fit well into the formulae-as-types paradigm, it seems fair to say that, since Tait's [117, 118] intensional semantics of Gödel's [69] *Dialectica* system of functionals of finite type, there has been enough indication that the framework would also be applicable to logics beyond the realm of intuitionism. Ultimately, the foundations of a 'functional' approach to logic are to be found in the work of Frege with his system of 'concept writing', not in that of Curry, or Howard, or indeed Heyting.

With the advent of *labelled* systems, as put forward by D. Gabbay in his forthcoming book on *Labelled Deductive Systems* [54], where the declarative unit is a labelled formula 't : **A**' (read 't labels **A**'), a logical system can now be seen as not simply a calculus of logical deductions on formulae, but a suitably harmonious combination of a functional calculus on the labels and a logical calculus on the formulae.

The role of the label is to provide additional information about the formula which is not of the same declarative nature as that of the formula itself. The label t in $t : \mathbf{A}$ can represent the degree of reliability of the item of data $\mathbf{A}$, or can be a λ-term representing a proof of $\mathbf{A}$, or as in the case of many valued logics, t can indicate the range of truth values of $\mathbf{A}$. Thus, depending on the logical system involved, the intuitive meaning of the labels vary. In querying databases, we may be interested in labelling the assumptions so that when we get an answer to a query we can indicate, via its label, from which part of the database the answer was obtained. Another area where labelling is used is temporal logic. We can time-stamp assumptions as to when they are true and, given those assumptions, query whether a certain conclusion will be true at a certain time. The consequence notion for labelled deduction is essentially the same as that of any logic: given the assumptions, does a conclusion follow?

The labels allow us to code meta-level information. For example, if we want to reason about 'our proof so far', we can either go to a meta-logic which names proofs and talks about them, or we can tag (label) formulae and propagate the tag, coding the necessary meta-level information in the tag. In computer science terms this would be identified as some sort of implementation. Of course, in *LDS* it is done logically. Thus, whereas in the traditional logical system the consequence relation is defined using proof rules on the formulae, in the *LDS* methodology the consequence relation is defined by using rules on both formulae and their labels. Formal rules are then established for manipulating labels and this allows for more scope in decomposing the various features of the consequence relation. The meta-level features coded by the extra tag aforementioned can be formally reflected in the algebra or logic of the labels and the object-level logical features can be reflected in the rules operating on the formulae.

7.2 Back to Frege

What we have presented in this monograph may be seen as a framework for studying the mathematical foundations of *Labelled Deductive Systems.* We can also regard it as an attempt at a reinterpretation of Frege's logical calculus where abstractors and functional operators work harmoniously together with logical connectives and quantifiers. In other words, the functional interpretation (sometimes referred to as the Curry–Howard–Tait interpretation) can be viewed in a wider perspective of a labelled deductive system which can be used to study a whole range of logics, including some which may not abide by the tenets of the intuitionistic interpretation (e.g. classical implicational logic, many-valued logics, etc.). The result is a labelled natural deduction system which we would like to see as a reinterpretation of Frege's 'functional' account of logic: it is as if the theory of functions of *Grundgesetze* is put together with the theory of predicates of *Begriffsschrift* in such a way that a formula is true (valid) if and only if a deduction of it can be constructed where the label contains no free variable (i.e. its proof-construction is a 'complete' object [49], meaning that the truth of the formula relies on no assumptions). The reinterpretation of Frege's functional interpretation in terms of an analysis of *deduction* gives rise to what we would call the *label-and-formula* notion of construction. (Architecture of the system: *Grundgesetze* alongside *Begriffsschrift*.)

7.3 The Functional Interpretation and Computer Science

Since the publication in the early–middle seventies of seminal papers by P. Martin-Löf [97, 99], J.-Y. Girard [61], P. Aczel [3], *et al.*, on the use of the Curry–Howard functional interpretation as a foundational basis for the semantics of programming languages, the logical foundations of computation, proof theory and type theory, there has been much interest in the topic within the computer science community. For historical reasons, perhaps, the interpretation has been mostly associated with the intuitionistic interpretation of logic and mathematics. Indeed, the well-established Curry–Howard interpretation, as it was originally formulated, is especially suitable for Heyting's intuitionistic predicate calculus.

With the characterization of labelled natural deduction systems, it is our objective to extend the functional interpretation to classical as well as non-classical (including modal) logics. One of the key ideas, already discussed in de Queiroz' [20] recent doctoral dissertation, is to adopt a semantic theory based on *convertibility* in the style of Tait [117, 118], which would be more general than the semantic theories based on canonical proofs (such as the one proposed by P. Martin-Löf [101, 102, 103]), allowing significantly more flexibility as to which logical system to choose as adequate for the particular application one might have in mind.

The extension of the functional interpretation to classical and non-classical logics shall be of great usefulness to some key topics in current computer science research, namely:

- the integration of functional and logic programming;
- the integration of techniques from λ-calculus, proof theory and the theory of abstract data types;
- the establishment of closer ties between model-theoretic and proof-theoretic approaches to automated deduction;
- the establishment of stronger connections between categorical and proof-theoretic techniques to the semantics of programming languages;
- the integration of discourse representation formalisms and categorial grammar with the more familiar (to computer scientists) 'computational' interpretation of types.

It shall also be of relevance to the integration of different research trends, as can be seen from the pertinence of each topic listed above: software engineering, automated deduction, theoretical computer science, natural language and computational linguistics.

Acknowledgements

Various people have contributed with their criticisms and observations, on the occasion of presenting the material in various places, including the workshops of ESPRIT Basic Research Action MEDLAR, and the Amsterdam-London exchange symposium in February 1992. We are particularly grateful to , Johan van Benthem, Thierry Boy de la Tour, Marcello D'Agostino, Ruth Kempson, Christoph Kreitz, Wilfried Meyer-Viol, Heinrich Wansing.

Research partially supported by Brazilian National Research Council, CNPQ ('Bolsa de Pesquisador 1C').

Ruy J. G. B. de Queiroz
Universidade Federal de Pernambuco, Brasil.

Dov M. Gabbay
King's College, London.

References

1. Alan R. Anderson and Nuel D. Belnap Jr. *Entailment. The Logic of Relevance and Necessity I.* Princeton University Press, Princeton, xxxii+541pp, 1975. With contributions by J. Michael Dunn and Robert K. Meyer.
2. Samson Abramsky. Computational interpretations of linear logic. Theoretical Computer Science, volume 111(1–2):3–57, 1993.
3. Peter H. G. Aczel. The strength of Martin-Löf's intuitionistic type theory with one universe. In S. Miettinen and J. Väänänen, editors, *Proceedings of the Symposia in Mathematical Logic, Oulu, 1974, and Helsinki, 1975*, volume 2 of *Technical Reports*, pages 1–32. Department of Philosophy, University of Helsinki, 1977.
4. Peter H. G. Aczel. Frege structures and the notions of proposition, truth and set. In J. Barwise, H.-J. Keisler, and K. Kunen, editors, *The Kleene Symposium*, volume 101 of *Studies in Logic and The Foundations of Mathematics*, pages 31–59, Amsterdam, xx+425pp, 1980. North-Holland Publishing Co. Proceedings of the Symposium held in June 18–24, 1978, at Madison, Wisconsin, USA.
5. Peter H. G. Aczel. Term declaration logic and generalized composita. In *Sixth Annual IEEE Symposium on Logic in Computer Science (LICS '91)*, pages 22–30. IEEE Press, 1991. Proceedings of the Symposium held July 15–18 1991, in Amsterdam, The Netherlands.
6. Arnon Avron. The Semantics and Proof Theory of Linear Logic. *Theoretical Computer Science*, 57:161–184, 1988.
7. Jon Barwise. *The Situation in Logic*, volume 17 of *CSLI Lecture Notes*. Center for the Study of Language and Information, Stanford, xvi+327pp, 1989.
8. Jon Barwise and Dov M. Gabbay. Situating Labelled Entailments. Typescript, Dept of Computing, Imperial College, 1997
9. Errett Bishop. *Foundations of Constructive Analysis. McGraw-Hill series in Higher Mathematics.* McGraw-Hill Book Company, New York, xiv+371pp, 1967.

10. Michael Barr and Charles Wells. *Category Theory for Computing Science. Prentice Hall International Series in Computer Science.* Prentice Hall, New York, xv+432pp, 1990.
11. Brian F. Chellas. *Modal Logic. An Introduction.* Cambridge University Press, Cambridge, xii+295pp, 1980.
12. Haskell B. Curry. Functionality in combinatory logic. *Proceedings of the National Academy of Sciences of USA*, 20:584–590, 1934.
13. Haskell B. Curry. *A Theory of Formal Deducibility*, volume 6 of *Notre Dame Mathematical Lectures.* Notre Dame University Press, Notre Dame, Indiana, xi+129pp, third printing (1966) of second (1957) edition, 1950.
14. Anjolina G. de Oliveira. *Transformations Between Proofs in Labelled Natural Deduction via Term Rewriting.* MSc Thesis. Departamento de Informática, Universidade Federal de Pernambuco, April 1995. (In Portuguese)
15. Anjolina G. de Oliveira and Ruy J. G. B. de Queiroz. A new basic set of proof transformations (abstract). *Logic Colloquium'95*, Haifa, Israel, August 1995.
16. Ruy J. G. B. de Queiroz. Note on Frege's notions of definition and the relationship proof theory vs. recursion theory (Extended Abstract). In *Abstracts of the VIIIth International Congress of Logic, Methodology and Philosophy of Science. Vol. 5, Part I*, pages 69–73, Moscow, 1987. Institute of Philosophy of the Academy of Sciences of the USSR.
17. Ruy J. G. B. de Queiroz. A proof-theoretic account of programming and the rôle of reduction rules. *Dialectica*, 42(4):265–282, 1988.
18. Ruy J. G. B. de Queiroz. The mathematical language and its semantics: to show the consequences of a proposition is to give its meaning. In Paul Weingartner and Gerhard Schurz, editors, *Reports of the Thirteenth International Wittgenstein Symposium 1988*, volume 18 of *Schriftenreihe der Wittgenstein-Gesellschaft*, pages 259–266, Vienna, 304pp, 1989. Hölder-Pichler-Tempsky. Symposium held in Kirchberg/Wechsel, Austria, August 14–21 1988.
19. Ruy J. G. B. de Queiroz. Normalization and the semantics of use (abstract). *Journal of Symbolic Logic*, 55:425, 1990. Abstract of a paper presented at *Logic Colloquium '88* held in Padova, Italy, August 23–30 1988.
20. Ruy J. G. B. de Queiroz. *Proof theory and computer programming. The Logical Foundations of Computation.* PhD thesis, Department of Computing, Imperial College, University of London, February 1990.
21. Ruy J. G. B. de Queiroz. Meaning as grammar *plus* consequences. *Dialectica*, 45(1):83–86, 1991.
22. Ruy J. G. B. de Queiroz. *Grundgesetze* alongside *Begriffsschrift* (abstract). In *Abstracts of Fifteenth International Wittgenstein Symposium*, pages 15–16, 1992. Symposium held in Kirchberg/Wechsel, August 16–23 1992.
23. Ruy J. G. B. de Queiroz. Normalization and language-games. *Dialectica*, 48(2):83–125.
24. Ruy J. G. B. de Queiroz and Dov M. Gabbay. The functional interpretation and the sequent calculus. Technical report, Department of Computing, Imperial College, Draft April 20, 1992.
25. Ruy J. G. B. de Queiroz and Dov M. Gabbay. Labelled natural deduction. Technical report, Department of Computing, Imperial College, Draft April 20, 1992.
26. Ruy J. G. B. de Queiroz and Dov M. Gabbay. The functional interpretation of the existential quantifier. *Bulletin of the Interest Group in Pure and Applied Logics* 3(2–3):243–290, 1995. (Presented at *Logic Colloquium '91*, Uppsala, August 9–16 1991. Abstract *JSL* 58(2):753–754, 1993.)
27. Ruy J. G. B. de Queiroz and Dov M. Gabbay. Equality in labelled deductive systems and the functional interpretation of propositional equality. In *Proceedings of the 9th Amsterdam Colloquium*, P. Dekker and M. Stockhof (editors), ILLC/Department of Philosophy, University of Amsterdam, pp. 547–566.
28. Ruy J. G. B. de Queiroz and Thomas S. E. Maibaum. Proof theory and computer programming. *Zeitschrift für mathematische Logik und Grundlagen der Mathe-*

matik, 36:389–414, 1990.
29. Ruy J. G. B. de Queiroz and Thomas S. E. Maibaum. Abstract data types and type theory: theories as types. *Zeitschrift für mathematische Logik und Grundlagen der Mathematik*, 37:149–166, 1991.
30. Ruy J. G. B. de Queiroz and Michael B. Smyth. Induction rules for non-inductive types in type theory. Technical report, Department of Computing, Imperial College, 1989. Presented at the *Fifth British Colloquium for Theoretical Computer Science*, Royal Holloway and Bedford New College, Egham, Surrey, UK, April 11–13 1989. Abstract to appear in *EATCS*.
31. Michael A. E. Dummett. Truth. *Proceedings of the Aristotelian Society (new series)*, 59:141–162, 1959.
32. Michael A. E. Dummett. *Frege: Philosophy of Language*. Duckworth, London, xiii+708pp, second (1981) edition, 1973.
33. Michael A. E. Dummett. The philosophical basis of intuitionistic logic. In H. E. Rose and J. C. Shepherdson, editors, *Logic Colloquium '73*, volume 80 of *Studies in Logic and The Foundations of Mathematics*, pages 5–40, Amsterdam, viii+513pp, 1975. North-Holland. Proceedings of the Colloquium held in Bristol, UK, 1973.
34. Michael A. E. Dummett. *Elements of Intuitionism*. Series *Oxford Logic Guides*. Clarendon Press, Oxford, xii+467pp, reprinted (1985) edition, 1977. With the assistance of Roberto Minio.
35. Michael A. E. Dummett. Comments on Professor Prawitz's paper. In G. H. von Wright, editor, *Logic and Philosophy*, Series *Entretiens of the International Institute of Philosophy*, pages 11–18. Martinus Nijhoff Publishers, The Hague, viii+84pp, 1980. Symposium held in Düsseldorf, August 27 – September 1 1978.
36. Michael A. E. Dummett. *Frege: Philosophy of Mathematics*. Duckworth, London, xiii+331pp, 1991.
37. Michael A. E. Dummett. *The Logical Basis of Metaphysics*. Duckworth, London, xi+355p, 1991. Revised account (1989) of *The William James Lectures* given at Harvard University in 1976.
38. Jens E. Fenstad, editor. *Proceedings of the Second Scandinavian Logic Symposium*, volume 63 of *Studies in Logic and The Foundations of Mathematics*. North-Holland, Amsterdam, viii+405pp, 1971. Proceedings of the Symposium held in Oslo, June 18–20 1970.
39. Luis Fariñas del Cerro and Andreas Herzig. Modal deduction with applications in epistemic and temporal logics. Research report, LSI-IRIT, Toulouse, 60pp, 1990.
40. Kit Fine. *Reasoning with Arbitrary Objects*, volume 3 of *Aristotelian Society series*. Basil Blackwell, Oxford, viii+220pp, 1985.
41. Frederic B. Fitch. Natural deduction rules for obligation. *American Philosophical Quarterly*, 3:27–38, 1966.
42. Frederic B. Fitch. Tree proofs in modal logic. *Journal of Symbolic Logic*, 31:152, 1966. Abstract of a paper presented at a meeting of the Association for Symbolic Logic in conjunction with the American Philosophical Association, at Chicago, Illinois, 29–30 April 1965.
43. Melvin Fitting. An epsilon-calculus system for first-order S4. In *Lecture Notes in Mathematics*, pages 103–110. Springer-Verlag, 1972.
44. Melvin Fitting. A modal logic epsilon-calculus. *Notre Dame Journal of Formal Logic*, 16:1–16, 1975.
45. Melvin Fitting. *Proof Methods for Modal and Intuitionistic Logics*, volume 169 of *Synthese Library. Studies in Epistemology, Logic, Methodology and Philosophy of Science*. D. Reidel, Dordrecht, viii+555pp, 1981.
46. Melvin Fitting. Modal logic should say more than it does. In J.-L. Lassez and G. Plotkin, editors, *Computational Logic. Essays in Honor of Alan Robinson*. MIT Press, Cambridge, MA. 1989.
47. Gottlob Frege. *Begriffsschrift, eine der arithmetischen nachgebildete Formelsprache des reinen Denkens*. Verlag von Louis Nebert, Halle, 1879. English translation 'Begriffsschrift, *a formula language, modeled upon that of arithmetic, for pure*

thought' in [123], pages 1–82.

48. Gottlob Frege. *Die Grundlagen der Arithmetik. Eine logisch mathematische Untersuchung über den Begriff der Zahl.* Verlag Wilhelm Koebner, Breslau, 1884. English translation *The Foundations of Arithmetic* by J. L. Austin, 2nd rev. ed. 1978, Basil Blackwell, Oxford.
49. Gottlob Frege. Funktion und Begriff. *Proceedings of the Jena Medical and Scientific Society*, 1891. English translation 'Function and Concept' (by Peter Geach) published in [95], pages 137–156.
50. Gottlob Frege. *Grundgesetze der Arithmetik. Begriffsschriftlich abgeleitet. I.* Verlag von Hermann Pohle, Jena, 1893. Reprinted in volume 32 of *Olms Paperbacks*, Georg Olms Verlagsbuchhandlung, Hildesheim, 1966, XXXII+254pp. Partial English translation in [53].
51. Gottlob Frege. *Grundgesetze der Arithmetik. Begriffsschriftlich abgeleitet. II.* Verlag von Hermann Pohle, Jena, 1903. Reprinted in volume 32 of *Olms Paperbacks*, Georg Olms Verlagsbuchhandlung, Hildesheim, 1966, XVI+266pp. Partial English translation in [55].
52. Gottlob Frege. Logic in mathematics, 1914. In [79], pages 203–250.
53. Montgomery Furth, editor. *The Basic Laws of Arithmetic. Exposition of the System.* University of California Press, Berkeley and Los Angeles, lxiv+143pp, 1964. Partial English translation of Gottlob Frege's *Grundgesetze der Arithmetik.*
54. Dov M. Gabbay. *Labelled Deductive Systems, Volume I – Foundations.* Oxford University Press. First Draft 1989. Current Draft, 465pp., May 1994, 1994. Published as MPI-I-94-223, Max-Planck-Institut für Informatik, Im Stadtwald D-663123 Saarbrücken, Germany.
55. Peter Geach and Max Black, editors. *Translations from the Philosophical Writings of Gottlob Frege.* Basil Blackwell, Oxford, x+228pp, 3rd (1980) edition, 1952.
56. Dov M. Gabbay and Ruy J. G. B. de Queiroz. An attempt at the functional interpretation of the modal necessity, 1991. First Draft Mar 11, 1991. Presented at MEDLAR 18-month Workshop, Torino, Italy, Apr 27–May 1, 1991. Published in the *MEDLAR Deliverables PPR2, 1991.*
57. Dov M. Gabbay and Ruy J. G. B. de Queiroz. Extending the Curry–Howard–Tait intepretation to linear, relevant and other resource logics (abstract). *Journal of Symbolic Logic*, 56(3):1139–1140, 1991. Presented at the *Logic Colloquium '90*, Helsinki, July 15–22, 1990.
58. Dov M. Gabbay and Ruy J. G. B. de Queiroz. Extending the Curry-Howard intepretation to linear, relevant and other resource logics. *Journal of Symbolic Logic*, 57(4):1319–1365, 1992.
59. Gerhard Gentzen. Untersuchungen über das logische Schliessen. *Mathematische Zeitschrift*, 39:176–210 and 405–431, 1935. English translation 'Investigations into Logical Deduction' in [116], pages 68–131.
60. Dov M. Gabbay and Ian Hodkinson. An axiomatization of the temporal logic with until and since over the real numbers. *Journal of Logic and Computation*, 1(2):229–259, 1990.
61. Jean-Yves Girard. Une Extension de l'Interpretation de Gödel à l'Analyse, et son Application à l'Elimination des Coupures dans l'Analyse et la Théorie des Types, 1971. In [38], pages 63–92.
62. Jean-Yves Girard. Linear Logic. *Theoretical Computer Science*, 50:1–102, 1987.
63. Jean-Yves Girard. Towards a geometry of interaction. In J. W. Gray and A. Scedrov, editors, *Category Theory in Computer Science and Logic*, volume 92 of *Contemporary Mathematics*, pages 69–108. American Mathematical Society, 1989. Proceedings of the Symposium held in 1987, Boulder, Colorado.
64. Jean-Yves Girard. A new constructive logic: classical logic. *Mathematical Structures in Computer Science*, 1:255–296, 1991.
65. Dov M. Gabbay and Ruth M. Kempson. Labelled abduction and relevance reasoning. In *Proceedings of the Workshop on Non-Standard Queries and Non-Standard Answers*, Toulouse, July 1991.

66. Dov M. Gabbay and Ruth M. Kempson. Natural-language content and information flow: a proof-theoretic perspective. In P. Dekker, editor, *Proceedings of the 8th Amsterdam Colloquium on Formal Semantics*, 1992.
67. Jean-Yves Girard, Yves Lafont, and Paul Taylor. *Proofs and Types*, volume 7 of *Cambridge Tracts in Theoretical Computer Science*. Cambridge University Press, Cambridge, xii+175pp, reprinted with minor corrections (1990) edition, 1989.
68. Kurt Gödel. Eine Interpretation des intuitionistischen Aussagenkalküls. *Ergebnisse eines mathematischen Kolloquiums*, 4:39–40, 1933. English translation 'An interpretation of the intuitionistic propositional calculus' in *The Philosophy of Mathematics*, edited by J. Hintikka, Oxford University Press, 1969.
69. Kurt Gödel. Über eine bisher noch nicht benützte Erweiterung des finiten Standpunktes. *Dialectica*, 12:280–287, 1958. English translation 'On a hitherto unexploited extension of the finitary standpoint' in *Journal of Philosophical Logic*, 9:133–142, 1980.
70. Andrej Grzegorczyk. Some relational systems and the associated topological spaces. *Fundamenta Mathematicae*, 60:223–231, 1967.
71. John Guttag. Abstract data types and the development of data structures. *Communications of the ACM*, 20(6):396–404, June 1977.
72. Ian Hacking. What is logic? *Journal of Philosophy*, LXXVI(6):285–319, 1979.
73. David Hilbert and Paul Bernays. *Grundlagen der Mathematik I*, volume XL of *Die Grundlehren der mathematischen Wissenschaften*. Verlag von Julius Springer, Berlin, XII+471pp, 1934. Reprinted by Edwards Brothers, Ann Arbor, Michigan, 1944.
74. David Hilbert and Paul Bernays. *Grundlagen der Mathematik II*, volume L of *Die Grundlehren der mathematischen Wissenschaften*. Verlag von Julius Springer, Berlin, XII+498pp, 1939. Reprinted by Edwards Brothers, Ann Arbor, Michigan, 1944.
75. Arend Heyting. Die formale Regeln der intuitionistische Logik. *Sitzungsberichte der preussischen Akademie von Wissenschaften (physicalischmathematische Klasse)*, pages 42–56, 1930.
76. Arend Heyting. *Intuitionism. An Introduction.* Series *Studies in Logic and the Foundations of Mathematics*. North-Holland, Amsterdam, viii+133pp, 1956.
77. K. Jaakko Hintikka. Quantifiers vs. Quantification Theory. *Linguistic Inquiry*, 5:153–177, 1974.
78. K. Jaakko Hintikka. Game-theoretical semantics: insights and prospects. In J. Hintikka and J. Kulas, editors, *The Game of Language*, volume 22 of *Synthese Language Library*, pages 1–31. D. Reidel, Dordrecht, xii+319pp, 1983.
79. Hans Hermes, Friedrich Kambartel, and Friedrich Kaulbach, editors. *Gottlob Frege. Posthumous Writings*. Basil Blackwell, Oxford, XIII+288pp, 1979. Transl. by Peter Long and Roger White.
80. William A. Howard. The formulae-as-types notion of construction. In J. P. Seldin and J. R. Hindley, editors, *To H. B. Curry: Essays on Combinatory Logic, Lambda Calculus and Formalism*, pages 479–490. Academic Press, London, xxv+606pp, 1980. Privately circulated notes, 1969, only later published in Curry's *Festschrift*.
81. Gerard Huet, editor. *Logical foundations of functional programming*. Series *UT Year of Programming*. Addison-Wesley, Reading, Mass., xvi+491pp, 1990.
82. Stephen C. Kleene. On the interpretation of intuitionistic number theory. *Journal of Symbolic Logic*, 10:109–124, 1945.
83. Hans Kamp and Uwe Reyle. *From Discourse to Logic*. Kluwer, 1993.
84. Saul A. Kripke. A completeness theorem in modal logic. *Journal of Symbolic Logic*, 24:1–14, 1959.
85. Saul A. Kripke. Semantic analysis of modal logic. I: normal propositional calculi. *Zeitschrift für mathematische Logik und Grundlagen der Mathematik*, 9:67–96, 1963.
86. Saul A. Kripke. Semantical analysis of modal logic II: non-normal modal propositional calculi. In J. W. Addison, Leon Henkin, and Alfred Tarski, editors, *The*

Theory of Models, pages 206–220. North-Holland, Amsterdam, 1965.
87. Albert C. Leisenring. *Mathematical Logic and Hilbert's ε-Symbol*. A volume of *University Mathematical Series*. MacDonald Technical and Scientific, London, ix+142pp, 1969.
88. Clarence Irving Lewis and Cooper Harold Langford. *Symbolic Logic*. The Century Co., New York, second (with Dover, New York, 1959) edition, 1932.
89. M. H. Löb. Solution of a problem of Leon Henkin. *Journal of Symbolic Logic*, 20:115–118, 1955.
90. Paul Lorenzen. *Einführung in die operative Logik und Mathematik*, volume LXXVIII of *Die Grundlehren der mathematischen Wissenschaften*. Springer-Verlag, Berlin, iv+298pp, 1955.
91. Paul Lorenzen. Ein dialogisches Konstruktivitätskriterium. In *Infinitistic Methods*, Oxford, 362pp, 1961. Pergamon Press. Proceedings of the Symposium on the Foundations of Mathematics (International Mathematical Union and Mathematical Institute of the Polish Academy of Sciences) held in Warsaw, 2–9 September 1959.
92. Paul Lorenzen. *Normative Logic and Ethics*, volume 236* of *B.I-Hochschultaschenbücher. Systematische Philosophie*. Bibliographisches Institut, Mannheim/Zürich, 89pp, 1969.
93. Joachim Lambek and Philip J. Scott. *Introduction to Higher-order Categorical Logic*, volume 7 of *Cambridge Studies in Advanced Mathematics*. Cambridge University Press, Cambridge, ix+293pp, 1986.
94. Saunders MacLane. *Categories for the Working Mathematician*, volume 5 of *Graduate Texts in Mathematics*. Springer-Verlag, New York, IX+262pp, 1971.
95. Brian McGuinness, editor. *Gottlob Frege. Collected Papers on Mathematics, Logic and Philosophy*. Basil Blackwell, Oxford, viii+412pp, 1984. Translated by Max Black, V. H. Dudman, Peter Geach, Hans Kaal, E.-H. W. Kludge, Brian McGuinness, R. H. Stoothoff.
96. D. H. Mellor, editor. *Foundations: Essays in Philosophy, Logic, Mathematics and Economics / by F. P. Ramsey*. Series *International library of psychology, philosophy and scientific method*. Routledge & Kegan Paul, London, viii+287pp, 1978.
97. Per Martin-Löf. A theory of types. Report 71-3, Department of Mathematics, University of Stockholm, 1971. 57pp. February 1971, revised October 1971.
98. Per Martin-Löf. About models for intuitionistic type theories and the notion of definitional equality. In S. Kanger, editor, *Proceedings of the Third Scandinavian Logic Symposium*, Series *Studies in Logic and The Foundations of Mathematics*, pages 81–109, Amsterdam, 1975. North-Holland. Symposium held in 1973.
99. Per Martin-Löf. An intuitionistic theory of types: predicative part. In H. E. Rose and J. C. Shepherdson, editors, *Logic Colloquium '73*, volume 80 of *Studies in Logic and The Foundations of Mathematics*, pages 73–118, Amsterdam, viii+513pp, 1975. North-Holland. Proceedings of the Colloquium held in Bristol, UK, in 1973.
100. Per Martin-Löf. Constructive mathematics and computer programming. In L. J. Cohen, J. Łos, H. Pfeiffer, and K.-P. Podewski, editors, *Logic, Methodology and Philosophy of Science VI*, Series *Studies in Logic and The Foundations of Mathematics*, pages 153–175, Amsterdam, xiii+738pp, 1982. North-Holland. Proceedings of the International Congress held in Hannover, August 22–29 1979.
101. Per Martin-Löf. *Intuitionistic Type Theory*. Series *Studies in Proof Theory*. Bibliopolis, Naples, iv+91pp, 1984. Notes by Giovanni Sambin of a series of lectures given in Padova, June 1980.
102. Per Martin-Löf. On the meanings of the logical constants and the justifications of the logical laws. In C. Bernardi and P. Pagli, editors, *Atti degli incontri di logica matematica. Vol. 2*, Series *Scuola di Specializzazione in Logica Matematica*, pages 203–281. Dipartimento di Matematica, Università di Siena, 1985.
103. Per Martin-Löf. Truth of a proposition, evidence of a judgement, validity of a proof. *Synthese*, 73:407–420, 1987. Special Issue on *Theories of Meaning*, Guest Editor:

Maria Luisa Dalla Chiara, collecting articles originally presented as contributions to the conference 'Theories of Meaning', organised by the Florence Center for the History and Philosophy of Science, Firenze, Villa di Mondeggi, June 1985.

104. Richard Montague. Universal grammar. *Theoria*, 36:373–398, 1970. Reprinted in [120], pages 222–246.

105. Bengt Nordström, Kent Petersson, and Jan M. Smith. *Programming in Martin-Löf's Type Theory. An Introduction*, volume 7 of *The International Series of Monographs on Computer Science*. Clarendon Press, Oxford, x+221pp, 1990.

106. Giuseppe Peano. *Arithmetices principia, nova methodo exposita*. Turin, 1889. English translation *The principles of arithmetic, presented by a new method* published in [123], pages 83–97.

107. Axel Poigné. Basic category theory. In S. Abramsky, D. Gabbay, and T. Maibaum, editors, *Handbook of Logic in Computer Science. Vol. I*. Oxford University Press, Oxford, 1992.

108. Dag Prawitz. *Natural Deduction. A Proof-Theoretical Study*, volume 3 of *Acta Universitatis Stockholmiensis. Stockholm Studies in Philosophy*. Almqvist & Wiksell, Stockholm, 113pp, 1965.

109. Dag Prawitz. Ideas and results in proof theory, 1971. In [38], pages 235–307.

110. Dag Prawitz. Meaning and proofs: on the conflict between classical and intuitionistic logic. *Theoria*, XLIII:2–40, 1977.

111. Dag Prawitz. Intuitionistic kogic: a philosophical challenge. In G. H. von Wright, editor, *Logic and Philosophy*, Series *Entretiens of the International Institute of Philosophy*, pages 1–10, The Hague, viii+84pp, 1980. Martinus Nijhoff Publishers. Proceedings of the Symposium held in Düsseldorf, August 27 – September 1 1978.

112. Frank P. Ramsey. The foundations of mathematics. *Proceedings of the London Mathematical Society, Ser. 2*, 25:338–384, 1925. Reproduced in [96], pp. 152–212.

113. Henrik Sahlqvist. Completeness and correspondence in the first and second-order semantics for modal logic. In S. Kanger, editor, *Proceedings of the Third Scandinavian Logic Symposium*, pages 110–143. North-Holland, 1975. Symposium held in Uppsala, 1973.

114. Robert Seely. Locally Cartesian closed categories and type theory. *Mathematical Proceedings of the Cambridge Philosophical Society*, 95:33–48, 1984.

115. Hans D. Sluga. *Gottlob Frege*. Series *The Arguments of the Philosophers*. Routledge & Kegan Paul, London, xi+203pp, 1980.

116. Manfred Egon Szabo, editor. *The Collected Papers of Gerhard Gentzen*. Series *Studies in Logic and The Foundations of Mathematics*. North-Holland, Amsterdam, xiv+338pp, 1969.

117. William W. Tait. Infinitely long terms of transfinite type. In J. N. Crossley and M. A. E. Dummett, editors, *Formal Systems and Recursive Functions*, Series *Studies in Logic and The Foundations of Mathematics*, pages 176–185, Amsterdam, 320pp, 1965. North-Holland. Proceedings of the *Logic Colloquium '63*, held in Oxford, UK.

118. William W. Tait. Intensional interpretations of functionals of finite type I. *Journal of Symbolic Logic*, 32:198–212, 1967.

119. William W. Tait. Against intuitionism: constructive mathematics is part of classical mathematics. *Journal of Philosophical Logic*, 12:173–195, 1983.

120. Richmond H. Thomason, editor. *Formal Philosophy. Selected Papers of Richard Montague*. Yale University Press, New Haven and London, 1974.

121. Richmond H. Thomason and R. Stalnaker. Modality and reference. *Noûs*, 2:359–372, 1968.

122. Anne S. Troelstra and Dirk van Dalen. *Constructivism in Mathematics: An Introduction. Vol. II*, volume 123 of *Studies in Logic and The Foundations of Mathematics*. North-Holland, Amsterdam, xvii+535pp, 1988.

123. Jean van Heijenoort, editor. *From Frege to Gödel: A Source Book in Mathematical Logic. 1879-1931*. Series *Source Books in the History of the Sciences*. Harvard University Press, Cambridge, Massachussetts, xii+664pp, 1967.

A GENERAL REASONING SCHEME FOR UNDERSPECIFIED REPRESENTATIONS

ESTHER KÖNIG AND UWE REYLE

1 The Landscape of Underspecified Semantic Representations

Underspecified semantic representations have attracted increasing interest within computational linguistics. Several formalisms have been developed that allow to represent sentence or text meanings with that degree of specificity that is determined by the context of interpretation. As the context changes they must allow for (partial) disambiguation steps performed by a process of refinement that goes hand in hand with the construction algorithm. And as the interpretation of phrases often[1] relies on deductive principles and thus any construction algorithm must be able to integrate the results of deductive processes, any semantic formalism should be equipped with a deductive component that operates directly on its semantic forms.

We call a meaning ϕ of a representation formalism $\mathcal{L}$ *underspecified*, if it represents an ambiguous natural language sentence or text in a more compact manner than by a disjunction of all its readings. $\mathcal{L}$ is called *semantic* if its representations are model-theoretically interpretable or if it comes with a disambiguation device that turns underspecified representations into sets of model-theoretically interpretable representations.[2] If $\mathcal{L}$'s disambiguation steps produce representations of $\mathcal{L}$ only, then $\mathcal{L}$ is called *closed*. And if $\mathcal{L}$'s disambiguation device produces all possible refinements of any ϕ, then $\mathcal{L}$

[1]E.g. in order to apply nominal and temporal resolution, consistency checks, integration of world knowledge or other non-compositional interpretation principles.

[2]Note that the second disjunct requires that either the underspecified representations themselves or the disambiguation algorithm is subject to certain demands on wellformedness, as, e.g. the so-called 'free-variable constraint' ([13, 10]). Although we think that this is a very important distinction (in particular under computational aspects) we do not distinguish here between those formalisms which are provided with a check of such meta-level constraints directly for underspecified representations and those formalisms whose well-formedness test requires all the total disambiguations.

H.J. Ohlbach and U. Reyle (eds.), Logic, Language and Reasoning, 251–277.

is called *complete*. Completeness is, of course, dependent on the particular natural language fragment $\mathcal{L}$ is supposed to cover. In this paper we restrict ourselves to the fragment of simple sentences containing singular indefinite as well as quantified NPs,[3] relative clauses and negation. To give an example what completeness involves let us consider a sentence with three quantified NPs with underspecifed scoping relations. Then $\mathcal{L}$ must be able to represent all $2^{3!} = 64$ refinements, i.e. partial and complete disambiguations of this sentence. For many formalisms the question whether they are complete wrt. a particular fragment, or not, is not decided yet. We, therefore, take a very liberal view and interpret 'complete' more in the sense of 'intended to be comlete' than in the sense of a precise characterization of expressive power. A formalism $\mathcal{L}$ is called *proper* if it is closed and complete. It is *c-deductive* (or 'classically deductive') if there is an inference mechanism for the disjunction of fully specified formulae the underspecified formula is supposed to represent. The formalism is called *u-deductive* if it is equipped with a deductive component that operates directly on the underspecified forms. If the deduction on the underspecified formulae can be merged with disambiguation steps, it is named *cu-deductive.* Table 1 gives a classification of some underspecified formalisms according to these properties. **LFG** stands for the linear logic approach to LFG semantics [7]. **MG** means Montague Grammar [8]. **MRS** is the Minimal Recursion Semantics of [5]. Quasi Logical Forms **QLF** and underspecification has been explored in [2]. For Underspecified Discourse Representation Structures **UDRS** see [15]. **USDL** is one of the formalisms which have been described in the section on underspecification in [3]. **UL** is the *U(nderspecified) L(ogic)*, we present in this paper. As can be judged from the available literature, almost all formalisms are semantic. The completeness property will be discussed subsequently for each formalism. Obviously, all the 'semantic' formalisms are classically deductive, but only **UDRSs** and **UL** are u-deductive. And only **UL** is cu-deductive.

The underspecified logic **UL** is a pair consisting of a proper underspecified semantic representation formalism $\mathcal{L}$, and a deductive component that directly operates on these structures. For the purpose of this paper and also for the sake of comparison we have split up the representations ϕ of a formalism $\mathcal{L}$ into three components, B, C, and D. M specifies the building blocks of the representation language, and C tells us how these building blocks are to be put together. D is the disambiguation device which implements the construction of the individual meaning representations from a meaning description $\langle B, C\rangle$. In the remainder of this section we present the different formalisms from the point of view of B, C, and D. See also Appendix A for an overview. Section 2 will then explain the deductive principles of our un-

[3]With the additional assumption that the interpretation of indefinite NPs is clause-bounded.

TABLE 1. Comparison of various underspecified formalisms with respect to some desirable logical properties.

	LFG	MG	MRS	QLF	UDRS	USDL	UL
semantic	yes	yes	no	yes	yes	yes	yes
closed	yes	yes	yes	yes	yes	yes	yes
complete	no	no	no	almost	yes	yes	yes
proper	no	no	no	almost	yes	yes	yes
c-deductive	yes	yes	no	yes	yes	yes	yes
u-deductive	no	no	no	no	yes	no	yes
cu-deductive	no	no	no	no	no	no	yes

derspecified logic **UL** and will show how these principles can be imported into the other underspecified semantic representation formalisms. The set of importable rules will of course depend on the properties of the particular formalisms.

1.1 Linear Logic Approach to LFG-semantics

In the case of [7]'s linear logic approach to LFG semantics, M consists of linear logic formulae built up from semantic projections (i.e. formulae of the form $h_\sigma \rightsquigarrow Y$ with h referring to an f-structure and Y being a variable or a formula of higher-order predicate logic). C reflects the hierarchical ordering of the underlying f-structure. The structure in (2) is the f-structure of the ambiguous sentence (1).

Every boy saw a movie. (1)

$$f : \begin{bmatrix} \text{SUBJ} \ g : \begin{bmatrix} \text{SPEC every} \\ \text{PRED boy} \end{bmatrix} \\ \text{OBJ} \ \ h : \begin{bmatrix} \text{SPEC a} \\ \text{PRED movie} \end{bmatrix} \\ \text{PRED see} \end{bmatrix} \quad (2)$$

The semantic projections associate the following meaning constructors for *every boy*, *a movie* and *saw*:

$$\left\{ \begin{array}{l} \forall G, R.(\forall x.g_\sigma \rightsquigarrow x \multimap G \rightsquigarrow R(x)) \multimap G \rightsquigarrow \mathtt{every(boy}, R), \\ \forall H, S.(\forall y.h_\sigma \rightsquigarrow y \multimap H \rightsquigarrow S(y)) \multimap H \rightsquigarrow \mathtt{a(movie}, S), \\ \forall x, y.g_\sigma \rightsquigarrow x \otimes h_\sigma \rightsquigarrow y \multimap f_\sigma \rightsquigarrow \mathtt{see}(x, y) \end{array} \right\} \quad (3)$$

D consists of a proof method for linear logic which in the case of (3) allows for two different formulae to be derived. If C only contains restrictions derived from f-structure, then the formalism is not complete. The incompleteness can be shown in a similar way as [9] does for the standard HPSG semantics (cf. [14]). E.g. for a verb with three complements SUBJect, OBJect, and IOBJect, one cannot state a constraint that SUBJ must have wide scope over OBJ while leaving open the scope relation between SUBJ and IOBJ. We must add additional constraints to the effect that certain proof steps are excluded. But this requires also that the proof theory of linear logic is made sensitive to this kind of constraint: a non-trivial extension (viz. the contribution by Crouch and van Genabith in [3]).

1.2 Underspecified Montague Grammar

For Montague Grammar, we take B to be sets of formulae of intensional logic, and C their c-structural relations. We thus consider the tree in (4) to be the underspecified representation of the ambiguous sentence (1).

(4)

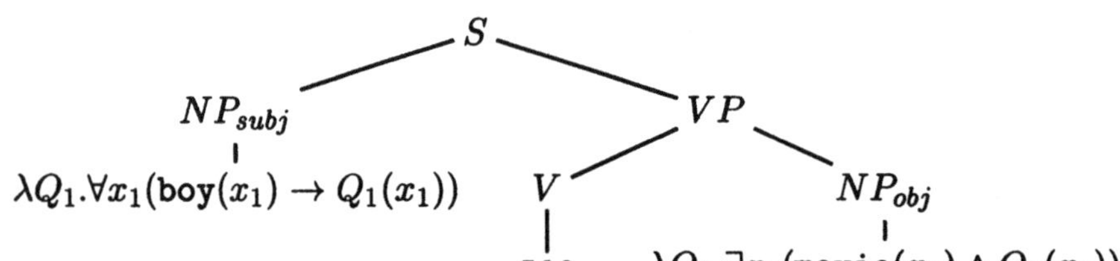

The disambiguation device D of this underspecified Montague Grammar formalism is given by the usual method of deriving sets of meanings on the basis of syntax trees via a quantifier storage mechanism. Applying D to representations like (4) results in structures of the same form, i.e. sets of formulae of intensional logic with empty C, as in

$$\left\{ \begin{array}{l} \forall x_1(\texttt{boy}(x_1)) \rightarrow \texttt{see}(x_1, \lambda Q.\exists x_2(\texttt{movie}(x_2) \wedge Q(x_2))), \\ \exists x_2(\texttt{movie}(x_2) \wedge \forall x_1(\texttt{boy}(x_1) \rightarrow \texttt{see}(x_1, \lambda Q.Q(x_2)))). \end{array} \right\} \quad (5)$$

Hence, this formalism is closed. If D is based on flat quantifier stores then the formalism is not complete, in the same respect as the above mentioned LFG approach. However, the expression of more fine-grained structural relations is possible if quantifier stores can be nested as it has been suggested by [12].

1.3 Minimal Recursion Semantics

The idea of a Minimal Recursion Semantics (MRS) has been brought up by Kay [11] and appeared in published form e.g. as [5]. A nested, i.e. 'recursive' formula is represented in a flat manner as the list of its structural pieces, which come with labels (handles) and whose argument positions are labels. Our previous example will look as in (6) in an HPSG-style feature structure representation (The feature names HANDEL and LISZT are used by [5] to designate the label of an expression and its list of building blocks, resp.)

$$\begin{bmatrix} \text{HANDEL} & \boxed{1} \vee \boxed{5} \\ \text{INDEX} & \boxed{8} \\ \text{LISZT} & \left\langle \begin{array}{l} \begin{bmatrix} \texttt{every\_rel} & \\ \text{HANDEL} & \boxed{1} \\ \text{BV} & \boxed{2} \\ \text{RESTR} & \boxed{3} \end{bmatrix}, \begin{bmatrix} \texttt{boy\_rel} & \\ \text{HANDEL} & \boxed{3} \\ \text{INST} & \boxed{2} \end{bmatrix}, \\ \begin{bmatrix} \texttt{some\_rel} & \\ \text{HANDEL} & \boxed{5} \\ \text{BV} & \boxed{6} \\ \text{RESTR} & \boxed{7} \end{bmatrix}, \begin{bmatrix} \texttt{movie\_rel} & \\ \text{HANDEL} & \boxed{7} \\ \text{INST} & \boxed{6} \end{bmatrix}, \begin{bmatrix} \texttt{see\_rel} & \\ \text{HANDEL} & \boxed{4} \\ \text{EVENT} & \boxed{8} \\ \text{ACT} & \boxed{2} \\ \text{UND} & \boxed{6} \end{bmatrix} \end{array} \right\rangle \end{bmatrix} \quad (6)$$

The flat representations are harder to read for humans (because the subformulae have to be substituted back mentally for the labels), but, according to the authors [5], they are easier to be processed automatically in a Machine Translation scenario. Note that there are no SCOPE features for the quantifiers `every_rel` and `some_rel`. This means that scope has been left unspecified, i.e. the hierarchical relations of the subformulae have been specified only partially.

In our schematization, the building blocks B of an MRS are the 'pieces of formulae' and the structural constraints C are given by the usage of the labels plus some logically motivated conditions, e.g. the semantics of the verbal complements must outscope the verb semantics.

To our knowledge, there is neither a formal semantics for the MRS language nor a scoping algorithm for the derivation of fully specified formulae.

1.4 Quasi Logical Forms

The relevant kernel of the language of Quasi Logical Forms (QLFs) [2, 4] comprises pairs of scope orderings and predicate/argument structures

`ScopeOrdering:PredArgStructure` (7)

where `ScopeOrdering` is a list of labels, and `PredArgStructure` consists of a predicate name and labelled argument terms with the shape

`term(Label,Features,Restrictor,Scope,Index).` (8)

`Scope` and `Index` may be variables. The earlier example sentence corresponds to the QLF in (9)

$$[]:\mathtt{see}\left(\begin{array}{l}\mathtt{term(+g,<type=q,lex=every>,boy,?Q,?X),}\\ \mathtt{term(+h,<type=q,lex=a>,movie,?P,?R).}\end{array}\right) \tag{9}$$

The empty list of scope orderings means that no scope restrictions apply among the arguments of the verb semantics. The wide scope reading of the universal quantifier will be fixed by

$$\mathtt{[+g,+h]}:\mathtt{see}\left(\begin{array}{l}\mathtt{term(+g,<type=q,lex=every>,boy,?Q,?X),}\\ \mathtt{term(+h,<type=q,lex=a>,movie,?P,?R).}\end{array}\right) \tag{10}$$

In QLFs, the structural constraints C are made up from two kinds of information:

1. the scope orderings;
2. the inverse embedding relations of the verb semantics and the argument semantics.

The second kind of information is required, since, logically, the quantified expressions for `every` and `a` on the argument positions must outscope the representation for the verb `see(?X,?R)` in order to avoid dangling variables.

For QLFs, there is both, a direct interpretation method, and a scoping algorithm D [1] which produces higher-order logic formulae. Most of the scopings, which the scoping algorithm can produce, can be expressed alternatively by refinements of the scoping orderings. In [4], it is sketched how the language of scope orderings can be made complete by admitting nested scope orderings.

1.5 Underspecified Discourse Representation Structures

Reyle's Underspecified Discourse Representation Structures (UDRS) [15] uses a set of labelled DRSs

$$B = \left\{ \begin{array}{l} l_0 : \Box, \\ l_1 : \boxed{\begin{array}{l} \boxed{\begin{array}{l} x_1 \\ \mathtt{boy}(x_1)\end{array}} \Rightarrow l_{12} : \Box \end{array}}, \ l_2 : \boxed{\begin{array}{l} x_2 \\ \mathtt{movie}(x_2)\end{array}}, \\ l_3 : \boxed{\begin{array}{l} \\ \mathtt{see}(x_1,x_2)\end{array}} \end{array} \right\} \tag{11}$$

which are partially ordered by possible scoping relations

$$C = \{l_3 \leq l_{12}, l_3 \leq l_2, l_1 \leq l_0, l_2 \leq l_0\}. \tag{12}$$

The disambiguation device D consists of consistently adding conditions that restrict the partial order, such as $l_1 \leq l_2$ (which gives the wide scope reading for the indefinite).

1.6 USDL

A **USDL** representation [3] is a set of equations between partial semantic representations. Naturally, this set can be devided into a set B of building blocks (e.g. the left set in (13)) and a set C of structural constraints (the right set in (13)).

$$\left\{ \begin{array}{l} X_1 = \texttt{every\_boy'}@L_{x_1}(X_4) \\ X_2 = \texttt{a\_movie'}@L_{x_2}(X_6) \\ X_3 = \texttt{see}(x_1, x_2) \end{array} \right\} \quad \left\{ \begin{array}{l} X_0 = C_1(X_1) \\ X_0 = C_2(X_2) \\ X_4 = C_3(X_3) \\ X_6 = C_4(X_3) \end{array} \right\} \tag{13}$$

An expression like $X_1 = \texttt{every\_boy'}@L_{x_1}(X_4)$ is an abbreviation for labelling the application of $\lambda Q.[\forall x_1(\texttt{boy'}(x_1) \rightarrow Q(x_1))]$ to $L_{x_1}(X_4)$ with the name X_1. Due to the granularity of the structural constraints, the formalism is complete according to our standards.

The disambiguation device of **USDL** is higher-order unification, restricted to *linear* formulae. This fact suggests a rather close connection to the disambiguation method in *linear* logic based **LFG** semantics, possibly along the lines of the Curry-Howard isomorphism between formulae and types.

2 UL – Underspecified Logic

The next section presents a general formalism that subsumes the above mentioned ones. Our policy is to keep the formalism as neutral as possible with respect to the particular shape of B and C. Its design is mainly dependent on the objective of being proper, semantic and (c)u-deductive. First, the general issues of such a formalism are discussed, before defining the details of its syntax and semantics.

2.1 The Ambiguity Connective and the Consequence Relation

Suppose a hearer is confronted with an utterance that is non-committal on some point of intense interest to him. Then he may well identify a number of alternatives to complete the utterance with the information that he considers essential. But he is unable to choose among them as long as this bit of information is not provided. On the basis of this intuition the semantic meaning of ambiguous utterances A is adequately represented by a (partial) function $[\![A]\!] : \mathcal{K} \rightarrow \{[\![A_i]\!]\}_{i=1,\ldots,n}$ from contexts $\mathcal{K}$ to the set of fully specified readings $\{[\![A_i]\!]\}_{i=1,\ldots,n}$ of A. As not all contexts do provide sufficient information to identify exactly one reading we may identify fully specified readings $[\![A_i]\!]$ with constant functions $[\![A_i]\!] : \mathcal{K} \rightarrow \{[\![A_i]\!]\}$ and generalize $[\![A]\!]$ to functions mapping contexts $\kappa \in \mathcal{K}$ to functions from contexts to meanings of less ambiguous expressions. We thus assume that the underlying logical formalism is proper. To see what syntactical devices we need to guarantee properness consider (14), (15), and (16).

James knows Jeeves. (14)

He smokes. (15)

James knows Jeeves. He smokes. (16)

Pronouns as well as proper names may have more than one possible reference, leading to ambiguities in (14), (15), and (16). The problem is that when (14) and (15) are combined to (16), their ambiguities do not multiply out. To see this suppose the domain of individuals consists of four people, $\{a, b, c, d\}$ of which $\{a, b\}$ are bearers of the name *James* and $\{c, d\}$ bear the name *Jeeves*. Then (14) and (15) are both four times ambiguous, or have four *possible* disambiguations. The sentence *He smokes* is also four times ambiguous if uttered in the context of (14), as in (16). In this context, however, the pronoun *he* is *contextually bound*, or *restricted* by the constraint that its antecedent should be either James or Jeeves. We will use a coindexing device as in (17) and (18) to indicate contextual disambiguation.

James knows Jeeves$_1$. *He*$_1$ *smokes.* (17)

James$_1$ *knows Jeeves*$_2$. *He*$_{1/2}$ *smokes.* (18)

The effect of contextual restriction on possible disambiguations is that the possible disambiguations of simple sentences do not multiply out when they are contextually combined. Taken the contextual restriction in (17) we do not get 16 readings for the whole sequence in (16), but only four. More interesting is the contextual disambiguation in (18). Although any of $\{a, b, c, d\}$ may be the referent of the pronoun *he* we only get eight readings for (18). It is important to note that this kind of contextual restriction on possible disambiguation is at work for all kinds of ambiguities. The sentences (19), (20), and (21) are sample cases of ambiguities that do not involve quantifier scope.

I like squash. (19)

The students get £ 100. (20)

Firemen are available. (21)

The sentence *It tastes wonderful* expresses a (post-hoc partial) restriction to (19) that excludes the interpretation of *squash* as a sport. If (20) is uttered with the floating quantifier *each* then the collective reading is excluded. And the existential reading of (21) may be forced by adding an ellipsis construction like *and their chief too.*[4]

As contextual disambiguation applies to all kinds of ambiguities the coindexing device must be equally flexible. Consider (22) (taken from [16]).

$$\frac{\begin{array}{c}\textit{If [the students get £100]}_i \textit{ then they buy books.}\\ \textit{[The students get £100]}_j.\end{array}}{\textit{The students buy books.}} \tag{22}$$

According to the most natural interpretation the two occurrences of

$$\textit{The students get £100.} \tag{23}$$

are taken to mean the same thing, i.e. i is taken to be equal to j. Under this coindexing constraint the meaning of the premise of (22) is given by (25) not by (24), where A_1 represents the first and A_2 the second reading of the second sentence of (22).

$$((A_1 \to B) \vee (A_2 \to B)) \wedge (A_1 \vee A_2) \tag{24}$$

$$((A_1 \to B) \wedge A_1) \vee ((A_2 \to B) \wedge A_2). \tag{25}$$

Note that *The students buy books* and *They buy books* must also be correlated in (25). Otherwise the argument would not be sound (under the assumption that the distributive reading of buying books (we mean: distributive with respect to the set of students) is not logically equivalent to the collective reading).

Before we go on let us make a small remark on non-monotonicity. The choice a context makes among different readings may be subject to revision, as shown in (26) and (27).

$$\begin{array}{c}\textit{James enters the room.}\\ \textit{When he smokes Mary gets angry.}\end{array} \tag{26}$$

$$\begin{array}{c}\textit{James enters the room.}\\ \textit{When he smokes Lord Leicester wants to have a brandy.}\end{array} \tag{27}$$

Ambiguity and context change thus result in non-monotonicity. This does, however, not affect the problem of ambiguous consequence we are discussing

[4]Proposals for the parallel disambiguation of quantifiers in the context of coordination and elliptical construction have been made in [1] and [6].

in this section. The reason is the following: we take a set of underspecified representations to be given as the result of interpreting – say – a piece of text. In particular we assume that contextual disambiguations relevant for the understanding of the text have been made by the interpreter. That is we assume the data to be decorated with a fixed set of indices that express the contextual choices made by the interpreter. Given this kind of data, we want to know what can be derived from it.

2.1.1. *The Ambiguity Connective,* $\sharp$

Let $\sharp$ be an operator that represents A's ambiguity between (possibly ambiguous) sentences A_1 and A_2 by $A_1 \sharp A_2$. We have seen that any attempt to represent the interpretation of $\sharp$ by a function $[\![A_1 \sharp A_2]\!]$ is doomed to failure, because its interpretation does not take contextual disambiguation into account. It must thus be parametrized by contexts κ to $[\![A_1 \sharp A_2]\!]^\kappa$. What are the properties of $[\![\cdot]\!]^\kappa$?

First of all, it has to guarantee that the $\sharp$-operator distributes over negation.[5] The ambiguity in (16) is present in exactly the same way in *James doesn't know Jeeves. He doesn't smoke.* This means that

$$[\![\neg(A_1 \sharp A_2)]\!]^\kappa = [\![(\neg A_1)\sharp(\neg A_2)]\!]^\kappa \quad \text{for any } \kappa \tag{28}$$

For conjunction and implication, $\odot \in \{\wedge, \rightarrow\}$, the case is more complicated, because they are binary and thus must respect (mutual) contextual restrictions between the formulae, A and B, they combine. If contextual constraints affect A and B, then the whole product set

$$\delta(A \odot B) := \{[\![A_1 \odot B_1]\!], [\![A_1 \odot B_2]\!], [\![A_2 \odot B_1]\!], [\![A_2 \odot B_2]\!]\} \tag{29}$$

is no longer available. The set is restricted to pairs

$$\delta^\kappa(A \odot B) := \{[\![A_{\iota_1} \odot B_{\nu 1}]\!] \ldots, [\![A_{\iota_n} \odot B_{\nu n}]\!]\} \tag{30}$$

that are admitted by the constraint set κ expressing coindexations between (sub-)phrases of A and B.[6] This means that the interpretation function $[\![\cdot]\!]^\kappa$ must satisfy the following property for two-place connectives $\odot \in \{\wedge, \rightarrow\}$.

$$[\![(A_1 \sharp A_2) \odot (B_1 \sharp B_2)]\!]^\kappa = [\![(A_{\iota_1} \odot B_{\nu 1})\sharp \ldots \sharp(A_{\iota_n} \odot B_{\nu n})]\!]^\kappa \tag{31}$$

$\delta^\kappa(A)$ is a *disambiguation operation* that respects contextual restrictions within A. We may assume that the contextual constraints, κ, are given

[5]We restrict ourselves to cases here where the presence of negation doesn't increase the set of possible readings as, e.g. in *John doesn't admire any linguist*, which is ambiguous, whereas *John admires any linguist* is not.

[6]In the following $\wedge$ is the dynamic (left-associative) conjunction operation on formulae with the intuition that the first argument presents the context in which the second argument is asserted.

as sets of equations, or membership relations, indicating coreferentiality of certain term expressions, or, more generally, correlatedness of phrase meanings. Consider again (17) and (18). Assume that A, B, C and D unambiguously refer to the individuals a, b, c and d, respectively. Then (32) corresponds to (17) and (33) to (18). (34) is no possible disambiguation.

$$\begin{aligned}&\delta^{\{[He]=[James]\}}(\textit{James knows Jeeves. He smokes.})\\ &= \left\{\begin{array}{l}\textit{A knows C. A smokes., B knows C. B smokes.,}\\ \textit{A knows D. A smokes., B knows D. B smokes.}\end{array}\right\}\end{aligned} \tag{32}$$

$$\begin{aligned}&\delta^{\{[He]\in\{[James],[Jeeves]\}\}}(\textit{James knows Jeeves. He smokes.})\\ &= \delta^{\{[He]=[James]\}}(\textit{James knows Jeeves.He smokes.})\\ &\cup \left\{\begin{array}{l}\textit{A knows C. C smokes., B knows C. C smokes.,}\\ \textit{A knows D. D smokes., B knows D. D smokes.}\end{array}\right\}\end{aligned} \tag{33}$$

$$\begin{aligned}&\delta^{\{[He]\in\{[James],[Jeeves]\}\}}(\textit{James knows Jeeves. He smokes.})\\ &= \{\textit{A knows C. B smokes.}\}\end{aligned} \tag{34}$$

Consider again (22) the data of which are abbreviated here as (35). Let A_1 and A_2, B_1 and B_2 be the two readings of the sentences A and B, respectively. (36) makes this explicit. If we assume that the only contextual disambiguation between A and B concerns the binding of *they* by *the students* then (36) is equivalent to (37) by applying (31) to (35)'s antecedent. (37) is equivalent to (38) if we assume that no contextual disambiguation occurs between the two occurrences of A. And (37) is equivalent to (39) if we assume that the two occurrences of A are co-indexed.

$$(A \to B) \wedge A \tag{35}$$

$$(A_1 \sharp A_2) \to (B_1 \sharp B_2)) \wedge A_1 \sharp A_2) \tag{36}$$

$$((A_1 \to B_1) \sharp (A_1 \to B_2) \sharp (A_2 \to B_1) \sharp (A_2 \to B_2)) \wedge A_1 \sharp A_2) \tag{37}$$

$$\begin{aligned}&((A_1 \to B_1) \wedge A_1) \sharp ((A_1 \to B_1) \wedge A_2)\\ &\sharp ((A_1 \to B_2) \wedge A_1) \sharp ((A_1 \to B_2) \wedge A_2)\\ &\sharp ((A_2 \to B_1) \wedge A_1) \sharp ((A_2 \to B_1) \wedge A_2)\\ &\sharp ((A_2 \to B_2) \wedge A_1) \sharp ((A_2 \to B_2) \wedge A_2)\end{aligned} \tag{38}$$

$$\begin{aligned}&((A_1 \to B_1) \wedge A_1) \sharp ((A_1 \to B_2) \wedge A_1)\\ &\sharp ((A_2 \to B_1) \wedge A_2) \sharp ((A_2 \to B_2) \wedge A_2)\end{aligned} \tag{39}$$

2.1.2. *The Consequence Relation*

Suppose a reasoning step is performed by an agent in order to make some information B explicit that is contained in his mental state, A, only implicitly. Then B automatically inherits all the contextual restrictions attached to the information bits it is derived from. Consider once more (22) with the coindexing constraints given in (40)

$$\frac{\begin{array}{c}\textit{If [[the students]}_1 \textit{ get £100]}_2 \textit{ then [[they]}_1 \textit{ buy books]}_3.\\ \textit{[[The students]}_1 \textit{ get £100]}_2.\end{array}}{\textit{[[The students]}_i \textit{ buy books]}_j.} \tag{40}$$

Given the kind of forward reasoning performed by an agent then the conclusion B in (40) must carry the same indices as the consequent of the implication, i.e. the index i of the conclusion must be set equal to 1, and j to 3. This co-indexing is an essential part of the derivation, because B is completely disambiguated by the contextual constraints imposed by what the agent knows, i.e. the data A, it is derived from. In technical terms, $|\delta^\kappa(B)| = |\delta^\kappa(A \wedge B)|$ for all κ.

A case to be distinguished from this one arises, for example, if some person, P_1, asks a question B to some other person, P_2. For reasons of dialogue cohesion there will be certain contextual restrictions between the question B and the representation A of P_2's knowledge.[7] But there may be ambiguities in the query that P_2 cannot resolve. Suppose the query B is *Do they*$_1$ *buy books?* and A corresponds to the data in (40). Here the interpretation of the pronoun *they* in B is correctly bound to the NP [*The students*]$_1$ by co-indexation. Now P_2 may well – by the kind of forward reasoning described above – derive [[*they*]$_1$ *buy books*]$_3$. But the question is, under which circumstances will he give the answer *yes* to P_1's question? There are two cases to be distinguished here. P_1 may not be aware of the ambiguity in neither [[*they*]$_1$ *buy books*]$_3$ nor in the representation B of P_1's question. In this case he will consider the two representations to be equivalent and give a positive answer. So, let us assume he knows about [[*they*]$_1$ *buy books*]$_3$'s ambiguity. Then he must conceive the possibility that the meanings of his and P_1's occurrence may diverge. This will in fact be the case if P_1 actually had a collective meaning in mind but did not make this explicit in the way he formulated the query. And if P_2 contextually disambiguates [[*they*]$_1$ *buy books*]$_3$ to the distributive reading, because he came to learn more about the students practices, the correlation between the antecedent and consequent of the implication in (35), and the absolute amount of money the students actually get. Being aware of this possible divergence

[7]Among them restrictions concerning the interpretation of proper names, pronouns, tenses and so on.

P_2 will not give the answer *yes*. If he is sure about the distributive reading he will instead give an answer that provides the disambiguating information to P_1's query, namely *They each bought books*. And if his own representation is ambiguous as well then he may make this explicit by answering *Either they each bought books or they bought books together*. This last answer shows that P_2's representation of the ambiguous [[*they*]$_1$ *buy books*]$_3$ is equivalent to the disjunction of its disambiguations (disjunction modulo contextual constraints within A, that is). The first answer indicates that on the other hand P_2 represents P_1's ambiguous query as equivalent to the conjunction of its disambiguations.

Thus, if B is ambiguous between B_1 and B_2, then $B_1 \sharp B_2 \models B_1 \sharp B_2$ is true if all of $B_1 \models B_1$, $B_1 \models B_2$, $B_2 \models B_1$, and $B_2 \models B_2$ are. Only if the two occurrences of B are coindexed, i.e. if P_2 knows that his and P_1's B mean the same thing, then $B_i \models B_i$ is true if $B_1 \models B_1$ and $B_2 \models B_2$ is. Hence both scenarios we discussed, the forward reasoning and the dialogue case, may be subsumed by the following general definition of ambiguous consequence.

$$A \models^{\kappa} B \text{ iff } \text{ for all } \delta^{\kappa} \;\; \delta^{\kappa}(A) \models \delta^{\kappa}(B). \tag{41}$$

The following versions of reflexivity, monotonicity and transitivity hold for $\models^{\kappa}$.

THEOREM 2.1

Reflexivity: $A \wedge B \models^{\kappa} B$ *iff for all* $B_i, B_j \in \delta^{\kappa}(B)$ $A \wedge B \models^{\kappa} B_i \leftrightarrow B_j$.
Monotonicity: If $A \models^{\kappa} B$ *and* $\kappa \subseteq \kappa'$, *then* $A \wedge A' \models^{\kappa'} B$.
Transitivity: If $A \models^{\kappa} B$, $B \models^{\kappa'} C$ *and for all* $B_i, B_j \in \delta^{\kappa \cup \kappa'}(B)$
it holds that $A \wedge B \models^{\kappa \cup \kappa'} B_i \leftrightarrow B_j$,
then $A \models^{\kappa \cup \kappa'} C$.

2.2 The Language of UL

Subsequently, the project of a general language of underspecification will be carried out in more detail.
The *signature* of **UL** consists of the following disjoint sets

1. a set of operators $\exists$, $\forall$, $\rightarrow$, $\sharp$ (disambiguation), $\Box_1$, $\Box_2$, ... (indices);
2. a set of (first-order) terms $\mathbf{t}_1$, $\mathbf{t}_2$, ..., which include a set of variables x_1, x_2, ...
3. a set of predicate symbols $\perp$ (false), P_1, P_2, ...;
4. and a set of labels l_1, l_2, ...

The *syntax* of *underspecified formulae* in **UL** is defined subsequently. Note that we distinguish between *underspecified formulae* and *underspecification forms*. A basic underspecified formula is an underspecification form which is labelled with a contextual *index*. In this way, one can make sure that underspecification comes always together with a contextual parameter, which could serve to disambiguate it. Of course, the coindexing of underspecification forms makes only sense, if there is a reasonable amount of 'similarity' among the coindexed material, e.g. sharing of labels.

Atomic formula:
If $\mathtt{t}_1, \ldots, \mathtt{t}_n$ are terms and P is a predicate symbol which requires n arguments then
$P(\mathtt{t}_1, \ldots, \mathtt{t}_n)$ is an *atomic formula.*

Partial formula:
If u is an underspecified formula, and x is a variable then

1. $l_1 : \neg l_2$
2. $l_1 : u \rightarrow l_2$
3. $l_1 : \exists x.u(x) \wedge l_2$
4. $l_1 : \forall x.u(x) \rightarrow l_2$

are *partial formulae.* For a partial formula, l_1 is called the *label* of the partial formula, and l_2 is its *embedded label.*

Set B of building blocks:
B consists of labelled underspecified and partial formulae.

Basic underspecification form:
If B is a set of buildings blocks, C is a set of relational constraints then $\langle B, C\rangle$ is a *basic underspecification form.*

Complex underspecification form:
If v and v' are underspecification forms, then $v \sharp v'$ is a *complex underspecification form.*

Basic underspecified formula:
If v is an underspecification form and $\square_i$ is an index then
$\square_i(v)$ is a *basic underspecified formula.*

Complex underspecified formula:
If u_i is an atomic formula, or a basic or complex underspecified formula then

1. $\neg u$	4. $u_1 \rightarrow u_2$
2. $u_1 \wedge u_2$	5. $\exists x.u$
3. $u_1 \vee u_2$	6. $\forall x.u$

are *complex underspecified formulae.*

Labelled underspecified formula:
If u is an underspecified formula then $l : u$ is a *labelled underspecified formula.*

Concerning the *relational constraints*, note that the partial formulae themselves induce structural constraints, as well. E.g. the constraint $l_2 \leq l_1$ could have been derived from the partial formula $(l_1 : \forall x.u(x) \rightarrow l_2)$.

We call the set C in an underspecification form $\langle B, C\rangle$, the *explicit constraints*, and the constraints which are derived from the elements of B the *implicit constraints.* The *constraints of* $\langle B, C\rangle$, i.e. union of the explicit and the implicit relational constraints of an underspecification form $\langle B, C\rangle$, must satisfy at least the following conditions

1. The constraints must form an asymmetric relation.
2. They must ensure that all variables are bound by some quantifier. I.e. if a variable occurs in some formula, this formula must be required to be subordinate to the partial formula which contains the corresponding quantifier.
3. The minimal elements must be underspecified formulae (not partial ones).

A constraint set is *total* iff it is a total order.

Furthermore, we demand that all the constraint sets of the basic underspecification forms in a complex underspecification form $v\sharp v'$ must be mutually incompatible.

The basis of the *disambiguation device* D is given by a disambiguation function δ which maps a basic underspecification form $v := \langle B \cup \{l : p\}, C\rangle$ onto an (underspecified) formula

$$p\,[l'/\langle B, C\,[l/l']\rangle] \quad \text{where } l' \text{ is an embedded label of } p$$

if the set of constraints $\{l_i \leq l \mid l_i \text{ occurs in } v\}$ is consistent with the constraints of v. The disambiguation function δ must be total, i.e. it must provide for a value for any underspecification form.

The dependency on contextual constraints will be realized as a dependency on a value assignment to the indices of an underspecified formula. Such a *disambiguation assignment* α is a function from indices $\Box_i$ onto constraint sets C. A *disambiguation function* δ^α *with associated disambiguation assignment* α is a restriction of a disambiguation function δ to those values which are compatible with a given disambiguation assignment α, i.e.

$$\delta^\alpha(\Box_i(\langle B, C\rangle)) := \begin{cases} \delta(\langle B, C\rangle) & \text{if } \alpha(\Box_i) \text{ is compatible with the} \\ & \quad \text{(recursively embedded) constraints of } \delta(\langle B, C\rangle) \\ \bot & \text{otherwise.} \end{cases} \tag{42}$$

A disambiguation assignment α' is called a *refinement* of another disambiguation assignment α, if for all indices $\Box_i$ $\alpha'(\Box_i) \models \alpha(\Box_i)$. A disambiguation assignment is called *total* if all its values are total.

Disambiguation functions δ^α with total assignments α can also disambiguate complex underspecification forms:

$$\delta^\alpha(\Box_i(v_1\sharp v_2)) := \begin{cases} \delta^\alpha(\Box_i(v_1)) & \text{if } \delta^\alpha(\Box_i(v_1)) \neq \bot \\ \delta^\alpha(\Box_i(v_2)) & \text{if } \delta^\alpha(\Box_i(v_2)) \neq \bot \\ \bot & \text{otherwise.} \end{cases} \tag{43}$$

Whenever no explicit reference to the labels is required, labels can be omitted:

long form	short form
$l_1 : \neg l_2$	$\neg$
$l_1 : u \rightarrow l_2$	$u \rightarrow$
$l_1 : \exists x.u(x) \wedge l_2$	$\exists x.u(x)$.

For example let us assume that $\forall x_1.\mathtt{boy}(x_1)$, $\exists x_2.\mathtt{movie}(x_2)$ and $\mathtt{see}(x_1, x_2)$[8] denote the meanings of *every boy*, *a movie*, and *saw*, respectively, in sentence (1). This means that an NP-meaning consists of the specification of its quantifier and its restrictor. We do not make any further stipulations which would be specific to any individual semantic theory.

2.3 The Logic of UL

In this section, the notion of satisfiability of underspecified formulae will be defined and the rule of Generalized Modus Ponens will be introduced.

DEFINITION 2.2 (Satisfiability of **UL** formulas)

Let $\mathcal{M}$ be a first-order model in the usual manner with interpretations $P_{\mathcal{M}}$ and $\mathtt{t}_{\mathcal{M}}$ for predicate symbols P and for terms $\mathtt{t}$ respectively. A formula u is satisfiable in a first-order model $\mathcal{M}$ if for all disambiguation functions δ^α there exists a total refinement α' of α such that one of the following cases applies:

1. $\mathcal{M}, \delta^{\alpha'} \models P(\mathtt{t}_1, \ldots, \mathtt{t}_n)$ if $\langle \mathtt{t}_{1\mathcal{M}}, \ldots, \mathtt{t}_{n\mathcal{M}} \rangle \in P_{\mathcal{M}}$. ($\mathtt{t}_1, \ldots, \mathtt{t}_n$ are ground terms)
2. $\mathcal{M}, \delta^{\alpha'} \models \Box_i(\langle B, C \rangle)$ if $\mathcal{M}, \delta^{\alpha'} \models \delta^{\alpha'}(\Box_i(\langle B, C \rangle))$
3. $\mathcal{M}, \delta^{\alpha'} \models \Box_i(v \sharp v')$ if $\mathcal{M}, \delta^{\alpha'} \models \delta^{\alpha'}(\Box_i(v \sharp v'))$
4. (a) $\mathcal{M}, \delta^{\alpha'} \models \neg u$ if $\mathcal{M}, \delta^{\alpha'} \not\models u$,
   (b) $\mathcal{M}, \delta^{\alpha'} \models u_1 \wedge u_2$ if $\mathcal{M}, \delta^{\alpha'} \models u_1$ and $\mathcal{M}, \delta^{\alpha'} \models u_2$,
   (c) $\mathcal{M}, \delta^{\alpha'} \models u_1 \vee u_2$ if $\mathcal{M}, \delta^{\alpha'} \models u_1$ or $\mathcal{M}, \delta^{\alpha'} \models u_2$,
   (d) $\mathcal{M}, \delta^{\alpha'} \models u_1 \rightarrow u_2$ if $\mathcal{M}, \delta^{\alpha'} \not\models u_1$ or $\mathcal{M}, \delta^{\alpha'} \models u_2$,
   (e) $\mathcal{M}, \delta^{\alpha'} \models \exists x.u$ if for some ground term $\mathtt{t}$ $\mathcal{M}, \delta^{\alpha'} \models u\,[x/\mathtt{t}]$,
   (f) $\mathcal{M}, \delta^{\alpha'} \models \forall x.u$ if for every ground term $\mathtt{t}$ $\mathcal{M}, \delta^{\alpha'} \models u\,[x/\mathtt{t}]$.

Note that $\sharp$ is unlike disjunction because it respects the contextual constraints given by the indices $\Box_i$. Furthermore, the condition that the constraint sets which occur in a complex underspecification form $v \sharp v'$ be mutually incompatible, guarantees for the distributivity (28) of the $\sharp$-operator

[8] We ignore tense in this paper.

since eventually at most one 'reading' will be chosen during the evaluation against a total disambiguation assignment. For example, let v be the reading which is picked by the (total) disambiguation assignment α, then

$$\begin{aligned}&\mathcal{M},\delta^\alpha \models \neg\Box_i(v\sharp v') \text{ iff } \mathcal{M},\delta^\alpha \not\models \Box_i(v)\,\sharp\,\Box_i(v')\\ &\text{iff } \mathcal{M},\delta^\alpha \not\models \delta^\alpha(\Box_i(v)\,\sharp\,\Box_i(v')) \text{ iff } \mathcal{M},\delta^\alpha \not\models \Box_i(v) \text{ iff } \mathcal{M},\delta^\alpha \models \neg\Box_i(v)\\ &\text{iff } \mathcal{M},\delta^\alpha \models \langle\neg\Box_i(v),\{\}\rangle \text{ iff } \mathcal{M},\delta^\alpha \models \delta^\alpha(\langle\neg\Box_i(v),\{\}\rangle\sharp\langle\neg\Box_i(v'),\{\}\rangle)\\ &\text{iff } \mathcal{M},\delta^\alpha \models \langle\neg\Box_i(v),\{\}\rangle\sharp\langle\neg\Box_i(v'),\{\}\rangle.\end{aligned} \tag{44}$$

In order to formulate side conditions on our deduction rules, we need the notion of *polarity* of a subformula, following [16]. The polarity of a formula is defined if its relative scope to all negations or monotone decreasing quantifiers in the whole formula is fixed: The formula has positive (resp. negative) polarity if it is in the scope of an even (resp. odd) number monotone decreasing quantifiers. Otherwise, its polarity is undefined.

Based on the above semantics, the following inference rule is sound, if the partial formula $\forall x.P(x)$ is restricted to have positive polarity.

$$\frac{\begin{array}{c}\Box_i(\langle\{\forall x.P(x)\}\cup B, C\rangle)\\ P(\mathtt{t})\end{array}}{\Box_i(\langle B, C\rangle)\,[x/\mathtt{t}]} \tag{45}$$

THEOREM 2.3 *If* $\forall x.P(x)$ *has positive polarity, then the rule (45) is sound.*

Proof: We have to show that

$$\begin{array}{c}\text{If } \mathcal{M},\delta^\alpha \models \Box_i(\langle\{\forall x.P(x)\}\cup B, C\rangle) \text{ and } \mathcal{M},\delta^\alpha \models P(\mathtt{t})\\ \text{then } \mathcal{M},\delta^\alpha \models \Box_i(\langle B, C\rangle)\,[x/\mathtt{t}].\end{array} \tag{46}$$

By assuming $\mathcal{M},\delta^\alpha \models P(\mathtt{t})$ and applying the semantic definitions, we get

$$\begin{array}{c}\forall\ \delta^\alpha \text{ it must hold that}\\ \text{if } \mathcal{M},\delta^\alpha \models \delta^\alpha(\Box_i(\langle\{\forall x.P(x)\}\cup B, C\rangle))\\ \text{then } \mathcal{M},\delta^\alpha \models \delta^\alpha(\Box_i\langle B, C\rangle)\,[x/\mathtt{t}].\end{array} \tag{47}$$

We show (47) by an induction on all possible disambiguated formulae (cf. [15, p.176])

$$\delta^\alpha(\Box_i(\langle\{\forall x.P(x)\}\cup B, C\rangle))$$

which corresponds simultaneously to an induction on

$$\delta^\alpha(\Box_i\langle B, C\rangle)\,[x/\mathtt{t}]$$

due to the coindexation of both formulae. The induction on possible disambiguated formulae amounts to an induction on the number e of partial formulae 'above' the formula $\forall x.P(x)$ in a disambiguated formula (i.e. the 'embedding level' of the partial formula).

$e = 0$: clear, since this is the usual format of the Generalized Modus Ponens.
$e = j - 1$: Assume that

$$\begin{array}{l} \text{if } \mathcal{M}, \delta^\alpha \models \delta^\alpha(\Box_i(\left\langle \begin{array}{l} \{l_j : \forall x.P(x) \rightarrow l_{j2}\} \cup B, \\ \{l_j \leq l_{1,\dots,j-1}\} \cup C \end{array} \right\rangle)) \\ \text{then } \mathcal{M}, \delta^\alpha \models \delta^\alpha(\Box_i \langle B, \{l_j \leq l_{1,\dots,j-1}\} \cup C \rangle) \, [x/\mathtt{t}] \end{array} \tag{48}$$

$e = j$: We have to show that

$$\begin{array}{l} \text{if } \mathcal{M}, \delta^\alpha \models \delta^\alpha(\Box_{i'}(\left\langle \begin{array}{l} \{l_j : \forall x.P(x) \rightarrow l_{j2}\} \cup B, \\ \{l_j \leq l_{1,\dots,j-1,j+1}\} \cup C \end{array} \right\rangle)) \\ \text{then } \mathcal{M}, \delta^\alpha \models \delta^\alpha(\Box_{i'} \langle B, \{l_j \leq l_{1,\dots,j-1,j+1}\} \cup C \rangle) \, [x/\mathtt{t}]. \end{array} \tag{49}$$

Since there is no restriction on B in (48), a formula which is embedded under l_j can be removed without harm:

$$\begin{array}{l} \text{if } \mathcal{M}, \delta^\alpha \models \delta^\alpha(\Box_i(\left\langle \begin{array}{l} \{l_j : \forall x_1.P_1(x_1) \rightarrow l_{j2}\} \cup B \\ \backslash \{l_k : \exists x_2.P_2(x_2) \wedge l_{k2}\} \\ \{l_j \leq l_{1,\dots,j-1}, l_k \leq l_j\} \cup C \end{array} , \right\rangle)) \\ \text{then } \mathcal{M}, \delta^\alpha \models \delta^\alpha(\Box_i \left\langle \begin{array}{l} B \setminus l_k : \exists x_2.P_2(x_2) \wedge l_{k2}, \\ \{l_j \leq l_{1,\dots,j-1}, l_k \leq l_j\} \cup C \end{array} \right\rangle) \, [x_1/\mathtt{t}] \end{array} \tag{50}$$

However, then

$$\text{if } \mathcal{M}, \delta^\alpha \models \exists x_2 \left(\begin{array}{l} P_2(x_2) \\ \wedge \; \delta^\alpha(\Box_i(\left\langle \begin{array}{l} \{l_j : \forall x_1.P_1(x_1) \rightarrow l_{j2}\} \cup B \\ \backslash \{l_k : \exists x_2.P_2(x_2) \wedge l_{k2}\} \\ \{l_j \leq l_{1,\dots,j-1}, l_k \leq l_j\} \cup C \end{array} , \right\rangle)) \end{array} \right)$$

then

$$\mathcal{M}, \delta^\alpha \models \exists x_2 \left(\begin{array}{l} P_2(x_2) \\ \wedge \; \delta^\alpha(\Box_i \left\langle \begin{array}{l} B \setminus l_k : \exists x_2.P_2(x_2) \wedge l_{k2}, \\ \{l_j \leq l_{1,\dots,j-1}, l_k \leq l_j\} \cup C \end{array} \right\rangle) \end{array} \right) [x_1/\mathtt{t}] \tag{51}$$

Similarly for other, monotone increasing partial formulae. ∎

Consider again (1). It can be understood as saying that each of the boys saw a different movie or that all of them saw the same movie. If in addition to (1) we know that Kevin is a boy, then we should be able to derive the fact that Kevin saw a movie – irrespective of what reading the first sentence is intended to have. The fact that Kevin is a boy is represented by `boy(kevin)`. An application of Generalized Modus Ponens (GMP) looks as follows.

$$\frac{\begin{array}{c}\Box_i(\langle\{\forall x_1.\mathtt{boy}(x_1), \exists x_2.\mathtt{movie}(x_2), \mathtt{see}(x_1,x_2)\}, C\rangle)\\ \mathtt{boy}(\mathtt{kevin})\end{array}}{\Box_i(\langle\{\exists x_2.\mathtt{movie}(x_2), \mathtt{see}(\mathtt{kevin},x_2)\}, C\rangle).} \tag{52}$$

As the conclusion is not ambiguous any more there is no problem with correctness here. But assume we had an ambiguous relative clause ϕ attached to *movie* (e.g. *in which he didn't like every actor*) then we need to guarantee correctness by contextual restrictions. Modus Ponens must, therefore, mark the consequent and the clause it derives from as correlated.

Inferences in the context of negation will be discussed later, in Section 4.

3 Instantiations of UL

In order to be more specific, let us look at particular instantiations of our abstract logical language. Since none of them provides for a coindexation mechanism, we add the indices $\Box_i$ of **UL** to each of the formalisms under consideration.

3.1 LFG

In [7]'s linear logic approach, Modus Ponens manipulates meaning constructors:

$$\frac{\begin{array}{c}\Box_i(\{\forall \mathrm{G}, S.(\forall x.g_\sigma \rightsquigarrow x \multimap \mathrm{G} \rightsquigarrow S(x)) \multimap \mathrm{G} \rightsquigarrow \mathtt{every}(P,S)\} \cup \phi)\\ P(\mathtt{t})\end{array}}{\Box_i(\{g_\sigma \rightsquigarrow \mathtt{t}\} \cup \phi)} \tag{53}$$

For example, if we instantiate the first premiss with the meaning constructor for *every boy* and the second premiss with $\mathtt{boy}(\mathtt{kevin})$, we get that the subject of the derived sentence is Kevin, i.e. $g_\sigma \rightsquigarrow \mathtt{kevin}$.

Soundness can be proved by mapping the LFG linear logic formulae to **UL** formulae[9] via a translation function τ_{LL}, where $\langle Q, Op\rangle \in \{\langle\exists,\wedge\rangle, \langle\forall,\rightarrow\rangle\}$:

$$\begin{array}{l}\tau_{LL}(\forall \mathrm{G}, S.(\forall x.g_\sigma \rightsquigarrow x \multimap \mathrm{G} \rightsquigarrow S(x)) \multimap \mathrm{G} \rightsquigarrow Q(R,S))\\ \quad := l_g : Qx.R(x)\; Op\; l_S\end{array} \tag{54}$$

We may go even further and formulate deduction rules that operate on f-structures themselves, as suggested by the map τ_{FS}:

$$\tau_{FS}\left(\left[\ \mathrm{G}\, g : \begin{bmatrix}\mathrm{SPEC}\ Q\\ \mathrm{PRED}\ R\end{bmatrix}\right]\right) := l_g : Qx.R(x)\; Op\; l_S \tag{55}$$

[9]For an in depth investigation of mappings among underspecified semantic representation formalisms, see the contribution by Crouch and van Genabith in [3].

An application of Modus Ponens to (2) would then result in the instantiation of

$$\left[\text{SUBJ}\, g : \left[\begin{array}{l}\text{SPEC } \mathbf{every}\\ \text{PRED } \mathbf{boy}\end{array}\right]\right] \tag{56}$$

to $\left[\text{PRED}\, \mathtt{kevin}\right]$ yielding

$$f : \left[\begin{array}{lll}\text{PRED } \mathbf{see} & & \\ \text{SUBJ} & g : & \left[\text{PRED}\, \mathtt{kevin}\right]\\ \text{OBJ} & h : & \left[\begin{array}{l}\text{SPEC } \mathtt{a}\\ \text{PRED } \mathtt{movie}\end{array}\right]\end{array}\right] \tag{57}$$

3.2 MG

In the case of MG, the Generalized Modus Ponens looks as follows

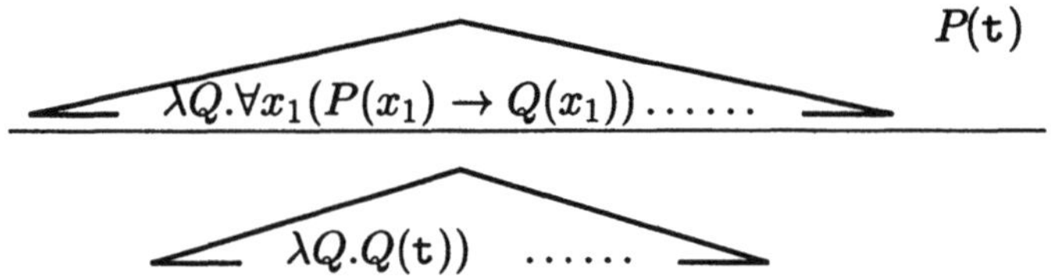

(58)

An application of Modus Ponens to $\lambda Q.\forall x_1(\mathbf{boy}(x_1) \rightarrow Q(x_1))$ in (4) and $\mathbf{boy}(\mathtt{kevin})$ yields a tree that results from (4) by replacing $\lambda Q.\forall x_1(\mathbf{boy}(x_1) \rightarrow Q(x_1))$ with $\lambda Q.Q(\mathtt{kevin})$. Applying D to it yields

$$\left\{\begin{array}{l}\mathbf{see}(\mathtt{kevin}, \lambda Q.\exists x_2(\mathtt{movie}(x_2) \wedge Q(x_2))),\\ \exists x_2(\mathtt{movie}(x_2) \wedge \mathbf{see}(\mathtt{kevin}, \lambda Q.Q(x_2))).\end{array}\right\} \tag{59}$$

Traditionally, we get (59) by applying non-ambiguous Modus Ponens to the members of (5). For extensional verbs like *see*, both formulae in (59) are equivalent to

$$\exists x_2(\mathtt{movie}(x_2) \wedge \mathbf{see}(\mathtt{kevin}, x_2)).$$

3.3 Minimal Recursion Semantics

The example (6) of a MRS representation allows us to apply the Generalized Modus Ponens as the substitution

$$\left[\begin{array}{ll}\mathtt{every\_rel} & \\ \text{HANDEL} & \boxed{1}\\ \text{BV} & \boxed{2}\\ \text{RESTR} & \boxed{3}\end{array}\right], \left[\begin{array}{ll}\mathtt{boy\_rel} & \\ \text{HANDEL} & \boxed{3}\\ \text{INST} & \boxed{2}\end{array}\right] \Longrightarrow \left[\begin{array}{ll}\mathtt{boy\_rel} & \\ \text{HANDEL} & \boxed{1}\\ \text{INST} & \boxed{2}\end{array}\right], \left[\begin{array}{ll}\mathtt{kevin\_rel} & \\ \text{HANDEL} & \boxed{1}\\ \text{INST} & \boxed{2}\end{array}\right] \tag{60}$$

based on the map

$$\tau_{MRS}\left(\begin{bmatrix}\texttt{Q\_rel} & \\ \text{HANDEL} & \boxed{g} \\ \text{BV} & \boxed{2} \\ \text{RESTR} & \boxed{3}\end{bmatrix}, \begin{bmatrix}\texttt{R\_rel} & \\ \text{HANDEL} & \boxed{3} \\ \text{INST} & \boxed{2}\end{bmatrix}\right) := l_g : Qx.R(x)\ Op\ l_S \qquad (61)$$

3.4 Quasi Logical Forms

For the QLF representation, the scheme of the Generalized Modus Ponens is instantiated e.g. as

$$\frac{\begin{array}{c}\texttt{[]:see}\left(\begin{array}{l}\texttt{term(+g,<type=q,lex=every>,boy,?Q,?X),}\\ \texttt{term(+h,<type=q,lex=a>,movie,?P,?R)}\end{array}\right)\\ \texttt{[]:boy(kevin1)}\end{array}}{\texttt{[]:see}\left(\begin{array}{l}\texttt{[]:boy(kevin1),}\\ \texttt{term(+h,<type=q,lex=a>,movie,?P,?R)}\end{array}\right)} \qquad (62)$$

This is justified by the map

$$\tau_{QLF}(\texttt{term(+g,<type=q,lex=Q>,R,?S,?X)}) := l_g : Qx.R(x)\ Op\ l_S \qquad (63)$$

3.5 UDRSs

The Generalized Modus Ponens for UDRSs reads

$$\frac{\begin{array}{c}\left\langle\left\{ l_g : \boxed{\boxed{\begin{array}{l}x\\ P(x)\end{array}} \Rightarrow l_{g2} : \Box}\right\} \cup B, C\right\rangle\\ P(\texttt{t})\end{array}}{\langle B, C\rangle\,[x/\texttt{t}]} \qquad (64)$$

The map from UDRSs on **UL** -formulae can be developed on the basis

$$\begin{array}{l}\tau_{UDRS}(l_g : \boxed{\boxed{\begin{array}{l}x\\ R(x)\end{array}} \Rightarrow l_S : \Box}) := l_g : \forall x.R(x) \rightarrow l_S\\ \tau_{UDRS}(l_g : \boxed{\begin{array}{l}x\\ R(x)\end{array}}) := l_g : \exists x.R(x) \wedge l_S\end{array} \qquad (65)$$

3.6 USDL

Finally, let us adapt the scheme of Generalized Modus Ponens to **USDL**.

$$\frac{\begin{array}{c}\{X_1 = \texttt{every\_P'@}L_x(X_4)\} \cup \phi \\ P(\texttt{t})\end{array}}{\{X_1 = \lambda Q.Q(\texttt{t})\texttt{@}L_{\texttt{t}}(X_4)\} \cup \phi.} \tag{66}$$

4 Negation

As mentioned earlier, an application of GMP (Generalized Modus Ponens) is only correct in the absence of negations, or monotone decreasing quantifiers.

John is a politician.
At least one problem preoccupies every politician. (67)

At least one problem preoccupies John.

John is a politician.
Few problems preoccupy every politician. (68)
$\not\vdash$ *Few problems preoccupy John.*

John is a politician.
Every politician doesn't sleep. (69)
$\not\vdash$ *John doesn't sleep.*

The examples in (67), (68), and (69) show that GMP may only be applied to an element γ of B, if there is no disambiguation of $\langle B, C\rangle$ (by D) that assigns γ narrow scope with respect to (an odd number of occurrences of) negations and monotone decreasing quantifiers, i.e. to cases where γ has positive polarity.

Thus the occurrence $\forall x.\texttt{boy}(x)$ in (1) has positive polarity, whereas it has negative polarity in (70) and indefinite polarity in (71).

Few mothers believed that every boy saw a movie. (70)

Every boy didn't see a movie. (71)

Subsequently, polarities, $+, -$, or i (indefinite), will be superscripted to the labels of the partial formulae. Now, let us concentrate on (71). We may split its completely underspecified representation

$$\langle\left\{\begin{array}{c} l_{1^i} : \forall x_1.\texttt{boy}(x_1) \rightarrow l_{12}, \\ l_{2^i} : \exists x_2.\texttt{movie}(x_2) \wedge l_{22}, \\ l_{3^+} : \neg l_{31}, \\ l_4 : \texttt{see}(x_1, x_2) \end{array}\right\}, \{l_4 \leq l_{12,22,31}\}\rangle \tag{72}$$

into two underspecified representations in which $\forall x.\mathtt{boy}(x)$ has definite polarity ($\{l_4 \leq l_{12,22,31}\}$ abbreviates the set $\{l_4 \leq l_{12}, l_4 \leq l_{22}, l_4 \leq l_{31}\}$.)

$$\langle \left\{ \begin{array}{c} l_{1^+} : \forall x_1.\mathtt{boy}(x_1) \rightarrow l_{12}, \\ l_{2^i} : \exists x_2.\mathtt{movie}(x_2) \wedge l_{22}, \\ l_{3^+} : \neg l_{31}, \\ l_4 : \mathtt{see}(x_1, x_2) \end{array} \right\}, \{l_3 \leq l_{12}, l_4 \leq l_{12,22,31}\} \rangle \tag{73}$$

$$\langle \left\{ \begin{array}{c} l_{1^-} : \forall x_1.\mathtt{boy}(x_1) \rightarrow l_{12}, \\ l_{2^i} : \exists x_2.\mathtt{movie}(x_2) \wedge l_{22}, \\ l_{3^+} : \neg l_{31}, \\ l_4 : \mathtt{see}(x_1, x_2) \end{array} \right\}, \{l_1 \leq l_{31}, l_4 \leq l_{12,22,31}\} \rangle \tag{74}$$

(73) represents the set of meanings of (71) in which the negation has narrow scope with respect to the universal quantifier, and (74) the set of meanings in which it has wide scope. These two descriptions can be combined into one formula with the help of the connective $\sharp$. The introduction rule (and simultaneously the elimination rule) for the $\sharp$-operator in (75) states that an underspecified formula may be replaced by two $\sharp$-connected underspecified formulae for the same set B of building blocks with orthogonal constraint sets $C \cup C_1$ and $C \cup C_2$ which together describe exactly the set of readings admitted by C.

$$\frac{\Box_i(\langle B, C\rangle)}{\Box_i(\langle B, C \cup C_1\rangle) \sharp \Box_i(\langle B, C \cup C_2\rangle)} \text{ if } \models C \cup C_1 \text{ iff } \not\models C \cup C_2 \tag{75}$$

It is the $\sharp$-operator which makes the formalism *cu-deductive*. Disambiguation steps and inference steps can alternate, while always producing meaningful formulae. On this basis (72) can be rewritten equivalently by (76) as a combination of an underspecified formula where the negation has narrow scope wrt. the universally quantified NP in the left subformula and wide scope in the right subformula.

$$\begin{array}{l} \langle \left\{ \begin{array}{c} l_{1^+} : \forall x_1.\mathtt{boy}(x_1) \rightarrow l_{12}, \\ l_{2^i} : \exists x_2.\mathtt{movie}(x_2) \wedge l_{22}, \\ l_{3^+} : \neg l_{31}, \\ l_4 : \mathtt{see}(x_1, x_2) \end{array} \right\}, \{l_3 \leq l_{12}, l_4 \leq l_{12,22,31}\} \rangle \\ \sharp \langle \left\{ \begin{array}{c} l_{1^-} : \forall x_1.\mathtt{boy}(x_1) \rightarrow l_{12}, \\ l_{2^i} : \exists x_2.\mathtt{movie}(x_2) \wedge l_{22}, \\ l_{3^+} : \neg l_{31}, \\ l_4 : \mathtt{see}(x_1, x_2) \end{array} \right\}, \{l_1 \leq l_{31}, l_4 \leq l_{12,22,31}\} \rangle \end{array} \tag{76}$$

What can we infer from (76) resp. (72)? By the usual equivalence transformations, the universal quantifier with negative polarity can be replaced by an existential one with positive polarity.

$$\langle\left\{\begin{array}{c} l_{1^+} : \forall x_1.\mathtt{boy}(x_1) \to l_{12}, \\ l_{2^i} : \exists x_2.\mathtt{movie}(x_2) \wedge l_{22}, \\ l_{3^+} : \neg l_{31}, \\ l_4 : \mathtt{see}(x_1, x_2) \end{array}\right\}, \{l_3 \leq l_{12}, l_4 \leq l_{12,22,31}\}\rangle$$

$$\sharp\langle\left\{\begin{array}{c} l_{1^+} : \exists x_1.\mathtt{boy}(x_1) \wedge l_{12}, \\ l_{2^i} : \exists x_2.\mathtt{movie}(x_2) \wedge l_{22}, \\ l_{3^+} : \neg l_{31}, \\ l_4 : \mathtt{see}(x_1, x_2) \end{array}\right\}, \{l_3 \leq l_{12}, l_4 \leq l_{12,22,31}\}\rangle \quad (77)$$

We can weaken $l_{1^+} : \exists x_1.\mathtt{boy}(x_1) \wedge l_{12}$ to $l_{1^+} : \exists x_1.\mathtt{boy}(x_1) \to l_{12}$ and realize now that the right subformula of $\sharp$ follows from the left one. Hence, we conclude

$$\langle\left\{\begin{array}{c} l_{1^+} : \exists x_1.\mathtt{boy}(x_1) \to l_{12}, \\ l_{2^i} : \exists x_2.\mathtt{movie}(x_2) \wedge l_{22}, \\ l_{3^+} : \neg l_{31}, \\ l_4 : \mathtt{see}(x_1, x_2) \end{array}\right\}, \{l_3 \leq l_{12}, l_4 \leq l_{12,22,31}\}\rangle \quad (78)$$

This means

There exists an x with the property that
if x is a boy then x didn't see a movie (79)

is a consequence of (71).

5 Possible Extensions

For the sake of simplicity we have restricted ourselves to scope ambiguities. We have to admit coindexed atomic formulae in order to represent correlated lexical meanings. For example, one might want to express that a lexically ambiguous word like *plant* means *factory* all over a text, instead of a designating living things like flowers in some occurrences and factories in other occurrences. Hence, the syntax of **UL** will be extended by one more kind of basic underspecified formulae:

$$\Box_k(P(\mathtt{t}_1, \ldots, \mathtt{t}_n)). \quad (80)$$

The satisfiability condition will be analogous to the existing ones for basic underspecified formulae.

In order to account for plural ambiguities like the one in (22), one needs to extend the disambiguation device by rules which introduce new formula material when a plural is disambiguated to a collective reading or to a distributive reading etc. Similarly, the disambiguation device has to consider interaction postulates which allow the meaning representation of an indefinite NP in an embedded underspecified formula, e.g. in a relative clause, to get wide scope over parts of the meaning representation of the matrix clause, as e.g. in the example taken from [15]

Every student to whom every professor recommends
a certain book which the student has already read is lucky. (81)

6 Conclusion

We presented a general inference scheme for an underspecified semantic representation formalism **UL** and showed how this inference scheme can be specialized for some existing underspecified representation languages like the linear logic formulae in LFG semantics, the Quasi Logical Forms of the SRI Core Language Engine, Minimal Recursion Semantics, Underspecified Discourse Representation Structures, and USDL. The benefit is that the logical properties of the inference scheme can be investigated on an abstract level, and are then 'inherited' by those formalisms which fulfill the necessary requirements. For example, the novel property of the **UL** formalism of being cu-deductive, i.e. of allowing for the alternation of disambiguation steps and proper inference steps, could be ported to those underspecified semantic representation formalisms which come with a well-defined disambiguation method.

Since we assume a very general language for expressing the structural constraints in an underspecified representation, there is enough room for extensions. For example, the structural constraints could mirror as well syntactic conditions as well as semantic requirements on the set of fully disambiguated representations.

Appendix

A Overview of Formalisms

The following table gives an overview of the instantiation of the language of buildings blocks B, the structural constraints C, and the disambiguation mechanism D by various underspecified formalisms.

LFG/Linear Logic

B e.g. $\forall \mathrm{G}, R.(\forall x.g_\sigma \rightsquigarrow x \multimap \mathrm{G} \rightsquigarrow R(x)) \multimap \mathrm{G} \rightsquigarrow \mathsf{every}(\mathsf{boy}, R)$

C derived from f-structures;

D Linear Logic deduction.

Montague Grammar

B e.g. $\lambda Q \forall x(boy(x) \rightarrow Q(x))$;

C derived from syntactic structures;

D scoping mechanism for the quantifier store.

Minimal Recursion Semantics

$$B \text{ e.g. } \begin{bmatrix} \texttt{every\_rel} & \\ \text{HANDEL} & \boxed{1} \\ \text{BV} & \boxed{2} \\ \text{RESTR} & \boxed{3} \end{bmatrix}, \begin{bmatrix} \texttt{boy\_rel} & \\ \text{HANDEL} & \boxed{3} \\ \text{INST} & \boxed{2} \end{bmatrix}$$

C derived from the uses of labels + well-formedness constraints;

D (not worked out).

Quasi Logical Forms

B e.g. `term(+g,<type=q,lex=every>,boy,?Q,?X)`;

C explicit scope orderings + well-formedness constraints;

D scoping mechanism for the embedded quantifiers.

Underspecified Discourse Representations

$$B \text{ e.g. } l_1 : \boxed{\boxed{\begin{array}{l} x \\ \mathsf{boy}(x) \end{array}} \Rightarrow l_{12} : \square}$$

C e.g. $\{l_{verb} \leq l_{12}, l_1 \leq l_{\top}\}$ + well-formedness constraints;

D disambiguation algorithm.

USDL

B e.g. $X_1 = \texttt{every\_boy'}@L_{x_1}(X_4)$;

C e.g. $\{X_{top} = C_5(X_1), X_4 = C_6(X_{verb})\}$;

D linear higher-order unification.

University of Stuttgart, Germany.

References

1. Hiyan Alshawi, editor. *The Core Language Engine.* ACL-MIT Press Series in Natural Languages Processing. MIT Press, Cambridge, Mass., 1992.
2. Hiyan Alshawi and Richard Crouch. Monotonic semantic interpretation. In *Proceedings of ACL*, pages 32–39, Newark, Delaware, 1992.
3. Robin Cooper, Dick Crouch, Jan van Eijck, Chris Fox, Josef van Genabith, Jan Jaspars, Hans Kamp, Manfred Pinkal, Massimo Poesio, and Steve Pulman. Building the framework. Deliverable 15, FraCaS, LRE 62-051, University of Edinburgh, 1996.

4. Robin Cooper, Dick Crouch, Jan van Eijck, Chris Fox, Josef van Genabith, Jan Jaspars, Hans Kamp, Manfred Pinkal, Massimo Poesio, Steve Pulman, and Espen Vestre. Describing the approaches. Deliverable 8, FraCaS, LRE 62-051, University of Edinburgh, 1994. URL: `ftp://ftp.cogsci.ed.ac.uk /pub/FRACAS/del8.ps.gz`.
5. Ann Copestake, Dan Flickinger, Rob Malouf, Susanne Riehemann, and Ivan A. Sag. Translation using minimal recursion semantics. In *Proceedings of the Sixth International Conference on Theoretical and Methodological Issues in Machine Translation*, University of Leuven, Belgium, 1995.
6. Richard Crouch. Ellipsis and quantification: A substitutional approach. In *Proceedings of the 6th Meeting of the Association for Computational Linguistics, European Chapter*, Dublin, 1995.
7. Mary Dalrymple, John Lamping, Fernando C.N. Pereira, and Vijay Saraswat. A deductive account of quantification in LFG. In Makoto Kanazawa, Christopher J. Pinon, and Henriette de Swart, editors, *Quantifiers, Deduction and Context*. CSLI, Stanford, Ca., 1995.
8. David R. Dowty, Robert E. Wall, and Stanley Peters. *Introduction to Montague Semantics*. Reidel, Dordrecht, Holland, 1981.
9. Anette Frank and Uwe Reyle. Principle based semantics for HPSG. In *Proceedings of the 6th Meeting of the Association for Computational Linguistics, European Chapter*, Dublin, 1995.
10. Jerry Hobbs and Stuart M. Shieber. An algorithm for generating quantifier scopings. *Computational Linguistics*, 13:47–63, 1987.
11. Martin Kay. Machine translation. Lecture Notes, Institut für Maschinelle Sprachverarbeitung, University of Stuttgart, June 1993.
12. William R. Keller. Nested cooper storage: the proper treatment of quantification in ordinary noun phrases. In Uwe Reyle and Rohrer Christian, editors, *Natural Language Parsing and Linguistic Theories*, pages 432–437. Reidel, Dordrecht, 1988.
13. Fernando C.N. Pereira. Categorial semantics and scoping. *Computational Linguistics*, 16(1):1–10, 1990.
14. Carl Pollard and Ivan A. Sag. *Head Driven Phrase Structure Grammar*. University of Chicago Press, Chicago, 1994.
15. Uwe Reyle. Dealing with ambiguities by underspecification: Construction, representation, and deduction. *Journal of Semantics*, 10(2):123–179, 1993.
16. Uwe Reyle. On reasoning with ambiguities. In *Proceedings of the 6th Meeting of the Association for Computational Linguistics, European Chapter*, Dublin, 1995.

DEDUCTIVE SYSTEMS AND CATEGORIES IN LINGUISTICS

JOACHIM LAMBEK

1 Introduction

Deductive systems appear as soon as one attempts a formal presentation of the grammar of a natural language. To see this one merely has to think of syntactic or semantic *types* as *formulas* and of grammatical *derivations* as *deductions*. Although categories had originally been introduced by Eilenberg and Mac Lane [13] to deal with problems in algebra and topology, they soon found applications in other branches of mathematics, particularly in logic and computer science. I don't wish to argue that categories should play a role in linguistics, but rather that they already do. Indeed, a deductive system turns into a category, as soon as one pays proper attention to the notion of equality between arrows. However, it is not just plain categories that enter linguistics, but certain structured categories and certain generalized categories. We shall begin by looking at three kinds of deductive systems and then introduce categorical considerations.

(1) The *syntactic calculus* developed by the author [20] had antecedents in Ajdukiewicz [1] and Bar-Hillel [2].

(2) *Context-free grammars* were formally introduced into linguistics by Chomsky [10]. They play a special role in the description of formal languages, see e.g. Ginsburg [15]. They are essentially equivalent to Gentzen-style deductive systems (but without Gentzen's structural rules), see Szabo [43].

(3) *Production grammars*, also known as *rewrite systems*, are essentially the same as Chomsky's [10] *generative grammars* (of type 0), though they have as an antecedent in mathematics the *semi-Thue systems*, see e.g. Kleene [19]. Productions are so named in Rosenbloom [42] and Brainerd [7]. I have pushed for them in [24, 27] as the most appropriate tool for handling natural languages.

H.J. Ohlbach and U. Reyle (eds.), Logic, Language and Reasoning, 279–294.

(4) The *categorical imperative* induces us to view the syntactic calculus as a *biclosed monoidal* or *residuated* category [22, 23], a context-free grammar as a *multicategory* [23, 24], and a production grammar as a *strictly monoidal* category, see Hotz [17] and Benson [4, 5].

(5) *Two-dimensional categories*, or 2-categories, appear as soon as one views the free monoid generated by the vocabulary as a one-object category, but their introduction becomes imperative when this monoid is replaced by the free category generated by a graph, following an idea of Brame [8, 9].

(6) *Semantics as a functor* first appears in Benson [4] it may be viewed as the process of imposing Gentzen's three structural rules. The ideas advocated here have also been developed by Dougherty [12].

Although we cannot do justice here to all attempts to apply category theory to linguistics, we should mention Barr [3], Nelson [37], Eytan [14] and applications to the semantics of natural languages by Gonzalo Reyes and his collaborators, see Reyes [39], La Palme *et al.* [41] and Macnamara and Reyes [35]. Moreover, Christian Houzel has lectured on categories in linguistics at the University of Paris, though apparently he never published his ideas.

2 Syntactic Calculus

The simplest kind of deductive system, which I have called a *Lawvere-style* deductive system in [28], deals with deductions of the form $f : A \to B$, where A and B are formulas in some formal language and f is a process for deducing B from A. However, traditionally, logicians were more interested in deducibility than in the actual deduction. Since deducibility is reflexive and transitive, we must stipulate the identity deduction

$$1_A : A \to A$$

for each formula A and a rule for obtaining new deductions from old:

$$\frac{f : A \to B \quad g : B \to C}{gf : A \to C}.$$

By the *syntactic calculus* we understand a Lawvere-style deductive system with binary operations $\otimes$, / (*over*) and \ (*under*) between formulas satisfying the following axioms and rules of inference:

$$\begin{aligned}(A \otimes B) \otimes C \quad &\leftrightarrow \quad A \otimes (B \otimes C),\\ A \otimes B \to C \ \text{ iff } \ &A \to C/B,\\ A \otimes B \to C \ \text{ iff } \ &B \to A\backslash C.\end{aligned}$$

Usually also a nullary operation I is admitted, satisfying

$$A \otimes I \leftrightarrow A \leftrightarrow I \otimes A\,.$$

One may furthermore introduce the usual lattice operations $\top$, $\wedge$, $\perp$ and $\vee$ satisfying

$$C \to \top, \quad \perp \to C,$$
$$C \to A \wedge B \text{ iff } C \to A \text{ and } C \to B,$$
$$A \vee B \to C \text{ iff } A \to C \text{ and } B \to C.$$

One easily obtains a number of theorems and derived rules of inference, e.g.

$$B \otimes (B\backslash A) \to A \qquad (A/B) \otimes B \to A, \qquad \text{(Ajdukiewicz)}$$

$$B \to A/(B\backslash A), \qquad B \to (A/B)\backslash A, \qquad \text{(type raising)}$$

$$\frac{A \to B \quad C \to D}{A \otimes C \to B \otimes D}, \qquad \frac{A \to B \quad C \to D}{A/D \to B/C}, \qquad \text{(functoriality)}$$

$$A\backslash(B/C) \leftrightarrow (A\backslash B)/C, \quad (C/B)/A \leftrightarrow C/(A \times B), \qquad \text{(associativity)}$$
$$(A/B) \otimes (B/C) \to A/C, \qquad \text{(composition)}$$
$$A/B \to (A/C)/(B/C). \qquad \text{(Geach)}$$

The syntactic calculus resembles the intuitionistic propositional calculus; in fact, it reduces to the latter if one imposes the following three axioms:

$$A \otimes B \to B \otimes A \qquad \text{(interchange)}\ ,$$
$$A \to A \otimes A \qquad \text{(contraction)}\ ,$$
$$A \to I \qquad \text{(weakening)}\ .$$

These are Gentzen's three structural rules, disguised as axioms, in a Lawvere-style deductive system. If they hold, one can prove:

$$A/B \leftrightarrow B\backslash A\ ,$$
$$A \otimes B \leftrightarrow A \wedge B\ ,$$
$$I \leftrightarrow T\ ,$$

and one usually writes $B\backslash A$ as $B \Rightarrow A$.

While all three of these axioms are absent from the syntactic calculus, weakening is absent from relevance logic, contraction is absent from BCK logic and both are absent from linear logic. In the presence of $\otimes$, interchange follows from contraction and weakening.

The linguistic application of the syntactic calculus, called *categorial grammar*, is based on the following idea. To each word in the dictionary one assigns a *syntactic type* made up from basic syntactic types

$$\mathbf{s} = \text{(declarative) sentence}\ ,$$
$$\mathbf{n} = \text{name}\ ,$$

and perhaps others, with the help of the operations $\otimes$, / and \, and perhaps I and the lattice operations. (We shall not discuss quantifiers here, see [25].)

For example, the dictionary might list:

$$\begin{aligned} \textit{Jane, time} &\rightarrow \mathbf{n}\,, \\ \textit{flies} &\rightarrow \mathbf{n}\backslash\mathbf{s}\,, \\ \textit{slowly} &\rightarrow \mathbf{s}\backslash\mathbf{s}\,, \\ \textit{he, she} &\rightarrow \mathbf{s}/(\mathbf{n}\backslash\mathbf{s})\,, \\ \textit{likes} &\rightarrow \mathbf{n}\backslash(\mathbf{s}/\mathbf{n}) \leftrightarrow (\mathbf{n}\backslash\mathbf{s})/\mathbf{n}\,, \\ \textit{him, her} &\rightarrow (\mathbf{s}/\mathbf{n})\backslash\mathbf{s}\,. \end{aligned}$$

One can then prove, with the help of the syntactic calculus:

$$\begin{aligned} \textit{time flies slowly} &\rightarrow (\mathbf{n} \otimes (\mathbf{n}\backslash\mathbf{s})) \otimes (\mathbf{s}\backslash\mathbf{s}) \\ &\rightarrow \mathbf{s} \otimes (\mathbf{s}\backslash\mathbf{s}) \\ &\rightarrow \mathbf{s} \end{aligned}$$

and

$$\begin{aligned} \textit{she flies} &\rightarrow (\mathbf{s}/(\mathbf{n}\backslash\mathbf{s})) \otimes (\mathbf{n}\backslash(\mathbf{s}) \\ &\rightarrow \mathbf{s}\,. \end{aligned}$$

Similarly one shows:

$$\textit{Jane likes her} \rightarrow \mathbf{s}\,,$$

where *her* cannot be replaced by *she*.

Note that 'formulas' have now become 'types', hence the slogan: *formulas as types*. We shall avoid the older word 'category' in place of 'type', as it might lead to confusion with the categories of Eilenberg and Mac Lane [13], about which we shall speak later.

The intuitionistic propositional calculus may also be applied to linguistics, but its formulas should then be regarded as *semantic types*, as is implicit in Curry [11]. This idea was further developed by Montague [36]. Thus $\mathbf{s}$ will be replaced by the type $\mathbf{t}$ of truth values, $\mathbf{n}$ by the type $\mathbf{e}$ of entities and $\mathbf{n}\backslash\mathbf{s}$ as well as $\mathbf{s}/\mathbf{n}$ by the type $\mathbf{e} \Rightarrow \mathbf{t}$ of all functions from the set of entities to the set of truth values.

3 Context-free Grammars

The most popular kinds of grammar are the context-free ones. The derivations in such a grammar have the form

$$f : A_1 \cdots A_n \rightarrow A_{n+1},$$

when the A_i are elements of some vocabulary, say the vocabulary of English enlarged by technical grammatical terms such as Subj (= subject), Pred (= predicate), Adv (= adverb) and Sent (= sentence). We have taken the liberty of reversing the usual direction of the arrow, thus taking the point of view of the hearer rather than that of the speaker. Thus reversed, a context-free derivation is formally the same as a deduction (sequent), as proposed by Gentzen for intuitionistic logic, yet not subject to his structural rules (see below).

We shall postulate reflexivity

$$1_A : A \to A;$$

but transitivity is now replaced by what Gentzen calls the *cut rule*:

$$\frac{f : \Lambda \to A \quad g : \Gamma A \Delta \to B}{g \,\&\, f : \Gamma\Lambda\Delta \to B},$$

where the capital Greek letters Λ, Γ and Δ denote strings of words, possibly empty strings.

The rules of English grammar must now be stated in the form of axioms, e.g.

$$\begin{aligned} \textit{Jane, time, she} &\to \text{Subj}\ , \\ \textit{flies} &\to \text{Pred}\ , \\ \textit{slowly} &\to \text{Adv}\ , \\ \text{Subj Pred} &\to \text{Sent}\ , \\ \text{Sent Adv} &\to \text{Sent}\ . \end{aligned}$$

Thus we have, for example:

$$\textit{she flies} \to \text{Subj } \textit{flies} \to \text{Subj Pred} \to \text{Sent}\ ,$$

or also

$$\textit{she flies} \to \textit{she} \text{ Pred} \to \text{Subj Pred} \to \text{Sent}\ .$$

In some sense, to be made precise later, these two derivations should be viewed as equivalent.

Context-free grammars are also useful as an auxiliary device for introducing the operations of the syntactic calculus or of intuitionistic logic (alias 'semantic calculus'). In fact, it was for the latter that Gentzen had originally proposed such a deductive system, though he also imposed three structural rules:

$$\frac{f : \Gamma A B \Delta \to C}{f^i : \Gamma B A \Delta \to C} \qquad \text{(interchange)},$$

$$\frac{f : \Gamma A A \Delta \to B}{f^c : \Gamma A \Delta \to B} \qquad \text{(contraction)},$$

$$\frac{f : \Gamma\Delta \to B}{f^w : \Gamma A \Delta \to B} \qquad \text{(weakening)}.$$

We remind the reader that these structural rules are not operative in the syntactic calculus.

The operations of the syntactic calculus may be introduced into a context-free grammar as follows. (In case of the tensor product, this was first done by Bourbaki [6].)

$$\frac{f : \Gamma AB\Delta \to C}{f^\S : \Gamma(A \otimes B)\Delta \to C}, \qquad \mathbf{m}_{AB} : AB \to A \otimes B,$$

$$\mathbf{e}_{AB} : (A/B)B \to A, \qquad \frac{f : \Gamma B \to A}{f^* : \Gamma \to A/B},$$

$$\mathbf{e}'_{AB} : B(B\backslash A) \to A, \qquad \frac{f : B\Gamma \to A}{f^\dagger : \Gamma \to B\backslash A},$$

$$\frac{f : \Gamma\Delta \to A}{f^\# : \Gamma I \Delta \to A}, \qquad \mathbf{i} : \ \to I.$$

Note that the string before the arrow in the last axiom is empty. Similar axioms and introduction rules for the lattice operations go back to Gentzen, e.g.:

$$\begin{array}{l} \mathbf{p}_{AB} : A \wedge B \to A, \\ \mathbf{q}_{AB} : A \wedge B \to B, \end{array} \qquad \frac{f : \Lambda \to A \quad g : \Lambda \to B}{< f, g >: \Lambda \to A \wedge B}.$$

With the help of the cut rule, we may replace the axioms $\mathbf{m}_{AB}$, $\mathbf{e}_{AB}$, $\mathbf{e}'_{AB}$, $\mathbf{p}_{AB}$ and $\mathbf{q}_{AB}$ by introduction rules as follows:

$$\frac{f : \Gamma \to A \quad g : \Delta \to B}{f\mu g : \Gamma\Delta \to A \otimes B},$$

$$\frac{f : \Lambda \to B \quad g : \Gamma A \Delta \to C}{g\varepsilon f : \Gamma(A/B)\Lambda\Delta \to C},$$

$$\frac{f : \Lambda \to B \quad g : \Gamma A \Delta \to C}{g\varepsilon' f : \Gamma\Lambda(B\backslash A)\Delta \to C},$$

$$\frac{f : \Gamma A \Delta \to C}{f_p : \Gamma(A \wedge B)\Delta \to C},$$

$$\frac{f : \Gamma B \Delta \to C}{f_q : \Gamma(A \wedge B)\Delta \to C}.$$

After these replacements have been carried out, one can prove, following Gentzen, that the cut rule is now redundant for the freely generated syntactic calculus: each derivation may be replaced by one whose construction does not involve a cut. The proof is even easier in the absence of the structural rules, see [20, 26]. Since each rule of inference now introduces an operation, either on the left or on the right of the arrow, one has a procedure for deciding whether, for given Γ and A, there is a derivation $\Gamma \to A$ (in fact, for finding all such derivations), by working backwards, e.g.:

$$\frac{\dfrac{1_A : A \to A \qquad 1_B : B \to B}{1_A \varepsilon_{AB} 1_B : (A/B)B \to A}}{(1_A \varepsilon_{AB} 1_B)^\dagger : B \to (A/B)\backslash A} \, .$$

The existence of a decision procedure led Chomsky to conjecture that the categorial grammar, obtained by assigning syntactic types to words in the dictionary, is in fact equivalent to a context-free grammar. After some unsuccessful attempts by various people to prove this, the conjecture was finally proved by Pentus [38], as long as the only operations admitted are $\otimes$, $/$ and $\backslash$. However, Kanazawa [18] showed that, with the help of the lattice operations, one could go further. In fact, the best known example of a language that cannot be described by a context-free grammar is the intersection of two languages that can be so described, and Kanazawa showed that such a language can be described by a categorial grammar which counts $\wedge$ among its operations.

4 Production Grammars

There is some doubt whether context-free or categorial grammars are powerful enough to describe natural languages. In fact, it seems that this has been 'proved' for Swiss German and perhaps also for Dutch, see Maclachlan [32]. It is clear, however, that every natural language may be described by a production grammar, since such grammars are capable of generating all recursively enumerable sets of symbols.

A *production grammar* $\mathcal{G} = (\mathcal{V}, \mathcal{V}_t, \mathcal{V}_i, \mathcal{P})$ consists of a set $\mathcal{V}$, the *vocabulary*, two subsets of $\mathcal{V}$, the *terminal* and *initial* vocabularies $\mathcal{V}_t$ and $\mathcal{V}_i$ respectively, and a finite subset $\mathcal{P}$ of $\mathcal{V}^* \times \mathcal{V}^*$, the set of *productions*, where $\mathcal{V}^*$ is the free monoid generated by $\mathcal{V}$.

With each pair $p = (\Gamma, \Delta)$ of $\mathcal{P}$ there is associated a derivation $p : \Gamma \to \Delta$. Other derivations are obtained from these and the reflexive axioms $1_\Gamma : \Gamma \to \Gamma$ with the help of transitivity

$$\frac{f : \Gamma \to \Delta \quad g : \Delta \to \Lambda}{g \circ f : \Gamma \to \Lambda}$$

and the substitution rule

$$\frac{f : \Gamma \to \Delta}{\Phi f \Psi : \Phi\Gamma\Psi \to \Phi\Delta\Psi} \, .$$

The latter is equivalent to

$$\frac{f : \Gamma \to \Delta \quad f' : \Gamma' \to \Delta'}{f'f : \Gamma'\Gamma \to \Delta'\Delta},$$

as is seen by putting $f'f = f'\Delta \circ \Gamma' f$ and $\Phi f \Psi = 1_\Phi f 1_\Psi$.

We shall obey the usual convention concerning the direction of the arrow, the reverse of that in Sections 1 and 2, thus looking at a production grammar from the point of view of the speaker. For example, $\mathcal{V}_t$ might be the set of English words (including inflected forms) and $\mathcal{V}_i = \{S, Q, C, \cdots\}$ might consist of a few symbols denoting types of sentences, say S = statement, Q = question, C = command, etc. In addition to the subsets $\mathcal{V}_t$ and $\mathcal{V}_i$, $\mathcal{V}$ contains as elements certain grammatical terms, such as

$$\begin{aligned}
\text{Subj} &= \text{subject}\ , \\
\text{Pred} &= \text{predicate}\ , \\
NP_3 &= \text{third person singular noun-phrase}\ , \\
P_3 &= \text{third person singular}\ , \\
VP &= \text{verb-phrase}\ , \\
T_1 &= \text{present tense}\ , \\
+s &= \text{the third person singular morpheme}\ .
\end{aligned}$$

Among the productions there might be the following:

$$\begin{aligned}
S &\to \text{Subj Pred}\ , \\
\text{Subj} &\to NP_3\ P_3\ , \\
\text{Pred} &\to T_1\ VP\ , \\
NP_3 &\to \mathit{time}\ , \\
VP &\to \mathit{fly}\ , \\
P_3\ T_1\ V &\to V + s \text{ (for any regular verb } V)\ , \\
V + s &\to \begin{cases} V es \text{ when } V \text{ ends in a sibilant or } o\ , \\ \cdots ies \text{ when } V = \cdots y \text{ and } \cdots \text{ ends in a consonant}\ , \\ V s \text{ otherwise} \end{cases}
\end{aligned}$$

Here is how we would expect to generate the sentence *time flies*:

$$\begin{aligned}
S &\to \text{Subj Pred} \\
&\to NP_3\ P_3\ \text{Pred} \\
&\to \mathit{time}\ P_3\ \text{Pred} \\
&\to \mathit{time}\ P_3\ T_1\ VP \\
&\to \mathit{time}\ P_3\ T_1\ \mathit{fly} \\
&\to \mathit{time}\ \mathit{fly} + s \\
&\to \mathit{time}\ \mathit{flies}\ .
\end{aligned}$$

I believe that this is more or less how people produce sentences in practice. On the other hand, the following derivation would unnecessarily burden the speaker's short term memory:

$$\begin{aligned} S &\to \text{Subj Pred} \\ &\to \text{Subj } T_1\, VP \\ &\to NP_3\; P_3\; T_1\; VP \\ &\to NP_3\; P_3\; T_1 \text{ fly} \\ &\to NP_3\; fly + s \\ &\to NP_3\; flies \\ &\to time\ flies\ . \end{aligned}$$

Still, from a mathematical point of view, these two derivations should be identified.

Production grammars resemble the deductive systems introduced by Gentzen to tackle classical logic. However, when Gentzen wrote

$$(*) \qquad ABC \to DE\ ,$$

he had in mind

$$A \wedge B \wedge C \to D \vee E\ .$$

In the absence of his structural rules, this should really be

$$A \otimes B \otimes C \to D \oplus E\ ,$$

where $\oplus$ is the DeMorgan dual of $\otimes$. But, in a production grammar, $(*)$ stands for

$$A \otimes B \otimes C \to D \otimes E\ ,$$

So we are really dealing with an exceptional situation in which $\otimes$ and $\oplus$ coincide. In mathematics, this happens when we deal with finite dimensional vector spaces.

5 The Categorical Imperative.

A slight extension of the fragment of English considered in Section 3 would produce not only

$$S \to time\ flies$$

but also

$$C \to time\ flies\ ,$$

meaning the command: 'measure the speed of flies'. There is no ambiguity here, the two derivations having different sources, as long as we distinguish between S and C. But now consider the declarative sentence:

experience teaches spiders that time flies.

This can be construed as a declarative sentence in two different ways, as is seen by looking at the passives

spiders are taught by experience that time flies

and

spiders that time flies are taught by experience.

Perhaps the most famous such example with built in ambiguity is due to Chomsky [10], who points out the necessity to distinguish two derivations

$$NP \to \textit{ the shooting of the hunters},$$

depending on whether '*the hunters*' is the subject or the object of the verb '*hunt*'.

Considerations such as these lead us to the question as to when two deductions in a deductive system or two derivations in a grammar should be identified. In a Lawvere-style deductive system the answer is clear: one stipulates

$$f1_A = f, \quad 1_B f = f, \quad (hg)f = h(gf) ,$$

for all $f : A \to B$, $g : B \to C$ and $h : C \to D$, thus turning the deductive system into a *category*, in the sense of Eilenberg and Mac Lane [13].

Similar equations may be imposed on the intuitionistic proposition calculus, turning it into a *bicartesian closed category*, see e.g. Lambek and Scott [29]. Restricting oneself to the positive intuitionistic propositional calculus, which omits the operations $\perp$ and $\vee$, one obtains a *Cartesian closed category*, in the sense of Lawvere [30], which may also serve as a categorical version of Curry's [11] semantic calculus.

When we introduce appropriate equations into the syntactic calculus, we obtain what I have elsewhere [22, 23] called a *biclosed monoidal* or *residuated* category. However, already for a monoidal category, it is not obvious what the appropriate equations should be. For example, following Mac Lane [33], one should postulate that the two derivations

$$((A \otimes B) \otimes C) \otimes D \to A \otimes (B \otimes (C \otimes D)) ,$$

one via $(A \otimes B) \otimes (C \otimes D)$ and the other via

$$(A \otimes (B \otimes C)) \otimes D \to A \otimes ((B \otimes C) \otimes D) ,$$

coincide. This is what is meant by saying that the associativity isomorphism

$$\alpha_{ABC} : (A \otimes B) \otimes C \to A \otimes (B \otimes C)$$

is *coherent.*

One need not worry about coherence if one stipulates that $(A \otimes B) \otimes C = A \otimes (B \otimes C)$, in which case one obtains a *strictly monoidal category.* That a production grammar should be viewed as a strictly monoidal category was already asserted by Hotz [17] and Benson [4, 5], who interpreted the derivation $ABC \to DE$, say, as a morphism $A \otimes B \otimes C \to D \otimes E$. Alternatively, we may view a production grammar as a special kind of 2-category, see e.g. Mac Lane [34], in which the 1-arrows are elements of a free monoid. We shall return to this viewpoint in the next section.

Imposing appropriate equations between derivations in a context-free grammar, one obtains a *multicategory,* see e.g. [24]. Given

$$f : \Lambda \to A,\ g : \Gamma A \Delta \to B,\ h : \Phi B \Psi \to C\ ,$$

one stipulates

$$g \mathbin{\&} 1_A = g,\quad 1_A \mathbin{\&} f = f$$

and

$$(h \mathbin{\&} g) \mathbin{\&} f = h \mathbin{\&} (g \mathbin{\&} f)\ .$$

But, if

$$k : \Phi A \Theta A' \Psi \to C\ ,\quad f' : \Lambda' \to A'\ ,$$

one stipulates

$$(k \mathbin{\&} f') \mathbin{\&} f = (k \mathbin{\&} f) \mathbin{\&} f'\ .$$

If operations $\otimes$, $/$, $\backslash$ etc. are introduced into a multicategory rather than into a category, giving rise to what may be called a *residuated multicategory* it is quite clear what equations should be postulated, namely the equations which guarantee that the following rules of inference are mutually inverse:

$$\begin{array}{rl}
f \mapsto f^{\S}\ , & g \mapsto g \mathbin{\&} \mathbf{m}_{AB}\ ;\\
f \mapsto f^{*}\ , & g \mapsto \mathbf{e}_{AB} \mathbin{\&} g\ ;\\
f \mapsto f^{\dagger}\ , & g \mapsto \mathbf{e}'_{AB} \mathbin{\&} g\ ;\\
f \mapsto f^{\#}\ , & g \mapsto g \mathbin{\&} \mathbf{i}\ ,\\
(f,g) \mapsto < f,g >\ , & h \mapsto (\mathbf{p}_{AB} \mathbin{\&} h,\ \mathbf{q}_{AB} \mathbin{\&} h)\ .
\end{array}$$

The coherence condition for the associativity of $\otimes$ can now be proved, hence need not be postulated, and the same is true for other conditions.

Similar equations should be postulated for the other lattice operations. Moreover, in a freely generated residuated multicategory, each derivation is now *equal* to one constructed without cuts. For details, see [24, 28].

6 Two-dimensional Categories

We note that both context-free grammars and production grammars involve free monoids generated by sets. If a production grammar is viewed as a strictly monoidal category, the objects are precisely the elements of the free monoid generated by the vocabulary. Now, a monoid is itself a one-object category, hence a production grammar may be viewed as a 2-category whose 1-arrows are strings of words and whose 2-arrows are derivations. Pursuing an idea of Michael Brame [8, 9], one is led to replace the free monoid generated by a set by the *free category generated by a graph.* This makes it even more imperative to view a production grammar as a 2-category.

For a 2-category one requires some further equations, as was first noticed for the 2-category of small categories. For example, one must postulate the *interchange law*, see Mac Lane [34]:

$$(g' \circ f')(g \circ f) = (g'g) \circ (f'f) \ ,$$

where $\circ$ denotes composition of 1-arrows and juxtaposition denotes the bifunctor:

$$\frac{f : \Gamma \to \Delta \quad f' : \Gamma' \to \Delta'}{f'f : \Gamma'\Gamma \to \Delta'\Delta} \ .$$

Brame's original idea was to replace the free monoid generated by the vocabulary by the free category generated by the graph whose arrows (oriented edges) are words and whose objects (nodes) are types. The way I read him, he would analyze the sentence *spiders time flies* as follows:

$$S \overset{spiders}{\longleftarrow} VP \overset{time}{\longleftarrow} NP \overset{flies}{\longleftarrow} 1 \ .$$

The oriented edges

$$\begin{aligned} spiders &: VP \to S \ , \\ time &: NP \to VP \ , \\ flies &: 1 \to NP \end{aligned}$$

are to be listed in the dictionary, meaning that

'*flies*' is a noun-phrase,

'*time*' operates on a noun-phrase to produce a verb-phrase,

'*spiders*' operates on a verb-phrase to produce a sentence.

It was Brame's intention that, aside from the dictionary entries, no further grammatical rules should be postulated. Of course, such a Brame grammar is equivalent to a very simple categorial grammar with dictionary entries:

$$\begin{aligned} spiders &\to S/VP \ , \\ time &\to VP/NP \ , \\ flies &\to NP \ . \end{aligned}$$

If Brame's idea is extended to production grammars, one may also look at derivations $\Gamma \to \Delta$, where Γ and Δ are arrows in the free category generated by the dictionary graph, for example:

$$spiders\ time\ flies \ \to \ flies\ are\ timed\ by\ spiders\ .$$

This opens up the possibility of presenting Chomskyan transformations as 2-arrows, analogous to the natural transformations in the 2-category of small categories. This is just a vague suggestion, but it might bear looking into.

Brame's idea may also be applied to the syntactic calculus by allowing the basic types to be arrows and permitting operations between arrows as follows: if $f : \alpha \to \gamma$, $g : \beta \to \gamma$ and $h : \alpha \to \gamma$, one admits

$$\begin{aligned} g \otimes f &: \alpha \to \gamma\ , \\ h/f &: \beta \to \gamma\ , \\ g\backslash h &: \alpha \to \beta \end{aligned}$$

such that

$$\mathrm{Hom}\ (g \otimes f, h) \cong \mathrm{Hom}\ (g, h/f) \cong \mathrm{Hom}\ (f, g\backslash h)\ .$$

One then obtains what may be called a *residuated bicategory.*

While linguistic applications of this idea have yet to be explored, residuated bicategories abound in mathematics (and may even have been studied abstractly by Bénabou). For example, think of the arrow $f : \alpha \to \beta$ as a bimodule ${}_\beta f_\alpha$, where α and β are rings or small additive categories, $\otimes$ is the usual tensor product and / and \ denote Hom-functors, as in my book on ring theory [21]. Lawvere's [31] generalized bimodules will provide other examples. It should be emphasized that the present discussion involves 'bicategories' rather than '2-categories', because the tensor product of bimodules is only associative up to coherent isomorphism.

There is yet another way in which one may try to turn the categories appearing in linguistics into 2-categories. Instead of allowing the objects (types) to become arrows as above, one could introduce arrows between arrows (derivations). This seems to be the strategy favoured in computer science when one replaces the arrows of a Cartesian closed category by terms of the lambda calculus and then studies reductions of one lambda term into another, see for instance Lambek and Scott [29].

7 Semantics as a Functor

It had already been pointed out by Benson [4] that *semantics* may be viewed as a structure preserving functor from a strictly monoidal category to a Cartesian category, that is, a category endowed with finite Cartesian products. In the same way, one may look at a semantics functor from the residuated category of syntactic types to the cartesian closed category of semantic types. This means, having interpreted each type A by an object $[A]$, one should interpret $A \otimes B$ by $[A] \times [B]$ and both A/B and $B\backslash A$ by $[B] \Rightarrow [A]$, here usually written $[A]^{[B]}$. Presumably, the type $\mathbf{s}$ of affirmative sentences should be interpreted as the subobject classifier of a topos.

Inasmuch as the internal language of a Cartesian closed category is the typed lambda calculus, I believe that this idea is implicit in Montague [36] semantics. One might take the topos in question to be of the form $\mathrm{Set}^{\mathcal{P}}$, where $\mathcal{P}$ is a poset made up of time and modality, indicating how the past can develop into many possible futures.

More sophisticated topos models have been studied by Gonzalo and Marie LaPalme Reyes [40] and Reyes [39].

If $[A]$ is taken to be an object of the given category, the square brackets behave like one of Girard's [16] modalities.

McGill University, Canada.

References

1. Kasimierz Ajdukiewicz. Die syntaktische Konnexität. *Studia Philosophica*, 1:1–27, 1937. Translated in S. McCall, Polish Logic 1920-1939, Clarendon Press, Oxford 1967.
2. Yehoshua Bar-Hillel. A quasiarithmetical notation for syntactic description. *Language*, 29:47–58, 1953.
3. Michael Barr. The theory of theories as a model of syntactic acquisition. *Theoretical Linguistics*, 5:261–274, 1978.
4. David B. Benson. Syntax and semantics: a categorical view. *Information and Control*, 17:145–160, 1970.
5. David B. Benson. The basic algebraic structures in categories of derivations. *Information and Control*, 28:1–29, 1975.
6. Nicolas Bourbaki. *Algèbre multilinéaire*. Paris, 1948.
7. Barron Brainerd. *Introduction to the mathematics of language study*. Elsevier, New York, 1971.
8. Michael Brame. Recursive categorical syntax and morphology I. *Linguistic Analysis*, 14:265–287, 1984.
9. Michael Brame. Recursive categorical syntax and morphology II. *Linguistic Analysis*, 15:137–176, 1985.
10. Noam Chomsky. *Syntactic Structures*. Mouton, The Hague, 1957.
11. Haskell B. Curry. Some logical aspects of grammatical structure. In R. Jacobson, editor, *Structure of Language and its Mathematical Aspects.*, number 12 in Proc. Symposia Applied Mathematics., Providence, 1961. American Mathematical Society.
12. Daniel J. Dougherty. Closed categories and categorial grammar. *Notre Dame Journal of Formal Logic*, 34:36–49, 1993.
13. Samuel Eilenberg and Saunders MacLane. General theory of natural equivalences. *Trans. Amer. Math. Soc.*, 58:231–254, 1945.
14. Michel Eytan. *Semantique doctrinale appliquée*. PhD thesis, Université René Descartes, Paris, 1982.

15. Seymour Ginsburg. *The Mathematical Theory of Context Free Languages*. McGraw-Hill, New York, 1966.
16. Jean-Yves Girard. Linear logic. *J. Theoretical Computer Science*, 50:1–102, 1987.
17. Günter Hotz. Eindeutigkeit und Mehrdeutigkeit formaler Sprachen. *Elektronische Informationsverarbeitung und Kybernetik*, 2:235–247, 1966.
18. Makoto Kanazawa. The Lambek calculus enriched with additional connectives. *Journal of Logic, Language and Information*, 1:141–171, 1992.
19. Stephen Cole Kleene. *Introduction to Metamathematics*. Van Nostrand, New York, 1952.
20. Joachim Lambek. The mathematics of sentence structure. *Amer. Math. Monthly*, 65:154–169, 1958. reprinted in: Buszkowski *et al.*, editors, *Categorial Grammar*, Benjamin, Amsterdam 1988, pp 153-172.
21. Joachim Lambek. *Lectures on Rings and Modules*. Ginn, Waltham Mass. 1966: Chelsea, New York 1976, 1986, 1966.
22. Joachim Lambek. Deductive systems and categories I. *J. Math. Systems Theory*, 2:278–318, 1968.
23. Joachim Lambek. Deductive systems and categories II. *Springer LNM86*, pages 76–122, 1969.
24. Joachim Lambek. Multicategories revisited. *Contemporary Mathematics*, 92:217–239, 1989.
25. Joachim Lambek. On a connection between algebra, logic and linguistics. *Diagrammes*, 22:59–75, 1989.
26. Joachim Lambek. Logic without structural rules. In Došen et al., editor, *Substructural Logics*, pages 179–206. Oxford University Press, 1992.
27. Joachim Lambek. Production grammars revisited. *Linguistic Analysis*, 23:205–225, 1993.
28. Joachim Lambek. What is a logical system? In Dov M. Gabbay, editor, *What is a logical System?* Oxford University Press, 1994.
29. Joachim Lambek and Philip J. Scott. *Introduction to Higher Order Categorical Logic*. Cambridge University Press, Cambridge, 1986.
30. F. William Lawvere. Adjointness in foundations. *Dialectica*, 23:281–296, 1969.
31. F. William Lawvere. Metric spaces, generalized logic and closed categories. *Rend. del Sem. Mat. e Fis. di Milano*, 43:135–166, 1973.
32. A. MacLachlan. The Chomsky hierarchy and natural languages: an overview. Technical report, Report from the department of Math. and Stat. 93-12, McGill University, 1993.
33. Saunders Mac Lane. *Natural Associativity and Commutativity*. Rice University Studies 49, 1963.
34. Saunders Mac Lane. *Categories for the Working Mathematician*. Springer Verlag, New York, 1971.
35. John Macnamara and Gonzalo E. Reyes (eds.). *The Logical Foundations of Cognition*. Oxford University Press, Oxford, 1994.
36. Richard Montague. Formal Philosophy. In Richmond H. Thomason, editor, *Formal philosophy*. Yale University Press, New Haven, 1974.
37. Evelyn Nelson. Categorical and topological aspects of formal languages. *Math. Systems Theory*, 13:255–273, 1980.
38. Matti Pentus. Lambek grammars are context free. *Proceedings of the 8th LICS Conference*, pages 429–433, 1993.
39. Gonzalo E. Reyes. A topos theoretic approach to reference and modality. *Notre Dame Journal of Formal Logic*, 32:359–391, 1991.
40. Gonzalo E. Reyes and Marie LaPalme Reyes. *The Logic of Kinds*. Manuscript, Université de Montréal, 1987.
41. Marie LaPalme Reyes, John Macnamara, Gonzalo E. Reyes, and Houman Zolfaghari. Count nouns, mass nouns and their transformations. Technical report, Rapports de recherche du département de Math. et de Stat., Université de Montréal #374, 1994.

42. Paul C. Rosenbloom. *The Elements of Mathematical Logic.* Dover, New York, 1950.
43. Manfred Egon Szabo, editor. *The Collected Papers of Gerhard Gentzen.* North Holland, Amsterdam, 1969.

TOWARDS A PROCEDURAL MODEL OF NATURAL-LANGUAGE INTERPRETATION CROSSOVER: A CASE STUDY

RUTH KEMPSON

1 Crossover: Preliminaries

This paper describes preliminary work on a model of natural language in which the dichotomy between syntactic and semantic algebras (assumed by linguists since Lewis [39]) is replaced by a system which defines inference over the pair of semantic and syntactic information expressed as label and formula. In this system, the left-right projection of on-line language interpretation provides the sole concept of structure defined for natural-language strings; and syntactic structure is defined as a proof structure through which interpretation is progressively built up. The logic assumed is a labelled deductive system in which syntactic, semantic and control devices are defined together (cf. Gabbay [15]). The motivation underpinning this approach to natural language is the aim of modelling the process whereby given a sequence of words, a hearer incrementally combines syntactic/semantic/pragmatic information to yield an overall interpretation of a string relative to the particular context in which it is intended to be interpreted. Context-dependent aspects of interpretation are modelled as abductive run-time choices from input specifications which constrain but do not fully determine the assigned interpretation. The overall process of interpretation is then defined to be sensitive not merely to lexically encoded bottom-up syntactic/semantic information but also to the way in which interpretation is built up on a left-right basis.[1] The model is introduced on a

[1]This work is part of a project with Dov Gabbay on the development of a processing model for natural language understanding (EPSRC award no: GRK67397, ESRC award no:R000-23-2069). Some of the central ideas are due to him, and I am grateful to him for ongoing support. I am also grateful for comments and suggestions to Wilfried Meyer-Viol, Rodger Kibble, Asli Göksel, Andrew Simpson, Mark Steedman, Stavroula Tsiplakou, Jiang Yan, and to the responses from various audiences to whom this material has been presented. An extended version of this article is appearing in the Journal of Linguistics. Volume 34:1, pp. 73-143

H.J. Ohlbach and U. Reyle (eds.), Logic, Language and Reasoning, 295–334.

case-study basis. I take a linguistic phenomenon assumed to be a syntactic phenomenon and currently granted to be a mystery (Postal [52]), and show how the mystery dissolves, looked at from the perspective of the model being developed. The approach is then tested against its ability to provide cross-language explanations.

The mystery in question is the interaction between the construal of *wh*-expressions and anaphora resolution – the so-called 'cross-over' phenomenon. Crossover has been principally studied within the Principles and Parameters framework (Chomsky [5, 6], Higginbotham [24, 25], Koopman and Sportiche [36], Lasnik & Stowell [38], Postal [52], and many others). Originally brought to light by Postal [51], the phenomenon is invariably subdivided into two discrete phenomena, strong and weak crossover. Strong crossover is the restriction universally displayed across all languages which prohibits a pronoun from being interpreted as dependent on a c-commanding operator if it both precedes and c-commands the trace bound by that operator:

(1) *Who$_i$ did he$_i$ think that Bill liked e_i?

This phenomenon is said to be due to a principle C effect, the trace being required to be free (not coindexed) with respect to any c-commanding expression in an A-position (Chomsky [6]). Weak crossover on the other hand is a restriction that has to be separately stated to preclude a pronoun from being bound by some c-commanding operator in the presence of a following trace, because the pronoun does not c-command that trace and so could not give rise to any principle C effects arising from co-indexation with it:[2]

(2) *Who$_i$ did his$_i$ mother ignore e$_i$?

Lasnik and Stowell [38] observe that weak crossover appears to need a semantic-based solution, as there are environments meeting the required specification where weak crossover effects nevertheless fail to emerge; and these seem to be associated with some concept of referentiality attributed to the *wh*-operator:

(3) John, who$_i$ his$_i$ mother had regularly ignored e_i, fell ill during the exam period.

[2]In the first GB account of weak crossover (Chomsky [6]), a parallel is drawn between such *wh*-examples as these and:

(i) His$_i$ mother ignores every child$_i$

but, as observed by Postal, this parallelism seems less striking in the light of the problems he adduces (cf. Postal [52], and also Williams [56]). For the most part, I shall focus exclusively on examples involving *wh*-constructions (though cf. Section 3.1.1). I shall also focus principally on accounts of crossover in the Principles and Parameters framework, since it is in this framework that these phenomena have received most attention (though cf. the discussion of Dowty [10] in Section 3.1).

(4) Which$_i$ of the team did the judge put away without his$_i$ mother being able to see e_i?
(5) Sam$_i$ is all too easy for his$_i$ mother to ignore e_i.

Indeed according to Lasnik and Stowell such cases provide evidence of a third phenomenon, weakest crossover, in which a *wh* empty category acts not as a name defined to be free, but as a form of pronominal. Weak crossover effects are said to arise in a weak crossover configuration only if the operator binding pronoun and trace is a true quantifier phrase. As demonstrated by Postal however [52], not only does this 'semantic' solution have to be hedged around with additional syntactic restrictions (for example that if the quantifier phrase is in a PP, the PP as a whole must inherit quantificational properties from its subparts), but there are cases where despite a relation of reference, weak crossover effects none the less remain in force:

(6) *It was Carl$_i$ that$_i$ a picture of him$_i$ fell on e_i

There are also cases involving quantification that do not give rise to weak crossover effects:

(7) Most of the patients, in particular all those who$_i$ their$_i$ doctor had said e_i were harmless were allowed home.

There is, furthermore, a parallelism between weak crossover phenomena and what Postal calls 'secondary strong crossover' which remains entirely unexplained under the Lasnik & Stowell account, for despite the fact that these constitute 'strong crossover examples', the pronoun being in an argument position c-commanding the *wh*-trace and being c-commanded by the the *wh*-phrase as a whole, the crossover restriction gets suspended under exactly the same circumstances as the weak crossover effect:

(8) *Whose$_i$ mother$_j$ did you tell him$_i$ e_j had refused the operation?
(9) John, whose$_i$ mother$_j$ I told him$_i$ e_j had refused the operation, was very upset.
(10) *Whose$_i$ exam results$_j$ was he$_i$ certain e_j would be better than anyone else's?
(11) John, whose$_i$ exam results$_j$ he$_i$ had been certain e_j would be better than anyone else's, failed dismally.

The conclusion Postal arrived at was that the weak crossover phenomenon remains deeply mysterious.

In this paper, I propose an analysis of crossover phenomena within the outlined interpretation-as-deduction framework that enables us to explain the data without sub-division into weakest, weak or strong crossover. The account emerges as a consequence of general properties of the reasoning system and the assumption that interpretation is built up on a left-right basis. The general pattern builds on the well-established model of parsing-as-deduction over logical type specifications (Pereira [49], [50], Moortgat [42], König [35], Morrill et al [45], Hepple [23]), but uses the added richness provided by making essential use of labels introduced into the logical language. Within this general perspective, analyses of *wh*-expressions, pronominals and relative clauses are presented in terms of the dynamics of the particular labelled deductive system, each of these analyses being set up as a direct reflection of the process involved in interpreting these structures in context. In Section 3, I show that the analyses of the individual phenomena when taken together are sufficient to predict the array of crossover effects without any ancillary construction-specific principle.

2 A Labelled Deductive System for Natural Language Interpretation — $\mathbf{LDS}_{NL}$

The framework to be introduced (cf. Gabbay and Kempson 26 [18], Gabbay, Kempson & Pitt [20], Kempson [32, 33], Gabbay, Kempson & Meyer-Viol [19] in preparation) is one in which utterance interpretation is modelled as an inferential yet structure-building process which allows integration of encoded constraints on interpretation and pragmatic processes such as anaphora resolution (cf. Fodor, [14], Sperber and Wilson [55], Marslen-Wilson and Tyler [40], Kempson [30], Partee [48]). Utterance interpretation is defined as a task of reasoning in a Labelled Deductive System which leads to progressive assignment of structure to a string on a left-right basis. LDS is a very general logic methodology within which information about the proof instead of being informally recorded alongside the proof, as in a Fitch style of natural deduction proof, is incorporated into it as part of the logical system (Gabbay [15, 16]). All rules, e.g., Modus Ponens are defined over pairs of label-plus-formula (= a declarative unit), the set of premisses $\{\alpha : P, \beta : P \rightarrow Q\}$ leading to conclusion $\mu(\beta, \alpha) : Q$ with μ some operation on the labels relative to some extra restriction $\Psi(\alpha, \beta)$ holding between α and β :

(12) $\rightarrow_\mu$ Elim: (μ-Modus Ponens)

$$\frac{\alpha : P \qquad\qquad \beta : P \rightarrow Q}{\mu(\beta, \alpha) : Q} \quad \text{if } \Psi(\alpha, \beta).$$

The labels provide a local record of any information relevant to individual steps of the proof – assumptions used, pertinent semantic information as required, extra control devices further restricting the applicability of the rules, etc. The general methodology is to blur syntactic and semantic characterizations of inference by absorbing semantic (or other meta-level) information into the syntax of the system. The particular interpretation of the labels depends on the particular LDS.

In the LDS for natural language interpretation (LDS_{NL}) being developed by Gabbay & Kempson (Gabbay & Kempson [18], Kempson [32, 33]), semantic constructs are given a proof-theoretic format. The formulae are the logical types **e**, **t**, **e**→ **t**, **e**→ (**e**→ **t**),... corresponding to the syntactic categories NP, S, intransitive verb, transitive verb, and so on.[3] The labels to such formulae are the concepts expressed by words, expressions of a predicate logic. The language for the labels is as follows. It contains predicate constants 'sing', 'see', 'smile', etc. names 'John', 'Mary', etc. a set of logical proper names, $m_{21}, m_{22}, \ldots$ and the connectives $\&, \vee, \rightarrow, \neg$. It has a number of restricted variables of the form $x, \Psi(x)$, each variable having an associated restriction on its range. These restricted variables are of two types, free and dependent. The latter are expressed as $f_\alpha(\beta)$, where α expresses the range of the function f, β its domain (cf. Section 2.2). There are also specialized place-holding meta-variables, $\mathbf{u}_{pro}, \mathbf{u}_{the}, \mathbf{u}_{wh}, \ldots$ which are subject to a CHOOSE operation which instantiates them with other expressions of the labelling algebra (see below). These expressions combine together through operations such as functional application, coordination etc. to give rise to complex labels. Databases of premisses may also serve as labels (see below). There is an analogous vocabulary for temporal reference with constants and variables denoting units of time, relations such as '>' holding between them, and meta-variables ranging over such temporal expressions. And, finally, there are control labels whose purpose is to impose extra constraints on inference such as imposing ordering on the premisses (see below).

The initial premisses are projected from individual lexical specifications. The operation on the labels for every step of Modus Ponens is functional application. These steps project databases of premisses of the form (13):

(13) { John:**e** smile:**e**→ **t** smile(John):**t** }.

The operations of the semantics (functional application) are thereby represented in the proof system alongside structural information while nevertheless retaining their separate identity as expressions of the labelling algebra. The effect is that for every step of Modus Ponens which eliminates type-structure in the formulae, conversely structure is increasingly built up in the labels.[4]

[3]In an alternative format, Meyer-Viol and Kempson (in preparation) reverse the label-formula relation, treating the logical type as the label to a lambda formula, incremental semantic compilation of the formula being driven by steps of deduction defined over the labels.

[4]The pattern is an extension of the type-deduction familiar in categorial grammar formalisms, with semantic and syntactic information explicitly twinned as label and formula (cf. independent developments using such label-formula pairs by Moortgat [43], Morrill [46], Oehrle [47]).

The novelty of the general LDS framework is not merely in the blurring of semantic and syntactic distinctions resulting from presenting label and formula together, but in the freedom it provides to incorporate into the system extra restrictiveness in the inferences allowed without defining extra connectives – through the use of control labels.[5] For example, in developing LDS_{NL} (Gabbay & Kempson [18]), we define the concept of subject proof-theoretically, with a restriction $\mathcal{S}$ which when added to a label for a premise guarantees that that premise is to be used last in a minimal sequence of steps of Modus Ponens leading to a conclusion of type **t**. This definition of subject is the proof-theoretic analogue to Dowty's [9] semantic characterization ($\mathcal{S}$ is a particular instance of Ψ in μ-Modus Ponens as defined above). A simple example is (14):

(14) { (John,$\mathcal{S}$):**e** saw:**e**→ (**e** →**t**) Mary:**e**},

a database of premisses from which two steps of Modus Ponens may establish the proposition expressed, with 'saw(Mary)(John)' as label to the formula **t**:

saw(Mary):**e**→ **t**
saw(Mary)(John):**t**.

Each clause is seen as a database conforming to the format of (14) – a set of premisses containing one 'major' premise β:$\mathbf{e}^n \rightarrow \mathbf{t}$, and a requisite number of 'minor' premisses of the form γ:**e**, from which the outcome must be α:**t** for some α which is only established through the words as they are presented. The words provide the assumptions, a pair of label and formula making up a single declarative unit, but they may also project additional information about how the database and its structure are to be projected – e.g. inflection determines which premise is to bear the annotation $\mathcal{S}$ identifying it as subject (to be used last in the local database). Monoclausal structure is then projected, as interpretation, from information given in the lexicon and steps of Modus Ponens in a conditional logic with labels. In this system everything is labelled, so each database bears a label. Databases take the form 's:A', where 'A' is a set of declarative units and the label 's' provides extra information about the inferential properties of the collection of premisses as a unit. Following Gabbay and Finger's development (Finger & Gabbay [12]) of a natural deduction system for temporal logic with t_1:A taken to stand for 'A is true at time t_1', the database label is construed as specifying the time at which the conclusion derived from the

[5]Gabbay and de Queiroz [17] demonstrate that sub-structural logics can be expressed as nested families of classical/intuitionistic logics agreeing on the language of the formulae but varying according as the associated labelling algebra dictates increasingly severe restrictions. This methodology stands in contrast to the non-modular approach of e.g. Linear Logic, in which the set of inferences licensed by such a more restricted sub-structural logic is captured through additional connectives, without labels.

set of premisses is projected as holding, i.e. for some labelled database, $s : \Delta, s : \Delta \vdash (s : \phi) : \mathbf{t}$. This label is projected from inflection on the verb. All information projected by the words that gives rise to this database configuration must be used once and once only, a restriction that means the logic has the restrictiveness of linear logic.

For a system to provide a base for modelling how utterances are assigned an interpretation, there is one essential ingredient over and above the projection of structure for a string: it needs a reconstruction of the phenomenon of context-dependence. All items that display such context-dependent effects need to be analysed as presenting a relatively weak input specification, their actual interpretation being selected by a choice mechanism based on other information made available in the interpretation process. The labels provide a natural basis for representing this concept of under-determined information, for we can say a word projects a fully specified type but an incomplete label, with a control specification constraining restrictions on how the label may be completed. In the case of pronouns I shall say that pronouns project information about their logical type but otherwise only a distinguished meta-variable (here listed as $\mathbf{u}_{he}$)[6] with a constraint on identifying the label to be paired with the type $\mathbf{e}$. The constraint on the instantiation function θ that identifies this label requires that if the pronoun is projected in a database as label to a minor premise (of the form α:$\mathbf{e}$), then its value may not be selected from a label to another minor premise in the same constellation of minor premisses (in more orthodox terms, a pronoun as argument to a verb (or adjective) may not be identified with any co-argument in the same complete functional complex). The identification of this label is part of the process of establishing the overall goal α:$\mathbf{t}$ but has to be chosen from other labels of appropriate type that are accessible from within the proof structure.[7] The effect is to define Principle B of the binding theory as a lexically encoded constraint on processing.

2.1 *Wh*-expressions and Goal-directed Deduction

If we add an explicit element of goal-directedness to this very simple system, we find a pattern remarkably similar to that displayed by *wh*-structures. The conditional logic proof (15) displays the relation of a stated goal on some proof structure to the elements of that proof (the proof format is a 'metabox' format in which all subroutines and their associated goals are explicitly represented):

[6]Meta-variables will be invariably represented in bold face.

[7]I assume that actual choices involve pragmatic factors such as relevance (cf. Sperber and Wilson [55]). In this paper I am concerned only with the limits within which such choices are made.

(15) $P \rightarrow (Q \rightarrow R) \vdash Q \rightarrow (P \rightarrow R)$

1	$\alpha : P \rightarrow (Q \rightarrow R)$	GOAL $Q \rightarrow (P \rightarrow R)$
2	$\beta : Q$	GOAL $P \rightarrow R$
3	$\gamma : P$	GOAL R
4	$\alpha(\gamma) : Q \rightarrow R$	Modus Ponens
5	$\alpha(\gamma)(\beta) : R$	Modus Ponens
6	$\lambda x.\alpha(x)(\beta) : P \rightarrow R$	$\rightarrow$ Intro.
7	$\lambda y \lambda x.\alpha(x)(y) : Q \rightarrow (P \rightarrow R)$	$\rightarrow$ Intro.

(The labels in this proof are arbitrary – they provide a name for the premise which can be used to display how the premisses applied. Introduction involves retraction of an assumption and is not used in this paper.)

Two properties of goal-directed reasoning stand out:

(A) to specify some goal and elements to be contained in that goal does not constitute achieving it or having those elements. P and Q are not elements of the proof in (15) in virtue of being elements of the goal: no more are R, and indeed Q→ (P→ R). Such elements only become parts of the proof as they are assumed or derived;

(B) the properties of the goal, together with the rules of the system, determine the form of all subordinate routines to be set up in achieving that goal. Hence the goal is reflected in the subordinate routines within the tree structure that represents the path to be followed in getting to that goal. Here in a conditional logic system, the effect is that each subformula explicitly represented in the goal is carried down within the representation of subordinate goals until the subroutine of assuming or deriving the subformula succeeds in achieving the step involving that subformula necessary for the eventual success in achieving the overal goal. Displayed as a tree structure, the subformulae of the overall goal will duly be carried down to subordinate routines (goals and extra assumptions) as the steps needed to establish that overall goal are progressively set out.[8]

These two properties display a striking parallelism with *wh*-expressions in *wh*-initial languages such as English. The initial *wh*-element is exceptional

[8]The conditional logic proof in (15) is strongly goal-directed in that the form of the goal fully dictates the set of steps necessary to achieve a goal. But the concept of goal is applicable to an array of cases ranging from such fully specified goals over weaker forms of goal-directedness in which the form of the goal only partially dictates the set of steps needed to achieve the goal. In all cases, we would however expect the patterns A,B.

as a constituent in not being visible for e.g. purposes of anaphoric dependency from the point of its occurrence in the string. It only becomes visible once the gap is established. This is the basic crossover pattern:

(16) *Who$_i$ does Joan think that he$_i$ worries e_i is sick?
(17) Who$_i$ does Joan think e_i worries he$_i$ is sick?
(18) *Who$_i$ does Joan think that his$_i$ mother worries e_i is sick?
(19) 19 Who$_i$ does John think e_i worries his$_i$ mother is sick?

Secondly, the information about that initial *wh*-element gets carried down clause by clause until the point at which the gap is constructed (and no further). This is the basic reconstruction pattern.

(20) Which pictures$_j$ of herself$_i$ did Bill say Sue$_i$ was planning to sell e_j
(21) *Which pictures$_j$ of herself$_i$ did Bill say John was planning to sell e_j to prevent Sue$_i$ giving away e_j?

If following the pattern of goal-directed proofs, we analyse *wh* as an initiator of a specifically goal-directed task – the task of constructing a 'gap' as assumption – we will predict this distribution in a principled way, for the information about the nature of the goal as expressed by the *wh*-expression will be automatically transferred through the configuration database by database (= clause by clause) until the point at which the requirement can be realized. We might present this schematically in the metabox format as in (22):

(22)

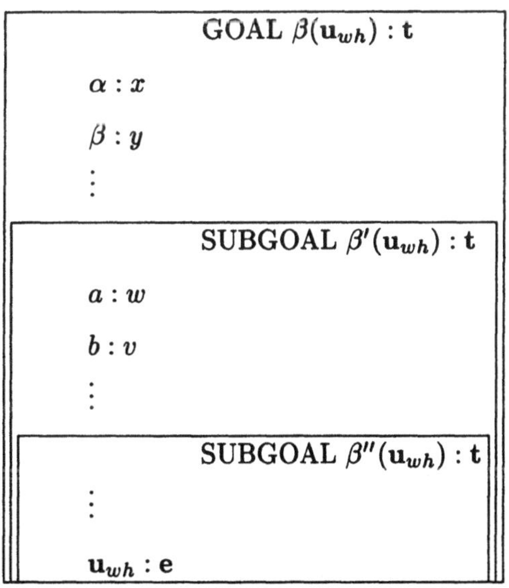

β, β' and β'' are schemas indicating that the target to be achieved is a conclusion of type **t** in which one of the premisses to be constructed at some point in the derivation is of the form $\mathbf{u}_{wh}$:**X** (to be read 'gap labelling type

X'). The $\mathbf{u}_{wh}$ (= 'gap') is a distinguished variable as label to a formula of type **e** that must be constructed at some point during the interpretation process in order for the goal specification to be met, and the string licensed as wellformed. (This meta-variable $\mathbf{u}_{wh}$ is instantiated either by the hearer in answer to the question, or, in relative clauses, by a unification device creating identity between labels, of which more later – Section 2.2). The constraint on the system requiring all premisses to be used will require that $\beta(\mathbf{u}_{wh})$ is not itself carried down to subgoals $\beta'(\mathbf{u}_{wh})$ or to $\beta''(\mathbf{u}_{wh})$ for the words in each database must be used once and once only, so each such schematic goal specification must be discrete. Nevertheless the overall goal and subgoals have an essential shared element – viz. the need at some point to construct the assumption corresponding to the gap.

To adopt this pattern in analysing *wh*-structures we shall require clause-sized units not merely to contain premisses combining together by Modus Ponens but also to contain a goal specification that is not part of the database, but a look-ahead device dictating what has to be done. In the declarative case, the goal will simply be to deduce α:**t**, but *Who does John like*, for example, will get projected as in (23), with the initial *wh*-expression projecting the requirement of the enriched goal specification that some premise with the label $\mathbf{u}_{wh}$ be constructed and through steps of Modus Ponens yield a conclusion containing an occurrence of $\mathbf{u}_{wh}$ in its accompanying label (here $\beta(\mathbf{u}_{wh})$:**t** is satisfied by like($\mathbf{u}_{wh}$)(John):**t**):

(23)

$s_1 < s_{now}$:

GOAL $\vdash \beta(\mathbf{u}_{wh}) : \mathbf{t}$
John, $\mathcal{S}$: **e**
like : $\mathbf{e} \to (\mathbf{e} \to \mathbf{t})$
$\mathbf{u}_{wh}$: **e**
like($\mathbf{u}_{wh}$) : $\mathbf{e} \to \mathbf{t}$
like($\mathbf{u}_{wh}$)(John) : **t**

$\mathcal{S}$, recall, is the annotation marking the premise John:**e** as to be used last in reaching this target. This format could equivalently be set out in a tree configuration, in which nodes are distinguished according to the role entries in the database play in the proof process: g for a goal node, L for a label node, F for a formula node, d for a node for a declarative unit, n for a database node, gn the node for its goal, Ln, the node for its label, Lgn the node for the label to its goal, Fgn the node for the formula in its goal, d^1n the node for the first declarative unit in the database at n, Ld^1n the node for its label, Fd^1n the node for its formula, d^2n the node for the second declarative unit in the database at n, etc:

(24) $[_n\ [_{Ln}\ \mathrm{s}_1, \mathrm{s}_1 < s_{now}\ ] \quad [_{gn}\ [_{Lgn}\ \mathrm{f}(\mathbf{u}_{wh})\ ]\ [_{Fgn}\ \mathbf{t}]]$

$[_{d^1 n}\ [_{Ld^1 n}\ \text{John}, S]\ [_{Fd^1 n}\ \mathbf{e}]]$

$[_{d^2 n}\ [_{Ld^2 n}\ \text{like}]\ [_{Fd^2 n}\ \mathbf{e} \rightarrow (\mathbf{e} \rightarrow \mathbf{t})\ ]]$

$[_{d^3 n}\ [_{Ld^3 n}\ \mathbf{u}_{wh}]\ [_{Fd^3 n}\ \mathbf{e}]]_n]$

Lexical specifications are stated using this node vocabulary, as this allows an explicit definition of the incremental way in which the proof tree is progressively built up node by node, with all extra restrictions imposed by an individual word able to be expressed in terms of required modifications to some partial tree.[9]

However, I shall retain the metabox format in this paper, as this graphically distinguishes goal specifications, entries in the database and database labels.

As we can see from (23) (and (24)), *wh*-expressions in *wh*-initial languages have a distinguished status according to this analysis. They do not themselves project an assumption in thc database, but merely impose the constraint that an assumption with a special label $\mathbf{u}_{wh}$:**e** be constructed at some point within the given proof domain that they initiate. Where that assumption is to be constructed is dictated by the sequence of words themselves. If both the goal specification and the requirement that information projected by all words be used once and once only are to be met, there must be just one point at which the requirements imposed by a word are not met by its neighbouring elements. In (23), this is exemplified by *like*, which there being no following expression in the string requires at this juncture the construction of some assumption *:**e**, for arbitrary *, a requirement that is met by assuming $\mathbf{u}_{wh}$:**e** at this point (cf. fn.9).

If in some initial string of words, the point at which the gap must be constructed does not arise, and the words present, rather, the need for a nested database labelling **t** as with the propositional attitude predicates *say, think*, etc. for which databases are required of the form:

(25) s_1 : | .... think : $\mathbf{t} \rightarrow (\mathbf{e} \rightarrow \mathbf{t})$ (s_2 : [ ]) : **t** |

the principal goal as yet unrealized in s_1 will have to be analogously carried down to s_2 if it is to be met. The goal to the database s_2 is indeed a joint consequence of the overall goal constraining the step of deduction in s_1 and the set of premisses so far entered into s_1. Hence, we get the pattern

[9]For example *like* is specified (presuming on a pointer $\uparrow$ that keeps track of the parsing process which rests at some arbitrary point d^i in database x) as:

$[_x[_{d^{i+1}x}[_{Ld^{i+1}x}\text{like}][_{Fd^{i+1}x}\mathbf{e} \rightarrow (\mathbf{e} \rightarrow \mathbf{t})]][_{d^{i+2}x}[_{Ld^{i+2}x}\ {\uparrow}{*}][_{Fd^{i+2}x}\mathbf{e}]]]$.

The addition of any such specification to the database is a process of unification. * is an empty Prolog-like variable, with a restriction that the very next addition be a process of unification replacing this variable. In this paper all details of lexical specifications are omitted. This process has been implemented in Sicstus Prolog (cf. Finger, Kibble, Gabbay and Kempson [13]).

analogous to (15) in the interpretation of (26), which can be set out as a series of steps as the building up of a database, as (26′):

(26) Who do you think Bill likes?

26′

s_1 :

1		GOAL $\beta(\mathbf{u}_{wh} : \mathbf{t})$
2, 3	you, $\mathcal{S}$: **e**	
3	think : $\mathbf{t} \rightarrow (\mathbf{e} \rightarrow \mathbf{t})$	
s_2 :		
4		GOAL $\beta'(\mathbf{u}_{wh}) : \mathbf{t}$
5, 6	Bill, $\mathcal{S}$: **e**	
6	like : $\mathbf{e} \rightarrow (\mathbf{e} \rightarrow \mathbf{t})$	
7	$\mathbf{u}_{wh}$: **e**	
8	like($\mathbf{u}_{wh}$) : $\mathbf{e} \rightarrow \mathbf{t}$	
9	like($\mathbf{u}_{wh}$)(Bill) : **t**	
10	think(s_2 : like($\mathbf{u}_{wh}$)(Bill)) : $\mathbf{e} \rightarrow \mathbf{t}$	
11	think(s_2 : like($\mathbf{u}_{wh}$)(Bill))(you) : **t**	

The characterization $\beta(\mathbf{u}_{wh})$ in the goal specification with some meta-level function on '$\mathbf{u}_{wh}$' stands for the restriction that the set of premisses must contain a minor premise of the form '$\mathbf{u}_{wh}$:**x**' which by steps of Modus Ponens will lead to the conclusion **t** with an occurrence of $\mathbf{u}_{wh}$ in the accompanying label. The particular lexical specification of *who* restricts this to being some minor premise of the form $\mathbf{u}_{wh}$:**e**. In (26′), steps of Modus Ponens (steps 8–9) duly derive first the conclusion to the innermost database (as label to some formula) and then the final conclusion 'think(s_2:like($\mathbf{u}_{wh}$)(Bill))(you):**t**'. To achieve the step of Modus Ponens as indicated at line 10, an extension of Modus Ponens is necessary to allow its application at step 10 in (26′), for we have a whole database deriving the type '**t**' at step 9. What is needed is to ensure that it is only the conclusion derived from this database that gets carried up by Modus Ponens and not the database-structure itself. This is achieved by defining a so-called 'meta-declarative unit' where a meta-declarative unit is defined as any database of premisses Δ that uniquely prove some conclusion α :A, a concept that applies equally to 'Mary:**e**', which trivially proves itself, and some database of premisses that non-trivially proves some conclusion. Modus Ponens is then defined in the neutral terms of meta-declarative units:[10]

[10]Uniqueness of the derivation for a given interpretation of the string is guaranteed by annotations such as '$\mathcal{S}$' = 'use last', and the fact that unlike categorial grammars,

(27) **Modus Ponens (for meta-declarative units)**

$$\frac{\Delta : A \qquad\qquad \Delta' : A \rightarrow B}{\beta(\alpha) : B}$$

where $\Delta \vdash \alpha : A, \Delta' \vdash \beta : A \rightarrow B$

The change in the revised Modus Ponens format is that the input to the rule is in terms of meta-declarative units, the output as before in terms of a declarative unit whose label solely reflects the semantic mode of combination. With this extension, an interpretation is assigned to *Who do you think Bill likes?* and in so doing, the goal $\beta(\mathbf{u}_{wh})$:**t** imposed as a target by the initial word in the string is fulfilled.

The overall perspective thus adopted is that interpretation is the target-driven incremental projection of database configurations from the input provided by lexical specifications through to a conclusion satisfying the imposed target as goal. This analysis of *wh*-initial expressions as a goal specification, rather than a straightforward premise contributing its interpretation from its position in the string will be the heart of the analysis that follows.

2.2 Linked Databases and Relative Clause Adjunction

This system so far has assumed only that databases of premisses may be nested one within the other. Within this limitation, I have outlined an account of how clausal interpretations are projected, how the phenomenon of anaphora resolution can be modelled, and how *wh*-questions are interpreted. To be able to address the problem of crossover however, we shall need to add an account of relative clause modification – as clause adjunct structures. For this the LDS notation provides a model taken directly from the LDS for modal logic (Gabbay and Finger [12]), in which accessibility relations between locally distinct databases of premisses are defined through labels assigned to databases. One such relation licensed in the present system is that of LINK. This yields a pair of databases that are independent apart from the result of a rule unifying a label in the one database with that contained within the other. Diagrammatically in (28), t_1' is unified with t_1, t_3'' unified with t_3, the result is a database with two further databases linked to it:

(28)

$$s : \boxed{t_1 : A_1, t_2 : A_2, t_3 : A_3}$$

$$\text{Unify } t_1'/t_1 \uparrow \qquad\qquad \uparrow \text{ Unify } t_3''/t_3$$

$$s' : \boxed{t_1' : B_1, t_2' : B_2, t_3' : B_3} \qquad s'' : \boxed{t_1'' : C_1, t_2'' : C_2, t_3'' : C_3}$$

---

no processes of type-lifting or composition of functions are allowed in this system. The definition of meta-declarative unit and the consequent revision of Modus Ponens are due to Dov Gabbay.

It is this pattern of linked databases that we adopt for relative clause modification.

From an informal processing perspective, relative clauses are optional clausal units of information that can be added in to some host clause:

(29) John who adores Mary left Sue.
(30) The idiot who adores Mary left Sue.

The relative clause in (29) for example simply provides a means of adding extra information about *John* into the database projected from the string *John left Sue.* Following the pattern of (28) we might schematically represent (29) (ignoring tense) as:

(31)

s_1 : | John:e, leave:e $\rightarrow$ (e$\rightarrow$ t), Sue:e |

↑

s_2 : | who/John:e, adore:e $\rightarrow$ (e$\rightarrow$ t), Mary:e |

The critical property of such linked databases is that they are independent save only for forced agreement with respect to two labels. The relation LINK itself is defined for a distinguished meta-variable '$\mathbf{u}_{wh}$' as:

$\text{LINK}(\Delta(t_i : A), \Delta'(\mathbf{u}_{wh} : A), \mathbf{u}_{wh}/\mathbf{t}_i) \Leftrightarrow$
For database Δ containing at least one occurrence of '$\mathbf{t}_i$: A', Δ' a distinct database containing at least one occurrence of '$\mathbf{u}_{wh}$:A' in which '$\mathbf{u}_{wh}$' is a place-holding meta-variable, 't_i' unifies with '$\mathbf{u}_{wh}$', 't_i' as the most general unifier replacing all occurrences of '$\mathbf{u}_{wh}$'

(The restriction that there be a unique host '$\mathbf{t}_i$:A' and a unique assumption licensed by the presence of the *wh*-expression is a consequence of the general restriction that words project information to be used uniquely.)

This informal account of relatives can be modelled directly by analysing who_{rel} as having a lexical specification that initiates the building of a new database with a goal specification that an assumption of the form $\mathbf{u}_{wh}$:e be constructed, with the extra restriction that that variable be unified with the label in the host database from which the construction of the new database is initiated. Such a lexical specification will induce the sequence of steps in (29′) in processing *John who adores Mary*, steps 3–5 fulfilling the target specification imposed at step 2:

29′ s_1 : | 1 John : e GOAL α : t

↑

s_2 :		
2		GOAL $\beta'(\mathbf{u}_{wh}) : \mathbf{t}$
		LINK $(s_1(\text{John} : \mathbf{e}), s_2(\mathbf{u}_{wh} : \mathbf{e}), \mathbf{u}_{wh}/\text{John})$
3	$\mathbf{u}_{wh}/\text{John}, \mathcal{S} : \mathbf{e}$	
4	adores : $\mathbf{e} \rightarrow (\mathbf{e} \rightarrow \mathbf{t})$	
5	Mary : $\mathbf{e}$	

Agreement on the verb indicating the presence of a premise annotated as '$\mathcal{S}$', in combination with the restriction imposed at step 2 by the initial *wh*-expression, guarantees that an assumption '$\mathbf{u}_{wh}$:**e**', with '$\mathbf{u}_{wh}$' unified with 'John', is constructed as the first entry in the database. (Unless such an assumption is constructed, the goal specification with the additional LINK restriction will not be fulfilled; and it must be constructed at this point, otherwise the annotation defining the subject projected by agreement will have no premise available to bear the annotation.) A database deriving 'adore(Mary)($\mathbf{u}_{wh}$/John)' thus results, meeting the initially imposed target specification. The projection of interpretation for (29) is then completed with the addition of the two further premisses 'leave: $\mathbf{e} \rightarrow (\mathbf{e} \rightarrow \mathbf{t})$', 'Sue:**e**', the identification of 'John:**e**' as to be annotated as $\mathcal{S}$ (= 'use last') in virtue of agreement on *leave*, and the resulting interpretation of (29) is 'leave(Sue)(John) & adore(Mary)(John)'.

Unconventionally, this analysis has treated non-restrictive relatives as displaying the basic relative-clause pattern. The distinction between restrictive and non-restrictive construals of a relative clause emerges merely as a result of different choices made in the process of projecting an interpretation. The non-restrictive serves to provide additional information about an entity, previously identified, exactly as with names, and the projection of the required linked database takes place after that process of identification has taken place. The restrictive relative on the other hand contributes to the identification of the entity denoted by enabling a recursively complex restriction to be projected, and the projection of the linked database configuration must therefore feed that identification process.

Implicit in this move, as we shall see, is the assumption that determiners are like anaphoric elements and project a place-holding meta-variable to be replaced. There is a considerable body of evidence supporting this analysis. The determiner *the*, for example, can be construed in a number of ways, directly analogous to a pronominal, with uses that range over indexically fixed reference, anaphorically determined reference, bridging cross-reference, bound-variable, and E-type (amongst others), as displayed by:

(32) The door was jammed.
(33) I couldn't get out of the car. The door was jammed.

(34) Every house I put on the market I have checked to make sure that the house is problem-free.
(35) Every car gets the door jammed occasionally.

Reflecting this, the definite article is analysed in terms of some input meta-variable with choice of a value for that variable to be made in context (cf. Section 2). The primary difference between definite NPs and pronouns lies in the predicate that is projected by the nominal content associated with the determiner as the restriction on the instantiation of that variable. *The door*, for example, is said to project a meta-variable '$\mathbf{u}_{the}$' whose value has to be chosen in context given the restriction that the value to be selected must denote an individual bearing the property 'door' – represented as '$\mathbf{u}_{the}$, door($\mathbf{u}_{the}$)'.[11]

Where such a predicate is complex, the predicate will have to be built up through the projection of a linked database structure; and this linkage is effected through unification of the *wh*-meta variable with the meta-variable projected by the determiner. There is one additional step in the projection of the restrictive complex-predicate construal: an extra step of inference is required, which collapses the created linked database to create the complex restriction added to the meta-variable in the initial database. In (30′), displaying a restrictive construal of (30), this step takes place at step 8:

30 The idiot who adores Mary left Sue.
30′

s_1 :

1		GOAL $\vdash \alpha : \mathbf{t}$
2	($\mathbf{u}_{the}$ idiot($\mathbf{u}_{the}$)), $\mathcal{S}$: **e**	
9	($\mathbf{u}_{the}$, idiot($\mathbf{u}_{the}$) & adore(Mary)($\mathbf{u}_{the}$)), : **e**	
10	CHOOSE $\mathbf{u}_{the} = m_{21}$ (for m_{21} a selected logical proper name)	
11	leave : $\mathbf{e} \rightarrow (\mathbf{e} \rightarrow \mathbf{t})$	
12	Sue : **e**	
13	leave(Sue) : $\mathbf{e} \rightarrow \mathbf{t}$	
14	leave(Sue)(m_{21}, idiot(m_{21}) & adore(Mary)(m_{21})) : **t**	

[11] This representation '$\mathbf{u}_{the}$, door($\mathbf{u}_{the}$)' is an abbreviation for a condition on the instantiation of the unrestricted variable $\mathbf{u}_{the}$ that its value $\theta\mathbf{u}_{the}$ must be some α such that 'door(α)' holds. With this interpretation in mind, I allow systematic representation of meta-variables both with and without restrictions. There may be restrictions on such meta-variables as to the set of labels that they may range over. However in the case of the determiner *the* and its associated meta-variable $\mathbf{u}_{the}$, the value of this meta-variable can be any label twinned with the formula 'e' meeting the condition given in the specified restriction. Note that on this analysis, nouns do not project a pair of label and formula but are defined directly as providing a restriction on some independently projected variable.

↑

s_2 :

3	GOAL $\vdash \beta(\mathbf{v}_{wh}) : \mathbf{t}$
	LINK $s_1((\mathbf{u}_{the}, \text{idiot}(\mathbf{u}_{the})) : \mathbf{e}).s2(\mathbf{v}_{wh} : \mathbf{e}), \mathbf{v}_{wh}/(\mathbf{u}_{the}, \text{idiot}(\mathbf{u}_{the}))$
4	$\mathbf{v}_{wh}/(\mathbf{u}_{the}\ \text{idiot}(\mathbf{u}_{the})), \mathcal{S} : \mathbf{e}$
5	$\text{adore} : \mathbf{e} \to (\mathbf{e} \to \mathbf{t})$
6	$\text{Mary} : \mathbf{e}$
7	$\text{adore(Mary)} : \mathbf{e} \to \mathbf{t}$
8	$\text{adore(Mary)}(\mathbf{v}_{wh}/\mathbf{u}_{the}, \text{idiot}(\mathbf{u}_{the})) : \mathbf{t}$

The inference rule required applies to a pair of a declarative unit and a database containing a label unified with the label of that declarative unit, and is defined over meta-declarative units:

(36) The Link Inference Rule

$$\frac{\Delta \vdash (\mathbf{c}_i, \varphi(\mathbf{c}_i)) : \mathbf{e}, \Delta' \vdash \Psi(\mathbf{v}_{wh}) : \mathbf{t}}{(\mathbf{c}_i, \varphi(\mathbf{c}_i)\ \&\ \Psi(\mathbf{c}_i)) : \mathbf{e}}$$

provided that

LINK $\Delta((\mathbf{c}_i, \varphi(\mathbf{c}_i)) : \mathbf{e}), \Delta'(\mathbf{v}_{wh} : \mathbf{e}), \mathbf{v}_{wh}/(\mathbf{c}_i, \varphi(\mathbf{c}_i))$

for $\mathbf{c}_i$, $\mathbf{v}_{wh}$ meta-variables ranging over expressions of type $\mathbf{e}$

The rule (36) follows the same format as the earlier extension of Modus Ponens. The input to the rule is in terms of meta-declarative units – the declarative unit trivially proving $\mathbf{c}_i, \varphi(\mathbf{c}_i) : \mathbf{e}$, and the database Δ' proving $\Psi(\mathbf{v}_{wh}) : \mathbf{t}$. The output is a labelled formula, the label specifying the predicate established from the conclusion $\Psi(\mathbf{v}_{wh}) : \mathbf{t}$ derived from Δ'.[12]

In (30), once having constructed a complex predicate by use of this rule at step 8, a value for the meta-variable '$\mathbf{u}_{the}$,idiot($\mathbf{u}_{the}$) & adore(Mary)($\mathbf{u}_{the}$)' can be selected; and some logical proper name uniquely denoting an individual meeting the description 'idiot($\mathbf{u}_{the}$) & adore(Mary)($\mathbf{u}_{the}$)' is duly chosen (or constructed) as the value of the meta-variable at step 9.

The alternative non-restrictive construal of (30) is displayed in (30″) (note that the process of identification takes place at step 2, prior to the projection of the linked database):

30 The idiot, who adores Mary, left Sue.

[12]This process serves to guarantee that variables introduced and free within a given proof domain are only available in that proof domain (witness the ungrammaticality of **Every man, who I admire, ignores me*). Cf. fn.11 for the systematic variation in representation of meta-variables as restricted and unrestricted.

30″

s_1 :

1		GOAL $\alpha : \mathbf{t}$
2	$\mathbf{u}_{the}$, idiot($\mathbf{u}_{the}$) : $\mathbf{e}$	
3	CHOOSE $\mathbf{u}_{the}$ = Alan	
10	leave : $\mathbf{e} \rightarrow (\mathbf{e} \rightarrow \mathbf{t})$	
11	Sue : $\mathbf{e}$	
12	leave(Sue) : $\mathbf{e} \rightarrow \mathbf{t}$	
13	leave(Sue)(Alan) : $\mathbf{t}$	

↑

s_2 :

4		GOAL ($\mathbf{u}_{wh}$) : $\mathbf{t}$
	LINK(s_1(Alan : $\mathbf{e}$), s_2($\mathbf{u}_{wh}$: $\mathbf{e}$), $\mathbf{u}_{wh}$/Alan)	
5	$\mathbf{u}_{wh}$/Alan : $\mathbf{e}$	
6	adore : $\mathbf{e} \rightarrow (\mathbf{e} \rightarrow \mathbf{t})$	
7	Mary : $\mathbf{e}$	
8	adore(Mary) : $\mathbf{e} \rightarrow \mathbf{t}$	
9	adore(Mary)($\mathbf{u}_{wh}$/Alan) : $\mathbf{t}$	

The difference between non-restrictive and restrictive relative clause construals can thus be straightforwardly modelled as a difference in the order of choices the hearer makes in projecting an interpretation for the string.

Central to this account has been the analysis of the definite article as an anaphoric element projecting a meta-variable for which a process of identification is required. More surprisingly, indefinites, which also allow both restrictive and non-restrictive construals of relative clause modification, can be analysed as displaying a form of anaphoric dependence in their interpretation. The analysis is an extension of the use of Skolem terms to capture the construal of an existentially quantified expression within the scope of some other (universally) quantified expression. In this higher-order notation, the force of the existential is represented without any quantifier as a functionally dependent element for some fixed function mapping the variable bound by the universal quantifier onto some new variable. The interpretation of (37)

(37) Every professor refereed a book

as

37 a. ($\forall$ x, professor(x)) ($\exists$ y, book(y)) referee(y)(x)

can be expressed by use of the Skolem term '$f_{book}(x)$' where f_{book} is a function applying to the set of professors to yield the book that each refereed:[13]

37 a'. ($\forall$ x, professor (x)) referee(f_{book}(x))(x).

The form of the name '$f_{book}(x)$' explicitly records that it is dependent on x (and ranges over books). With the use of labels, and in particular database labels, we are able to extend this analysis to wide scope construal of existential expressions by allowing such Skolem terms to be defined over database labels. This will provide a representation of the wide scope construal of (37), viz.

37 b. ($\exists$ y, book(y)) ($\forall$ x, professor(x)) referee(y)(x)

as (37b')

37 b'. s_i : ($\forall$ x, professor(x)) referee($f_{book}(s_i)$)(x)

with '$f_{book}(s_i)$' as a name recording dependency on the database label. In words: 'Some function f_{book} applying to some world-time unit s_i yields a book such that everyone read it'. If we then allow in an element of underdeterminacy in the lexical specification for indefinites, we can give the indefinite article a unitary characterization as invariably projecting a meta-variable that ranges over dependent names $f_{N'}(\alpha)$, being a place-holding meta-variable for the domain of the function, the restriction N' on its range being projected by the associated nominal content, and the value of α having to be chosen by the hearer as part of the process of utterance interpretation exactly analogous to anaphoric resolution. It is this choice that will give rise to the array of interpretations for the indefinite. In building linked database configurations, such a choice – just as with definite noun phrases – can be made before or after the projection of any linked database structure. Hence the ambiguity of (38) according as the relative clause is restrictively or non-restrictively construed:

(38) A mistake that I should have spotted remained in the final manuscript.

The analysis can then be generalized, and all quantified expressions will be said to project a meta-variable for which a value has to be chosen, with some choices being trivial. For example, for a regular quantifier such as *every*, any new free (individual) variable can be chosen with the locality restriction that its occurrence is licensed only within the proof domain into which it is introduced. The projection of NP meanings for singular NPs is therefore able to be uniform: all determiners will be said to project a meta-variable of some distinguished form with varying constraints as to what type of label must be used to instantiate that meta-variable (a free variable in the case

[13]The restriction on the variables is specified only in the first occurrence of the variable for legibility reasons.

of *every*, a dependent name in the case of the indefinite, either of these or names in the case of *the* and pronouns).[14] All variables are restricted, and the accompanying nominal expressions project a restriction on this name/variable. The recursively specified addition of further restrictions is achieved through the construction of linked databases defined to contain a label unified with the label projected by the determiner/name. In case the linkage is through a meta-variable projected by a determiner, there is an inference rule projecting the conclusion of the linked database as an additional restriction on the variable to be selected.

It is worth noting that this analysis is a straightforward and natural extension of the proof-theoretic methodology already adopted. In predicate logic proofs, the first step of inference for a quantified formula is a step eliminating the quantifer, replacing it with a name. In analysing determiners as mapping onto names of one kind or another, the intermediate step of extracting the quantified expression out of its position in the clausal sequence to create a predicate logic structure as the input to interpretation is simply being by-passed. No step of quantifier raising is needed, and no account of indefinites as highly exceptional determiners need be invoked (contra Reinhart [53, 54]).[15]

This account of quantification is transparently nothing more than a statement of what is required, and a full account of the dependencies intrinsic to these terms (building on the work of e.g Fine [11], Meyer-Viol [41]) remains to be given. In particular an account of the semantics of $\mathbf{u}_{wh}$ will need to be formally defined (cf. Higginbotham [26]). However, in giving an analysis of *wh* in terms of a meta-variable $\mathbf{u}_{wh}$ to be instantiated by expressions of the labelling algebra (rather than directly by their model-theoretic counterparts), one problem will be side-stepped. The analysis of the natural language expression itself will be weaker than any given type of answer and will not need to distinguish between regular questions and

[14] In rounding out the details of this account, restrictions on anaphoric dependency on such constructed names will have to be stated in terms of the varying availability of those names. Free names (corresponding to universal quantification) are available only within the proof domain into which they are introduced. Dependent names (corresponding to existential quantification) are available within the domain into which the element on which they are dependent is introduced. Hence the restriction that quantifier scope variation for non-existential quantifiers is clause internal, and the converse freedom of indefinites to give rise to very wide-scope effects, as displayed in (i)–(ii):

(i) A porter thought that everyone was in the building.

(ii) Everyone thought that a friend of Sue's was in the building.

[15] I do not address the problem of plurals in this paper, so leave on one side the need for set-naming devices for plural quantification and the issues these raise (though cf. Alechina & van Lambalgen [1], van Lambalgen [37], for the development of a proof-theoretic account for generalized quantifiers).

what have been called 'functional questions' (Groenendijk & Stokhof [22]). On the more orthodox accounts in which the semantics of *wh*-questions is specified directly as a mapping from the natural language expression onto the set of model-theoretic counterparts of possible answers, a functional question must be distinguished from a question which is to receive a simple answer, since answers to the former must take the form of a function from individuals to individuals (of type '⟨e, e⟩') and so cannot be reduced to a single individual (of type 'e'). According to the present analysis in which *wh* is a place-holding meta-variable instantiated by an expression of the labelling algebra, the problem does not arise. The prediction of question-answer pairs such as (39):

(39) Who does every student ignore? Her tutor.

arises from the possibility of substituting in place of the assumption '$\mathbf{u}_{wh}$:e' already projected in the database '{x, student(x):e, ignore:e→(e→ t), $\mathbf{u}_{wh}$:e}' the replacement premise triggered by *her tutor* which I represent as '$f_{tutor}(\mathbf{u}_{pro})$:e'. With *every* projecting a new variable free within the proof domain into which it introduces a premise, such a choice of value as answer to the question allows '$\mathbf{u}_{pro}$' contained in '$f_{tutor}(\mathbf{u}_{pro})$' to be subsequently identified as 'x, student(x)' within the domain in which 'x, student(x):e' is licensed, giving rise to the database configuration:[16]

{x, student(x):e, ignore:e→ (e→t), (f_{tutor}(x):e}.

The sentence boundary between question and answer is irrelevant: the process of interpreting the answer involves substituting the premise projected by the lexical items comprising the answer in place of the meta-variable '$\mathbf{u}_{wh}$:e'. And in the case of an answer such as *her tutor* in (39), such a substitution contains a (pronominal) meta-variable, which is itself subject to subsequent interpretation. Nothing idiosyncratic to this kind of answer needs to be specified in the interpretation projected by the *wh*-question itself. The analysis of *wh* thus can remain simple without need of any tiered binding as proposed by Chierchia [4].

3 Crossover – The Basic Pattern

This proof-theoretic format for presenting natural-language interpretation provides an unexpected bonus. Not merely does it capture the incremental way in which interpretation may be assigned to natural language strings, but in addition it shows potential for explaining data classically presumed to require strictly syntactic explanation. In particular we shall see that the

[16]As before I suppress all occurrences of the restriction on the variable x except the first.

analyses now set up for *wh*-expressions, anaphora and relative clauses – each of which is independently motivated – provide explanations of the array of crossover data without any auxiliary construction-specific principles.

Three distinguishing features of the proposed LDS framework will provide the backbone of the explanation:

(1) The model is defined on the assumption that projection of interpretation should be procedural in the sense of projecting not merely the content of component parts, but also how that content is incrementally built up on a left-to-right basis.
(2) The format chosen to reconstruct this assumption is goal-directed labelled deduction. The goal-directedness provides the basis for maintaining a basic linearity account while allowing systematic exceptions.
(3) The model reconstructs at each stage the asymmetry between some provided input specification and the choices that are made to fix the interpretation. The process of interpretation thereby reflects the partial nature of information at any stage during that process. It is the different information that is available to the hearer in making interim choices that determine the different crossover effects. In *wh*-questions, the hearer is confronted with two tasks:

    (i) constructing the proof to a certain pre-set format (to wit that it contain some 'gap' in its set of labels) and

    (ii) identifying the value of the variable projected by the pronoun.

Taking English as the starting point, the general pattern – from a discourse processing perspective – is that the *wh*-expression is invisible for purposes of establishing anaphoric dependency until after the gap is constructed. The goal-driven model gives an immediate answer to this problem: the initial *wh*-expression projects a goal specification. As a mere target on the outcome, it is but a license for some assumption to be constructed later, and cannot be visible for anaphoric dependency. There is no question of the information projected by the *wh*-expression violating or not violating any locality restriction associated with the pronominal for it does not constitute information presented by the database. Once the assumption necessary to meeting that target has however been constructed, it is visible for purposes of anaphoric dependency within the domain within which that assumption has been constructed. Hence the pattern (16)–(19) (repeated here):

(16) *Who$_i$ does Joan think that he$_i$ worries e_i is sick?
(17) Who$_i$ does Joan think e_i worries he$_i$ is sick?
(18) *Who$_i$ does Joan think that his$_i$ mother worries e_i is sick?
(19) Who$_i$ does John think e_i worries his$_i$ mother is sick?

The pattern is not a language-particular stipulation. To the contrary, the pattern is exactly that of natural deduction proofs. A conditional conclusion of the form $Q \rightarrow R$ does not itself constitute the making of the assumption Q. But once the assumption Q has been made and entered in the proof as a premise, its status within the proof is that of any other assumption in the proof.

This preliminary analysis of the crossover phenomenon predicts successfully a number of the distributional properties listed in Lasnik & Stowell. It predicts the basic required pattern of '*wh* – gap – pronoun' rather than '*wh* – pronoun – gap' as above. It also predicts (along with the linking account of Hornstein [27]) that parasitic gap constructions should not constitute a counter example to the present account, as indeed they do not, because the pronoun follows the initial 'true' *wh*-gap, so under this account the pattern of gap followed by pronoun is preserved:[17]

(40) Which book_i did John read e_i after having given it_i to Mary to review e_i

It further predicts that easy-to-please constructions will not give rise to crossover effects, because the subject of the construction preceding the pronominal is certainly available to provide a value for the pronominal, again not withstanding the following gap:

(41) John_i is easy for his_i friends to please e_i

On this account, it is not so much whether a gap follows the pronominal, but whether a gap precedes the pronominal. Moreover it correctly allows the prediction that a pronoun following a gap may only be construed as identical to the gap if the locality restriction of the pronoun is satisfied. Hence (42)

(42) *Who_i does Sue know e_i gave him_i a rise?

[17]Some account is of course needed to explain why additional 'parasitic' elements are licensed at all. The framework here allows the possibility of multiple instances of $\mathbf{u}_{wh}$:e once the initial uniquely licensed assumption is constructed through strategies such as anaphora resolution. In particular these complex clausal adverbials suggest an analysis as a composite database acting as label to the database projected by the main clause, and as such project a domain within which secondary instances of such a variable would be licensed. But I leave this issue unresolved. I also leave unresolved the reported contrast in some dialects between negative-marked clauses which allow parasitic resumptive pronouns, as in (i) and clauses with no negation, which do not, as in (ii):

(i) Which cake_i did the woman who didn't like it_i refuse to eat e_i?

(ii) Which cake_i did the woman who liked it_i refuse to eat e_i?

No such contrast is unacceptable in my dialect, and both are unacceptable. The phenomenon of those dialects would have to be explained along lines set out for French (cf. s. 3.1.1).

3.1 Crossover and Relative Clauses

Nonetheless the account as it stands is clearly too strong. The analysis in terms of *wh* as merely a goal specification suggests that the restriction prohibiting the sequence '*wh*-pronoun gap' should always be illicit across strong and weak crossover constructions, and for all *wh*-initial languages. Yet this is clearly false, given the range of constructions in which in English it doesn't hold, of which non-restrictive relatives remain as the primary case:

(3) John, who$_i$ his$_i$ mother had regularly ignored e_i, fell ill during the exam period.

Moreover not only are there many languages with *wh*-in situ constructions for which no weak crossover effects are displayed (cf. Georgopoulos [21]), but there may be variation between closely related languages such as English and French. Strong crossover effects are universal, but weak crossover effects are highly variable. In the French/English difference, it is in relative clauses in particular that there is variation as to whether or not crossover effects can be suppressed. The analysis apparently needs to predict that sometimes the relative pronoun is visible and interacts with pronoun construal, but sometimes it is not.

To achieve this result, all we need to do is put together the LDS_{NL} accounts of *wh*, pronominal and relative clauses. I take the strong crossover data first, which do not vary across languages:

(43) *John$_i$ who$_i$ Sue thinks he$_i$ worries e_i will fail, left.

The properties involved in processing *wh*-initial constructions, non-restrictive construals and pronouns all interact. (1) A *wh*-initial expression in English does not project a premise in its own right, so the hearer has the task of constructing the appropriate assumption at some suitable point. (2) The non-restrictive construal of a relative however results in the goal specification that has to be met being fully identified as to the type and content of the premise. All that is missing is the information as to what role in the database it is to play. (3) Pronouns have an associated locality constraint that if they project a minor premise of the form $\mathbf{u}_{pro}$:**e**, then in what is otherwise a free process of construal, the value of $\mathbf{u}_{pro}$ may not be chosen from a discrete label of type **e** occurring in the database under construction (= not an argument in its own clause).[18] The hearer therefore has the task

[18]Given the incremental node description of meta-boxes as defined in Section 1, this restriction can be stated as in the lexical specification of *he*:

$$[_x[_{d^{i+1}x}[_{Ld^{i+1}x}\mathbf{u}_{pron}, \text{male}(\mathbf{u}_{pron}), \mathcal{S}, \theta\mathbf{u} = \alpha, \alpha \notin \{y \mid y = Ld^j x, Fd^j x = \mathbf{e}\}][_{Fd^{i+1}x}\mathbf{e}]]].$$

However for the purposes of this paper, I retain the simpler presentation displayed in (43′).

of deciding on an antecedent for the pronoun in the knowledge that it must not include any premise of the form α:**e** in its own database. These three properties combine to preclude any dependence of the pronominal on the variable projected by the *wh* in (43). For example, processing of the *wh* in (43) induces the building of the new database for the relative clause s_2 with new target. So processing of *John* followed by *who* leads to:

$$s_1 : \boxed{\text{John}_1 : \mathbf{e} \qquad \text{GOAL } \alpha : \mathbf{t}}$$

$$\uparrow$$

$$s_2 : \boxed{\begin{array}{l} \hfill \text{GOAL}(\mathbf{u}_{wh}) : \mathbf{t} \\ \text{LINK}(s_1(\text{John}_1 : \mathbf{e}), s_2(\mathbf{u}_{wh} : \mathbf{e}), \mathbf{u}_{wh}/\text{John}_1) \\ \vdots \end{array}}$$

As a target specification, this information about the GOAL will get carried down the tree until such point as the assumption associated with the gap can be constructed. The target that *wh* imposes cannot be satisfied in s_2 given the presence of premisses {Sue:**e**, think:**t**→ (**e**→ **t**)} projected by the words *Sue* and *think*:

$$s_1 : \boxed{\text{John}_1 : \mathbf{e} \qquad \text{GOAL } \alpha : \mathbf{t}}$$

$$\uparrow$$

$$s_2 : \boxed{\begin{array}{l} \hfill \text{GOAL}(\mathbf{u}_{wh}) : \mathbf{t} \\ \text{LINK}(s_1(\text{John}_1 : \mathbf{e}), s_2(\mathbf{u}_{wh} : \mathbf{e}), \mathbf{u}_{wh}/\text{John}_1) \\ \text{Sue} : \mathbf{e} \\ \text{think} : \mathbf{t} \rightarrow (\mathbf{e} \rightarrow \mathbf{t}) \end{array}}$$

So having processed the sequence *John who Sue thinks* the database s_3 will be constructed in accordance with the type specification encoded in *think*. In the light of the overall goal not yet being satisfied, this further database will duly be assigned the subgoal

$$s_3 : \boxed{\begin{array}{c} \text{SUBGOAL } \beta'(\mathbf{u}_{wh}) : \mathbf{t} \\ \text{LINK}(s_1(\text{John}_1 : \mathbf{e}), s_3(\mathbf{u}_{wh} : \mathbf{e}), \mathbf{u}_{wh}/\text{John}_1) \end{array}}$$

so the task in s_3 is to construct a premise of the form John_1:**e**. (The carrying down of this target as a task to be completed in s_3 is essential if the overall goal of providing a conclusion to s_2 that contains the label $\mathbf{u}_{wh}/\text{John}_1$ is to be realized.) But now the completeness of the label $\mathbf{u}_{wh}/\text{John}_1$ means that the premise $\mathbf{u}_{wh}/\text{John}_1$:**e** is fully determined as a premise playing a part in that database. The assumption does not have to be constructed – on the contrary it is given. All that has to be clarified is what its role should be. The information encoded by *he* is that the label to be assigned the premise projected by the pronoun must not be that of any label of type **e** in the database currently under construction. But, given the goal specification, this cannot include John_1:**e**. for John_1:**e** is required to be a premise in s_3. Hence the dependence of *he* on *John* is precluded, even though there is no local relation between the two expressions in the surface string. The clash induced as a principle B effect is caused by the transfer of fully identified information down through the proof tree. Hence the unacceptability of (43) on the indicated interpretation. I give the full specification of the interpretation for (43) as (43′):

43′

s_1 : $\text{John}_1 : \mathbf{e}$ GOAL $\alpha : \mathbf{t}$

$\text{leave} : \mathbf{e} \rightarrow \mathbf{t}$

↑

s_2 : GOAL $\beta(\mathbf{u}_{wh}) : \mathbf{t}$

$\text{LINK}(s_1(\text{John}_1 : \mathbf{e}), s_2(\mathbf{u}_{wh} : \mathbf{e}), \mathbf{u}_{wh}/\text{John}_1)$

$\text{Sue} : \mathbf{e}$

$\text{think} : \mathbf{t} \rightarrow (\mathbf{e} \rightarrow \mathbf{t})$

s_3 : GOAL $\beta'(\mathbf{u}_{wh}) : \mathbf{t}$

$\text{LINK}(s_1 1(\text{John}_1 : \mathbf{e}), s_3(\mathbf{u}_{wh} : \mathbf{e}), \mathbf{u}_{wh}/\text{John}_1)$

$\mathbf{u}_{he}, (\text{if } \mathbf{u}_{pro} : \mathbf{e} \in s_i, \text{CHOOSE } \theta(\mathbf{u}_{pro}) \neq \alpha, \alpha : \mathbf{e} \in s_i) : \mathbf{e}$

$\text{CHOOSE } \mathbf{u}_{he} \neq \text{John}_1$

$\text{worry} : \mathbf{t} \rightarrow (\mathbf{e} \rightarrow \mathbf{t})$

s_4 : GOAL $\beta'(\mathbf{u}_{wh}) : \mathbf{t}$

$\text{LINK}(s_1(\text{John}_1 : \mathbf{e}), s_4(\mathbf{u}_{wh} : \mathbf{e}), \mathbf{u}_{wh}/\text{John}_1)$

$\mathbf{u}_{wh}/\text{John}_1 : \mathbf{e}$

$\text{fail} : \mathbf{e} \rightarrow \mathbf{t}$

Should either the selected antecedent or the pronoun itself not be an argument, no such clash will arise and the relevant interpretation for (44) will be licensed:

(44) Joan_i, who her_i mother worries about e_i all the time, is sick again.

As before, the identification of the goal specification induced by the *wh*-expression as a completely identified premise Joan:**e** has the effect of rendering it visible for purposes of anaphoric dependency. But in these cases one or other of antecedent or pronoun is not a premise (= argument) and so the locality principle intrinsic to the pronominal is satisfied. The interpretation process is displayed in (44′). At the point at which the value of the pronominal is fixed, there is nothing to debar the instantiation of $\mathbf{w}_{pro}$ as 'Joan' given its projection as a mere sub- part of the premise '$\mathrm{f}_{mother}(\mathbf{w}_{pro})$: e' projected from his mother: [19]

44′ s_1 :

1 Joan : **e**	GOAL $\alpha : \mathbf{t}$
	↑

s_2 :

2		GOAL $\beta'(\mathbf{u}_{wh}) : \mathbf{t}$
	LINK $(s_1(\text{Joan} : \mathbf{e}), s_2(\mathbf{u}_{wh} : \mathbf{e}), \mathbf{u}_{wh}/\text{Joan})$	
3	$f(\mathbf{w}_{pro}), \text{mother}(f(\mathbf{w}_{pro})) : \mathbf{e}$	
	for $\mathbf{w}_{pro} : \mathbf{e} \in s_i$, CHOOSE $\theta(\mathbf{w}_{pro}) \neq \alpha, \alpha : \mathbf{e} \in s_i$	
4	worry-about : $\mathbf{e} \rightarrow (\mathbf{e} \rightarrow \mathbf{t})$	
5	$\mathbf{u}_{wh}/\text{Joan} : \mathbf{e}$	

Hence the acceptability of (44). An exactly similar explanation will extend to the extended strong crossover data of (9),(11) (repeated here):

(9) John, whose$_i$ mother$_j$ I told him$_i$ e_j had refused the operation, was very upset.

(11) John, whose$_i$ exam results$_j$ he$_i$ had been certain e_j would be better than anyone else's, got a nasty shock when they$_j$ came out.

This predicts the following patterns. (45) should pattern in parallel with (46):

(45) The bastard$_i$ who$_i$ his$_i$ mother worshipped...

(46) The bastard$_i$ worships his$_i$ mother.

In (46) the premise projected by *the bastard* is a minor premise of the database. In (45), it is AS THOUGH the premise projected by *the bastard* were a minor premise of the database projected from the relative clause, for the goal specification of the database initiated by *who* is fully identified as containing a premise with the very same label as projected by *the bastard.* In both cases then, the information projected from *the bastard* can be used to complete the incomplete label projected by *his*, *his* not being a premise of the form '$\mathbf{u}_{pro}$:e' in the database (it is '$\mathrm{f}_{mother}(\mathbf{u}_{pro})$' that is). Similarly, (11) should pattern with (47):

(47) John's exam result gave him a nasty shock.

[19] The characterization of the genitive is informal at best.

In both (11) and (47) the pronoun corresponds to a premise of the form '$\mathbf{u}_{pro}$:**e**' in the database, but neither *whose* (identified as 'John's') in (11) nor John's in (47) correspond to any such premise. It is '$\mathrm{f}_{exam-result}$(John):**e**' that constitutes the premise. Hence the locality restriction on the pronoun cannot cause a clash with the '$\mathbf{u}_{wh}$' meta-variable identified as 'John' in (11) despite the guarantee of its presence at some point in the same database.[20] On the other hand, *wh* in a PP adjunct will be predicted to disallow an interpretation with a following dependent subject pronoun as in (48), because PP adjuncts containing a name cannot provide an antecedent for a following subject pronominal:

(48) *Joan$_i$, behind whom$_i$ she$_i$ looked e_i nervously, coughed.
(49) *Behind Joan$_i$. she$_i$ looked nervously.

Notice that the analysis does not turn on the expression being in any semantic sense referential (denoting a fixed individual in the model). What is critical is whether the *wh*-expression is construed *ab initio* as projecting a fully identified logical expression whether as a variable or as a name. Hence the analysis provides a semantic base to the explanation without narrowing it to the concept of referentiality.

In this simple interaction of procedures for dynamically building an interpretation comprising a relative and a pronoun, we have, then, the main body of data – the unvarying imposition of strong crossover effects (due either to the locality condition on the pronominal in the relative clause or to the invisibility of the goal specification), the presence of weak crossover effects in questions (there being no linkage to fix the assumption to be constructed), and the lack of weak crossover effects or extended strong crossover effects when the construal of the relative clause is non-restrictive (because the locality condition on the pronominal is satisfied).

This asymmetry between non-restrictive relatives and questions is not predicted by a categorial account of crossover given by Dowty [10] (using assumptions of Hepple [23]). In this account, *wh*-expressions are defined as combining with an open propositional function which is arrived at by discharging some previously constructed assumption. The interpretation of pronouns likewise, modelling the construal of pronominals as bound variables, is assumed to involve an extra assumption that has to be discharged, but the discharge process does not involve change of type assignment. This

[20]For many speakers, sentences such as (11) are not as acceptable as the weak crossover cases such as (44). There is interference here from the fact that in any cases, as in (11), such complex possessive relative structures can almost invariably be expressed more directly given this construal of the pronoun. e.g. for (11) as (i):

(i) John$_i$, who$_i$ had been certain that his$_i$ exam results would be better than anyone else's, failed dismally.

Where no such more direct means is acceptable, as in (9), judgments are more solid.

analysis predicts correctly that only the order *wh*-gap-pronoun is licensed, since the assumption created by the pronoun can be discharged by identity with the assumption corresponding to the gap, which can then be discharged in the presence of the *wh*-expression, with which it combines. In the reverse order, however, the pronoun would discharge the gap, but this would leave nothing to be discharged by the *wh*-operator since the pronoun being merely identified with some antecedent wouldn't alter its type as is required for the *wh*-operator. The order *wh*-pronominal-gap is duly precluded – for both weak and strong crossover. So far, so good. But the account is general to all instances of *wh*, and the asymmetry between weak and strong crossover effects in non-restrictive relatives will not be predicted. On the categorial analysis in which indexical interpretations of a pronoun and principle B effects are dismissed as pragmatic and not part of the analysis (Hepple [23]), there is nothing, correctly, to prevent *the* man_i, who_i his_i *mother thinks* e_i *is happy* via indexical identification of his with the independently identified referent of the man, but this leaves no way of precluding the strong crossover case, (50):

(50) *the man_i, who_i Sue thinks he_i knows e_i is happy,...

Here, equally, once the constructed assumption is discharged in the presence of the higher type *wh*-expression, there is nothing subsequently to preclude the pronoun from being indexically identified with the referent of the man.

Despite similarities between the present analysis of *wh* and the categorial analysis, the former being a parsing analogue of the latter, it is notable that the proposed analysis of the crossover phenomena does not carry over straightforwardly to the categorial account. The meta-level characterization of *wh* as projecting a goal to be achieved makes explicit reference to the way in which information within a goal specification is progressively transferred down through the tree from one database to another as the projection of the proof structure progressively unfolds. It is this added reference to intermediate stages in the process of projecting the required semantic content on which the preclusion of strong crossover data, and contrarily the context-sensitivity of weak crossover effects, depends. The Hepple/Dowty/Morrill account, in contrast, reflecting merely the semantic result, involves solely a constructed assumption and its subsequent discharge – in other words a variable-binding phenomenon. There is no reference to the process involved in achieving that result, and so no means of predicting the availability of information about the licensed assumption at any intermediate stages in the projection of interpretation. The partial context-sensitivity of the phenomenon is accordingly not predicted.

3.1.1. *Restrictive Relatives and the French Puzzle*

The final piece still to be provided in the crossover puzzle as presented by Postal [52] is the problem posed by restrictive relative clauses, where the data are somewhat unclear even in English, and do not appear to exist in closely related languages. Sometimes the data suggest that a restrictive relative is not a structure that licenses anaphoric dependency on the structure projected by the *wh*-expression, sometimes they suggest that it does. The prediction made by this analysis is in principle clearcut, but we have a margin of doubt that mirrors the unclarity and displayed parametric variation in the data. According to the analysis so far set out, if the projection of content for a relative clause involves building a linked database to be used to provide a restriction on some variable projected by the determiner, as in restrictive relative clauses, the process of linkage will involve unifying (= identifying) one place-holding meta-variable with another as displayed in (30′) above: the determiner projects from the lexicon a place-holding meta-variable, udet, the *wh*-expression involves constructing a premise with meta-variable u_{wh} as label. By definition meta-variables are placeholders and have no content of their own. A meta-variable in a goal specification indicates a target to be met, an assumption of appropriate form to be constructed. It is not an entry in the database. Unifying one meta-variable with another will accordingly not change the status of the meta-variable *vis a vis* the new database to be constructed: the appropriate assumption remains to be constructed, it is not an entry in the database in its own right. Restrictive relative clauses should therefore give rise to weak crossover effects exactly as in questions, because the goal specification in the database in question specifies what sort of assumption is to be constructed – it does not itself provide it. We predict correctly the unacceptability of the indicated interpretation in (6) (repeated here):

(6) ?*Every student who_i his_i mother had regularly ignored e_i fell ill during the exam period.

Cf also:

(51) ?*No-one who_i her_i mother helped e_i got a distinction.

However we still face the puzzle as to why a closely related language such as French should systematically fail to give rise to weak crossover effects in relative clauses (a phenomenon which, incidentally, Dowty's categorial analysis would provide no basis for explaining):

(52) le médicin que_i sa_i femme a présenté e_1 a cette infirmière charmante... (Postal [52])
'the doctor that his wife introduced to that delightful nurse.'

(53) Chaque $enfant_i$ que_i son_i père a récupéré e_i aux grilles de l'école est dans de bonnes mains.
'Each child that his father picked up from the school gates is safe.'

(54) chaque $enfant_i$ à qui sa_i mère a donné un livre ...
'each child to whom his mother gave a book ...'

As (53)–(54) demonstrate, this phenomenon is entirely independent of the referential properties of the determiner, for it is available for all types of determiner.

The clue to the analysis required comes from the structure made available at the point at which the value of a label for the pronoun is chosen.[21] I take (55):

(55) The woman who her mother ignored e_i failed

and display the configuration at the point at which choice of value for projected by *her* is made:

$$s_1 : \quad 1\ \mathbf{u}_{the}\text{woman}(\mathbf{u}_{the}) : \mathbf{e} \qquad \text{GOAL } \alpha : \mathbf{t} \qquad \uparrow$$

$$s_2 : \quad 2 \qquad \text{GOAL } \beta'(\mathbf{v}_{wh}) : \mathbf{t}$$

$$\text{LINK } (s_1(\mathbf{u}_{the}, \text{woman}(\mathbf{u}_{the}) : \mathbf{e}), s_2(\mathbf{v}_{wh} : \mathbf{e}), \mathbf{v}_{wh}/\mathbf{u}_{the}, \text{woman}(\mathbf{u}_{the}))$$

$$3 \quad f(\mathbf{w}_{pro}), \text{mother}(f(\mathbf{w}_{pro})) : \mathbf{e}$$

$$\vdots$$

The next step required, step 4, is to identify $\mathbf{w}_{pro}$ so that the label projected from *her mother* can be fully identified. By analysis, the meta-variable specified as required by the target is certainly invisible. But what of '$\mathbf{u}_{the}$, woman($\mathbf{u}_{the}$)' specified in s_1? This too is a meta-variable, without as yet a fixed value; but it is entered in the database s_1 as an incomplete label to a premise. Can it count as a value from this position? There are two possible answers. One is that premisses in a database are only available for identification purposes when their entry into the database is complete whether by lexical or constructed assumption. This is in effect a semantically driven criterion. An entry in a database only constitutes an entry if either it is name-like with a fixed semantic value, or some assignment of values to variables has given it a value protem. By this criterion, the premise of the form '$\mathbf{u}_{the}$, woman($\mathbf{u}_{the}$):e' in s_1 is not complete because the very purpose of the linked database currently under construction is to provide a further restriction on the placeholding meta-variable $\mathbf{u}_{the}$ so that a value for the meta-variable can duly be made. The process of providing that restriction has been initiated by the *wh* by the building of s_2 but it is not at this

[21] I am presuming on a linearity account of anaphoric dependence. Cf. Kempson, [30, 31], and Williams [56] for a defence of this position. In this framework, such an account can be upheld because VP adjuncts such as in *With his bag on his shoulder, John left the house*, are analysed as parasitic on a premise of the type $\mathbf{e} \rightarrow \mathbf{t}$, a function from a label of type $\mathbf{e} \rightarrow \mathbf{t}$ to a new label ADV(β):$\mathbf{e} \rightarrow \mathbf{t}$. Accordingly, the choice of any pronominal contained within an adverbial is delayed until the entry in the database of suitable $\beta : \mathbf{e} \rightarrow \mathbf{t}$, for the presence of $\beta : \mathbf{e} \rightarrow \mathbf{t}$ is an essential prerequisite to the adjunct forming a wellformed label of the logical system.

stage complete; and so there is no complete assumption in the database s_1 to count as providing a value for the pronoun's meta-variable in s_2. Hence by this criterion the label '$\mathbf{u}_{the}$, woman($\mathbf{u}_{the}$)' cannot provide a value for the pronoun. This predicts that weak crossover effects are maintained in restrictive relative clauses.

However there is an alternative criterion. Suppose the very presence of $\mathbf{u}_{the}$ in s_1, albeit only an interim computational device, is taken to be sufficient for providing a value for $\mathbf{w}_{pro}$ in s_2, independent of $\mathbf{v}_{wh}$ (with which it will ultimately be identical). If such a choice procedure is licensed, it will be used as a purely syntactic means of identifying the label to be assigned the pronoun, and in consequence there will be no weak crossover effects, for the pronoun will be identified with a place-holding meta-variable $\mathbf{u}_{the}$ that turns out to be the unifier replacing $\mathbf{v}_{wh}$. Given the subtlety of this distinction, one might well expect unclarity in the data with speaker variation according as the process of instantiating the variable for the pronoun is or is not semantically driven. Moreover one might expect parametric variation across languages according as languages allow meta-variables to be instantiated by syntactic devices independent of the content assigned to that device.

This analysis leads to a further prediction. If French is a language that licenses such a purely syntactic base from which to identify the pronominal using the variable introduced by the determiner, we should predict that weak crossover effects in French are not symmetrical in questions and relatives. While in relative clauses weak crossover effects will be completely absent given the presence of the determiner head, *wh*-questions which lack any such determiner head with its consequent LINK procedure should provide no basis for identifying the pronoun, and weak crossover effects are predicted to re-emerge. This prediction is partially confirmed though the data are clouded by an additional property distinguishing French and English. The interpretation of *wh*-relatives and *wh*-questions are certainly not symmetric. In French relatives, as we have seen, potential crossover effects are completely suppressed, and a pronominal intervening between the *wh*-relatives and the point at which its assumption is constructed fully licenses an interpretation as dependent on the variable projected by the determiner and linked to the gap. In *wh*-questions, there is much greater hesitancy over acceptability judgments with questions so construed:

(56) *Qui$_i$, si'il ya quelqu'un, crois-tu que sa$_i$ mère a appelé e_i?
'Who, if there is someone, do you think his mother called?'

(57) *Qui$_i$ sa$_i$ vie satisfait?
Who does her life satisfy?

(58) ?Quel enfant$_i$ crois tu que sa$_i$ mère a appelé e_i
'Which child do you think his mother called?'

(59) ??Qui$_i$ crois tu que sa$_i$ mère a appelé e_i
'Who do you think his mother called?'

(56)–(57) are unacceptable, and (58)–(59) are said to be marginally acceptable but only when contextualized to a scenario in which a fixed individual is presumed to be picked out by the *wh*-expression for whom the speaker needs identification.[22]

These data conform to the general prediction that on the one hand a *wh*-expression, unless in some sense fully identified, counts as a goal and is invisible, but on the other that if such a goal can independently be fully specified, it will count as visible for purposes of anaphoric dependency. In these cases, the person denoted by the *wh*-expression IS taken to be conceptually fully identified – all that is being asked for is how they are named. What appears to distinguish French and English is that French allows enrichment of *wh*-expressions to yield such referential uses; English apparently does not. This is a consequence of a common pattern – that where languages differ in the number of choices lexically encoded to cover a given interpretive range, the language with less choices will license correspondingly greater use of pragmatic enrichment in compensation. To cover a range of interpretations analogous to that made available by the two lexical items in English $what_{DET}/which_{DET}$, French licenses enrichment of the single *wh*-determiner *quel* (with its gender-specific counterpart *quelle*). Enrichment itself however is open-ended and subject only to the pragmatic requirement of leading to an interpretation compatible with any conditions imposed by the structure from which it is projected – hence the availability of a meta-linguistic type of construal for (58) which is not available in the analogous English examples. With this one added auxiliary statement, I take the general prediction to be confirmed.[23]

The analysis as so far given distinguishes *wh* from other binding operators in that amongst all determiners, only it encodes the projection of a goal specification. This has a consequence already suggested by Postal – that

[22]In a toddlers' playground for example, in which little Harry (whose identity is not known to the speaker) wanders away from his mother, settles in the sandpit at the feet of speaker and hearer, and is subsequently desperately called for by his mother, it is reported by some speakers to be acceptable for the speaker to turn to the hearer and utter (59) intending dependence of the pronominal *sa* on *qui*.

[23]This claim appears to face a counterexample with

(i) ?Quel auteur$_i$ sa$_i$ première poème a-t-elle jamais satisfait e_i ?

These 'complex inversion' structures involve duplication of the subject expression with a resumptive pronoun, the expression *sa première poème* acting as a topicalized expression, linked to the database-internal position through a relation of anaphoric dependency. Given the presence of the pronoun within a topicalized construction which is construed as independently identifiable, and the anaphoric dependence of that pronominal on the *wh*-expression, the only possible interpretation of the *wh* itself is as in (58)–(59), and the same explanation would apply here.

the generalisation between *wh*-phrases and quantifier phrases such as *every N* on which the original weak crossover account was based (Chomsky [6]) (achieved by s-structure movement in the first case and LF movement in the second) is spurious (cf. also Williams [56]). As Postal observes, the parallelism breaks down in French since French fails to manifest weak crossover effects in restrictive relative clauses but apparently fully retains them with quantified expressions. The contrast is between:

(60) le médicin que$_i$ sa$_i$ femme a présenté e_i a cette infirmière charmante

(Postal's (33b)) and (61) (Postal's fn.14.iv):

(61) *Sa$_i$ femme n'a présenté aucun homme$_i$ a cette infirmière charmante.

On this analysis, the structures in (62) and (63) are clearly distinct:

(62) *Who$_i$ does his$_i$ mother love e$_i$?
(63) *His$_i$ mother loves every man$_i$.

Initial *wh*-projects a look-ahead device requiring the construction of a premise at some point, but not itself providing such a premise. *Every*, in situ, projects a premise into the database. What they have in common is that they both lead to the introduction of a newly constructed variable (as label) into a given proof domain viz. the minimal domain for which some conclusion α:t is required. But they differ in the manner of its introduction. Pronouns on the other hand do not introduce a new variable. To the contrary, they rely on identifying the label to the premise they project from labels independently made available. In so far as there is parallelism between (62) and (63) it is due to different causes. The label to be associated with the premise projected by *his* in (62) cannot be identified as identical with that projected by *who* because initial *who* has the special property of not providing a database entry. The *his* in (63) cannot be identified with a variable projected by *every* for linearity reasons – the information projected from the lexical entry from *every* is not available in computing a value for the label to be projected by the pronoun. [24] Thus the similarity between (62) and (63) is not an argument for explaining the similar lack of indicated

[24]Hornstein ([27]) cites as evidence for a non-linear analysis in terms of linking the acceptability of

(i) His$_i$ mother introduced every child$_i$ to his$_i$ teacher

in which the first occurrence of *his* is said to be linked to the second, thus allowing the indicated interpretation through linking of the second pronoun to the variable (corresponding to the trace of the moved quantifier phrase). These data are, however, highly marginal, and are excluded by the present account. The borderline judgements can be explained in terms of the conflict between the first occurrence of *his*, which clearly indicates a construal of the pronominal that is independent of what follows, and the second occurrence, whose natural construal is dependent on the quantifier, with nothing to make salient the necessary disjointness of the two pronominals. Hence the tension and uncertain judgements.

anaphoric dependence in terms of identity of structure. Characterising the particular properties of *wh*-initial in a language such as English and its manifest parallelism with goal-directed reasoning tasks is sufficient. [25]

Confirmation of this natural-deduction style of analysis comes from what Postal took to be the additional puzzle posed by cleft constructions in French. Despite showing no weak crossover effects in relative clauses, French shows weak crossover effects as in English in cleft constructions. If the expression is referential, no crossover effects are displayed:

(64) C'est Jeanne$_i$ que sa$_i$ mère admire e_i
(65) It's John$_i$ that his$_i$ mother told me e_i was guilty

If the expression is a quantifier however, crossover effects emerge. Though quantified expressions can occur in a focussed position, they cannot be used for establishing anaphoric dependency:

(66) It's every Linguistics book that I asked you to catalogue – not every History book.
(67) *It's every student$_i$ that his$_i$ tutor expects to see e_i at midday
(68) *C'est quelqu'un d'autre$_i$ que sa$_i$ mère a présenté e_i
'It's someone else that his mother had introduced.'

This framework immediately suggests an explanation: focus constructions in French as in English are goal specifications – a specification of a premise, fully identified, to be used at some later point in the proof sequence. The parallelism between focus constructions and *wh*-expressions follows at once. A focussed element, like a *wh*-expression will not itself constitute an entry in the database. Unlike a *wh*-expression in questions, however, a focussed

[25]Contrary to Hornstein's claim [27] that a linear account of weak crossover cannot predict what he takes to be the uniform unacceptability of:

(i) ?*The AG's investigation of every senator$_i$ threatened his$_i$ career.

(ii) A part of every article$_i$ undermined it$_i$.

The account sketched here can in principle correctly predict that the data are not homogeneous. If we presume that the internal specification of the nominal modifying some determiner constitutes a discrete licensing domain for variables (matching Postal's concept of scope domain), we can nevertheless predict that the dependent name projected by an indefinite/definite determiner may be identified by anaphoric choice as a name dependent on the variable projected from within that complement specification. Such a choice will have the effect of creating an occurrence of that variable within the containing database as part of the new dependent name, thus making the variable available as a possible antecedent for the following pronominal. Cp (iii)–(v):

(iii) A representative from every firm$_i$ in turn had to present its$_i$ accounts.

(iv) The top of every bottle$_i$ was lying beside it$_i$.

(v) The scarf that every child$_i$ has to wear tomorrow is lying beside his$_i$ bed.

This interpretation is not available if the specifier position is filled as in (i). Though the details of the account remain to be spelled out in full, they strongly indicate that no purely structural account of quantifier-pronominal dependencies will be sufficient.

element will characteristically be a fully identified premise, in which case the only information lacking will be where in the combinatorial process it is to play a role. In such cases, it will be exactly like *wh* in non-restrictive relative construals – there will be some fully identified premise and some domain within which it has to be constructed. A pronoun in the clause following will therefore generally be able to identify with the label there listed as long as no locality clash with the pronominal ensues. This yields the data (64)–(65). Suppose however the expression in the goal involves a quantified expression, itself a license to construct a new name at the point of its introduction in the proof. We then predict that such an expression, while still a specification in the goal only, cannot provide a name for purposes of establishing anaphoric dependence, for it is not yet introduced into the proof. This predicts exactly the results we want. Quantifiers may occur as focussed elements, but they cannot be used as a source of anaphoric dependency from that position. Unlike in relative clauses, there is no premise in a database already set up to which the element in the goal corresponds: there is only the focus expression, and that itself constitutes the goal exactly as is *wh*. Hence the invisibility of an expression in the focus position unless fully identified, and in particular the invisibility of quantified expressions occurring in that position for purposes of anaphora resolution – even in French.

We thus finally reach a familiar conclusion – that *wh*-initial expressions are focus constructions, in all these cases corresponding to a meta-level instruction on the projection of interpretation in the form of a goal specification.[26]

4 Summary

In this paper I have taken a very simple and intuitive idea from natural deduction – that of having a goal to be achieved – and shown how it can

[26]The one genuine set of problems remaining of those posed by Postal is the apparent ability of focus-sensitive items to license dependency between pronominals and *wh*-expressions in questions, even in English:

(i) Who_i did even his_i mother ignore e_i?

This analysis however is I suggest not what is going on. Rather the pronominal is picking up its value from the argument structure of the nominal mother which inherently expresses a two-place relation between mother and child. Such information is straightforwardly manipulated for the purpose of establishing extra pragmatic implicatures, indeed such a choice depends on accessing information about the concepts expressed (cf. Sperber & Wilson [55]), but it is normally not accessible for purposes of establishing anaphoric dependence. However, words such as *even* encode information that encourages the hearer to pick out a larger context set of premisses than would be selected in interpreting the same sequence modulo the occurrence of even. Hence the potential for manipulating such information in cases such as (i). The relation between *who* and *his* in (i) as construed is therefore one of coreference not referential dependence.

be used within an LDS setting to resolve one current syntactic mystery. This concept, as part of a general framework for modelling natural language understanding, promises to provide a uniform explanation to the crossover phenomenon – analysing it as nothing more than the interaction of *wh*-construal, relative-clause construal and anaphora resolution. Standing back from the details, there have been three recurrent themes. First, the explanation of natural language interpretation involves the projection of structure in an order-sensitive way. Secondly, this interpretation process is a goal-directed task of natural deduction. Thirdly, the lexical content of natural language expressions may under-determine the interpretation to be assigned to them, and values for such elements have to be chosen online as part of the task of assigning labelled structure to the string. All of these properties are reflected in the proposed framework. In this system the projection of structure is not bottom to top, defined exclusively in terms of hierarchical configuration, but in terms of a left-to-right procedure in which the words provide both the component parts and information about how to build the structure in which those parts are combined together. Modelling this projection as a natural deduction process plays an essential role in providing this procedural perspective. The particular advantage of an LDS system in addition is that it enables syntactic, semantic and control elements to be expressed together through the simultaneous presence of label and formula as a single inferential unit. The consequence is that structure is not defined within some syntactic system whose internal properties are defined in terms exclusive to natural language structure. To the contrary, the motivation for all concepts and their particular applications come from other areas of reasoning. In particular, the phenomenon of long-distance dependence is not on this account some irreducible property of the computational device (as in Chomsky [8]), but a phenomenon directly explicable in terms of goal-directed reasoning. This strongly suggests that the system underpinning natural language is not some independent module within the cognitive system with no relation to other modules with which it interacts, but is rather a member of a family of logical systems each a purpose-directed specialisation sharing properties with neighbouring modules.

Finally, we return to the motivation for this particular labelled deductive system. The system is being set up to provide a model of the interpretation process. At its heart is a reconstruction of the under-specification intrinsic to the content of natural language elements and the modelling of how this is resolved. The claim, following Sperber and Wilson [55], is that such enrichment, and indeed the overall process of natural-language understanding of which it is a part, is an essentially structural process, building a representation of content from a mixed meta-level/object-level input. However, the resulting model provides a new dynamically projected concept of syntax for natural-language strings, suggesting a much closer link between competence and performance systems. The challenge that this account of crossover poses is whether this dynamic system projecting structure linearly from a sequence of words should replace in toto more orthodox concepts of syntax, interpretation, and understanding. This challenge cannot be properly explored without full definition of the logical system and detailed lexical specifications for particular language fragments (Gabbay, Kempson & Meyer-Viol in preparation). Nevertheless, this preliminary success in explaining crossover phenomena suggests that the answer may well be that it should.

School of Oriental and African Studies, London.

References

1. Natasha Alechina & Michael van Lambalgen. Correspondence and completeness for generalized quantifiers. In Kempson, R. (ed.) *Deduction and Language, Bulletin of the Interest Group in Pure and Applied Logics*, Vol.3. No.2-3, 167–90, 1995.
2. Gennaro Chierchia. Formal semantics and the grammar of predication. *Linguistic Inquiry* 16, 417-44, 1985.
3. Gennaro Chierchia. Functional *WH* and weak crossover. WCCFL 10.75-91, 1991.
4. Gennaro Chierchia. Questions with quantifiers. *Natural Language Semantics* 1; 181-234, 1992.
5. Noam Chomsky. 1977. *Essays on Form and Interpretation.* North-Holland, New York, 1977.
6. Noam Chomsky. *Lectures on Government and Binding.* Foris. Dordrecht, 1981.
7. Noam Chomsky. *Knowledge of Language.* Praeger, 1985.
8. Noam Chomsky. *On Minimalism.* MIT Press, Cambridge MA, 1995.
9. David R. Dowty. Thematic proto-roles and argument selection. *Language*, 67(3), pp. 547-615, 1991.
10. David R. Dowty. 'Variable-free' syntax, variable-binding syntax, the natural deduction Lambek calculus and the crossover constraint. In *Proceedings of the 1992 West Coast Conference on Formal Linguistics*, 1992.
11. Kit Fine. *Reasoning with Arbitrary Objects.* Blackwell, 1985.
12. Marcelo Finger & Dov M. Gabbay. Adding a temporal dimension to a logical system. *Journal of Logic, Language and Information* 1, 203-33, 1993.
13. Marcelo Finger, Rodger Kibble, Dov M. Gabbay & Ruth Kempson. Parsing natural language using LDS: a prototype. Presented at the 3rd Workshop on Logic, Language and Computation, Salvador, Bahia (Brazil), May 1996.
14. Jerry A. Fodor. *Modularity of Mind.* MIT Press, 1983.
15. Dov M. Gabbay. *Labelled Deductive Systems.* Oxford University Press, 1996.
16. Dov M. Gabbay. 'Classical vs non-classical logics (the universality of classical logic). In Gabbay, D. *et al. Handbook of Logic in Artificial Intelligence & Logic Programming: Vol.2 Deduction Methodologies* 359-500, 1994.
17. Dov M. Gabbay & Ruy de Queiroz. Extending the Curry-Howard-Tait interpretation to linear, relevant and other resource logics. *Journal of Symbolic Logic* 57(4): 1319-65, 1993.
18. Dov M. Gabbay & Ruth Kempson. Natural-language content: a proof-theoretic perspective. In *Proceedings of 8th Amsterdam Semantics Colloquium.* Amsterdam, 1992.

19. Dov M. Gabbay, Ruth Kempson & Wilfred Meyer-Viol. Labelled deduction for natural language understanding. In preparation.
20. Dov M. Gabbay, Ruth Kempson & Jeremy V. Pitt. Labelled abduction and relevance reasoning. In Demolombe & Imielinski, T. (eds.) *Non-standard Queries and Non-standard Answers* 155-96. Clarendon Press, Oxford, 1994.
21. Carol Georgopolous. Canonical government and the specifier parameter: an ECP account of weak crossover. *Natural Language and Linguistic Theory* 9, 1-48, 1991.
22. Jeroen Groenendijk & Marten Stokhof. *Studies on the Semantics of Questions and the Pragmatics of Answers*. Academisch Proefschrift. Amsterdam, 1984.
23. Mark Hepple. *The Grammar and Processing of Order and Dependency: A Categorial Approach*. Ph.D Edinburgh, 1990.
24. James Higginbotham. Pronouns and bound variables. Linguistic Inquiry 11, 679-708, 1980.
25. James Higginbotham. Logical form, binding and nominals. *Linguistic Inquiry* 14, 395-420, 1983.
26. James Higginbotham. On Semantics. *Linguistic Inquiry* 16, 547-93, 1985.
27. Norbert Hornstein. *Logical Form: The Grammar of Logical Form from GB to Minimalism*. Blackwell, 1995.
28. Jiang Yan. Logical dependency in Chinese quantification. SOAS Working Papers in Linguistics 3, 1993
29. Jiang Yan. *Quantification in Chinese: An LDS Perspective*. Ph.D London, 1995.
30. Ruth Kempson. Logical form: the language cognition interface. *Journal of Linguistics*, 1988.
31. Ruth Kempson. Anaphora: a pragmatic account. *Proceedings of the Relevance Theory Workshop*. Braga. Portugal, 1990.
32. Ruth Kempson. Ellipsis: a natural deduction perspective. In ed. Kempson, R. *Bulletin of The Interest Group of Pure and Applied Logics: special edition on Language and Deduction*. Vol.3, nos.2-3, 1995.
33. Ruth Kempson. Semantics, pragmatics, and natural-language interpretation. In Lappin, S. (ed.) *Handbook of Contemporary Semantic Theory*. Blackwell, 1995.
34. Ruth Kempson. Crossover: a dynamic perspective. In Jensen, S. (ed.) *SOAS Working Papers in Linguistics and Phonetics* 6, 1995.
35. Esther König. Parsing as natural deduction. In *Proceedings of the Annual Meeting of the Association for Computational Linguistics*. Vancouver, 1989.
36. Hilda Koopman & Dominic Sportiche. Variables and the bijection principle. *The Linguistic Review* 2, 139-60.
37. Michiel van Lambalgen. Natural deduction for generalized quantifiers. In van der Does, J. & van Eijk, J. (eds.) *Generalized Quantifier Theory and Applications*, CSLI, 1995.
38. Howard Lasnik & Tim Stowell. Weakest crossover. *Linguistic Inquiry* 22:687-720, 1991.
39. David K. Lewis. General Semantics. In Davidson, D. & Harman. G. (eds.) *Formal Semantics of Natural Language*. Reidel, 1972.
40. William Marlsen-Wilson & Lorraine Tyler. Central processes in speech understanding. *Philosophical Transactions of the Royal Society London B* 295: 317-32, 1981.
41. Wilfred Meyer-Viol. *Instantial Logic: An Investigation into Reasoning with Instances*. ILLC Dissertation Series 1995 -11. ILLC. Amsterdam, 1995.
42. Michael Moortgat, *Categorial Investigations*. Foris. Dordrecht, 1988.
43. Michael Moortgat. Labeled deductive systems for categorial theorem proving. in Dekker, P. & Stokhof, M. (eds.) *Proceedings of Eighth Amsterdam Colloquium*, 1992.
44. Michael Moortgat. Residuation in mixed Lambek systems. in Kempson, R. (ed.) *Language and Deduction. Special Issue of Bulletin of the Interest Group in Pure and Applied Logics*. Vol. 3, Nos. 2-3. London, 1995.
45. Glyn Morrill *et al.* Categorial deductions and structural operations. In Barry, G. & Morrill, G.(eds.) *Studies in Categorial Grammar*. Edinburgh Working Papers in Cognitive Science. Edinburgh, 1990.

46. Glyn Morrill. *Type-logical Grammar*. Kluwer, 1994.
47. Dick Oehrle. Term-labelled categorial type system. *Linguistics and Philosophy* 18, 1995.
48. Barbara Partee. Binding implicit variables in quantified contexts. In *Proceedings of the Chicago Linguistic Society* 25, 1989.
49. Fernando C. N. Pereira. Categorial semantics and scoping. *Computational Linguistics* 16, 1, 1-10, 1990.
50. Fernando C. N. Pereira. Deductive interpretation. In Klein, E. & Veltman, F. (eds.) *Natural Language and Speech* 117-34. Springer-Verlag. Berlin, 1991.
51. Paul Postal. *Cross-over Phenomena*. Holt, Rinehart & Winston, 1972.
52. Paul Postal.Remarks on weak crossover effects. *Linguistic Inquiry* 24, 539-556, 1993.
53. Tanya Reinhart. *Wh*-in situ: an apparent paradox. in Dekker, P. & Stokhof, M. *Proceedings of the Eigth Amsterdam Colloquium*, 1992.
54. Tanya Reinhart. Interface strategies. OTS Working Papers. Utrecht, 1995.
55. Dan Sperber & Deirdre Wilson. *Relevance: Communication and Cognition*. Blackwell. Oxford, 1986.
56. Edwin Williams. *Thematic Structure in Syntax*. MIT Press, 1994.

TRANSFORMATION METHODS IN LDS

KRYSIA BRODA, MARCELLO D'AGOSTINO AND ALESSANDRA RUSSO

1 Introduction

The methodology of *Labelled Deductive Systems* – or simply LDS[1] – is a unifying framework for the study of logics and their interactions. It was proposed by Dov Gabbay a few years ago in response to conceptual pressure arising from application areas, and has now become a large and influential research programme providing logicians, both pure and applied, with a common language and a common set of basic principles in which to express and to solve their problems.

For the theoretician LDS provides the foundations for a *taxonomy* of logics and brings out the common structure underlying different logical systems proposed for a variety of different purposes – whether philosophical, or mathematical, or practical. For the applied logician (e.g. the computer scientist or the engineer) it provides a powerful technique to develop logical systems tailored to the needs of a specific application (see [16]), so as to maximize the role of logic in applied research.

In the LDS approach the basic declarative units of a deductive process are not just formulae but *labelled formulae*, i.e. expressions of the form $x : A$ where A is a formula of a standard logical language and x is a term of a given labelling language. The labels refer to elements of a suitable algebraic structure that we call *the algebra of the labels* (or *labelling algebra*). In general, a labelled formula $x : A$ expresses a relation between a formula A and an element x of the labelling algebra. It is not necessary to commit to any particular assumption on the nature of this relation. On some occasions we can read $x : A$ 'semantically' as 'A holds *relative to* x', where x is an element of an appropriate space to which it makes sense to relativize the

[1]Ian Hodkinson reports that Gabbay's original choice was LSD, for 'Labelled Systems of Deduction'; the acronym LDS was adopted, to avoid ambiguities, when someone pointed out that LSD was already in use with a slightly different meaning.

H.J. Ohlbach and U. Reyle (eds.), Logic, Language and Reasoning, 335–376.

truth of the proposition A[2]. On others, the label x may identify a region of a *structured database* – whose structure is modelled by the labelling algebra –, and the labelled formula records meta-level information about the dependence of A on x. According to some interpretations (originating in [23]) A may also represent a *type* and the label x a *term* of type A. What counts here is that in a labelled deductive system, the deduction rules act on the labels as well as on the formulae, according to certain fixed rules of propagation which depend on the intepretation of the labelled formulae.

For example, given a language whose only connective is $\rightarrow$, the *Modus Ponens* (or $\rightarrow$-elimination) rule may be given a general formulation as

$$\frac{\begin{array}{c} x : A \rightarrow B \\ y : A \end{array}}{f(x,y) : B}$$

where x and y are terms of a given labelling algebra and f is a function (associated with Modus Ponens) giving the new label after the rule has been applied.

We can obtain an appropriate $\rightarrow$-introduction rule by inverting the $\rightarrow$-elimination, as follows:

$$\frac{\begin{array}{c} t : A \\ f(x,t) : B \end{array}}{x : A \rightarrow B} \quad \text{with } t \text{ atomic.}$$

To show $x : A \rightarrow B$ we assume $t : A$ (with a new atomic label t) and we must prove $z : B$, for a z such that $z = f(x,t)$.

Different fs or different labelling algebras yield different variants of Modus Ponens and, possibly, different logics. For example, if we take an arbitrary semigroup as the algebra of the labels, with $\circ$ as multiplication and $f(x,y) = x \circ y$, we have the rules:

$$\frac{\begin{array}{c} x : A \rightarrow B \\ y : A \end{array}}{x \circ y : B} \qquad \frac{\begin{array}{c} t : A \\ x \circ t : B \end{array}}{x : A \rightarrow B}$$

The interpretation of these rules depends on the interpretation of the labelling algebra. For instance, the terms of the semigroup may represent tasks and a labelled formula $x : A$ may mean that the task x *verifies* the proposition expressed by A. A 'composite' task $x \circ y$ can then be interpreted as meaning the result of performing task x followed by task y. To be even more concrete, one can think of atomic formulae as describing different states of a system M, and of labels as describing actions that can make M

[2]For instance a Kripke frame for modal logics; for an LDS approach to normal modal logics see [28].

change its state. A labelled formula $x : A$, with A atomic, means that the task x makes M change state from its initial state to the state described by A. A formula like A is then interpreted as *a type* and one can say that the task x is *of type* A. A conditional labelled formula $x : A \rightarrow B$ means that, whenever y is a task of type A, the composition of x and y, written as $x \circ y$, is a task of type B. From this point of view, x can be seen as a function and $\circ$ as functional composition. Under this interpretation the conditional rules given above become sound rules for reasoning about the system M and its state transitions.

Clearly, these rules are a (labelled) generalization of the traditional rules of natural deduction [24, 21, 27]. The latter – together with the standard notion of a natural deduction proof of A from a set of assumptions Γ – characterize a well-known logical system, i.e. *intuitionistic* implication. But what about the labelled rules? Once a suitable labelling algebra and the interpretation of the function f have been fixed, do they still characterize a *logic*?

Let's consider the toy-example outlined above. Here the language contains only a binary logical operator that resembles the standard (classical or intuitionistic) conditional. We can consider an 'empty task' 1, such that $1 \circ x = x \circ 1$ and agree that the empty task 'verifies' the initial state of M (that is to say, we have to do nothing to bring the system to its initial state). Let $\mathbf{T}$ be the class of all formulae which are 'verified' by the empty task (i.e. are always true at the initial state). Under this interpretation, $A \rightarrow A$ belongs to $\mathbf{T}$ and so does $(A \rightarrow B) \rightarrow ((C \rightarrow A) \rightarrow (C \rightarrow B))$. However, the formula $A \rightarrow (B \rightarrow A)$ does not. For a counterexample, think of tasks a and b verifying A and B respectively, but such that $a \circ b$ does *not* verify A. The *monotonicity* principle of classical and intuitionistic logic does not hold in this context. A little reflection shows that the same applies to the formulae $(A \rightarrow (A \rightarrow B)) \rightarrow (A \rightarrow B)$, embodying the *contraction* principle, and $(A \rightarrow (B \rightarrow C)) \rightarrow (B \rightarrow (A \rightarrow C))$, embodying the *permutation* principle. So, we have a 'semantics' and a labelled deductive system for it. But do we also have a 'logic'?

An answer to this question depends, of course, on our definition of 'logic'. Very seldom definitions are completely stable, and this is no exception. A 'logic' is usually identified with a *consequence relation*, i.e. a binary relation formalizing the intuitive notion of logical consequence. The traditional definition of consequence relation, which was first formulated explicitly by Tarski [31, 32], can be described as a relation $\vdash$ between sets of formulae and formulae satisfying the following conditions:

Identity	$A \vdash A$
Monotonicity	$\Gamma \vdash B$ implies $\Gamma, A \vdash B$

Transitivity $\Gamma, A \vdash B$ and $\Delta \vdash A$ implies $\Gamma, \Delta \vdash B$

where Γ and Δ are *sets* of formulae.

For instance, the system of intuitionistic implication can be shown to correspond to the smallest consequence relation closed under the following additional condition concerning the $\rightarrow$ operator:

Cond $\Gamma, A \vdash B$ iff $\Gamma \vdash A \rightarrow B$

The closure conditions in the definition of a consequence relation are all *structural* conditions, i.e. they do not involve any specific logical operator, but express basic properties of the notion of logical consequence[3]. The emergence of Relevance Logic (for an overview see [13]) showed the inadequacy of the traditional definition of consequence relation. If the system R proposed by Anderson and Belnap had to be called 'logic', this definition had to be amended. Let's see why.

The whole idea of Relevance Logic is that in a 'proper' deduction all the premisses *have to be used at least once* to establish the conclusion, so as to stop the validity of the notorious 'fallacies of relevance' such as the so-called 'positive paradox' $A \rightarrow (B \rightarrow A)$, or the *ex-falso quodlibet* principle $\neg A \wedge A \rightarrow B$, or its dual $B \rightarrow A \vee \neg A$. This criterion of use is sufficient to prevent the fallacies from being provable. For instance, in the typical natural deduction proof of the positive paradox the assumption B is discharged 'vacuously' by the application of the $\rightarrow$-introduction rule, i.e. it is not used in obtaining the conclusion of the subproof constituting the premiss of the rule application. If we restrict our notion of proof in such a way that all the assumptions have to be *used* in order to obtain the conclusion of the proof, such 'vacuous' applications of the $\rightarrow$-introduction rule are not allowed, since the subproof which occurs as premiss would not be a 'proper proof'. So, the standard proof is no longer an acceptable proof of the positive paradox, and it can be shown that no alternative proofs can be found. Similar considerations apply to the other fallacies.

The criterion of use is clearly of a 'metalevel' nature. It takes the form of a side-condition that restricts the application of the natural deduction rules. Let us consider the restricted ND-deducibility relation. Is it a consequence relation in the traditional sense? The answer is obviously 'no', because it does not satisfy Monotonicity, in that this condition would allow the addition of 'irrelevant' assumptions which are not used in deducing the conclusion. So, if we want to consider the system of relevant implication as

[3]This terminology goes back to Gentzen [21] and his distinction between structural and operational rules in the sequent calculi.

a logic, we have to drop Monotonicity from our definition of a consequence relation.

But this is not the whole story. In the system R, the definition of a 'relevant' deduction requires that every single *occurrence* of an assumption is used to obtain the conclusion. Now, let us write $\Gamma \vdash_R A$, where Γ is a finite sequence of formulae, to mean that there is an R-deduction of A using all the elements of Γ (which are *occurrences* of formulae). Consider the statement $A, A \vdash_R A$. This is not provable because there is no way of using both occurrences of A in the antecedent in order to obtain the conclusion, i.e. one of these two occurrences is 'irrelevant'. Therefore, while $A \vdash_R A$ is trivially provable, $A, A \vdash_R A$ is not. It becomes provable if we 'dilute' the criterion of use to the effect that *at least one* occurrence of each assumption needs to be used, as in the 'Mingle' system (see [13]). The trouble is that the distinction between these two different approaches to the notion of relevance cannot be expressed by the traditional notion of consequence relation as a relation between *sets* of formulae and formulae: there is no way to distinguish the set $\{A, A\}$ from the set $\{A\}$ and, therefore, between $A, A \vdash_R A$ and $A \vdash_R A$. To make this sort of distinction we need a finer-grained notion of consequence relation.

In fact, such a finer-grained notion was developed by Gentzen back in the mid 1930's, with his calculus of sequents [21]. In Gentzen's approach, a (single-conclusion[4]) consequence relation is axiomatized as a relation between a finite *sequence* of formulae and a formula. That such a relation holds between a sequence Γ and a formula A is expressed by the *sequent* $\Gamma \Longrightarrow A$. The axioms of Gentzen's system are given by all the sequents of the form $A \Longrightarrow A$. Gentzen specified also two sets of rules to derive new sequents from given ones, that he called *operational rules* and *structural rules*. While the first type of rules embodied, in his view, the meaning of the logical operators, the second type embodied the meaning of $\Longrightarrow$. These structural rules are the following:

Thinning $$\frac{\Gamma, \Delta \Longrightarrow B}{\Gamma, A, \Delta \Longrightarrow B}$$

Permutation $$\frac{\Gamma, A, B, \Delta \Longrightarrow C}{\Gamma, B, A, \Delta \Longrightarrow C}$$

Contraction $$\frac{\Gamma, A, A, \Delta \Longrightarrow B}{\Gamma, A, \Delta \Longrightarrow B}$$

[4]Gentzen also considered *multi-conclusion* consequence relations and showed how this more general notion was more convenient for the formalization of classical logic. See [21].

Cut $$\frac{\Gamma \Longrightarrow A \quad \Delta, A, \Lambda \Longrightarrow B}{\Delta, \Gamma, \Lambda \Longrightarrow B}$$

Here the Thinning rule corresponds to the Monotonicity condition of a consequence relation, while Cut corresponds to Transitivity, except that, in the sequent formulation, the structure on the left of $\Longrightarrow$ is a *sequence*, rather than a *set*, of formulae. Moreover, in Gentzen's system the role of the Identity condition is played by the assumption that every axiom $A \Longrightarrow A$ can be used at any step of a sequent proof. The remaining rules of Contraction and Permutation have the effect of making Gentzen's relation $\Longrightarrow$ deductively equivalent to the consequence relation $\vdash$ with *sets* as the first argument.

Gentzen's richer axiomatization provides the means of characterizing 'resource-sensitive' systems like R as 'substructural' consequence relations, i.e. logics for which the standard structural rules of Gentzen's axiomatization may not hold. In the case of R, the rule which is dropped is the Thinning rule, responsible for the arbitrary introduction of 'irrelevant' items in the antecedent of a sequent. After removing Thinning, one can consider a rule symmetric to the contraction rule:

Expansion $$\frac{\Gamma, A, \Delta \Longrightarrow B}{\Gamma, A, A, \Delta \Longrightarrow B}$$

The more 'liberal' Mingle system is then distinguished from R by the fact that it allows for this weaker version of the Thinning rule.

The discussion of Relevance Logic clearly brings out the general idea that variations in the notion of logical consequence correspond to variations in the allowed *structural rules* of a suitable sequent-based system, leaving the basic *operational* rules unchanged. With Girard's linear logic [22, 2] the 'substructural movement' reached its climax. linear logic completely rejects the 'vagueness' of traditional proof-theory concerning the use and manipulation of assumptions in a deductive process. A 'proper' proof is one in which every assumption is used *exactly once*. If a particular assumption A can be used *ad libitum*, this has to be made *explicit* by prefixing it with the 'storage' operator. This means that the Contraction rule is not sound in linear logic, since it says, informally, that a proof of B from two occurrences of A can be turned into a proof of B from one occurrence only of A. But this is impossible, unless A is one of those assumptions which can be used *ad libitum*, in which case we should prefix it with the storage operator. In the 'non-commutative' variant of linear logic – which was put forward in 1958 by Lambek [25] as a system intended for applications to mathematical linguistics – also the *order* in which assumptions are used becomes crucial. This means that the Permutation rule is also put in question.

We shall not dive further into the sea of substructural logics. For an overview of the subject the reader is referred to the historical reconstruction in [12]. In this paper we shall, instead, use a fragment of this family of logics as a case-study to illustrate a set of methods originating in the LDS program. In particular, we aim to illuminate the following aspects:

(1) By virtue of the extra power of labels and labelling algebras, typical of the LDS approach, traditional proof systems can be transformed so as to become applicable over a much wider territory whilst retaining a uniform structure. Different logics can be obtained by defining different labelling algebras, which therefore act as 'parameters', and the transition from one logic to another can be captured as a parameter-changing process which leaves the structure of deductions unchanged.[5] In LDS logical systems are not studied 'statically', in isolation, but 'dynamically', observing the process of their generation and their interaction (via modifications of the labelling algebras) on the basis of a fixed proof-theoretical hard-core.

(2) The labelling algebras can be 'discovered' by means of a systematic process which leads from a sequent-based definition of a consequence relation to a Kripke-style semantics for it which is then used to provide a suitable algebra of the labels. A crucial intermediate step of this process consists in applying the Lindenbaum–Tarski method to formulate the 'algebraic semantics' of a given class of consequence relations. This time-honoured approach to the study of logics has been recently discredited because of its lack of 'usefulness'. Here we show how algebraic semantics can be taken as an important step in the systematic development of a suitable LDS[6]

(3) The theory of LDS is not confined to the natural deduction presentation given in [19]. In the sequel we shall illustrate, in a concrete example, how LDS can be reformulated as a general theory of *Labelled Analytic Deduction* based on the system **KE**. This system is somewhat similar to the

[5]Although the idea of annotating formulae with labels is not new (in particular Anderson and Belnap used labels to keep track of 'relevance of assumptions'), LDS has developed this idea in a very general and systematic way, turning it into a research program which is gradually changing our basic 'logical perception'. As with the well-known pictures of Gestalt psychology, Gabbay is pushing us to see a 'duck' (an LDS) where we used to see only a 'rabbit' (a traditional system of deduction, possibly with restrictions).

[6]The material devoted to the semantics of substructural logics contained in this paper overlaps with the content of many other papers, in particular [10, 11], but also [2]. The Kripke-style 'semantics' of implication which arises in section 4.2, and which can be found in a number of papers, is nothing but a generalization of Urquhart's semantics for relevant implication described in [33]. For related semantic investigations into substructural logics see also [26], [29], and [1]. As is partially shown in the present paper, under certain circumstances, semantics for arbitrary consequence relations can be generated in a systematic way. On this problem see also [20, 14, 15, 18]. In our view, however, the main use of these ideas is in the development of a suitable LDS. 'Semantics', here, is only a by-product of this process and its distinction from 'syntax' is blurred by the LDS perspective.

method of analytic tableaux, but drastically departs from the latter in the fact that it is *not* cut-free. On the contrary, it hinges on an *analytic cut* rule which is the only branching rule, all the elimination rules being of a *linear* format. For an exposition of the **KE** system and of its advantages over the tableaux method see [8, 9] and [5].

(4) The **KE**-style analytic formulation can be used as an algorithm for finding a (labelled) natural deduction proof. While the natural deduction rules may be very suitable to present a proof *once it has been found*, the **KE** rules are more suitable when the problem is that of *finding* it.[7] We shall therefore describe an algorithm to turn a labelled **KE**-refutation into a labelled natural deduction proof. This algorithm consists in inverting the derivation process and showing that the refutation rules can be read as backward reasoning in a natural deduction system.

2 Consequence Relations

2.1 Preliminaries

We assume a propositional language $\mathcal{L}$ with two binary operators $\rightarrow$ and $\otimes$ and a constant $\top$. We shall use the upper case letters A, B, C, etc. as variables ranging over formulae of $\mathcal{L}$. The set of all well-formed formulae of $\mathcal{L}$ will be denoted by $\mathcal{F}$, and $\mathcal{F}^*$ will denote the set of all *sequences* of elements of $\mathcal{F}$. Sequences of formulae will be denoted by enclosing formulae, separated by commas, within square brackets, as in $[A, B, C]$. We shall use $[\,]$ to denote the empty sequence. We shall also use the capital Greek letters Γ, Δ, etc. as variables ranging over sequences of formulae. We shall often write $A, B, C, \ldots$ instead of $[A, B, C, \ldots]$ and Γ, Δ for the concatenation of sequences Γ and Δ.

DEFINITION 2.1 We take a *consequence* relation[8] $\vdash$ as a relation $\mathcal{F}^* \times \mathcal{F}^*$ satisfying:

Identity $$A \vdash A$$

Surgical Cut $$\frac{\Delta, A, \Lambda \vdash B \quad \Gamma \vdash A}{\Delta, \Gamma, \Lambda \vdash B}$$

[7]This conflict between *discovery* and *justification* is not a recent one. The Greek mathematician Pappus (3rd century A.D.) called the first process 'analysis' and the second process 'synthesis'. He agreed with the nowadays ATP community that the 'analysis' involves reasoning backwards, from the theorem that one wants to prove to the axioms (or the data), while the 'synthesis' requires us to retrace our steps, somehow inverting them, from the axioms to the theorem. (See [3] for a description of a goal directed approach to natural deduction.)

[8]For a general approach to *structured* consequence relations see [17].

As we mentioned in the introduction, the notion of consequence relation is traditionally defined in terms of sets rather than sequences. The two approaches are equivalent whenever $\vdash$ is closed under the following conditions:

$$S_1 \quad \frac{\Gamma, A, B, \Delta \vdash C}{\Gamma, B, A, \Delta \vdash C} \qquad\qquad S_2 \quad \frac{\Gamma, A, A, \Delta \vdash C}{\Gamma, A, \Delta \vdash C}$$

$$S_3 \quad \frac{\Gamma, A, \Delta \vdash C}{\Gamma, A, A, \Delta \vdash C}$$

So, in a consequence relation satisfying all the structural rules S_1–S_3, the sequences occurring on the left of the turnstile can be replaced by *sets*. In the traditional definition, consequence relations are also required to satisfy the following *monotonicity* condition:

$$S_4 \quad \frac{\Gamma \vdash C}{\Gamma, A \vdash C}$$

We shall refer to conditions S_1–S_4 as to the *structural rules.* A logical system is said to be *substructural* if one or more of these structural rules are not assumed as valid.

The first question related to 'substructural' logics concerns the logical interpretation of sequences of formulae. When all the structural rules are allowed, the antecedent of a sequent is equivalent to the conjunction of its elements. In substructural logics, this is no longer the case, and a sequence of formulae represents different 'conjunctions' depending on which structural rules are allowed. We shall use the symbol '$\otimes$' to denote any binary operator which satisfies the *minimal* condition for a conjunction operator, namely

$$(\mathrm{C}_\otimes) \qquad \Gamma, A, B, \Delta \vdash C \;\text{ iff }\; \Gamma, A \otimes B, \Delta \vdash C.$$

So we can always consider a sequent $\Gamma \vdash A$ as equivalent to $\otimes\Gamma \vdash A$, where $\otimes\Gamma$ denotes the $\otimes$-concatenation of the formulae in Γ.

We shall also denote by '$\rightarrow$' any binary operator satisfying the minimal condition for an implication operator, namely

$$(\mathrm{C}_\rightarrow) \qquad \Gamma, A \vdash B \;\text{ iff }\; \Gamma \vdash A \rightarrow B.$$

Finally, we shall use the constant $\top$ to represent the empty sequence in the antecedent of a sequent. So, the following holds for every consequence relation $\vdash$:

$C_\top$ $\qquad \Gamma, \Delta \vdash B$ iff $\Gamma, \top, \Delta \vdash B$.

2.2 Substructural Cconsequence Rrelations

The smallest consequence relation containing the operators $\rightarrow, \otimes, \top$ and closed under the structural rules S_1–S_4 corresponds to the $\{\rightarrow, \wedge, \top\}$ fragment of intuitionistic logic, where $\wedge$ corresponds to $\otimes$.

In the sequel we shall consider the class of subsystems of this fragment of intuitionistic logic obtained by restricting the allowed structural rules to any subset of $\mathbf{S} = \{S_1, S_2, S_3, S_4\}$. Such logics can be classified according to the set of operational conditions (defining the meaning of the logical operators) and the subset of structural rules that they satisfy. We shall use the notation LJ^γ_α, where γ is a set of logical operators (defined by conditions like $C_\otimes$, $C_\rightarrow$ and $C_\top$ above) and α is a subset of $\{1,2,3,4\}$, to indicate the smallest consequence relation closed under the appropriate conditions for the operators in γ (whatever they may be), and the structural rules in $\{S_i | i \in \alpha\}$. For every fixed γ, the set of logics $\{LJ^\gamma_\alpha | \alpha \in 2^{\mathbf{S}}\}$ represents a family that we call 'the LJ^γ family'. In this paper we shall restrict ourselvers to the logics of the $LJ^{\{\otimes,\rightarrow,\top\}}$ family and, therefore, we shall systematically omit the superscript. For instance $LJ_\emptyset$ will denote the smallest consequence relation of the $LJ^{\{\otimes,\rightarrow,\top\}}$ family which is not closed under any of the structural rules S_1–S_4, while $LJ_{\{1,3\}}$ will be the smallest consequence relation of this family closed under S_1 and S_3.

3 The Algebraic Approach

3.1 The Algebraic Interpretations of Sequents

The well-known method of Lindenbaum–Tarski provides a way of turning the operators of a logical system L into the operators of an abstract algebra $\mathcal{A}$, which can then be used to define a valuation system characteristic of L. Provided L contains as theorems:

L1 $\qquad A \rightarrow A$

L2 $\qquad (A \rightarrow B) \rightarrow ((B \rightarrow C) \rightarrow (A \rightarrow C))$.

Then one defines $A \cong B$ iff $A \rightarrow B$ and $B \rightarrow A$ are both theorems of L, and observes that $\mathcal{F}/\cong$, where $\mathcal{F}$ is the set of all well-formed formulae of the language of L, is partially ordered by the relation $\|A\| \leq \|B\|$ iff $A \rightarrow B$ is a theorem of L.

One can then consider the equivalence classes $\|A\|$, $\|B\|$ etc. as the set of truth-values of a valuation system, and define operations on this set corresponding to the logical operators of L as defined by the proof-theory.

This method was meant to provide a characterization of the set of theorems of a logical system which was more convenient than the Hilbert-style axiomatic presentation for the purpose of mathematical investigation. It was originally applied to classical and intuitionistic logic, but can be in principle applied to any logic (satisfying L1 and L2 above) in which the relation $\cong$ is a congruence. In an axiom system, the notion of *theoremhood* comes first and that of *derivability* second. Today we tend to characterize a logical system in terms of its derivability relation, rather than in terms of theoremhood, and fix the axiomatic properties of this relation in terms of sequents, namely objects of the form $A_1, \ldots, A_n \vdash B$.

To apply the Lindenbaum–Tarski method to abstract consequence relations constructed as sets of sequents closed under certain conditions, we have to clarify the logical role played by the extra-logical symbols, like the comma and the turnstile. We have seen in the previous section that the comma is naturally associated with the operator $\otimes$. We can then interpret the comma by means of a binary operation, say $\circ$, satisfying different properties depending on the allowed structural rules. In any case, since the comma is a list-constructor, this operation is required to satisfy at least the laws of a monoid, with identity element 1 representing the empty list.

As we have seen in the previous section, a sequent $\Gamma \vdash A$ is equivalent to $\otimes\Gamma \vdash A$. Once we have shifted to this interpretation of a sequent as a relation between two formulae, it is easy to see that in every sequent calculus closed under Identity and Surgical Cut, the turnstile denotes a quasi-ordering[9] relation on the set of formulae. We can, therefore, take the equivalence classes of formulae under the equivalence relation $\cong$, defined as $A \cong B$ iff $A \vdash B$ and $B \vdash A$. These equivalence classes are partially ordered by: $\|A\| \leq \|B\|$ iff $A \vdash B$.

Let us turn our attention to the logical operators. It is not difficult to see that C$\otimes$, together with Surgical Cut, implies:

$$\frac{A \vdash B \quad C \vdash D}{A \otimes C \vdash B \otimes D} \tag{1}$$

This means that the condition

$$x \leq y \text{ and } w \leq z \text{ implies } x \circ w \leq y \circ z \tag{2}$$

has to be satisfied, for all $x, y, w, z \in \mathcal{F}/\cong$.

As to implication, in all the consequence relations considered here, its behaviour is characterized by C$\rightarrow$:

(C$\rightarrow$) $$\Gamma, A \vdash B \text{ iff } \Gamma \vdash A \rightarrow B.$$

[9]We recall that a binary relation is a quasi-ordering if it is reflexive and transitive.

By using Surgical Cut and C→ it is easy to derive:

$$\Gamma \vdash A \to B \text{ iff } \forall\Delta(\Delta \vdash A \text{ implies } \Gamma, \Delta \vdash B). \tag{3}$$

First assume $\Gamma \vdash A \to B$. By C→, $\Gamma, A \vdash B$. Suppose $\Delta \vdash A$, then by (Surgical Cut), $\Gamma, \Delta \vdash B$. For the converse, assume $\forall\Delta(\Delta \vdash A$ implies $\Gamma, \Delta \vdash B)$. Hence, if Δ is the sequence $[A]$, we have, by Identity, $\Gamma, A \vdash B$ and, by $C_{\to}$, we can conclude $\Gamma \vdash A \to B$. □

In terms of the partial ordering $\leq$ defined above, this means that we can define an operator $\rhd$, corresponding to $\to$, such that for all $x, y, z, v \in \mathcal{F}/\cong$:

$$x \leq v \rhd z \text{ iff } \forall y(y \leq v \text{ implies } x \circ y \leq z). \tag{4}$$

Notice that this implies:

$$x \circ y \leq z \text{ iff } x \leq y \rhd z. \tag{5}$$

The study of the algebra outlined above allows for a sharp analysis of a wide variety of sub-systems of intuitionistic logic and provides a means of separating, within a given logic, what pertains to the minimal 'meaning' of the logical operators, expressed by invariant inference rules, and what pertains to our *procedural uses* of such operators, expressed by the changing properties of the relation $\leq$.

3.2 The Algebra of LJ-structures

In this section we describe the algebra which arises from the previous discussion. It should be obvious that this algebra is in no way more informative than the definition of a consequence relation in terms of sets of sequents. However, we shall see in the next sections that it is *heuristically useful*, in that it suggests developments which do not arise naturally from the consideration of sequents.

3.2.1. *LJ-structures*

DEFINITION 3.1 An LJ-structure is a structure $(M, \circ, \rhd, 1, \leq)$, where

1. $(M, \circ, 1)$ is a monoid with identity 1, i.e. it satisfies:
   (a) $x \circ (y \circ z) = (x \circ y) \circ z$
   (b) $1 \circ x = x \circ 1 = x$
2. $\leq$ is a partial ordering of M, i.e. is a reflexive, antisymmetric and transitive binary relation between its elements;
3. the operation $\circ$ is *order-preserving*, that is:

$$x_1 \leq y_1 \text{ and } x_2 \leq y_2 \text{ iff } x_1 \circ x_2 \leq y_1 \circ y_2.$$

4. For every x, y the set $\{z | z \circ x \leq y\}$ has a maximum element denoted by $x \rhd y$.

The binary operation $\circ$ may or may not satisfy (any combination of) the following additional axioms:

B$_1$ $x \circ y = y \circ x$
B$_2$ $x \leq x \circ x$
B$_3$ $x \circ x \leq x$
B$_4$ $x \circ y \leq x$.

We shall speak of *LJ_α-structures*, where α is a subset of $\{1, 2, 3, 4\}$, to denote LJ-structures such that the operation $\circ$ satisfies the additional conditions in $\{B_i | i \in \alpha\}$.

4 Semantic Consequence Relations

In this section we shall define two kinds of semantic consequence relations, which are associated with LJ-structures: *many-valued* consequence relations and *possibile-world* consequence relations. We show that the two characterizations coincide.

4.1 Many-valued Consequence Relations

DEFINITION 4.1 Let S be an LJ-structure. A *many-valued valuation*, or *MV-valuation* for short, over S is a function h from the well-formed formulae of $\mathcal{F}$ to the terms of S, satisfying:

1. $h(A \to B) = h(A) \rhd h(B)$
2. $h(A \otimes B) = h(A) \circ h(B)$
3. $h(\top) = 1$

DEFINITION 4.2 The *MV-consequence relation* associated with a class of LJ-structures $\mathbf{C}$, is the relation $\models^{MV}_{\mathbf{C}}$ between finite sequences of formulae and formulae defined as follows:

$A_1, \ldots, A_n \models^{MV}_{\mathbf{C}} B$ iff $h(A_1) \circ \cdots \circ h(A_n) \leq h(B)$ for every LJ-structure $S \in \mathbf{C}$ and every MV-valuation h over S.

A formula A is *valid* in $\mathbf{C}$ whenever $\top \models^{MV}_{\mathbf{C}} A$.

It should be clear that every LJ$_\alpha$-structure corresponds to the consequence relation LJ$_\alpha$ defined in the previous section, as stated in the following theorem:

THEOREM 4.3 $\Gamma \vdash_{\mathrm{LJ}_\alpha} A$ *iff* $\Gamma \models^{MV}_{\mathrm{LJ}_\alpha} A$.

Proof: It is easy to verify that $\models^{MV}_{\mathrm{LJ}_\alpha}$ satisfies Identity, Surgical Cut and $\mathrm{C}_{\rightarrow}$. Moreover, it satisfies also the structural rule S_i whenever $\leq$ satisfies B_i, for $i = 1, \ldots, 4$. This is sufficient to establish the soundness of $\vdash_{\mathrm{LJ}_\alpha}$ with respect to $\models^{MV}_{\mathrm{LJ}_\alpha}$. For the completeness, consider the equivalence relation $\cong$, defined as:

$$A \cong B \text{ iff } A \vdash_{\mathrm{LJ}_\alpha} B \text{ and } B \vdash_{\mathrm{LJ}_\alpha} A.$$

The set $\mathcal{F}/\!\cong$ is partially ordered by:

$$\|A\| \leq \|B\| \text{ iff } A \vdash_{\mathrm{LJ}_\alpha} B. \tag{6}$$

Consider the operation $\bullet$ on $\mathcal{F}/\!\cong$ defined by $\|A\| \bullet \|B\| = \|A \otimes B\|$. It follows from the properties of $\otimes$ that $\mathcal{F}/\!\cong$ is a monoid with identity $\|\top\|$. Moreover, it is easy to show that in every consequence relation:

$$A \vdash B \text{ and } C \vdash D \text{ implies } A \otimes C \vdash B \otimes D.$$

Hence, for every $x_1, x_2, y_1, y_2 \in \mathcal{F}/\!\cong$,

$$x_1 \leq y_1 \text{ and } x_2 \leq y_2 \text{ implies } x_1 \bullet x_2 \leq y_1 \bullet y_2$$

so that $(\mathcal{F}/\!\cong, \bullet)$ is coordinatewise ordered by $\leq$. Finally, for every $\|A\|$ and $\|B\|$ in $(\mathcal{F}/\!\cong, \bullet, \leq)$,

$$\|A \rightarrow B\| = \max\{z | z \bullet \|A\| \leq \|B\|\}.$$

For, suppose $\|C\| \bullet \|A\| \leq \|B\|$. Then

$$\begin{aligned} \|C\| \bullet \|A\| \leq \|B\| \quad & \text{iff} \quad \|C \otimes A\| \leq \|B\| \text{ iff } C \otimes A \vdash B \\ & \text{iff} \quad C, A \vdash B \text{ iff } C \vdash A \rightarrow B \text{ iff } \|C\| \leq \|A \rightarrow B\|. \end{aligned}$$

So $(\mathcal{F}/\!\cong, \bullet, \leq)$ is an LJ-structure where $\|A\| \rhd \|B\|$ is defined as $\|A \rightarrow B\|$. Moreover, it is easy to see that it is also an LJ_α-structure. Now, consider the function h defined by $h(A) = \|A\|$. By definition h is an MV-valuation. Now, suppose, $\Gamma \not\vdash_{\mathrm{LJ}_\alpha} A$. Then $\otimes\Gamma \not\vdash_{\mathrm{LJ}_\alpha} A$ and $\| \otimes \Gamma\| \not\leq \|A\|$ by (6). So, $\|B_1\| \bullet \cdots \bullet \|B_n\| \not\leq \|A\|$, where $B_1, \ldots, B_n$ are all the elements of Γ. Therefore, $h(B_1) \bullet \cdots \bullet h(B_n) \not\leq h(A)$ for some MV-valuation over some LJ_α-structure. □

4.2 Possible-world Consequence Relations

In this section we show how to translate a many-valued consequence relation, such as the one outlined in the previous section, into an equivalent 'possible-world' consequence relation. For this purpose we have to replace the plurality of truth-values with a plurality of 'worlds' or 'pieces of information' considered as the valuation space of a *two-valued* relative valuation (or 'forcing' relation).

DEFINITION 4.4 We define:

1. A *quasi-ordered monoid* is a structure $(M, \circ, 1, \preceq)$ such that $(M, \circ, 1)$ is a monoid with identity 1, and $\preceq$ is a quasi-ordering satisfying
$$x \preceq y \text{ and } v \preceq z \text{ implies } x \circ v \preceq y \circ z.$$
2. A *partially ordered monoid* is a quasi-ordered monoid where the relation $\preceq$ is a partial ordering.

DEFINITION 4.5 Let Q be a quasi-ordered monoid. A *PW-valuation over* Q is a two-argument function $\mathcal{F} \times Q \mapsto \{T, F\}$, where $\mathcal{F}$ is the set of formulae of the language, satisfying the following conditions:

1. $v(A, x) = T$ and $x \preceq y$ implies $v(A, y) = T$.
2. $v(A \to B, x) = T$ iff $\forall y, v(A, y) = T$ implies $v(B, x \circ y) = T$.
3. $v(A \otimes B, x) = T$ iff $\exists y, z,\ y \circ z \preceq x$ and $v(A, y) = T$ and $v(B, z) = T$.
4. $v(\top, x) = T$ iff $1 \preceq x$.

So, the quasi-ordering of the monoid behaves like an 'accessibility relation'. The monoid Q is also called the *valuation space* or the *frame* of the PW-valuation. Sometimes the pair (Q, v), where Q is a frame and v is a PW-valuation over Q, is called a *model*.

DEFINITION 4.6 A point z in the valuation space Q of a PW-valuation is *A-minimal* if $v(A, z) = T$ and $(\forall x \in Q)(v(A, x) = T$ implies $z \preceq z)$. We say that a PW-valuation over Q is *regular* if
$$(\forall A \in \mathcal{F})((\exists w \in Q) v(A, w) = T \text{ implies } (\exists z \in Q) z \text{ is } A\text{-minimal}).$$

DEFINITION 4.7 The *regular PW-consequence relation* $\models^{PW}_{\mathbf{C}}$, associated with a class $\mathbf{C}$ of quasi-ordered monoids, is the relation between sequences of formulae and formulae defined as follows:

$$A_1, \ldots, A_n \models^{PW}_{\mathbf{C}} B \text{ iff } v(B, x_1 \circ \cdots \circ x_n) = T \text{ whenever } v(A_i, x_i) = T \text{ for all } i = 1, \ldots, n, \text{ for every } Q \in \mathbf{C} \text{ and every regular PW-valuation } v \text{ over } Q.$$

A formula A is valid in $\mathbf{C}$ if $\top \models^{PW}_{\mathbf{C}} A$.

Consider a quasi-ordered monoid Q. Let $\simeq$ be the equivalence relation defined as
$$x \simeq y =_{\mathsf{def}} x \preceq y \text{ and } y \preceq x.$$

Let the operation $\bullet$ on $Q/\simeq$ be defined as follows: $\|x\| \bullet \|y\| =_{\mathsf{def}} \|x \circ y\|$. Obviously, $(Q/\simeq, \bullet)$ is a monoid with identity $\|1\|$. Moreover, we can define a partial ordering $\sqsubseteq$ on $(Q/\simeq, \bullet)$ as usual: $\|x\| \sqsubseteq \|y\|$ iff $x \preceq y$. It is easily checked that $(Q/\simeq, \bullet, \sqsubseteq)$ is a partially ordered monoid.

LEMMA 4.8 *Let v be a regular PW-valuation over a quasi-ordered monoid Q. Let $(Q/\simeq, \bullet, \sqsubseteq)$ be the associated partially ordered monoid. Moreover, let v' be the valuation function over $(Q/\simeq)$ defined as follows:*

$$v'(A, \|x\|) = T \quad \textit{iff} \quad v(A, x) = T.$$

Then v' is a regular PW-valuation, namely it satisfies all the conditions in Definition 4.5 and the regularity condition in Definition 4.6.

We now consider classes of quasi-ordered monoids satisfying any combination of the following axioms

$\mathbf{B}_1'$ $x \circ y \preceq y \circ x$ (commutativity)
$\mathbf{B}_2'$ $x \circ x \preceq x$ (contraction)
$\mathbf{B}_3'$ $x \preceq x \circ x$ (expansion)
$\mathbf{B}_4'$ $x \preceq x \circ y$ (monotonicity)

Notice that these conditions are obtained by inverting the conditions B_1-B_4 of page 347.

We shall also denote each class by $\mathbf{Q}_\alpha$, where α is, as usual, a subset of $\{1, 2, 3, 4\}$. We shall denote by $\models^{PW}_{\mathbf{Q}_\alpha}$ the regular consequence relation associated with the class $\mathbf{Q}_\alpha$ of quasi-ordered monoids.

COROLLARY 4.9 *Let $\models^{PW}_{\mathbf{Q}_\alpha}$ be the regular PW-consequence relation associated with the class $\mathbf{Q}_\alpha$ of quasi-ordered monoids, and let $\models^{PW}_{\mathbf{P}_\alpha}$ be the similar consequence relation associated with the subclass $\mathbf{P}_\alpha$ of $\mathbf{Q}_\alpha$ such that $\mathbf{P}_\alpha = \{Q \in \mathbf{Q}_\alpha | Q$ is partially ordered$\}$. Then*

$$\Gamma \models^{PW}_{\mathbf{Q}_\alpha} A \quad \textit{iff} \quad \Gamma \models^{PW}_{\mathbf{P}_\alpha} A.$$

Proof: The *only-if* direction is trivial, because if there is a PW-valuation v over a partially ordered monoid which falsifies $\Gamma \vdash A$ then, a fortiori, there is such a valuation over a quasi-ordered monoid. For the *if* direction, if v is a PW-valuation over a quasi-ordered monoid which falsifies $\Gamma \vdash A$, consider the valuation v' over $Q/\simeq$ defined in Lemma 4.8. Then v' is a PW-valuation over the partially ordered monoid $Q/\simeq$ and it is easy to see that v' falsifies $\Gamma \vdash A$. □

4.3 Correspondence with Many-valued Semantics

LEMMA 4.10 *Let v be a regular PW-valuation over a partially ordered monoid $(Q, \circ, \sqsubseteq)$ and let Q' be the set of all $x \in Q$ such that x is A-minimal for some $A \in \mathcal{F}$. Then*

$$(\forall x, y \in Q')(\exists w)(w = \min\{z | y \sqsubseteq z \circ x\}).$$

Moreover, Q' is closed under $\circ$.

Proof: If x, y are in Q', then x is A-minimal for some A and y is B-minimal for some B. Since v is regular, there is $w \in Q$ such that w is $A \to B$-minimal. By definition of PW-valuation, $v(B, w \circ x) = T$ and, since y is B-minimal, $y \sqsubseteq w \circ x$. Now, suppose $y \sqsubseteq z \circ x$. Since x is A-minimal, it follows that for all u such that $v(A, u) = T$, $y \sqsubseteq z \circ u$ and, since y is A-minimal, $v(A, z \circ u) = T$. Hence, by definition of a PW-valuation, $v(A \to B, z) = T$ and, since w is $A \to B$-minimal, $w \sqsubseteq z$. Moreover, it is easy to verify that Q' is closed under $\circ$. □

COROLLARY 4.11 *Let $(Q, \circ, \sqsubseteq)$ be a partially ordered monoid and v a regular valuation over it. The structure $(Q', \circ, \leq)$, where Q' is defined as in Lemma 4.10, and $x \leq y$ iff $y \sqsubseteq x$ is an LJ-structure.*

THEOREM 4.12 *For all finite sequences Γ and all formulae A:*

$$\Gamma \models^{MV}_{\mathbf{LJ}_\alpha} A \quad \textit{iff} \quad \Gamma \models^{PW}_{\mathbf{Q}_\alpha} A$$

Proof: Consider an LJ_α-structure $S = (M, \circ, \leq)$. We define an associated structure $S' = (M', \bullet, \subseteq)$ as follows:

- M' is the set of all *increasing sets* or *order filters* generated by the points x of S, namely the sets $\uparrow x = \{y | x \leq y\}$, according to the partial ordering $\leq$.
- $\bullet$ is defined as follows: $\uparrow x \bullet \uparrow y = \uparrow (x \circ y)$;
- $\subseteq$ is ordinary set inclusion.

It is easy to verify that S' is partially ordered monoid belonging to $\mathbf{Q}_\alpha$.

Now, given a many-valued valuation h over S, we define an associated PW-valuation v_h over S' as follows:

$$v_h(A, x) = T \quad \text{iff} \quad h(A) \geq x(\text{ i.e. } h(A) \in \uparrow x).$$

The reader can verify that v_h satisfies all the conditions for a *regular* PW-valuation.

Similarly, given a quasi-ordered monoid $S = (Q, \circ, \preceq)$ such that $S \in \mathbf{Q}_\alpha$, and a *regular* PW-valuation v over it, we can define an associated LJ_α-structure $S' = (Q', \circ, \leq)$. First we take the valuation v' over the partially ordered monoid $(Q/\simeq)$, defined as in Lemma 4.8. Then, we consider the function $h_v : \mathcal{F} \mapsto Q/\simeq$ defined as follows:

$$h_v(A) = \min\{z | v'(A, z) = T\}.$$

Let $h_v(\mathcal{F}) = \{h_v(A) | A \in \mathcal{F}\}$. It follows from Lemma 4.10 that $S' = (h_v(\mathcal{F}), \circ, \leq)$, where $\leq$ is the reverse of $\sqsubseteq$, is an LJ_α-structure. It is not difficult to verify that h_v satisfies all the conditions for an MV-valuation over the LJ_α-structure S', namely:

1. $h_v(A \otimes B) = h_v(A) \circ h_v(B)$
2. $h_v(A \to B) = h_v(A) \rhd h_v(B)$
3. $h_v(\top) = \|1\|$

The correspondence between MV and PW-valuations that we have just outlined is expressed by the following identities:

$$h_{v_h} = h \qquad\qquad v_{h_v} = v.$$

Now, for the if-direction of the theorem, suppose $h(A_1) \circ \cdots \circ h(A_n) \not\leq h(B)$ for some MV-valuation h over S. Then, it is easy to verify that $v_h(B, x_1 \circ \cdots \circ x_n) = F$ and $v_h(A_i, x_i) = T$ for all i. For the only-if direction, assume $v(B, x_1 \circ \cdots \circ x_n) = F$ and $v(A_i, x_i) = T$, for some regular PW-valuation v over a quasi-ordered monoid. Then it is easy to verify that $h_v(A_1) \circ \cdots \circ h_v(A_n) \not\leq h_v(B)$. □

It follows from Theorem 4.3 and Theorem 4.12 that:

COROLLARY 4.13 *For all finite sequences of formulae* Γ *and all formulae* A

$$\Gamma \vdash_{\mathbf{LJ}_\alpha} A \;\; \textit{iff} \;\; \Gamma \models^{MV}_{\mathbf{LJ}_\alpha} A \;\; \textit{iff} \;\; \Gamma \models^{PW}_{\mathbf{Q}_\alpha} A.$$

Forcing Notation. A PW-valuation v is the characteristic function of a relation between points of the valuation space and formulae of the language. This kind of relation is usually called a *forcing* relation and is denoted by $\Vdash$, the translation between the two formulations being the following: $v(A, x) = T$ iff $x \Vdash A$.

5 From PW-Semantics to LKE-refutations

We now show how the PW-semantics described in the previous section can be reformulated in terms of a *labelled refutation system.* This will take the form of a generalization of the classical tableau-like system **KE** (see [7]). The generalization involves shifting from signed formulae to *labelled* signed formulae, or LS-formulae, as basic units of a refutation, and so allows us to incorportate the semantics into the syntax, in the spirit of Gabbay's LDS approach.

5.1 The Implication Fragment

DEFINITION 5.1 Let $\mathcal{A}$ be an alphabet containing (i) denumerably many symbols 'a_1',..., 'a_n' called *atomic labels,* (ii) the symbol '$\circ$'. The set of labels is defined as the least set satisfying:

1. Every atomic label is a label.
2. If x and y are labels, $x \circ y$ is also a label.

It is called *the labelling language* and is denoted by $\mathcal{L}_L$. A *label* is any element of $\mathcal{L}_L$. A *labelled signed formula*, or *LS-formula* for short, is an expression of the form $TA : x$ or $FA : x$ where A is a well-formed formula and x is a label.

The interpretation of LS-formulae is as follows: $TA : x$ stands for $v(A, x) = T$ (or, equivalently, $h(A) \geq x$ in terms of MV-valuations) and $FA : x$ for $v(A, x) = F$ (or, equivalently, $h(A) \not\geq x$).

Implication Rules. It follows from Definition 4.5 that for every PW-valuation

$$v(A \to B, x) = T \text{ and } v(A, y) = T \text{ imply } v(B, x \circ y) = T, \qquad (7)$$

and that

$$v(A \to B, x) = F \text{ implies } \exists y, v(A, y) = T \text{ and } v(B, x \circ y) = F. \qquad (8)$$

Therefore, using the notation defined above, the following expansion rules are sound:

$$\frac{\begin{array}{l} TA \to B : x \\ TA : y \end{array}}{TB : x \circ y} \qquad \frac{FA \to B : x}{\begin{array}{l} TA : a \\ FB : x \circ a \end{array}} \qquad (9)$$

where a is a *new atomic* label.

We have seen that (Corollary 4.9) we can restrict our attention, without loss of generality, to regular PW-valuations of partially ordered monoids. So in such valuations if there is a point that verifies a formula A, than there is also the *least* point that verifies A. (This property corresponds to the regularity property of PW-valuations.) So, we can always identify the new atomic label a in the rule for $FA \to B : x$ with such a minimum x at which A is true. Therefore, in every subsequent application of the EF→ rule to a conditional with the same antecedent, we can *reuse* the same atomic label a, instead of introducing a new one. This amounts to allowing for the rule:

$$\frac{\begin{array}{l} TA : a \\ FA \to B : x \end{array}}{FB : x \circ a} \qquad (10)$$

when a is atomic.

These rules are the *universal* rules for the conditional operator: they hold for every consequence relation which contains an operator → satisfying the deduction theorem.

We observe that our relative valuations are *bivalent*, so that for all formulae A and all points x of the valuation space:

$$v(A,x) = F \quad \text{or} \quad v(A,x) = T. \tag{11}$$

This means that the following branching rule is allowed:

$$\frac{\qquad\qquad}{FA : x \mid TA : x}$$

where A is an arbitrary formula of the logical language and x is an *arbitrary* label of the labelling language.

The rules just described are tree-expansion rules similar to the classical tableau rules. A branch is *closed* when it contains both $TA : x$ and $FA : x$ for some formula A and some label x. The extra-conditions on the accessibility relation, which characterize each particular implication logic, can be expressed as *structural* rules by means of the same notation. All the rules are summarized in Table 1.

TABLE 1. Types of Rules

Operational	Fundamental	Structural	
		Permutation	Contraction
ET→ $\dfrac{TA \to B : x \quad TA : y}{TB : x \circ y}$	Clos $\dfrac{TA : x \quad FA : x}{\times}$	$\dfrac{TA : z \circ x \circ y \circ v}{TA : z \circ y \circ x \circ v}$	$\dfrac{TA : z \circ x \circ x \circ y}{TA : z \circ x \circ y}$
		Expansion	Monotonicity
EF→ $\dfrac{FA \to B : x \quad TA : a}{FB : x \circ a}$	PB $\dfrac{\qquad}{FA : x \mid TA : x}$	$\dfrac{TA : z \circ x \circ y}{TA : z \circ x \circ x \circ y}$	$\dfrac{TA : z \circ x \circ v}{TA : z \circ x \circ y \circ v}$

The operational rules are the *universal rules* for implication, derived from the general definition of a PW-valuation. The fundamental rules express the basic properties of our semantic notion of truth and falsity. The structural rules correspond to the properties of the accessibility relation expressed in axioms B_1'–B_4' of Section 4.

Derivations in these systems are *trees* of labelled signed formulae constructed according to the rules which characterize each system. A branch is *closed* if it ends with the symbol ×, otherwise it is *open*. A tree is *closed* when all its branches are closed. An *LKE-tree for* **F**, where **F** is a set of LS-formulae, is a tree whose nodes are either LS-formulae in **F** or are obtained by an application of one of the LKE-rules. A formula A is *provable* from the set of assumptions Γ if and only if there is a closed tree for the set $\{TB_1 : b_1, \ldots, TB_n : b_n, FA : b_1 \circ \cdots \circ b_n\}$, where $B_i \in \Gamma$, for $i = 1, \ldots, n$ and

$b_i \neq b_j$ whenever $B_i \neq B_j$. Notice that the system of classical implication is obtained by simply ignoring the labels.

In our set-up the difference between the various implication systems is reduced to the difference between the corresponding structural rules. Such structural rules are not very handy from a practical point of view and may be troublesome in the formulation of a decision procedure. However, it is not difficult to show (see [7]) that the application of the structural rules can be pushed down to end of each branch.

THEOREM 5.2 *Every closed tree $\mathcal{T}$ for* **F** *can be transformed into a closed tree $\mathcal{T}'$ such that, in each branch, no application of an operational rule follows an application of a structural rule.*

So, for all practical purposes, we can dispense with the structural rules altogether, provided that we replace the closure rule with the following one:

$$\text{(Clos)} \qquad \frac{\begin{array}{c} TA : x \\ FA : y \end{array}}{\times} \ \textit{provided}\ x \sqsubseteq y$$

where $\sqsubseteq$ is the partial ordering of the class $\mathbf{P}_\alpha$ of partially ordered monoids under consideration. In this formalization the difference between the various implication logics is reduced to a difference in the side-condition associated with the 'closure' rule. This side-condition can be easily checked in each case. Without any side-condition, the resulting system is, of course, classical implication. In Tables 2 and 3 we show some examples of refutations.

Notice how the right derivation of Table 2 fails for the logic $\mathrm{LJ}_\emptyset$ (i.e. with no structural rules), since $b \circ a \circ c \not\sqsubseteq a \circ b \circ c$ for some monoid in $\mathbf{P}_\emptyset$. Notice also how in Table 3 the leftmost derivation fails in $\mathrm{LJ}_{\{1\}}$ (and, *a fortiori*, in $\mathrm{LJ}_\emptyset$), since $a \circ b \circ b \not\sqsubseteq a \circ b$ for some monoid in $\mathbf{P}_{\{1\}}$, the derivation in the center one fails in $\mathrm{LJ}_{\{1,2\}}$ and its subsystems, since $a \not\sqsubseteq a \circ a$ for some monoid in $\mathbf{P}_{\{1,2\}}$ and the rightmost one fails in $\mathrm{LJ}_{\{1,2,3\}}$ and its subsystems, since $a \not\sqsubseteq a \circ b$ for some monoid in $\mathbf{P}_{\{1,2,3\}}$.

The reader can easily verify that any attempt to derive Peirce's law fails in all the implication systems except, of course, the one in which the labels are ignored, corresponding to classical logic.

Are the rules we have been described so far *complete* for every logic of the $\mathrm{LJ}^{\{\rightarrow\}}$ family? The answer is not utterly positive because of a disturbing exception, namely the family of logics which satisfy the structural rule S_3 (Expansion), but not its stronger version S_4 (Monotonicity). These logics, which include the well-known system of 'mingle' implication, are provably closed under the following structural rule:

$$\frac{\Gamma \vdash A \quad \Delta \vdash A}{\Gamma, \Delta \vdash A}$$

TABLE 2.

$\vdash_{\mathrm{LJ}_\emptyset} (A\to B)\to((C\to A)\to(C\to B))$	$\vdash_{\mathrm{LJ}_{\{1\}}} (A\to B)\to((B\to C)\to(A\to C))$
$F(A\to B)\to((C\to A)\to(C\to B)) : 1$	$F(A\to B)\to((B\to C)\to(A\to C)) : 1$
$TA\to B : a$	$TA\to B : a$
$F(C\to A)\to(C\to B) : a$	$F(B\to C)\to(A\to C) : a$
$TC\to A : b$	$TB\to C : b$
$FC\to B : a\circ b$	$FA\to C : a\circ b$
$TC : c$	$TA : c$
$FB : a\circ b\circ c$	$FC : a\circ b\circ c$
$TA : b\circ c$	$TB : a\circ c$
$TB : a\circ b\circ c$	$TC : b\circ a\circ c$
×	×
	for commutative frames

TABLE 3.

$\vdash_{\mathrm{LJ}_{\{2\}}} A\to(A\to B))\to(A\to B$	$\vdash_{\mathrm{LJ}_{\{3\}}} A\to(A\to A)$	$\vdash_{\mathrm{LJ}_{\{4\}}} A\to(B\to A)$
$F(A\to(A\to B))\to(A\to B) : 1$	$FA\to(A\to A) : 1$	$FA\to(B\to A) : 1$
$TA\to(A\to B) : a$	$TA : a$	$TA : a$
$FA\to B : a$	$FA\to A : a$	$FB\to A : a$
$TA : b$	$FA : a\circ a$	$TB : b$
$FB : a\circ b$	×	$FA : a\circ b$
$TA\to B : a\circ b$	for expansive frames	×
$TB : a\circ b\circ b$		for monotonic frames
×		
for contractive frames		

This rule can be justified 'semantically' as follows. Suppose there are two points x_1 and x_2 at which a formula A is true. By definition of valuation there is the *minimum* point at which A is true. Let a be such a minimum point. If the frame is expansive we have that $a \sqsubseteq a \circ a$. Therefore, since $a \sqsubseteq x_i$ for $i = 1, 2$, A is verified also by the point $x_1 \circ x_2$. The argument can, of course, be generalized to any finite numbers of points verifying A.

This means that in any logic satisfying Expansion there cannot be a valuation v such that, for some finite set $\{x_1, \ldots, x_n\}$ of points $v(A, x_i) = T$, for all i, but $v(A, x_1 \circ \ldots \circ x_n) = F$. Hence, a branch containing all $TA : x_i$ and $FA : x_1 \circ \cdots \circ x_n$ should be considered closed.

This problem can be overcome by assuming that frames are meet-semilattices, i.e. for every two points x and y there exists their meet $x \sqcap y$, and that truth is preserved under such meets, i.e. if both x and y verify A, then their meet $x \sqcap y$ also verifies A. This allows for a *more general* closure rule of which the previous one is just a special case:

$$\begin{array}{c} TA : x_1 \\ \vdots \\ TA : x_n \\ FA : y \\ \hline \times \end{array} \tag{12}$$

provided that $x_1 \sqcap \cdots \sqcap x_n \sqsubseteq y$. (Observe that in every expansive frame $x_1 \sqcap \cdots \sqcap x_n \sqsubseteq x_1 \circ \cdots \circ x_n$.) Alternatively, we can introduce a function h that picks up the least point, if any, that verifies a given formula A, and modify the closure rule as follows: a branch is closed whenever it contains $FA : y$, where y is such that $h(A) \sqsubseteq y$. (Observe that, whenever, a labelled signed formula of the form $TA : x$ belongs to the branch, we know that $h(A)$ is defined and $h(A) \sqsubseteq x$.)

We call LKE$_{\rightarrow}$ (for 'Labelled KE') the refutation system characterized by the operational rules for $\rightarrow$ plus the general closure rule (12). In fact, this is not a single refutation system, but a *family* of refutation systems which differ from each other only for the *algebra of the labels* used in checking the side-condition on the closure rule. This takes the form of a set of axioms $\mathcal{A}$, characterizing the class $\mathbf{Q}_\alpha$ of partially-ordered monoids under consideration (i.e. it will comprise the usual axioms for the partial ordering $\sqsubseteq$ plus a set of axioms characterizing the additional constraints identified by the subscript α). We shall use the notation $\mathcal{A}_\alpha$ to indicate the algebra of the labels corresponding to the class $\mathbf{Q}_\alpha$.

Accordingly, given any specific algebra of the labels $\mathcal{A}_\alpha$, we shall say that a branch of an LKE$_{\rightarrow}$-tree is *closed for* $\mathcal{A}_\alpha$ if (i) it contains a suitable set of premisses for the closure rule and (ii) the set of axioms of $\mathcal{A}_\alpha$ implies that the side-condition associated with this putative application of the closure rule is satisfied. Obviously, an LKE$_{\rightarrow}$*-tree* will be *closed for* $\mathcal{A}_\alpha$, if all its branches are closed for $\mathcal{A}_\alpha$.

It is not difficult to show (see [7] for the details) that

THEOREM 5.3 *For every finite* Γ, $\Gamma \vdash^{\{\rightarrow\}}_{\mathrm{LJ}_\alpha} A$ *if and only if there is a closed LKE*$_{\rightarrow}$*-tree for* $\mathcal{A}_\alpha$.

The implication fragments of some of the logics $\mathrm{LJ}^{\{\rightarrow\}}_\alpha$ are well-known logical systems. The correspondence is summarized in Table 4.

5.2 Variables in the Labels

Each label occurring in a tree is built up from atomic labels and the relevant operations of the labelling algebra. New atomic labels are introduced by applications of the rule EF$\rightarrow$, and the propagation of the labels is uniquely determined by the tree rules. By contrast, the rule PB is sound for *every*

TABLE 4.

$LJ^{\{\to\}}_{\emptyset}$	Lambek's right implication
$LJ^{\{\to\}}_{\{1\}}$	Girard's linear implication
$LJ^{\{\to\}}_{\{1,2\}}$	Anderson's and Belnap's relevant implication
$LJ^{\{\to\}}_{\{1,2,3\}}$	Mingle implication
$LJ^{\{\to\}}_{\{1,2,4\}}$	Direct implication
$LJ^{\{\to\}}_{\{1,2,3,4\}}$	Intuitionistic implication

choice of the label x and we only know that, for every valid sequent $\Gamma \vdash A$, *there exists* a set of choices for the labels generated by the application of PB which leads to a closed tree. It is, therefore, convenient in practice to apply the rule PB with a *variable* label x and postpone the evaluation of this variable until enough information is available. For this purpose we need some new notions.

DEFINITION 5.4 We enrich our labelling language with a denumerable set of *variables* denoted by $\gamma_1, \gamma_2, \gamma_3$ etc. A *label-scheme* α is a label containing variables. A *potential closure set* is a set of LS-formulae of the form $\{TA : \alpha_1, \ldots, TA : \alpha_n, FA : \beta\}$, where $\alpha_1, \ldots, \alpha_n, \beta$ are label-schemes. A *potentially closed* branch is a branch containing a potential closure set. A *tree $\mathcal{T}$* is *potentially closed* if all its branches are potentially closed.

Notice that a potentially closed branch may contain more than one potential closure set. So, every potentially closed branch ϕ determines a finite set I_ϕ of inequations, one for each potential closure set occurring in it. (Recall that, with the exception of the logics characterized by frames which are both non-monotonic and expansive, a closure set is always a pair $\{TA : \alpha, FA : \beta\}$, so that the inequations I_ϕ have the simple form $\alpha \sqsubseteq \beta\}$.) Therefore, the 'closure problem' for a potentially closed tree takes the following form:

> *Is there a set* $\mathbf{S}$ *of inequations such that* $\mathbf{S} \cap I_\phi \neq \emptyset$ *for all branches* ϕ*, and all the inequations in* $\mathbf{S}$ *have a simultaneous solution in a given algebra of the labels* $\mathcal{A}_\alpha$*?*

EXAMPLE 5.5 As a simple example of the use of variables in the labels Figure 1 shows a tree for $\vdash ((A \to A) \to B) \to ((B \to C) \to C)$. This is a potentially closed tree which is turned into a closed tree for all commutative labelling algebras under the substitution $\gamma = 1$. ∎

5.3 Rules for $\otimes$

The appropriate tree-expansion rules for $\otimes$ can be read off the valuation condition for $\otimes$ in Definition 4.5, namely

$$v(A \otimes B, x) = T \text{ iff } \exists y, z,\ y \circ z \sqsubseteq x \text{ and } v(A, y) = T \text{ and } v(B, z) = T.$$

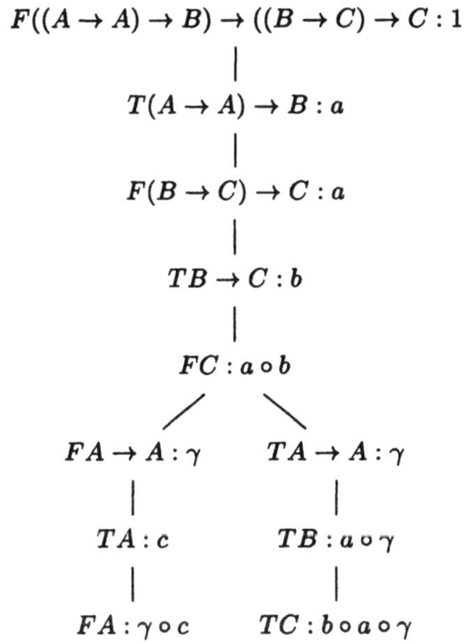

Figure 1.

Accordingly, the rules for $\otimes$ are as follows:

ET$\otimes$	EF$\otimes$	EF$\otimes$
$\dfrac{TA \otimes B : x}{TA : a \quad TB : x/a}$ with *a new*	$\dfrac{FA \otimes B : x \quad TA : y}{FB : x/y}$	$\dfrac{FA \otimes B : x \quad TB : y}{FA : x/y}$

where x/y denotes the *set* of points z such that $y \circ z \sqsubseteq x$. The consideration of a *set* of points as a label of a signed formula involves a reinterpretation of its intuitive meaning. If X is a set of points, an LS-formula $TA : X$ will mean that $v(A, x) = T$ for *some* $x \in X$, while the LS-formula $FA : X$ will mean that $v(A, x) = F$ for *all* $x \in X$. Indeed, we can re-interpret *all* the labels as sets of points. For this purpose it is sufficient to regard an atomic label a as shorthand for the singleton $\{a\}$. Moreover, if X and Y are sets of points, we can define $X \circ Y = \{x \circ y \mid x \in X \text{ and } y \in Y\}$. Under this interpretation the rules given above (as well as the implication rules) are easily seen to be sound. We can also formulate a 'liberalized' version of the ET$\otimes$ rule, as we did for the EF$\to$ rule, by assuming that the atomic label a denotes the *least* point at which A is true. Therefore, in every subsequent application of the ET$\otimes$ rule to an LS-formula of the form $TA \otimes C : x$, we can *reuse* the same atomic label a, instead of introducing a new one.

We call $\mathrm{LKE}_{\{\to,\otimes\}}$ the system which results by adding the three rules for $\otimes$ to the system $\mathrm{LKE}_{\to}$ considered in the previous section. An example of an $\mathrm{LKE}_{\{\to,\otimes\}}$-refutation is given in the next example.

EXAMPLE 5.6 The formula $(A \otimes B \to C) \to (A \to (B \to C))$ is valid in all frames. ■

$$F(A \otimes B \to C) \to (A \to (B \to C)) : 1$$
$$|$$
$$TA \otimes B \to C : a$$
$$|$$
$$FA \to (B \to C) : a$$
$$|$$
$$TA : b$$
$$|$$
$$FB \to C : a \circ b$$
$$|$$
$$TB : c$$
$$|$$
$$FC : a \circ b \circ c$$
$$\swarrow \qquad \searrow$$
$$FA \otimes B : x \qquad TA \otimes B : x$$
$$| \qquad\qquad\qquad |$$
$$FB : x/b \qquad TC : a \circ x$$

The right-hand branch of the above tree is closed under the substitution $x = b \circ c$. Since c is certainly an element of the set $(b \circ c)/b$, the left-hand branch is closed under the same substitution. □

Again it can be shown (see [7] for the details) that $\mathrm{LKE}_{\{\to,\otimes\}}$ is complete for all the logics of the $\mathrm{LJ}^{\{\to,\otimes\}}$ family.

5.4 Analytic Cut

All the elimination rules of the system $\mathrm{LKE}_{\{\to,\otimes\}}$ are *analytic* in the sense that they 'analyse' complex formulae by specifying the *consequences* of their truth and falsity in terms of the truth and falsity of their subformulae. On the other hand the branching rule PB (which is equivalent to the cut rule of the sequent calculus) can introduce *arbitrary formulae*.

In the classical **KE** system – with no labels – the use of the rule PB can be restricted to *analytic* applications, i.e. applications preserving the subformula property, without loss of completeness. Indeed, the choice of the 'cut formulae', namely the formulae introduced by an application of PB, can be even further restricted so that the resulting refutations follow a regular pattern or *canonical form* and can be found by means of a simple systematic procedure, similar to the usual tableau-completion procedure (see [8] for a detailed discussion of this point). Moreover, such canonical refutations are often essentially shorter than the corresponding tableau refutations and *never* significantly longer (see again [8] for the related technical results in terms of polynomial simulations).

This property extends also to the labelled version of **KE** that we are considering here: we can restrict the applications of PB to subformulae without loss of completeness, and define a refutation procedure which is a labelled generalization of the classical one. However, the use of *labelled* formulae introduces a further degree of freedom in the application of the PB rule: the choice of the label. So the problem of proof-search in our system depends crucially also on a new component: the choice of the labels associated with the PB-formulae. Investigating possible ways in which this extra-freedom can be restricted for each given algebra of the labels is the key to the decision problem for the corresponding substructural systems. Of course, fully mechanized proof-search can be achieved only when the complexity of the labels can be bound in one way or the other. This investigation goes beyond the limits of the present exposition, but the reader can find some clues in [7] and a more detailed discussion in the (forthcoming) Part II of the same article.

6 A Procedural Interpretation of the LKE-rules

In this section we describe an algorithm to find (labelled) natural deduction proofs for logics of the $LJ^{\{\rightarrow,\otimes\}}$ family, based on the $LKE_{\{\rightarrow,\otimes\}}$ system. It consists of inverting $LKE_{\{\rightarrow,\otimes\}}$-refutations, showing that the EF$\otimes$ and EF$\rightarrow$ rules can be read as backward reasoning in a natural deduction style. This algorithm could be useful in applications of substructural logics to all the areas where a human–oriented interface, based on natural deduction, is required.

6.1 Substructural Natural Deduction

Following the framework developed in [19, 3] we define in Table 5 six natural deduction (ND) inference rules which correspond to the tree-expansion rules for $\{\rightarrow,\otimes\}$[10], and to the closure and PB rules. The ND rules can be proved

[10] $\otimes$ has higher precedence than $\rightarrow$

to be complete and sound with respect to the LKE-rules (see Section 6.3). We will restrict our attention to regular PW-valuations, namely for every

$\boxed{\begin{array}{l} A : a \\ \vdots \\ B : x \circ a \end{array}}$ $A \to B : x \quad \to\mathcal{I}$ (i)	$\to\mathcal{E}\ \dfrac{\begin{array}{l} A \to B : x \\ A : y \end{array}}{B : x \circ y}$
$\boxed{\begin{array}{l} \vdots \\ A : \gamma_1 \end{array}}\ \boxed{\begin{array}{l} \vdots \\ B : \gamma_2 \end{array}}$ $A \otimes B : x \quad \otimes\mathcal{I}$ (ii)	$\otimes\mathcal{E}\ \dfrac{A \otimes B : x}{\begin{array}{l} A : a \\ B : b \end{array}}$ (i) (iii)
$\vdots$ $A : x$ $\vdots$ $A : y \quad \surd$ if $x \sqsubseteq y$	$\vdots$ $\boxed{\text{proof of lemma } A : \gamma}$ $A : \gamma$ Lemma $\vdots$ goal

(i) a is a *solo* parameter
(ii) $\gamma_1 \circ \gamma_2 \sqsubseteq x$
(iii) b is a *solo* parameter, and $a \circ b \sqsubseteq x$

TABLE 5. Natural Deduction Rules

formula A, if there is a point x that verifies A, then there is also the least point, called a, that verifies A. We will refer to this least point as the *A-characteristic* label.

The language of the ND system consists of two parts; a language for the labels, composed of the constant symbol 1, atomic parameters $\{a, b, c, \ldots\}$, variables $\{\gamma, \gamma_1, \gamma_2, \ldots\}$, the operator $\circ$ and the binary relation $\sqsubseteq$, and a standard set of wffs using the logical connectives $\{\to, \otimes\}$. A *solo* parameter is a label such that its atomic occurrences within a structural derivation label only the formula it is first introduced with, and which may occur (as atomic labels) repeatedly in the derivation whenever such a formula has to be re-introduced. Each characteristic label introduced in a closed LKE-tree is mapped into a solo parameter in the ND system. For simplicity the same names will be used. The partial ordering relation $\sqsubseteq$ of the underlying labelling algebra is mapped to the relation (of the same name) $\sqsubseteq$ in the ND system and similarly for the $\circ$ operator.

In the rule $\rightarrow\mathcal{I}$, the box can be seen as introducing a new context composed of the assumption $A : a$ together with the other sentences already assumed or derived. To deduce $A \rightarrow B : x$, it is needed to show that $B : x \circ a$ can be derived using the extra assumption $A : a$. The rule $\otimes\mathcal{E}$ corresponds to the ET$\otimes$ rule of LKE$_{\{\rightarrow,\otimes\}}$ system. However, to maintain a uniformity in the constraints on the labels the expression x/a used in Section 5.3 is replaced by the solo parameter b together with the imposed constraint $a \circ b \sqsubseteq x$. In the cases of $\rightarrow\mathcal{I}$ and $\otimes\mathcal{E}$, the parameter a is also a solo parameter. The rule $\rightarrow\mathcal{E}$ is the usual Modus Ponens adapted for labelled formulae. The inference rule $\otimes\mathcal{I}$ replaces the two EF$\otimes$ rules shown in Section 5.3. It is a symmetric rule which can be read as: 'to prove a formula $A \otimes B : x$, show $A : \gamma_1$ and $B : \gamma_2$', where the free variables γ_1 and γ_2 must satisfy the constraint $\gamma_1 \circ \gamma_2 \sqsubseteq x$. If $A : \gamma_1$ is already derived for some γ_1, then the rule reduces to the left EF$\otimes$ rule shown in Section 5.3. Analogously, if $B : \gamma_2$ is already derived for some γ_2, then the rule reduces to the right EF$\otimes$ rule. The variables γ_1 and γ_2 have to be instantiated by particular ground labels[11] which satisfy the required constraint $\gamma_1 \circ \gamma_2 \sqsubseteq x$ (see note (ii)). Depending on the logic in use, there may be different instantiation values. The $\surd$-rule recognizes when a particular goal has been achieved and corresponds to the closure rule, with the required constraint $x \sqsubseteq y$. As free variables may be introduced in the application of the $\otimes\mathcal{I}$ rule and of the Lemma rule, the labels x and y may be label-schemes. Therefore, suitable ground values have to be found so to satisfy the inequality $x \sqsubseteq y$.

Since LKE-trees make use of the semantic equivalent of the cut rule, namely the PB rule, we also need some way of including intermediate lemmas. This is achieved by the Lemma rule. Once a proof of $A : \gamma$ has been found, possibly with imposed or required constraints, then $A : \gamma$ can be used as an assumption in proving the main goal.

A proof of a theorem is constructed by working backwards from the formula (seen as a goal) using introduction rules and, as further labelled assumptions are introduced by $\rightarrow\mathcal{I}$, by working forwards using elimination rules. As in the classical case, declarative units assumed or derived in a particular context can be imported into a larger context (or inner subderivation). In general, a proof will impose some constraints and will require others to be satisfied. We call the imposed constraints *ICs* and the constraints to be satisfied *RCs*. *ICs* arise from applications of the $\otimes\mathcal{E}$ rule, and *RCs* occur with applications of $\otimes\mathcal{I}$ and $\surd$ rules. Unlike assumptions, *ICs* are global and are used to satisfy *RCs*. This is given in (13),

$$\bigwedge_{i=0}^{n} IC_i \rightarrow \bigwedge_{j=0}^{m} RC_j \qquad (13)$$

[11] We call a label-scheme *ground* when all its variables have been instantiated either to an atomic label or to a concatenation of atomic labels.

where the $\{IC_0, \ldots, IC_n\}$ are imposed constraints and the $\{RC_0, \ldots, RC_m\}$ are required constraints generated within a ND derivation. In order to satisfy (13) – find values for the variables (if any occurs in it) that make it true – the properties of the labelling algebra are used.

An example of a proof of a theorem in linear logic, which satisfies the commutative property of $\circ$, is given in Figure 2. In this example, the

1	$A : a$	
2	$A \otimes (A \otimes A \rightarrow B) : b$	
3	$A : a$	can be omitted
4	$A \otimes A \rightarrow B : c$	$\otimes\mathcal{E}$ $(a \circ c \sqsubseteq b)$
5	$\surd$ $(\gamma_1 = a)$	$\surd$ $(\gamma_2 = a)$
6	$A : \gamma_1$	$A : \gamma_2$
7	$A \otimes A : \gamma_3$	$\otimes\mathcal{I}$ if $\gamma_1 \circ \gamma_2 \sqsubseteq \gamma_3$
8	$A \otimes A : \gamma_3$	Lemma
9	$B : c \circ \gamma_3$	$\rightarrow\mathcal{E}$
10	$B : a \circ b$	$\surd$ if $c \circ \gamma_3 \sqsubseteq a \circ b$
11	$(A \otimes (A \otimes A \rightarrow B)) \rightarrow B : a$	$\rightarrow\mathcal{I}$
12	$A \rightarrow ((A \otimes (A \otimes A \rightarrow B)) \rightarrow B) : 1$	$\rightarrow\mathcal{I}$

Figure 2. An Example of Natural Deduction Proof

general constraint corresponding to (13) is given by

$$(a \circ c \sqsubseteq b) \rightarrow (a \circ a \sqsubseteq \gamma_3) \wedge (c \circ \gamma_3 \sqsubseteq a \circ b).$$

This is satisfied by choosing $\gamma_3 = a \circ a$ and using commutativity and the order-preserving property of $\circ$ (i.e. $a \circ c \sqsubseteq b \Rightarrow c \circ a \sqsubseteq b \Rightarrow c \circ a \circ a \sqsubseteq b \circ a \Rightarrow c \circ a \circ a \sqsubseteq a \circ b$).

6.2 LKE-rules as Natural Deduction

In this section we describe a procedural interpretation of the LKE-rules and show how an LKE-tree can be reformulated as a natural deduction proof. Let us interpret the labelled signed formulae as follows:

- $TA : x$ means that the labelled formula $A : x$ belongs to the data or is a provisional assumption.
  $FA : x$ means that the labelled formula $A : x$ is to be proven.

Under this interpretation, the rule EF$\rightarrow$ can be read as a *goal-reduction rule*:

$FA \rightarrow B : x$	prove $[A \rightarrow B : x]$
$TA : a$	assume $[A : a]$ and
$FB : x \circ a$	prove $[B : x \circ a]$

The rule (ET $\rightarrow$), on the other hand, is a *data-expansion rule* which processes the data via applications of the labelled version of Modus Ponens. It is easy to see that these two rules correspond exactly to the natural deduction rules $\rightarrow\mathcal{I}$ and $\rightarrow\mathcal{E}$ given in Section 6.1.

Let us see a simple example of how this interpretation works: consider the contraction axiom $(A \rightarrow (A \rightarrow B)) \rightarrow (A \rightarrow B)$. We show below its proof in LKE side by side with its procedural interpretation (recall that 1 is the empty data structure). Whether the LKE-tree closes depends on the

1.	$F(A \rightarrow (A \rightarrow B)) \rightarrow (A \rightarrow B) : 1$	prove $[(A \rightarrow (A \rightarrow B)) \rightarrow (A \rightarrow B) : 1]$
2.	$T(A \rightarrow (A \rightarrow B)) : a$	assume $[(A \rightarrow (A \rightarrow B)) : a]$ and
3.	$FA \rightarrow B : a$	prove $[A \rightarrow B : a]$
4.	$TA : b$	assume $[A : b]$ and
5.	$FB : a \circ b$	prove $[B : a \circ b]$
6.	$TA \rightarrow B : a \circ b$	$A \rightarrow B : a \circ b$ follows from 2 and 4
7.	$TB : a \circ b \circ b$	$B : a \circ b \circ b$ follows from 6 and 4
8.	$\times$ (if $a \circ b \circ b \sqsubseteq a \circ b$)	$\surd$ (if $a \circ b \circ b \sqsubseteq a \circ b$)

constraint $a \circ b \circ b \sqsubseteq a \circ b$ being satisfied by the labelling algebra. Similarly, in the procedural version (right-hand side), the deduction of $B : a\circ b\circ b$ from $B : a \circ b$ succeeds under the same condition. Closure is therefore the same as the $\surd$-rule of Section 6.1. We can therefore invert the above labelled analytic proof and turn it into a direct proof in which each application of a goal-reduction rule EF$\rightarrow$ becomes an $\rightarrow\mathcal{I}$ rule. An ND proof of the above theorem is given below.

1	$A \rightarrow (A \rightarrow B) : a$	
2	$A : b$	
3	$A \rightarrow B : a \circ b$	$\rightarrow\mathcal{E}$
4	$B : a \circ b \circ b$	$\rightarrow\mathcal{E}$
5	$B : a \circ b$	$\surd$ if $a \circ b \circ b \sqsubseteq a \circ b$
6	$A \rightarrow B : a$	$\rightarrow\mathcal{I}$
7	$(A \rightarrow (A \rightarrow B)) \rightarrow (A \rightarrow B) : 1$	$\rightarrow\mathcal{I}$

This is a correct proof in relevance logic but not in linear logic because the assumption A is used twice: first with $A \rightarrow (A \rightarrow B)$ to infer $A \rightarrow B$ and then with $A \to B$ itself to infer B. Indeed, to satisfy the RC $a \circ b \circ b \sqsubseteq a \circ b$ the contraction property is required.

This procedural interpretation can be extended to the PB, ET⊗ and EF⊗ rules. As for the PB rule, it is not difficult to see that its role is that of generating appropriate *lemmas* to be used in a proof. Consider an arbitrary application of the PB rule in a LKE-tree as shown in Figure 3. If

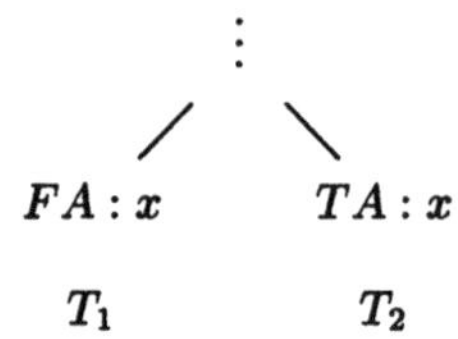

Figure 3.

the sub-tree T_1 is closed, then the 'goal' $FA : x$ succeeds. Hence the labelled formula $TA : x$ is provable and can be used as a lemma in the sub-tree T_2. So, this rule can be interpreted as follows: first generate as a sub-proof the proof of the lemma, which corresponds to the left sub-tree, and then add the statement of the lemma as an assumption below the subproof (in the same way as $TA : x$ is added by the PB rule in the right sub-tree). The proof of a lemma may lead to additional global RCs and ICs. These are propagated throughout the remaining ND proof.

We now consider an example in linear logic, in which use is made of the Lemma rule. The LKE-tree is given in Figure 1 of Example 5.5. The left sub-tree yields the following trivial natural deduction sub-proof with the single RC $c \sqsubseteq \gamma \circ c$. This is satisfied for $\gamma = 1$ as shown also in Example 5.5.

1	$A : c$	
2	$A : \gamma \circ c$	$\surd$ if $(c \sqsubseteq \gamma \circ c)$

3 $A \to A : \gamma$

The instantiation of the free variable γ is propagated in the proof by replacing any of its occurrences with 1. The complete corresponding natural deduction proof is shown in Figure 4. Here, the remaining RC $b \circ a \sqsubseteq a \circ b$ is satisfied, as the operator $\circ$ is commutative in the labelling algebra $\mathcal{A}_{\{1\}}$ corresponding to the class $\mathbf{Q}_{\{1\}}$ of partially-ordered monoids of linear logic.

As for the EF⊗ rule, the correspondence with the ND $\otimes\mathcal{I}$ rule is not quite immediate. In the $\text{LKE}_{\{\to,\otimes\}}$ system there are two sub-cases, depending on whether or not $TA : \gamma$ is already in the branch for some γ. In the first case (i.e. $TA : \gamma$ is already in the branch for some γ) the ND left sub-proof of the $\otimes\mathcal{I}$ rule succeeds immediately by the $\surd$ rule and the variable

1	$(A \to A) \to B : a$	
2	$B \to C : b$	
3	proof of $A \to A : \gamma$	
4	$A \to A : 1$	Lemma$\gamma = 1$
5	$B : a$	$\to\mathcal{E}$
6	$C : b \circ a$	$\to\mathcal{E}$
7	$C : a \circ b$	$\surd$ if $(b \circ a \sqsubseteq a \circ b)$
8	$(B \to C) \to C : a$	$\to\mathcal{I}$
9	$((A \to A) \to B) \to ((B \to C) \to C) : 1$	$\to\mathcal{I}$

Figure 4.

γ_1 is bound to γ. This leaves only the labelled formula $B : \gamma_2$ to be shown, with the RC $\gamma_1 \circ \gamma_2$ $(= \gamma \circ \gamma_2) \sqsubseteq x$. In the second case (i.e. $TA : \gamma$ is not in the branch) the EF$\otimes$ rule is preceded by the application of the PB rule on the LS-formula $TA : \gamma$. The left branch of this rule will provide a refutation proof of the formula $TA : \gamma$, whereas in the right branch the added assumption $TA : \gamma$ will allow the application of the EF$\otimes$. This case and its corresponding ND interpretation is shown in Figure 5. In the ND

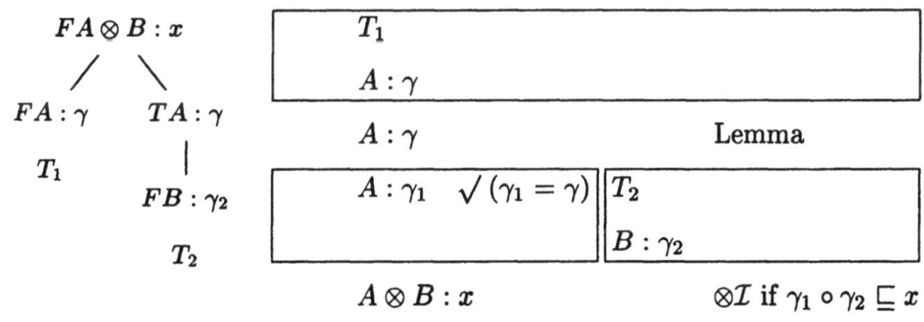

Figure 5. Translation of $\otimes$ rule

proof, the application of the $\otimes\mathcal{I}$ rule is equally preceded by the application of the Lemma rule (interpretation of the PB rule) as shown in Figure 5. Here the generation of the lemma $A : \gamma$ allows the left-hand subgoal of the $\otimes\mathcal{I}$ rule to be immediately satisfied by the application of the $\surd$ rule. The right-hand box of the $\otimes\mathcal{I}$ rule will include the right sub-tree of the PB rule application, and its context will be composed also of the lemma $A : \gamma$.

A more natural proof of $A \otimes B : x$ using the natural deduction rules includes the proof of $A : \gamma_1$ directly within the left-hand box of the rule $\otimes\mathcal{I}$. If $A : \gamma_1$ already exists, then the result is similar to that obtained using trees. If not, instead of using the Lemma rule, notice that the proof of $A : \gamma_1$ is not in any different context than the subsequent proof of $B : \gamma_2$, so its proof can be repeated if necessary and the two schemes are the same again.

6.3 The Translation Procedure

6.3.1. *Making Non-redundant LKE-proofs*
In order to translate a LKE-tree into a natural deduction proof the LKE-tree refutation should be non-redundant. Many steps may be made before it is known whether they are needed or not, so the resulting LKE-tree should be pruned, removing the steps which turn out to be unnecessary.

DEFINITION 6.1 A LKE-tree is *non-redundant* if each node[12] is used in some step – that is, either contributes to closure or one of its descendants does. The set of *used nodes* is the least set of nodes satisfying the following conditions:

- A node is *used* if at least one of its formulae is a premisse of a closure step.
- If n is a used node and m is a premisse of the rule that generated n then m is used.

Any LS-formula which belongs to a used node is also called a *used* LS-formula.

To generate a non-redundant LKE-tree we use the following procedure:

PNR – *Procedure to generate a non-redundant LKE-tree:*

1. Using the definition of a used node we can form the set of *used* nodes in a LKE-tree.
2. The remaining nodes are not used and may be deleted.
3. If one of the nodes of an application of the PB-rule is deleted then the whole sub-tree below it closes using the remaining nodes in that branch. So the PB application turns out to be unnecessary and the sub-tree beneath the other node of the PB application may be deleted. The procedure PNR is called recursively on the remaining tree.

[12]A node is either an assumption or the result of an LKE-rule application. So a node may contain more than one formula – e.g., the conclusion of the EF$\rightarrow$ rule.

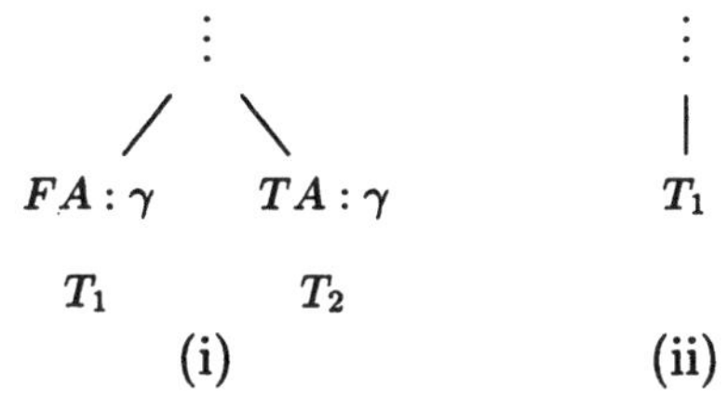

Figure 6.

For example, in Figure 6(i) suppose FA : γ is unused. Then the tree can be reduced to Figure 6(ii) by removing the right branch altogether, since T_1 can still be correctly closed.

Since PNR only deletes nodes it must eventually terminate giving a non-redundant tree. Note that by definition of used node the non-redundant tree will still close.

6.3.2. *Deriving a Natural Deduction Proof*

In this section we define the notion of a non-redundant proper LKE-tree and we describe a procedure called PND which generates from a given non-redundant proper LKE-tree a natural deduction proof. We also show that the procedure PND is correct, i.e. given a LKE-refutation there exists a corresponding ND proof generated by the PND interpretation. This will guarantee, together with the completeness of the LKE system, that the ND rules are also complete.

DEFINITION 6.2 A *proper closed* LKE-tree is a closed tree containing only ground labels and in which no atomic label introduced for the first time in a branch by an application of $\rightarrow\mathcal{I}$ or $\otimes\mathcal{E}$ occurs in the label of any PB-formula in the tree above that step.

LEMMA 6.3 *If a closed LKE-tree exists then a proper closed tree exists.*

Proof: This follows from the completeness of the 'unliberalized' rules[13] for EF$\rightarrow$ and ET$\otimes$, as shown in [7]. □

PND-*procedure to form a natural deduction proof from a proper non-redundant LKE-tree*

The proof procedure is applied step by step to a given LKE-refutation. Depending on the type of LKE rules used, a different action is taken:

Closure Use the $\surd$-rule. The closure constraint becomes a RC.

[13]As we have seen in Section 5.1 and Section 5.3 there are two versions of the rules EF$\rightarrow$ and EF$\otimes$. The 'unliberalized' version requires the atomic label a introduced by the rule application to be a new label (i.e. not occurring anywhere else in the tree).

PB Complete the translation of the left-hand (F branch) within a box and write the conclusion below the box (corresponds to the T-formula of the right-hand branch). Then complete the translation of the right-hand branch. (Propagate any substitution that is found for the free variables.)

ET→ The two premisses will already be in the proof and available to the current box, so apply $\rightarrow\mathcal{E}$.

EF→ Apply $\rightarrow\mathcal{I}$ and put the T-formula at the top and the F-formula at the bottom of the new box.

EF⊗ Suppose the major premisse of the step is $FA \otimes B : x$. Introduce two boxes. The left one will contain the proof of $TA : \gamma_1$. This can be obtained immediately by the $\surd$-rule using the T-premisse of the step which is of the form $TA : y$, and instantiating γ_1 to y. The right box will contain the proof of the conclusion of the LKE-rule, (i.e. $FB : x/y$), but with the label γ_2 and the RC $\gamma_1 \circ \gamma_2 \sqsubseteq x$.

ET⊗ Suppose the major premisse of the step is $TA \otimes B : x$. Add the two conclusions $TA : a$ and $TB : b$ and the IC $a \circ b \sqsubseteq x$. This IC implies $b \sqsubseteq x/a$, and $TB : b$ implies $TB : x/a$ as used in the LKE-tree.

The initial F-formula of the LKE-tree is at the bottom of the proof.

If PND is applied to the tree in Figure 7 then the natural deduction proof in Figure 2 is obtained. The left branch of the tree closes under the condition $a \sqsubseteq \gamma/a$, which is satisfied for $\gamma = a \circ a$. The right branch closes with the condition $(b/a) \circ a \circ a \sqsubseteq a \circ b$, which also holds by definition of $/$ and commutativity of $\circ$. The set of required constraints generated in the corresponding ND proof is $\{a \sqsubseteq \gamma_1, a \sqsubseteq \gamma_2, \gamma_1 \circ \gamma_2 \sqsubseteq \gamma_3, c \circ \gamma_3 \sqsubseteq a \circ b\}$. The only imposed constraint is $a \circ c \sqsubseteq b$. Instantiating γ_1 and γ_2 by a solves the first two required constraints and yields the remaing two constraints to become $a \circ a \sqsubseteq \gamma_3$ and $c \circ \gamma_3 \sqsubseteq a \circ b$. The first one is the satisfied by the same instantiation for γ_3 used in the LKE-tree, namely $\gamma_3 = a \circ a$. The second constraint $c \circ a \circ a \sqsubseteq a \circ b$ is satisfied by using the imposed constraint and the commutativity property of $\circ$ (i.e. $c \circ a \circ a \sqsubseteq a \circ c \circ a \sqsubseteq b \circ a \sqsubseteq a \circ b$).

A reverse translation method has been developed which shows that for a given natural deduction proof there exists a closed LKE-tree, so proving the soundness of the natural deduction proofs. This is described in [4, 6] where algorithms for solving label constraints are also given.

6.3.3. *Correctness of* PND

We show here that the PND procedure is correct – i.e given a LKE-tree there exists a corresponding ND proof generated by PND. To do so we will restrict ourselves, without loss of generality, to 'atomic closures' only and show first, as an intermediate result, that in such LKE-trees all the F-conclusions of EF→ rule applications are used. This will allow us to prove that there exists a resulting natural deduction proof.

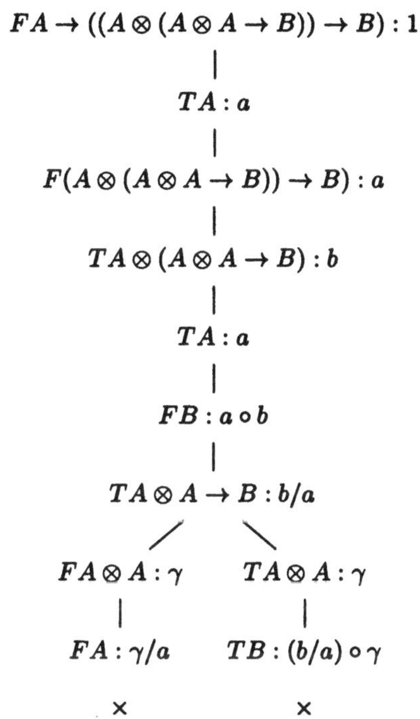

Figure 7. LKE-tree that gives rise to proof in Figure 2

DEFINITION 6.4 Let X be a formula. We define the *complexity* of X, written $C(X)$, as follows. $C(X) = 0$ if X is an atomic formula. $C(X) = 1 + \max\{C(Y),\ C(Z)\}$ if $X = Y op Z$, where Y and Z are formulae and *op* is one of the two binary operators $\rightarrow$, $\otimes$.

Let us say that an LKE-tree is *atomically closed* if the closure rule is restricted to atomic formulae.

LEMMA 6.5 *If there is a closed LKE-tree $\mathcal{T}$ for a set* $\mathbf{F}$ *of LS-formulae, then there is an atomically closed LKE-tree $\mathcal{T}'$ for* $\mathbf{F}$.

Proof: The proof is by induction on the complexity of the formulae used in the closure rule in a branch of a given LKE-tree.
Base Case: Trivial.
Inductive Step: Suppose by inductive hypothesis that a given LKE-tree in which all branch closures occur between formulae with complexity $\leq k$ can be reduced to a tree in which the branch closures occur between atomic formulae only. Consider a tree in which the maximum complexity of formulae involved in branch closures is $k+1$. There are two cases: (i) the closure formulae are of the form $TX \rightarrow Y : x$ and $FX \rightarrow Y : y$ with $x \sqsubseteq y$, or (ii) the closure formulae are of the form $TX \otimes Y : x$ and $FX \otimes Y : y$, with

$x \sqsubseteq y$. In both cases the closure can be reduced to be between formulae of complexity $\leq k$, namely the formula Y, as shown below. Then by the inductive hypothesis the resulting tree can be reduced to one in which branch closure occurs between atomic formulae.

(i)	$TX \to Y : x$	$TX \otimes Y : x$	(ii)
	$FX \to Y : y$	$FX \otimes Y : y$	
	$TX : a$	$TX : a$	
	$FY : y \circ a$	$TY : x/a$	
	$TY : x \circ a$	$FY : y/a$	

(Note that $TX : a$ may already have occurred.) The LKE-tree (i) closes because by hypothesis $x \sqsubseteq y$ and then $x \circ a \sqsubseteq y \circ a$. Analogously, in the LKE-tree (ii) the hypothesis $x \sqsubseteq y$ implies that $a \circ (x/a) \sqsubseteq y$, which is equivalent to $x/a \sqsubseteq y/a$. □

DEFINITION 6.6 The T-formula resulting from the application of an EF$\to$ rule to a LS-formula X is called the *T-conclusion* of X and the resulting F-formula is called the *F-conclusion* of X.

THEOREM 6.7 *Let $\mathcal{T}$ be a non-redundant proper LKE-tree. Every F-conclusion resulting from the application of the EF$\to$ rule to a used LS-formula is also used in $\mathcal{T}$.*

Proof outline: The proof is by contradiction. Let $\mathcal{B}$ be a branch of $\mathcal{T}$ in which the F-conclusion of a used LS-formula X is not used. Since X is used in $\mathcal{B}$ the T-conclusion must be used. Let $TA : a$ be the T-conclusion of X. Note that the formula $TA : a$ will not have previously been introduced in $\mathcal{B}$ by a EF$\to$ or ET$\otimes$ application, else again X would not have been used. Therefore, given also the fact that $\mathcal{T}$ is proper, the label a occurs in $\mathcal{B}$ only in formulae below the introduction of a. Applications of rules using $TA : a$ are either ET$\to$ or EF$\otimes$ steps. The first yields a T-formula with label of the form $y \circ a$, and the second an F-formula with label of the form y/a. These labels may subsequently appear in other terms as follows (x, y and z are labels):

1. $y \circ a$ (after ET$\to$)
2. $(y \circ a)/b$ (after ET$\to$, ET$\otimes$)
3. $x/(y \circ a)$ (after ET$\to$, EF$\otimes$)
4. $(x/(y \circ a))/(z \circ a)(= x/(y \circ a \circ z \circ a))$ (after ET$\to$, EF$\otimes$, ET$\to$, EF$\otimes$)
5. $x/(y \circ (z \circ a)/b)$ (after ET$\to$, ET$\otimes$, ET$\to$, EF$\otimes$).

Terms of the type shown in 1 and 2 label T-formulae, whereas terms of the form shown in 3-5 label F-formulae. In 3-5 a does not occur in x. For T-formulae a occurs in the numerator of the / operator and for F-formulae a occurs in the denominator of the / operator. Subsequent closure in a branch using such steps would involve labels in the following combinations:

- 1 with either 3,4, 5 or a term labelling an F-formula not including a
- 2 with either 3,4, 5 or a term labelling an F-formula not including a
- a term labelling a T-formula with no occurrence of a with either 3,4, 5 or another term labelling an F-formula with no a.

In the case of 1 with 3, for example, it is required to satisfy $w_1 \sqsubseteq w_2$, where $w_1 = U_1 a$ and $w_2 = x/(U_2 a)$ (where U_1 and U_2 are concatenations of labels). But since a does not occur in x, this cannot hold. We can reason analogously and reach a contradiction in all cases. Hence the F-conclusion of X is used in the branch $\mathcal{B}$. □

DEFINITION 6.8 Let $\mathcal{T}$ be a non-redundant LKE-tree. We define a *closure formula* in a branch $\mathcal{B}$ of $\mathcal{T}$ to be the atomic formula $FX : x$ used within the closure of $\mathcal{B}$. F_1 *directly supports* F_2 ($F_1 \ll F_2$) if it is the F-formula resulting from applying the EF$\rightarrow$ rule or the EF$\otimes$ rule to F_2. The *root* R of F_1 is the terminal formula of the (finite) chain $F_1 \ll \ldots \ll R$.

LEMMA 6.9 *Let $\mathcal{T}$ be a non-redundant proper LKE-tree. The root of the closure formula in a branch $\mathcal{B}$ is either the F-formula of the PB-step at greatest depth in $\mathcal{B}$ or, if there is no such formula, the initial F-formula.*

Proof outline: The proof is by induction on the number of PB-applications. **Base Case:** Suppose there are no PB-steps, then by Theorem 6.7 every F -formula in the single branch is used. Hence, in a closed branch $\mathcal{B}$, the chain beginning with the single closure formula must terminate at the initial F -formula.
Inductive Step: Suppose the lemma holds for $\leq k$ PB-steps. Let $\mathcal{T}$ be a non-redundant proper LKE-tree with $k+1$ PB-applications in a closed branch $\mathcal{B}_1$. Consider the PB-step at greatest depth in $\mathcal{B}_1$ (i.e. the PB-step in which the closure in its two branches uses no other PB-steps). $\mathcal{T}$ has the form shown in Figure 8(i), in which $\mathcal{B}_1$ is the rest of the branch above the considered application of PB.

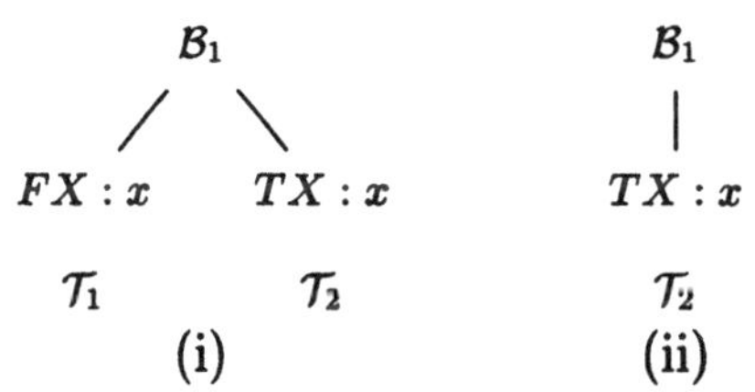

Figure 8.

If $FX : x$ is not the root of the closure formula in $\mathcal{T}_1$, then we can use Theorem 6.7 to show $FX : x$ is not used, which contradicts the non-redundancy of $\mathcal{T}$. The restricted LKE-tree of Figure 8(ii) uses k applications of PB and so, by the inductive hypothesis, the lemma holds. □

THEOREM 6.10 *Let $\mathcal{T}$ be a non-redundant proper LKE-tree. There exists an equivalent ND proof obtained by applying the* PND *procedure to $\mathcal{T}$.*

The above theorem can be proved by showing that the PND procedure allows each ND rule to be successfully applied. By Lemma 6.9 the closure below the root F-formula of an application of a PB rule in $\mathcal{T}$ uses only F-formulae derived from the root. Moreover, these used F-formulae form a chain starting from the closure formula. This guarantees that the PND procedure will generate a successful sub-proof within the box of the lemma rule, where the innermost goal corresponds to the closure formula and the last conclusion corresponds to the root formula (the chain of F-formulae appears in the ND sub-proof in the reverse order). Moreover, T-formulae will appear in the ND proof in the same order as in the tree, so the T-formulae necessary for the application of $\rightarrow\mathcal{E}$, $\otimes\mathcal{I}$ and $\surd$ rules will be available in the ND proof. Finally, our earlier discussion shows that the constraint implication (13) consisting of ICs introduced by $\otimes\mathcal{E}$ and RCs introduced by $\otimes\mathcal{I}$ and $\surd$ will be satisfied.

Krysia Broda and Alessandra Russo
Imperial College, London.

Marcello D'Agostino
Università di Ferrara, Italy.

References

1. Michele Abrusci. Phase semantics and sequent calculus for pure non-commutative classical linear propositional logic. *Journal of Symbolic Logic* 56:1403–1451, 1991.
2. Arnon Avron. The semantics and proof theory of linear logic. *Theoretical Computer Science* 57:161–184, 1988.
3. Krysia Broda, Susan Eisenbach, Hessam Khoshnevisan and Steve Vickers *Reasoned Programming*. Prentice Hall, 1994.
4. Krysia Broda, Marcelo Finger and Alessandra Russo. LDS-natural deduction for substructural logics. *Journal of the IGPL* 4:3:486–489, 1996. Extended Abstract. Full version in [6].
5. Krysia Broda, Marcello D'Agostino and Marco Mondadori A solution to a problem of Popper. To appear in *The Epistemology of Karl Popper*, Kluwer Academic Publishers, 1997.
6. Krysia Broda, Marcelo Finger and Alessandra Russo. LDS-natural deduction for substructural logics. *Journal of the IGPL*. Submitted, 1997.
7. Marcello D'Agostino and Dov M Gabbay. A generalization of analytic deduction via labelled deductive systems. Part I: Basic substructural logics. *Journal of Automated Reasoning* 13:243–281, 1994.

8. Marcello D'Agostino and Marco Mondadori. The taming of the cut. *Journal of Logic and Computation* 4:285–319, 1994.
9. Marcello D'Agostino. Are tableaux an improvement on truth-tables? *Journal of Logic, Language and Information* 1:235–252, 1992.
10. Kosta Dŏsen. Sequent systems and groupoid models I. *Studia Logica* 47:353–385, 1988.
11. Kosta Dŏsen. Sequent systems and groupoid models II. *Studia Logica* 48:41–65, 1989.
12. Kosta Dŏsen. A historical introduction to substructural logics. In Schroeder-Heister, Peter and Dŏsen, Kosta, editors, *Substructural Logics*. Oxford University Press. 1–31, 1993.
13. Michael J. Dunn. Relevance logic and entailment. In Gabbay, Dov M. and Guenthner, Franz, editors, *Handbook of Philosophical Logic*, volume III. Kluwer Academic Publishers. chapter 3, 117–224, 1986.
14. Dov M. Gabbay and Hans Jürgen Ohlbach. An algebraic fine structure for logical systems. Technical report, Department of Computing, Imperial College of Science, Technology and Medicine, 180 Queen's Gate, London, 1993.
15. Dov M. Gabbay and Hans Jürgen Ohlbach. From a Hilbert calclus to possible-world semantics. In Broda, Krysia, editor, *Proceedings of ALPUK Logic Programming Conference 1992*. Springer. Lecture Notes in Computer Science, 218–252, 1993.
16. Dov M. Gabbay. How to construct a logic for your application. In H. J. Ohlbach editor, *GWAI-92: Advances in Artificial Intelligence (LNAI 671)*. Springer. 1–30, 1992.
17. Dov M. Gabbay. General theory of structured consequence relations. In Schroeder-Heister, Peter and Dŏsen, Kosta, editors, *Substructural Logics*. Oxford University Press. 109–151, 1993.
18. Dov M. Gabbay. Classical versus non-classical logics. In Gabbay, Dov; Hogger, Chris; and Robinson, J. A., editors, *Handbook of Logic in AI and Logic Programming, Volume 2*. Oxford University Press, 1994.
19. Dov M. Gabbay. *Labelled Deductive Systems, Volume 1*. Oxford University Press, 1996.
20. Dov M. Gabbay. Fibred semantics and the weaving of logics, 1. To appear in *Journal of Symbolic Logic*, 1997.
21. Gerhard Gentzen. Unstersuchungen über das logische Schliessen. *Math. Zeitschrift* 39:176–210. English translation in [30], 1935.
22. Jean-Yves Girard. Linear logic. *Theoretical Computer Science* 50:1–102, 1987.
23. William A. Howard. The formulae-as-types notion of construction. In Seldin, J.P. and Hindley, J.R., editors, *To H.B. Curry: Essays on Combinatory Logics, Lambda Calculus and Formalism*. Academic Press, London, 1980.
24. Stanislaw Jaśkowski. On the Rules of Suppositions in Formal Logics. *Studia Logica* 1, 1934.
25. Joachim Lambek. The mathematics of sentence structure. *Amer. Math. Monthly* 65:154–169, 1958.
26. Hiroakira Ono. Semantics for substructural logics. In Schroeder-Heister, Peter, editor, *Substructural Logics*. Oxford University Press. 259–291, 1993.
27. Dag Prawitz. *Natural Deduction. A Proof-Theoretical Study*. Almqvist & Wilksell, Uppsala, 1965.
28. Alessandra Russo. Generalising propositional modal logic using labelled deductive systems. In *Applied Logic Series (APLS), 'Frontiers of Combining Systems, First International Workshop'*, volume 3. 57–73, 1996.
29. Giovanni Sambin. The semantics of pretopologies. In Schroeder-Heister, Peter, editor, *Substructural Logics*. Oxford University Press. 293–307, 1993.
30. Manfred Egon Szabo, editor. *The Collected Papers of Gerhard Gentzen*. North-Holland, Amsterdam, 1969.
31. Alfred Tarski. Fundamentale Begriffe der Methodologie der deduktiven Wissenschaften, I. *Monatshefte für Mathematik und Physik* 37:361–404, 1930.

32. Alfred, Tarski. Über einige fundamentale Begriffe der Metamathematik. *Comptes Rendus des Séances de la Société des Sciences et des Lettres de Varsovie* 23:22–29, 1930.
33. Alasdair Urquhart. Semantics for relevant logic. *The Journal of Symbolic Logic* 37:159–170, 1972.

LABELLED DEDUCTION IN THE COMPOSITION OF FORM AND MEANING

MICHAEL MOORTGAT

In the late Fifties, Jim Lambek has started a line of investigation that accounts for the composition of form and meaning in natural language in *deductive* terms: formal grammar is presented as a logic — a system for reasoning about the basic form/meaning units of language and the ways they can be put together into wellformed structured configurations. The reception of the categorial grammar logics in linguistic circles has always been somewhat mixed: the mathematical elegance of the original system [22] is counterbalanced by clear descriptive limitations, as Lambek has been the first to emphasize on a variety of occasions.

As a result of the deepened understanding of the options for 'substructural' styles of reasoning, the categorial architecture has been redesigned in recent work, in ways that suggest that mathematical elegance may indeed be compatible with linguistic sophistication. A careful separation of the logical and the structural components of the categorial inference engine leads to the identification of *constants* of grammatical reasoning. At the level of the basic rules of use and proof for these constants one finds an explanation for the *uniformities* in the composition of form and meaning across languages. Cross-linguistic *variation* in the realization of the form-meaning correspondence is captured in terms of *structural* inference packages, acting as plug-ins with respect to the base logic of the grammatical constants. Structural inferences are under the explicit control of lexically anchored licensing features. These features have the status of logical constants in their own right: they control the structural aspects of grammatical resource management in a way analogous to what the 'exponentials' (or 'modalities') of linear logic do for resource multiplicity. The reader is referred to [28] for a systematic presentation of these developments.

The categorial set-up sketched above presents a new challenge for the 'Parsing-as-Deduction' approach to natural language processing. Consider the question whether a string of words $w_1 \cdots w_n$ constitutes a wellformed expression of type B. Under the deductive view on grammatical computa-

H.J. Ohlbach and U. Reyle (eds.), Logic, Language and Reasoning, 377–399.

tion, this question is reformulated as the problem displayed in (1): given A_i as the logical 'parts-of-speech' for the words w_i, does the grammar logic allow the derivation of the conclusion B? In the original Lambek calculus, linear order is the only structural factor that affects derivability: assumptions can be faithfully represented as one-dimensional lists $A_1, \ldots, A_n$. But in the refined setting assumed here, the assumptions A_i are configured into a *structured database*: the precise configuration of the assumptions will determine which structural inferences are applicable and whether the goal formula B is derivable or not. Clearly, we cannot take the structure of the database as *given* in the statement of the parsing problem. Rather, we want to find a way of efficiently *computing* this structure in the process of grammatical deduction.

$$\begin{array}{lccccc} \text{PARSING} & w_1 & \cdots & w_n & & \\ \textit{as} & \vdots & & \vdots & & \\ \text{DEDUCTION} & \underbrace{A_1 \quad \cdots \quad A_n}_{\Gamma} & & & \vdash & B \end{array} \tag{1}$$

In §3 we provide a uniform algorithmic proof theory for the structure-sensitive style of grammatical reasoning in terms of Dov Gabbay's framework of Labelled Deduction. We'll see that this framework has exactly the right properties for dealing with the logical and the structural aspects of grammatical reasoning in a modular way. We start with a brief overview of the grammatical architecture assumed in this paper in §1. In §2 we give a linguistic illustration on the basis of a labelled Natural Deduction format that is useful for *displaying* proofs once they have been found, but that does not have the right properties for algorithmic proof *search* (parsing).

1 Grammatical Composition: Logic, Structure, and Control

In this paper, we consider a language of type formulas freely generated from a small number of atomic types $\mathcal{A}$ by means of the unary and binary connectives in (2). The binary $/, \bullet, \backslash$ are the familiar categorial product and slash connectives. The unary $\Diamond, \Box$ are the new control devices.

$$\mathcal{F} ::= \mathcal{A} \mid \Diamond\mathcal{F} \mid \Box\mathcal{F} \mid \mathcal{F}/\mathcal{F} \mid \mathcal{F}\bullet\mathcal{F} \mid \mathcal{F}\backslash\mathcal{F} \tag{2}$$

The categorial formula language is used to talk about the form-meaning units of language: 'signs', or 'grammatical resources', as we will call them here. An appropriate framework for reasoning about structured configurations of grammatical resources is *modal logic*: we base the models for the grammar logic on frames $F = \langle W, R^2, R^3 \rangle$. The domain W, in the case at hand, is the set of linguistic resources, and for each family of n-place connectives, we have an $n+1$-place 'accessibility relation' modelling the decomposition of a grammatical compound into its constituent part(s). This type of semantics has its ancestry in the Kripke models for relevant logics introduced in the Seventies by Routley and Meyer as pointed out in [5]. Frame based semantics for the extended type languages we consider in this paper is investigated in depth in [20] — see Van Benthem (this volume) for discussion.

As remarked in the introduction, we want to keep *logical* and *structural* aspects of the meaning of the constants apart. The categorial base logic is 'structurally uncommitted' in that it interprets $\Diamond$ and $\bullet$ as existential modal operators with respect to arbitrary binary and ternary composition relations R^2 and R^3. The constants $\Box$ and $/, \backslash$ are interpreted as the residuation duals of these existential modalities. See the interpretation clauses in (3).

$$\begin{aligned} V(\Diamond A) &= \{x \mid \exists y(R^2xy \ \&\ y \in V(A)\} \\ V(\Box A) &= \{y \mid \forall x(R^2xy \ \Rightarrow x \in V(A)\} \end{aligned} \quad (3)$$

$$\begin{aligned} V(A \bullet B) &= \{x \mid \exists y \exists z[R^3xyz \ \&\ y \in V(A) \ \&\ z \in V(B)]\} \\ V(C/B) &= \{y \mid \forall x \forall z[(R^3xyz \ \&\ z \in V(B)) \Rightarrow x \in V(C)]\} \\ V(A\backslash C) &= \{z \mid \forall x \forall y[(R^3xyz \ \&\ y \in V(A)) \Rightarrow x \in V(C)]\} \end{aligned}$$

The residuation laws of (4) capture the properties of $\Diamond, \Box$ and $/, \bullet, \backslash$ with respect to derivability. The residuation inferences, together with the reflexivity and transitivity of derivability, give the essential completeness result in the sense that $A \rightarrow B$ is provable iff $V(A) \subseteq V(B)$ for every valuation V on every frame F. Restricting our attention to the binary connectives, we have the completeness result of [5] for the calculus **NL** of [23]. For the language extended with unary connectives, see [27, 20].

$$\Diamond A \rightarrow B \quad \text{iff} \quad A \rightarrow \Box B \quad (4)$$

$$A \rightarrow C/B \quad \text{iff} \quad A \bullet B \rightarrow C \quad \text{iff} \quad B \rightarrow A\backslash C$$

The laws of the base logic hold universally, in the sense that they do not depend on structural properties of the composition relation. Cross-linguistic variation is obtained by adding to the base logic postulate packages regulating structural aspects of the grammatical resource management regime. Semantically, these postulates 'have a price': they introduce constraints on the interpretation of the grammatical composition relations R^2, R^3, see again [5, 20] for thorough discussion. An illustrative sample of structural postulates is displayed in (5) below.

$$\begin{array}{rcrc} C: & A \bullet B \to B \bullet A & K1: & \Diamond(A \bullet B) \to \Diamond A \bullet B \\ A: & (A \bullet B) \bullet C \leftrightarrow A \bullet (B \bullet C) & K: & \Diamond(A \bullet B) \to \Diamond A \bullet \Diamond B \\ C_\Diamond: & A \bullet \Diamond B \to \Diamond B \bullet A & MA: & (A \bullet_j B) \bullet_i C \to A \bullet_j (B \bullet_i C) \\ A_\Diamond: & (A \bullet B) \bullet \Diamond C \leftrightarrow A \bullet (B \bullet \Diamond C) & MC: & A \bullet_i (B \bullet_j C) \to B \bullet_j (A \bullet_i C) \end{array} \tag{5}$$

The postulates C and A on the left impose commutativity or associativity constraints on the interpretation of the composition relation R^3. Adding A to the base residuation logic produces the familiar associative Lambek calculus **L** of [22]; adding both the A and C postulates gives the Lambek-Van Benthem calculus **LP**. The postulates C and A change the resource management regime in a *global* fashion. In the presence of the unary modalities, one can consider refined options such as $C_\Diamond$ or $A_\Diamond$, where reordering or restructuring are not globally available, but have to be explicitly *licensed* by a structural control operator $\Diamond$. On the right, we have interaction postulates regulating the *communication* between the unary and binary multiplicatives (the weak and strong distributivity principles $K1$ and K), or between distinct binary multiplicatives (such as the weak distributivity principles of Mixed Associativity and Mixed Commutativity). These latter cases require a straightforward *multimodal* generalization of the architecture, with frames $F = \langle W, \{R_i^2\}_{i\in I}, \{R_j^3\}_{j\in J}\rangle$, where the indices keep *composition modes* apart. In §2.3, the reader will find an illustration of grammatical analysis in terms of modal control and interaction postulates like the above.

2 Labelling Proofs: Form and Meaning

We now present two systems of *labelled* deduction for the display of derivations in the extended categorial logics. As we have seen above, fine-tuning of categorial inference is obtained by considering mixed logics where interacting regimes of structural resource management are put together. Labelled presentations of the proof systems are particularly useful here: in line with the slogan of 'bringing the semantics into the syntax' the labelling systems allow explicit reference to the grammatical resources and the logical and structural aspects of their composition. On the meaning side, we have labelling in the sense of the 'formulas-as-types' program, producing 'semantic recipes' for categorial derivations. On the structural side, labelling can capture the configuration of linguistic resources in the form dimension, and the allowable structural manipulations of these configurations in the process of grammatical reasoning.

2.1 Categorical Combinators

The first system of labelled categorial deduction we consider is the categorical presentation of [24]. In the categorical presentation, deductions take the form of 'arrows' $f : A \to B$, where the proof label f codes a process of deducing B from A, i.e. a proof of the semantic inclusion $v(A) \subseteq v(B)$. For every type formula A, we have an identity arrow 1_A, capturing the reflexivity of derivability, and we have a rule of inference which from given proofs f and g produces a new proof $g \circ f$ for their sequential composition, thus capturing the transitivity of derivability. The pure residuation logic is then obtained by imposing the additional rules of inference of Definition 2.1, which establish the residuation laws for $\Diamond, \Box$ and $/, \bullet, \backslash$. One can now study equality of proofs in terms of appropriate categorical equations for the labelling system, cf. [25], and [37] for discussion in the context of combinatorial linear logic.

DEFINITION 2.1 *The pure logic of residuation: combinator proof terms* [24].

$$1_A : A \to A \qquad \frac{f : A \to B \quad g : B \to C}{g \circ f : A \to C}$$

$$\frac{f : \Diamond A \to B}{\mu_{A,B}(f) : A \to \Box B} \qquad \frac{g : A \to \Box B}{\mu^{-1}_{A,B}(g) : \Diamond A \to B}$$

$$\frac{f : A \bullet B \to C}{\beta_{A,B,C}(f) : A \to C/B} \qquad \frac{f : A \bullet B \to C}{\gamma_{A,B,C}(f) : B \to A\backslash C}$$

$$\frac{g : A \to C/B}{\beta^{-1}_{A,B,C}(g) : A \bullet B \to C} \qquad \frac{g : B \to A\backslash C}{\gamma^{-1}_{A,B,C}(g) : A \bullet B \to C}$$

One easily derives the arrows for the laws of left and right functional application, and their unary counterpart. See (6), where we write **app**$^\backslash$ for the proof $\gamma^{-1}(1_{B\backslash A})$, **app**$^/$ for $\beta^{-1}(1_{A/B})$, and **co-unit** for $\mu^{-1}(1_{\Box A})$.

$$\textbf{co-unit} : \Diamond\Box A \to A \quad \textbf{app}^/ : A/B \bullet B \to A \quad \textbf{app}^\backslash : B \bullet B\backslash A \to A \tag{6}$$

As examples of derived rules of inference, we have the Isotonicity laws for $\Diamond$ and $\bullet$. The $f \cdot g$ law is known as 'parallel composition' in the categorical setting, as contrasted with the 'sequential composition' of arrows $g \circ f$.

$$\frac{f : A \to B}{(f)^{\diamond} : \Diamond A \to \Diamond B} \qquad \frac{f : A \to B \quad g : C \to D}{f \cdot g : A \bullet C \to B \bullet D} \tag{7}$$

In (8), we give the derivation of $(f)^{\diamond}$ as $\mu^{-1}(\mu(1_{\Diamond B}) \circ f)$. For sequential composition, see [22].

$$\frac{f : A \to B \quad \dfrac{1_{\Diamond B} : \Diamond B \to \Diamond B}{\mu(1_{\Diamond B}) : B \to \Box\Diamond B}}{\dfrac{\mu(1_{\Diamond B}) \circ f : A \to \Box\Diamond B}{\mu^{-1}(\mu(1_{\Diamond B}) \circ f) : \Diamond A \to \Diamond B}} \tag{8}$$

Variation in grammatical resource management is obtained by adding to the pure residuation logic the required structural postulates, cf. (5) above. Each of these postulates, as an extra axiom schema, is labelled with its own primitive structural combinator.

The categorical labelling fully encodes a proof, both in its logical and in its structural aspects. As an example, we derive an implicational form of the Mixed Commutativity postulate MC from (5), dropping formula subscripts and composition mode indices for legibility. (f here stands for $\gamma^{-1}(\gamma(\beta^{-1}(1_{A/B}))\circ\gamma^{-1}(1_{C\backslash B}))$, as the reader will no doubt want to check.) Notice that the proof term is composed of a logical part (the residuation combinators γ, β and their inverses) and a structural component (the combinator **mc**).

$$\frac{\dfrac{\dfrac{\mathbf{mc} : C \bullet_i (A/_jB \bullet_j C\backslash_iB) \to A/_jB \bullet_j (C \bullet_i C\backslash_iB) \quad f : A/_jB \bullet_j (C \bullet_i C\backslash_iB) \to A}{f \circ \mathbf{mc} : C \bullet_i (A/_jB \bullet_j C\backslash_iB) \to A}}{\gamma(f \circ \mathbf{mc}) : A/_jB \bullet_j C\backslash_iB \to C\backslash_iA}}{\beta(\gamma(f \circ \mathbf{mc})) : A/_jB \to (C\backslash_iA)/_j(C\backslash_iB)} \tag{9}$$

2.2 Natural Deduction and Curry–Howard Labelling

In order to relate the categorical proof terms to the Curry-Howard-de Bruyn formulas-as-types interpretation, we now move to a Natural Deduction presentation, which we first consider in its unlabelled form. The arrows $f : A \to B$ are replaced by statements $\Gamma \vdash B$ representing a deduction of a formula B from a *structured database* of assumptions Γ. The structural 'packaging' of the resources is what distinguishes the categorial systems from linear logic — in the latter, the database can be seen as a *multiset* of assumptions: the occurrence aspect of the formulas matters, but not their further structuring.

To build a structured database of antecedent formulae, we need a language of structural connectives matching the language of logical connectives. This strategy goes back essentially to Belnap's Display Logic — see [12] for an up-to-date discussion of the substructural connections. We write (unary) $\langle\cdot\rangle$ for the structural counterpart of $\Diamond$, and (binary) $(\cdot \circ \cdot)$ for the structural connective corresponding to $\bullet$.

$$\mathcal{S} ::= \mathcal{F} \mid \langle\mathcal{S}\rangle \mid \mathcal{S} \circ \mathcal{S} \tag{10}$$

DEFINITION 2.2 *The pure residuation logic: Natural Deduction presentation. Introduction and Elimination rules for the constants. (Notation: $\Gamma[\Delta]$ for a structure Γ with a distinguished occurrence of a substructure Δ.)*

$$\frac{}{A \vdash A}\ (Ax)$$

$$\frac{\Gamma \vdash \Box A}{\langle\Gamma\rangle \vdash A}\ (\Box E) \qquad \frac{\langle\Gamma\rangle \vdash A}{\Gamma \vdash \Box A}\ (\Box I)$$

$$\frac{\Gamma \vdash A}{\langle\Gamma\rangle \vdash \Diamond A}\ (\Diamond I) \qquad \frac{\Delta \vdash \Diamond A \quad \Gamma[\langle A\rangle] \vdash B}{\Gamma[\Delta] \vdash B}\ (\Diamond E)$$

$$\frac{\Delta \vdash A \quad \Gamma \vdash A\backslash B}{\Delta \circ \Gamma \vdash B}\ (\backslash E) \qquad \frac{A \circ \Gamma \vdash B}{\Gamma \vdash A\backslash B}\ (\backslash I)$$

$$\frac{\Gamma \vdash B/A \quad \Delta \vdash A}{\Gamma \circ \Delta \vdash B}\ (/E) \qquad \frac{\Gamma \circ A \vdash B}{\Gamma \vdash B/A}\ (/I)$$

$$\frac{\Gamma \vdash A \quad \Delta \vdash B}{\Gamma \circ \Delta \vdash A \bullet B}\ (\bullet I) \qquad \frac{\Delta \vdash A \bullet B \quad \Gamma[A \circ B] \vdash C}{\Gamma[\Delta] \vdash C}\ (\bullet E)$$

It is not difficult to derive the Natural Deduction rules from the categorical formulation. Let us write $\Gamma^\natural$ for the formula that results from replacing the structural connectives $\circ$ and $\langle\cdot\rangle$ in Γ by their logical counterparts $\bullet$ and $\Diamond$. The Axiom case coincides in the two presentations. $(\Box I)$ and $(\Box E)$ become the residuation inferences μ and μ^{-1}, respectively. $(/I)$ and $(\backslash I)$ become the β and γ half of residuation. $(\Diamond I)$ and $(\bullet I)$ are the Monotonicity rules of inference (7) — derived rules of inference, as we saw above. For $(\backslash E)$, we have the derivation in (11) which composes Monotonicity with Application. The $(/E)$ case is similar.

$$\frac{\dfrac{f : \Delta^\natural \to A \quad g : \Gamma^\natural \to A\backslash B}{f \cdot g : (\Delta \circ \Gamma)^\natural \to A \bullet A\backslash B} \quad \mathbf{app}^\backslash : A \bullet A\backslash B \to B}{\mathbf{app}^\backslash \circ (f \cdot g) : (\Delta \circ \Gamma)^\natural \to B} \tag{11}$$

For ($\bullet E$), we have (12). We write $\pi(f)$ for the sequence of μ, β, γ inferences that isolate the target formula $(A \circ B)^\natural$ on the left hand side of the arrow, moving the context to the right hand side (notation $C \mid \Gamma^\natural$). At that point, we compose with the major premise g, and put the context back in place on the left hand side via π^{-1}. The ($\Diamond E$) case is similar.

$$\frac{\dfrac{g : \Delta^\natural \to A \bullet B \quad \dfrac{\dfrac{f : (\Gamma[A \circ B])^\natural \to C}{\vdots}}{\pi(f) : (A \circ B)^\natural \to C \mid \Gamma^\natural}}{\dfrac{\pi(f) \circ g : \Delta^\natural \to C \mid \Gamma^\natural}{\vdots}}}{\pi^{-1}(\pi(f) \circ g) : (\Gamma[\Delta])^\natural \to C} \tag{12}$$

Structural rules S, in the Natural Deduction presentation, take the form of inferences

$$\frac{\Gamma[\Delta'] \vdash A}{\Gamma[\Delta] \vdash A}\, S \tag{13}$$

where the formula equivalents $\Delta^\natural$ and $\Delta'^\natural$ of the structures Δ and Δ' match the left and right hand sides of a structural postulate $\sigma : A \to B$. Their derivation from the categorical presentation, then, follows the lines of (12), with the structural combinator axiom σ taking the place of the open premise g. As an illustration, consider (14), the Natural Deduction rule corresponding to the distributivity postulate $\Diamond(A \bullet B) \to \Diamond A \bullet \Diamond B$.

$$\frac{\Gamma[\langle\Delta_1\rangle \circ \langle\Delta_2\rangle] \vdash A}{\Gamma[\langle\Delta_1 \circ \Delta_2\rangle] \vdash A}\, K \tag{14}$$

Let us turn now to the more familiar decoration of Natural Deduction derivations with λ term annotation for the Curry–Howard–de Bruyn 'formulas-as-types' interpretation. Instead of the formula A, we take the labelled formula $t : A$ as the 'basic declarative unit'. Rules of inference manipulate both the formula and its label, and we build a recipe t for the construction of the meaning of the goal formula B out of a structured configuration of labelled assumptions $x_i : A_i$. The term decoration rules for the pure logic of residuation are given in (2.3). Introduction and Elimination of the implications correspond to functional abstraction and application, respectively. (We can collapse / and \ in the meaning dimension, coding the order requirements of these operators in the antecedent term structure.) Introduction and Elimination rules for • are associated with the pairing and projection operations. In an entirely analogous way, we have 'cap' and 'cup' operations for the Introduction and Elimination rules for $\Diamond$ and $\Box$. Substructural versions of the 'formulas-as-types' program, and of the relevant term equations, are studied in depth in [10, 38].

DEFINITION 2.3 *Natural deduction. Proof terms.*

$$\frac{}{x : A \vdash x : A}\ (Ax)$$

$$\frac{\Delta \vdash u : A \quad \Gamma \vdash t : A\backslash B}{\Delta \circ \Gamma \vdash (t\ u) : B}\ (\backslash E) \qquad \frac{x : A \circ \Gamma \vdash t : B}{\Gamma \vdash \lambda x.t : A\backslash B}\ (\backslash I)$$

$$\frac{\Gamma \vdash t : B/A \quad \Delta \vdash u : A}{\Gamma \circ \Delta \vdash (t\ u) : B}\ (/E) \qquad \frac{\Gamma \circ x : A \vdash t : B}{\Gamma \vdash \lambda x.t : B/A}\ (/I)$$

$$\frac{\Gamma \vdash t : A \quad \Delta \vdash u : B}{\Gamma \circ \Delta \vdash \langle t, u\rangle : A \bullet B}\ (\bullet I) \qquad \frac{\Delta \vdash u : A \bullet B \quad \Gamma[x : A \circ y : B] \vdash t : C}{\Gamma[\Delta] \vdash t[(u)_0/x, (u)_1/y] : C}\ (\bullet E)$$

$$\frac{\Gamma \vdash t : \Box A}{\langle\Gamma\rangle \vdash {}^{\vee}t : A}\ (\Box E) \qquad \frac{\langle\Gamma\rangle \vdash t : A}{\Gamma \vdash {}^{\wedge}t : \Box A}\ (\Box I)$$

$$\frac{\Gamma \vdash t : A}{\langle\Gamma\rangle \vdash {}^{\cap}t : \Diamond A}\ (\Diamond I) \qquad \frac{\Delta \vdash u : \Diamond A \quad \Gamma[\langle x : A\rangle] \vdash t : B}{\Gamma[\Delta] \vdash t[{}^{\cup}u/x] : B}\ (\Diamond E)$$

The Curry–Howard term decoration records the application of the *logical rules* of inference — the Elimination and Introduction rules for the connectives. The *structural rules* of resource management are not reflected in the Curry–Howard labelling: structural rules, schematically, manipulate a structural subterm of the antecedent, leaving the succedent formula annotation unaffected.

$$\frac{\Gamma[\Delta'] \vdash t : A}{\Gamma[\Delta] \vdash t : A}\ S \tag{15}$$

If we restrict the attention exclusively to the formula labels, we see a loss of information with respect to the categorical proof terms that faithfully encoded both the logical and structural aspects of a derivation. But of course, in the 'sequent style' Natural Deduction presentation, the antecedent has a term structure of its own, given by the structural operations $\circ$ and $\langle\cdot\rangle$, and structural rules manipulate this term structure.

2.3 Illustration: Crossed Dependencies

The components of the grammatical architecture proposed in the previous section are summarized below.

Logic. The core notions of 'grammatical composition' are characterized in terms of universal laws, independent of the structural properties of the composition relation. The operations of the base logic (introduction/elimination of the grammatical constants) provide the interface to a derivational theory of meaning via the Curry–Howard interpretation of proofs.

Structure. Packages of resource-management postulates function as 'plug-in' modules with respect to the base logic. They offer a logical perspective on structural variation, within languages and cross-linguistically.

Control. A vocabulary of control operators provides explicit means to fine-tune grammatical resource management, by imposing structural constraints or by licensing structural relaxation.

In order to illustrate the increased expressive power of the multimodal style of reasoning, we take a brief look at crossed dependencies in Dutch. As is well known, a proper analysis of the syntactic and semantic aspects of crossed dependencies is beyond the reach of strictly context-free grammar formalisms — beyond the reach of the standard Lambek calculus **L**, in the categorial case. The relevant phenomena are displayed in (16) below. As the examples (a) to (c) show, Dutch is a verb-final language: in their canonical position (the embedded clause), verbs look for their arguments to the left. Crossed dependencies arise in the presence of modal auxiliaries such as 'kunnen' ('can'), 'willen' ('want'). These auxiliaries select for an infinitival complement, but rather than consuming this complement in its entirety, they are prefixed to the clause-final infinitival *head* of their complement, 'skipping over' the arguments of the infinitival, if any. By connecting the infinitive 'plagen' to its direct object 'Alice' and the auxiliary 'wil' to its subject 'Tweedledum' in (e), one can see where the dependencies cross.

Consider the provisional type assignments in (16), where we write vp for $np\backslash s$. On the basis of (d), one could assign 'wil' the type vp/inf, so that it prefixes itself to its infinitival complement. But to obtain the combination 'wil plagen' with the transitive infinitive in (e), one would need a directionally crossed or mixed form of composition (schematically, $A/B, C\backslash B \Rightarrow C\backslash A$) which is invalid in **L**, as it violates the order sensitivity of the types involved. The grammatical example (e), in other words, is underivable given the **L** type assignments above. The sequence 'wil Alice plagen' in (f), on the contrary, *is* derivable, but it is *ungrammatical* in the embedded clausal context we are considering.

We can overcome these problems of overgeneration and undergeneration by moving to a multimodal setting, as shown in [29], and by exploiting

a als Alice slaapt (slaapt: $np\backslash s$) (16)
if Alice sleeps
b als Tweedledum Alice plaagt (plaagt: $np\backslash(np\backslash s)$)
if Tweedledum Alice teases ('if T teases A')
c of Alice Tweedledum gek vindt (vindt: $ap\backslash(np\backslash(np\backslash s))$)
whether Alice Tweedledum crazy considers
('whether A considers T crazy')
d als Alice wil slapen (wil: vp/inf, slapen: inf)
if Alice wants sleep ('if A wants to sleep')
e als Tweedledum Alice wil plagen (plagen: $np\backslash inf$, wil: ??)
if Tweedledum Alice wants tease ('if T wants to tease A')
f *if Tweedledum wil Alice plagen

the structural control devices $\Diamond, \Box$. The structural package in (17) makes a distinction between two binary modes. The regular combination of heads with their phrasal complements is realized by $\bullet_1$: subcategorizational requirements of the verbs in (16a-c), and to the transitive infinitive in (e), will be expressed in terms of the $\backslash_1$ implication. The head adjunction operation that gives rise to crossed dependencies is realized by $\bullet_0$: the type assignment for 'wil' in (16d-f) selects the infinitival complement in terms of $/_0$.

$$\begin{array}{ll} P1 & \Diamond_1(A \bullet_1 B) \to A \bullet_1 \Diamond_1 B \\ P2 & \Diamond_1 A \to \Diamond_0 A \\ P3 & \Diamond_0(A \bullet_0 B) \to \Diamond_0 A \bullet_0 \Diamond_0 B \\ P4 & A \bullet_1 (\Diamond_0 B \bullet_0 C) \to \Diamond_0 B \bullet_0 (A \bullet_1 C) \end{array} \quad (17)$$

Consider next the control component, expressed in terms of modalities $\Diamond_0$ and $\Diamond_1$, together with their residuals. In order to lexically anchor the structural control, we use a 'key-and-lock' strategy. Verbal elements (tensed and infinitival) are lexically typed with $\Box_0$ as their main connective — the transitive infinitive 'plagen', for example, would get the type assignment $\Box_0(np\backslash_1 inf)$. As we have seen above, subcategorizational requirements are expressed in terms of implications with respect to the composition modes $\bullet_1$ and $\bullet_0$. But before these implications can be used in a derivation, the $\Box_0$ lock has to be removed, by means of the basic reduction law $\Diamond_0\Box_0 A \to A$.

The role of the control devices $\Diamond_1$ and $\Diamond_0$ in (17) is to check whether the clause final verbal structure is indeed wellformed. We assign a complementizer like 'als' ('if') the type $sbar/_1\Box_1 s$, i.e. the goal type for an embedded clause is $\Box_1 s$. To prove that a structure Γ is of type $\Box_1 s$ amounts to proving $\langle\Gamma\rangle^1 \vdash s$ (via Box Introduction). Here our postulate package can start its work. $P1$ recursively inspects phrasal structure, and looks for the verbal head at the end. At the point where there is no more phrasal $\bullet_1$ structure to traverse, $P2$ switches the control to inspection of the verbal head constituent itself. This can either be a simple verb (which can then be directly unlocked by means of $\Diamond_0\Box_0 A \to A$), or it can be a compound verbal cluster, constructed by means of the head adjunction operation $\bullet_0$. In the latter case, $P3$ recursively checks whether the components of a verbal cluster are indeed verbs. Postulate $P4$ — a modally controlled version of mixed commutativity — undoes the crossed dependencies and makes sure that the phrasal complements that were skipped over can be consumed by means of $\backslash_1$ Elimination.

A Natural Deduction derivation for the verb phrase 'Alice wil plagen' of (16e) is given below. In (18), we focus on *structural* composition, dropping type formulae in the antecedent, and semantic labels in the succedent. Notice that the upper part of the derivation proceeds in 'bottom up' fashion from lexical type assignments purely in terms of *logical* inferences — Elimination rules for the implication and box connectives — producing the structure

$$\langle \text{wil}\rangle_0 \circ_0 (\text{Alice} \circ_1 \langle \text{plagen}\rangle_0)$$

The structural inferences $P1$–$P4$ mediate between this structure and the structure

$$\langle \text{Alice} \circ_1 (\text{wil} \circ_0 \text{plagen})\rangle_1$$

that is obtained in 'top down' fashion from the final conclusion by means of the logical $\Box_1$ Introduction inference.

$$\dfrac{\dfrac{\dfrac{\dfrac{\dfrac{\dfrac{\dfrac{\text{wil} \vdash \Box_0(vp/_0 inf)}{\langle \text{wil}\rangle_0 \vdash vp/_0 inf}\,\Box_0 E \qquad \dfrac{\text{Alice} \vdash np \qquad \dfrac{\text{plagen} \vdash \Box_0(np\backslash_1 inf)}{\langle \text{plagen}\rangle_0 \vdash np\backslash_1 inf}\,\Box_0 E}{\text{Alice} \circ_1 \langle \text{plagen}\rangle_0 \vdash inf}\,\backslash_1 E}{\langle \text{wil}\rangle_0 \circ_0 (\text{Alice} \circ_1 \langle \text{plagen}\rangle_0) \vdash vp}\,/_0 E}{\text{Alice} \circ_1 (\langle \text{wil}\rangle_0 \circ_0 \langle \text{plagen}\rangle_0) \vdash vp}\,P4}{\text{Alice} \circ_1 \langle \text{wil} \circ_0 \text{plagen}\rangle_0 \vdash vp}\,P3}{\text{Alice} \circ_1 \langle \text{wil} \circ_0 \text{plagen}\rangle_1 \vdash vp}\,P2}{\langle \text{Alice} \circ_1 (\text{wil} \circ_0 \text{plagen})\rangle_1 \vdash vp}\,P1}{\text{Alice} \circ_1 (\text{wil} \circ_0 \text{plagen}) \vdash \Box_1 vp}\,\Box_1 I \qquad (18)$$

In (19), we concentrate on the composition of *meaning*: we drop the structured antecedent database, and present just the succedent formulae with their Curry–Howard term labels. Observe that the modal auxiliary 'wil' has the required scope over the combination of the infinitive 'plagen' and its direct object 'Alice'.

$$\dfrac{\dfrac{\text{wil} : \Box_0(vp/_0inf)}{{}^{\vee}\text{wil} : vp/_0inf}\ \Box_0E \quad \dfrac{\text{alice} : np \quad \dfrac{\text{plagen} : \Box_0(np\backslash_1 inf)}{{}^{\vee}\text{plagen} : np\backslash_1 inf}\ \Box_0E}{({}^{\vee}\text{plagen alice}) : inf}\ \backslash_1E}{\dfrac{({}^{\vee}\text{wil}\ ({}^{\vee}\text{plagen alice})) : vp}{{}^{\wedge}({}^{\vee}\text{wil}\ ({}^{\vee}\text{plagen alice})) : \Box_1 vp}\ \Box_1 I}\ /_0E \tag{19}$$

Notice also that the distinction between $\Diamond_1$ and $\Diamond_0$ effectively imposes the constraint that the (standard Dutch) verb-raising cluster cannot contain phrasal compositions. Because we have $\Diamond_1 A \rightarrow \Diamond_0 A$, but not the other way around, the following attempt at deriving the ungrammatical (16f) '(als Tweedledum) wil Alice plagen' fails: $\Diamond_0$ does not distribute through a phrasal $\bullet_1$ configuration. This then solves the 'overgeneration' part of the problems with (16).

$$\dfrac{\dfrac{\text{wil} \vdash \Box_0(vp/_0inf)}{\langle\text{wil}\rangle_0 \vdash vp/_0inf}\ \Box_0E \quad \dfrac{\text{FAILS}}{\langle\text{Alice} \circ_1 \text{plagen}\rangle_0 \vdash inf}}{\dfrac{\dfrac{\dfrac{\langle\text{wil}\rangle_0 \circ_0 \langle\text{Alice} \circ_1 \text{plagen}\rangle_0 \vdash vp}{\langle\text{wil} \circ_0 (\text{Alice} \circ_1 \text{plagen})\rangle_0 \vdash vp}\ P3}{\langle\text{wil} \circ_0 (\text{Alice} \circ_1 \text{plagen})\rangle_1 \vdash vp}\ P2}{\text{wil} \circ_0 (\text{Alice} \circ_1 \text{plagen}) \vdash \Box_1 vp}\ \Box_1 I}\ /_0E \tag{20}$$

3 Proof Search and Labelling

The categorical and Natural Deduction formats are handy for *presenting* proofs once they have been found. But they do not provide an appropriate basis for automated proof *search*. In this section we consider two proof formats which *do* have an algorithmic interpretation: Gentzen sequent calculus and proof nets. Labelled presentations here make it possible to give a uniform presentation of the Curry–Howard 'derivational meaning' at the level of **LP**, i.e. at a level where multiplicity of assumptions matters but where one can abstract from the structural aspects of composition. Instead, these aspects are controlled via an appropriate structural labelling regime. A labelled Gentzen presentation can be seen as a first step towards a modular treatment of 'logic' and 'structure'. But the Gentzen format still suffers from spurious non-determinism, which can be effectively removed as soon as we move to a (labelled) proof net approach.

3.1 Labelled Gentzen Calculus

The relations between Natural Deduction and sequent calculus for resource logics are well-understood, syntactically and on the level of the Curry–

Howard interpretation, see for example [10, 11, 38]. The move from Natural Deduction to Gentzen sequent presentation requires that we reformulate all logical rules of inference in such a way that a connective is introduced in the *conclusion*, either in the antecedent (rules of use, left rules) or in the succedent (rules of proof, right rules). In the presence of a Cut Elimination result to the effect that the Cut rule in (21) does not increase the set of derivable theorems (or semantic recipes, modulo logical equivalence), one immediately obtains a procedure for decidable proof search based on systematic removal of connectives from conclusion to premises.

$$\frac{\Delta \Rightarrow u : A \quad \Gamma[x : A] \Rightarrow t : B}{\Gamma[\Delta] \Rightarrow t[u/x] : B}\ \text{Cut} \tag{21}$$

The Introduction rules in the Natural Deduction presentation have the required form — they can be taken over unchanged as rules of proof in the sequent calculus. The rules of use for $\Diamond$ and $\bullet$ are obtained from the Natural Deduction Elimination rules for these connectives by instantiating the major premise as the identity axiom.

$$\frac{\overline{A \bullet B \vdash A \bullet B}\ (Ax) \quad \Gamma[A \circ B] \vdash C}{\Gamma[A \bullet B] \vdash C}\ (\bullet E) \quad \rightsquigarrow \quad \frac{\Gamma[A \circ B] \Rightarrow C}{\Gamma[A \bullet B] \Rightarrow C}\ (\bullet L) \tag{22}$$

In the rules of use for $\Box$ and $/, \backslash$, we recognize compiled Cuts, on $\Diamond\Box A \Rightarrow A$, and on Application.

$$\frac{\Gamma[A] \Rightarrow B}{\Gamma[\langle \Box A \rangle] \Rightarrow B}\ \Box L \qquad \frac{\Delta \Rightarrow A \quad \Gamma[B] \Rightarrow C}{\Gamma[\Delta \circ A\backslash B] \Rightarrow C}\ \backslash L \qquad \frac{\Delta \Rightarrow A \quad \Gamma[B] \Rightarrow C}{\Gamma[A/B \circ \Delta] \Rightarrow C}\ /L \tag{23}$$

Considering proof search from a 'parsing as deduction' point of view, one notices an important difference between the case of the *associative* system **L**, and the generalized multimodal categorial logics that form the subject of this paper. Because of the global availability of Associativity in **L**, one can say that strong and weak generative capacity for this system coincide. Parsing a string $w_1 \cdots w_n$ as an expression of type B comes down to proving the sequent $A_1, \ldots, A_n \Rightarrow B$, where the A_i are types assigned to the lexical resources w_i, and where the antecedent is a 'flat' sequence of assumptions without hierarchical structure. In the general multimodal case, we need to know the structural configuration of the antecedent assumptions in terms of the structural connectives $\circ$ and $\langle\cdot\rangle$ with their mode indications. As remarked in the introduction to this paper, one cannot take the antecedent structuring as 'given' without trivializing the parsing problem. Rather, we have to find a proof format where the structure of the antecedent database is gradually 'discovered' in the proof process.

In the categorial literature, a variety of labelled sequent formulations have been proposed for this purpose — see, among others, [26, 34, 32, 16]. One considers labelled sequents $x_1 : A_1, \ldots, x_n : A_n \Rightarrow t : B$ where the antecedent is simply a multiset of labelled formulae, representing the lexical assumptions (seen as occurrences, i.e. with the x_i distinct), and where t is a structure label built from these x_i by means of the structural operations of the multimodal system one is dealing with. Below we present the structural labelling of [20], which is complete for the general multimodal architecture ($\Diamond$ and $\bullet$ plus their residuals, and structural rule packages relativized to composition modes). The syntax of the labelling system is given in (24). Definition 3.1 presents the labelled sequent rules. (We have slightly adapted the notation of [20]. Of course, one can also label the formulas with Curry–Howard terms for semantic interpretation — but we concentrate on the structural aspects here.)

$$\begin{array}{llll} \phi, \psi & \longrightarrow & x & \text{atomic labels} \\ & & \mathbf{un}(x, \phi) & \text{unary tree} \\ & & \mathbf{bin}(x, \phi, \psi) & \text{binary tree} \end{array} \tag{24}$$

DEFINITION 3.1 *Labelled Gentzen calculus: [20]. A structure label is called* proper *if all its atomic subterms are distinct. Notation: x, y, z for atomic structure terms, t, u, v for proper structure terms. Γ, Δ finite multisets of formulas decorated with atomic structure labels. $t[u \rightsquigarrow v]$ denotes the substitution of u for v in t.*

$$\frac{}{x : A \Rightarrow x : A}\ Ax$$

$$\frac{x : A, \Gamma \Rightarrow t : B}{y : \Box A, \Gamma \Rightarrow t[\mathbf{un}(x,y) \rightsquigarrow x] : B}\ \Box L \qquad \frac{\Gamma \Rightarrow \mathbf{un}(x,t) : A}{\Gamma \Rightarrow t : \Box A}\ \Box R$$

$$\frac{y : A, \Gamma \Rightarrow t : B}{x : \Diamond A, \Gamma \Rightarrow t[x \rightsquigarrow \mathbf{un}(x,y)] : B}\ \Diamond L \qquad \frac{\Gamma \Rightarrow t : A}{\Gamma \Rightarrow \mathbf{un}(x,t) : \Diamond A}\ \Diamond R$$

$$\frac{\Gamma \Rightarrow u : B \quad x : A, \Delta \Rightarrow t : C}{y : A/B, \Gamma, \Delta \Rightarrow t[\mathbf{bin}(x,y,u) \rightsquigarrow x] : C}\ /L \qquad \frac{z : B, \Gamma \Rightarrow \mathbf{bin}(x,t,z) : A}{\Gamma \Rightarrow t : A/B}\ /R$$

$$\frac{\Gamma \Rightarrow u : B \quad x : A, \Delta \Rightarrow t : C}{z : B\backslash A, \Gamma, \Delta \Rightarrow t[\mathbf{bin}(x,u,z) \rightsquigarrow x] : C}\ \backslash L \qquad \frac{y : B, \Gamma \Rightarrow \mathbf{bin}(x,y,t) : A}{\Gamma \Rightarrow t : B\backslash A}\ \backslash R$$

$$\frac{y : A, z : B, \Gamma \Rightarrow t : C}{x : A \bullet B, \Gamma \Rightarrow t[x \rightsquigarrow \mathbf{bin}(x,y,z)] : C}\ \bullet L \qquad \frac{\Gamma \Rightarrow t : A \quad \Delta \Rightarrow u : B}{\Gamma, \Delta \Rightarrow \mathbf{bin}(x,t,u) : A \bullet B}\ \bullet R$$

The above rules of labelled deduction represent the pure residuation logic. Recovery of the configuration of the antecedent in terms of unary $\langle\cdot\rangle$ and binary $\circ$ structural operations, and the underlying pairs and triples for the composition relations R^2 and R^3 in the semantics, is straightforward. Structural rules, in this presentation, translate into labelling rules

$$\frac{\Gamma \Rightarrow t[u'] : A}{\Gamma \Rightarrow t[u] : A} \tag{25}$$

replacing a subterm u by a structural alternative u', where u and u' are the labelling versions of the left- and righthand sides of a structural postulate $A \rightarrow B$. Below the distributivity principle K as an illustration.

$$(K) \quad \Diamond(A \bullet B) \rightarrow \Diamond A \bullet \Diamond B$$

$$u : \mathbf{un}(x, \mathbf{bin}(y, t', t'')) \qquad u' : \mathbf{bin}(x, \mathbf{un}(y', t'), \mathbf{un}(y'', t''))$$

Let us evaluate the labelled Gentzen presentation from the perspective of algorithmic proof search. On the formula level, the format allows 'backward chaining' search (elimination of connectives) on the basis of a goal formula and an antecedent multiset of lexical resources. But the flow of information on the level of the structure labels is at odds with the backward chaining regime, and requires destructive term manipulations in the rules that have $t[u \rightsquigarrow v]$ in the conclusion. This problem makes the labelled Gentzen format suboptimal for the purposes of 'parsing-as-deduction' for the general multimodal categorial framework. (Of course, there is also the problem of spurious non-determinism in rule application order characteristic for naive sequent proof search. But this problem can be tackled by adding 'procedural control', as shown in the categorial literature by [19, 15, 14], and by [17, 1, 4], among others, in the context of 'linear' refinements of Logic Programming.)

3.2 Labelled Proof Nets

In this section, we consider labelled versions of the 'proof nets' of Linear Logic as an optimalization of sequent proof search. Proof nets can be decorated with Curry–Howard λ-term labelling in a straightforward way, as shown in [36, 13]. In order to capture the syntactic fine-structure of systems more discriminating than **LP**, and multimodal architectures with structural inference packages, we complement the semantic labeling with *structural* labeling.

The construction of a proof net corresponding to a sequent $\Gamma \Rightarrow B$ can be presented as a three stage process. The first stage is deterministic and consists in unfolding the formula decomposition tree for the A_i antecedent terminal formulae of Γ and for the goal formula B. The unfolding has to keep track of the antecedent/succedent occurrences of subformulae: we work with signed formulae, and distinguish $(\cdot)^\bullet$ (antecedent) from $(\cdot)^\circ$ (succedent) unfolding, corresponding to the sequent rules of use and proof for the connectives. We also distinguish two types of decomposition steps: $\exists$-type decomposition for the $\Diamond$L,$\bullet$L, /R,\R rules, and $\forall$-type decomposition corresponding to the $\Box$L,/L,\L,$\bullet$R rules. (For the binary connectives, these are one-premise and two-premise inferences, respectively.)

DEFINITION 3.2 *Formula decomposition.*

$$\frac{(A)^\bullet \quad (B)^\circ}{(A/B)^\bullet}\,\forall \qquad \frac{(B)^\bullet \quad (A)^\circ}{(A/B)^\circ}\,\exists \qquad \frac{(B)^\circ \quad (A)^\bullet}{(B\backslash A)^\bullet}\,\forall \qquad \frac{(A)^\circ \quad (B)^\bullet}{(B\backslash A)^\circ}\,\exists$$

$$\frac{(A)^\bullet \quad (B)^\bullet}{(A\bullet B)^\bullet}\,\exists \qquad \frac{(B)^\circ \quad (A)^\circ}{(A\bullet B)^\circ}\,\forall$$

$$\frac{(A)^\bullet}{(\Diamond A)^\bullet}\,\exists \qquad \frac{(A)^\circ}{(\Diamond A)^\circ}\,\forall \qquad \frac{(A)^\bullet}{(\Box A)^\bullet}\,\forall \qquad \frac{(A)^\circ}{(\Box A)^\circ}\,\exists$$

We call the result of the unfolding a *proof frame*. The second stage, corresponding to the Axiom case in the Gentzen presentation, consists in linking the signed atomic formulae (literals) with opposite polarity marking. We call an arbitrary linking connecting the leaves of the proof frame a *proof structure*. Not every proof structure corresponds to a sequent derivation. The final stage is to perform a wellformedness check on the proof structure graph in order to identify it as a *proof net*, i.e. a structure which effectively corresponds to a sequent derivation. For the checking of the well-formedness conditions, there are various alternatives for Girard's original 'long trip' condition, which (in the case of the binary connectives) checks the graph for connectedness and acyclicity. We do not discuss these checking procedures here, but move on to labelled versions of the proof net format.

The proof net version of Curry–Howard labelling is presented in Definition 3.3.

DEFINITION 3.3 *Formula decomposition with Curry–Howard terms for* **LP** *meaning composition. We use x,y,z (t,u,v) for object-level variables (terms), M,N for meta-level (search) variables. The search variables are instantiated in establishing the axiom links. Newly introduced object-level variables and metavariables in the rules below are chosen fresh.*

$$\textit{Axiom links} \quad \overline{t:(A)^\bullet \quad M:(A)^\circ} \quad \overline{M:(A)^\circ \quad t:(A)^\bullet} \quad \textit{with } M:=t$$

$$\frac{(t\ M):(A)^\bullet \quad M:(B)^\circ}{t:(A/B)^\bullet}\,\forall \qquad \frac{x:(B)^\bullet \quad N:(A)^\circ}{\lambda x.N:(A/B)^\circ}\,\exists$$

$$\frac{M:(B)^\circ \quad (t\ M):(A)^\bullet}{t:(B\backslash A)^\bullet}\ \forall \qquad \frac{N:(A)^\circ \quad x:(B)^\bullet}{\lambda x.N:(B\backslash A)^\circ}\ \exists$$

$$\frac{(t)_0:(A)^\bullet \quad (t)_1:(B)^\bullet}{t:(A\bullet B)^\bullet}\ \exists \qquad \frac{N:(B)^\circ \quad M:(A)^\circ}{\langle M,N\rangle:(A\bullet B)^\circ}\ \forall$$

$$\frac{{}^\cup t:(A)^\bullet}{t:(\Diamond A)^\bullet}\ \exists \qquad \frac{M:(A)^\circ}{{}^\cap M:(\Diamond A)^\circ}\ \forall \qquad \frac{{}^\vee t:(A)^\bullet}{t:(\Box A)^\bullet}\ \forall \qquad \frac{M:(A)^\circ}{{}^\wedge M:(\Box A)^\circ}\ \exists$$

Proof nets, as we have considered them so far, give a geometric representation for the occurrence-sensitivity of **LP** derivability, but they ignore structural aspects of well-formedness. Roorda shows that one can impose a further geometric criterion of planarity on the axiom linkings to capture the order-sensitivity of the **L** refinement. It is not clear, however, how such a geometric approach would generalize to the general multimodal architectures we have been studying in this paper, where typically a base logic is combined with variable packages of structural postulates.

In order to obtain a general algorithmic proof theory for the multimodal systems, we now complement the Curry–Howard labelling with a system of *structure* labeling, that serves the same purpose as the antecedent structuring in the (labelled) Natural Deduction presentation of Definition 2.3. The labeling regime of Definition 3.4 is related to proposals in [16, 33, 34], but makes adjustments to accommodate the multimodal architecture in its full generality.

DEFINITION 3.4 *Structure labels: syntax. The labeling system uses atomic formula labels x and structure labels $\langle\sigma\rangle, (\sigma\circ\tau)$, for the $\forall$ formula decomposition nodes. For the $\exists$ nodes, we use* auxiliary *labels: expressions that must be rewritten to structure/formula labels under the residuation reductions of Definition 3.5.*

σ,τ	$\longrightarrow$	x	*(atoms)*	$(\sigma\circ\tau)$	*(constructor $\bullet$)*
		$\langle\sigma\rangle$	*(constructor $\Diamond$)*	${}^\triangleleft(\sigma)$	*(left-destructor $\bullet$)*
		$\lfloor\sigma\rfloor$	*(destructor $\Diamond$)*	$(\sigma)^\triangleright$	*(right-destructor $\bullet$)*
		$\lceil\sigma\rceil$	*(goal $\Box$)*	$x\backslash\sigma$	*(goal $\backslash$)*
				σ/x	*(goal $/$)*

DEFINITION 3.5 *Labelled formula decomposition: structure labels and residuation term reductions (boxed)* REDEX $\succ$ CONTRACTUM. *We use x,y,z (t,u,v) for object-level formula (structure) labels, Γ,Δ for meta-level search variables. Newly introduced formula labels and metavariables in the rules below are chosen fresh.*

$$\frac{(t \circ \Delta) : (A)^\bullet \quad \Delta : (B)^\circ}{t : (A/B)^\bullet}\,\forall \qquad \frac{x : (B)^\bullet \quad \Gamma : (A)^\circ}{\Gamma / x : (A/B)^\circ}\,\exists$$

$$\boxed{(t \circ x)/x \succ t}$$

$$\frac{\Delta : (B)^\circ \quad (\Delta \circ t) : (A)^\bullet}{t : (B\backslash A)^\bullet}\,\forall \qquad \frac{\Gamma : (A)^\circ \quad x : (B)^\bullet}{x \backslash \Gamma : (B\backslash A)^\circ}\,\exists$$

$$\boxed{x\backslash(x \circ t) \succ t}$$

$$\frac{{}^{\triangleleft}(t) : (A)^\bullet \quad (t)^{\triangleright} : (B)^\bullet}{t : (A \bullet B)^\bullet}\,\exists \qquad \frac{\Delta : (B)^\circ \quad \Gamma : (A)^\circ}{(\Gamma \circ \Delta) : (A \bullet B)^\circ}\,\forall$$

$$\boxed{({}^{\triangleleft}(t) \circ (t)^{\triangleright}) \succ t}$$

$$\frac{\lfloor t \rfloor : (A)^\bullet}{t : (\Diamond A)^\bullet}\,\exists \quad \frac{\Gamma : (A)^\circ}{\langle \Gamma \rangle : (\Diamond A)^\circ}\,\forall \qquad \frac{\langle t \rangle : (A)^\bullet}{t : (\Box A)^\bullet}\,\forall \quad \frac{\Gamma : (A)^\circ}{\lceil \Gamma \rceil : (\Box A)^\circ}\,\exists$$

$$\boxed{\langle \lfloor t \rfloor \rangle \succ t} \qquad \boxed{\lceil \langle t \rangle \rceil \succ t}$$

The basic residuation reductions in Definition 3.5 are dictated by the identities for complex formulae $\Diamond A, \Box A, A \bullet B, A/B, B\backslash A$. Structural postulates $A \to B$ translate to reductions $\sigma(B) \succ \sigma(A)$, where $\sigma(\cdot)$ is the structure label translation of a formula. The reduction for the distributivity postulate K is given as an illustration in (26). Notice that both residuation reductions and structural postulate reductions are asymmetric, capturing the asymmetry of the derivability relation.

$$\Diamond(A \bullet B) \to \Diamond A \bullet \Diamond B \quad \overset{\sigma(\cdot)}{\rightsquigarrow} \quad (\langle t \rangle \circ \langle u \rangle) \succ \langle (t \circ u) \rangle \tag{26}$$

The structural labelling is used in parsing in the following way. To determine whether a string $x_1 \cdots x_n$ can be assigned the goal type B on the basis of a multiset of lexical assumptions $\Gamma = x_1 : A_1, \ldots, x_n : A_n$, one considers the multiset of signed labelled literals resulting from the formula decomposition of the $x_i : (A_i)^\bullet$ in Γ and of the goal formula $\Delta : (B)^\circ$. One resolves literals with opposite signature, unifying the search variable decorating $(p)^\circ$ against the label decorating $(p)^\bullet$, making sure in the matching that the proof net conditions of acyclicity and connectedness are respected. Notice that the labelling regime is set up in such a way that the $(p)^\circ$ literals are always decorated with unstructured search variables: unification at the axiom linkings, in other words, is simple one-sided matching. Labels can be rewritten under the residuation and/or structural postulate rewritings. We require the label assigned to the goal type B to be *normal*, in the sense that all residuation redexes must be reduced. We say that the input string $x_1 \cdots x_n$ is parsed if it is the yield of the normalized structure label assigned to the goal type B.

Illustration
For an illustration of the proof net labelled deduction system, we now return to the multimodal head adjunction analysis of §2.3. The relevant structural postulates are repeated below with the corresponding term rewriting rules for the structural labelling system.

$$\begin{array}{llr} P1 & \Diamond_1(A \bullet_1 B) \rightarrow A \bullet_1 \Diamond_1 B & t \circ_1 \langle u \rangle_1 \succ \langle t \circ_1 u \rangle_1 \\ P2 & \Diamond_1 A \rightarrow \Diamond_0 A & \langle t \rangle_0 \succ \langle t \rangle_1 \\ P3 & \Diamond_0(A \bullet_0 B) \rightarrow \Diamond_0 A \bullet_0 \Diamond_0 B & \langle t \rangle_0 \circ_0 \langle u \rangle_0 \succ \langle t \circ_0 u \rangle_0 \\ P4 & A \bullet_1 (\Diamond_0 B \bullet_0 C) \rightarrow \Diamond_0 B \bullet_0 (A \bullet_1 C) & \langle t \rangle_0 \circ_0 (u \circ_1 v) \succ u \circ_1 (\langle t \rangle_0 \circ_0 v) \end{array} \tag{27}$$

We try to derive the type $\Box_1(np\backslash_1 s)$ for the string 'Alice wil plagen' on the basis of the lexical type assignments in (28). (We have expanded the atomic formula inf of our earlier assignments to $np\backslash_1 is$ in order to illustrate the unfolding of higher-order types — one can read is as 'infinitival clause'.)

$$\begin{array}{ll} \text{Alice} & := np \\ \text{plagen} & := \Box_0(np\backslash_1(np\backslash_1 is)) \\ \text{wil} & := \Box_0((np\backslash_1 s)/_0(np\backslash_1 is)) \end{array} \tag{28}$$

The labelled literals resulting from the formula unfolding for the lexical resources and for the goal formula $\Box_1(np\backslash_1 s)$ is given in (29). We use upper case italic for the unknowns associated with succedent (goal) literals; upper case Roman for the fresh atomic formula labels for hypothetical assumptions.

$$\begin{array}{lll} & (np)^\bullet & \text{Alice} \\ C & (is)^\circ & \\ & (np)^\bullet & \mathrm{P} \\ H & (np)^\circ & \\ & (s)^\bullet & (H \circ_1 (\langle \text{wil} \rangle_0 \circ_0 \mathrm{P}\backslash_1 C)) \\ K & (np)^\circ & \\ N & (np)^\circ & \\ & (is)^\bullet & (N \circ_1 (K \circ_1 \langle \text{plagen} \rangle_0)) \\ P & (s)^\circ & \\ & (np)^\bullet & \mathrm{Q} \end{array} \tag{29}$$

The step-by-step decomposition tree for the higher-order type assignment for the verb 'wil' is presented in (30).

$$\dfrac{\dfrac{\dfrac{H : (np)^\circ \quad H \circ_1 (\langle \mathrm{wil}\rangle_0 \circ_0 \mathrm{P}\backslash_1 C) : (s)^\bullet}{\langle \mathrm{wil}\rangle_0 \circ_0 \mathrm{P}\backslash_1 C : (np\backslash_1 s)^\bullet}\forall \quad \dfrac{\mathrm{P} : (np)^\bullet \quad C : (is)^\circ}{\mathrm{P}\backslash_1 C : (np\backslash_1 is)^\circ}\exists}{\langle \mathrm{wil}\rangle_0 : ((np\backslash_1 s)/_0(np\backslash_1 is))^\bullet}\forall}{\mathrm{wil} : (\Box_0((np\backslash_1 s)/_0(np\backslash_1 is)))^\bullet}\forall \tag{30}$$

The resolution steps (axiom linkings) and the term rewritings leading up to the identification of the goal label are presented below.

$$\begin{array}{rcl}
K^{np} & = & \mathrm{Alice} \\
N^{np} & = & \mathrm{P} \\
C^{is} & = & (\mathrm{P} \circ_1 (\mathrm{Alice} \circ_1 \langle \mathrm{plagen}\rangle_0)) \\
H^{np} & = & \mathrm{Q} \\
P^{s} & = & (\mathrm{Q} \circ_1 (\langle \mathrm{wil}\rangle_0 \circ_0 \mathrm{P}\backslash_1(\mathrm{P} \circ_1 (\mathrm{Alice} \circ_1 \langle \mathrm{plagen}\rangle_0)))) \\
\mathrm{Goal} & = & \lceil \mathrm{Q}\backslash_1(\mathrm{Q} \circ_1 (\langle \mathrm{wil}\rangle_0 \circ_0 \mathrm{P}\backslash_1(\mathrm{P} \circ_1 (\mathrm{Alice} \circ_1 \langle \mathrm{plagen}\rangle_0))))\rceil_1 \\
& & \rightsquigarrow \quad (\mathrm{Res}\backslash), (\mathrm{Res}\backslash) \\
& & \lceil \langle \mathrm{wil}\rangle_0 \circ_0 (\mathrm{Alice} \circ_1 \langle \mathrm{plagen}\rangle_0))\rceil_1 \rightsquigarrow \quad (P4) \\
& & \lceil \mathrm{Alice} \circ_1 (\langle \mathrm{wil}\rangle_0 \circ_0 \langle \mathrm{plagen}\rangle_0))\rceil_1 \rightsquigarrow \quad (P3) \\
& & \lceil \mathrm{Alice} \circ_1 \langle \mathrm{wil} \circ_0 \mathrm{plagen}\rangle_0\rceil_1 \rightsquigarrow \quad (P2) \\
& & \lceil \mathrm{Alice} \circ_1 \langle \mathrm{wil} \circ_0 \mathrm{plagen}\rangle_1\rceil_1 \rightsquigarrow \quad (P1) \\
& & \lceil \langle \mathrm{Alice} \circ_1 (\mathrm{wil} \circ_0 \mathrm{plagen})\rangle_1\rceil_1 \rightsquigarrow \quad (\mathrm{Res}\Box) \\
& & (\mathrm{Alice} \circ_1 (\mathrm{wil} \circ_0 \mathrm{plagen}))
\end{array} \tag{31}$$

As was the case with the labelled sequent presentation of §3.1, from the labelling information one can reconstruct the structural configuration of the antecedent database, and a sequent or Natural Deduction representation of the proof from the axiom linkings in (31).

The labelled proof net format discussed here is the basis for Richard Moot's theorem prover Grail, a general grammar development environment for multimodal categorial systems. We refer the interested reader to [30, 31] for information on the design of the Grail system — specifically, for discussion of efficient strategies to combine structural label rewriting with checking of proof net conditions.

4 Conclusion

The quinquagenarian who is the subject of this Festschrift has edited a volume [8] raising the question 'What is a logical system?' An answer is suggested in an interview he has given for the Computational Linguistics Magazine TA! The interview appears under the title '*I* am a logic' [9]. Following up on this suggestion, we have compared a number of proof formats for grammatical reasoning from a 'Labelled Deduction' point of view. Labelled presentations of the categorial proof theory turn out to be attractive: they naturally accommodate the required modular treatment of logical and structural aspects of grammatical resource management.

Utrecht Institute of Linguistics, The Netherlands.

References

1. Jean-Marc Andreoli. Logic programming with focussing proofs in Linear Logic. *Journal of Logic and Computation* 2(3), 1992.
2. Johan van Benthem. Correspondence theory. In D. Gabbay and F. Günthner (eds.) *Handbook of Philosophical Logic. Vol II*, Dordrecht, 167–247, 1984
3. Wojciech Buszkowski. Completeness results for Lambek syntactic calculus. *Zeitschrift für mathematische Logik und Grundlagen der Mathematik.* 32, 13–28, 1986.
4. Iliano Cervesato, Joshua S. Hodas and Frank Pfenning. Efficient resource management for Linear Logic proof search. In *Proceedings International Workshop on Extensions of Logic Programming*, Leipzig, 1996.
5. Kosta Došen. A brief survey of frames for the Lambek calculus. *Zeitschrift für mathematische Logik und Grundlagen der Mathematik* 38, 179–187, 1992.
6. Kosta Došen & Peter Schröder-Heister (eds.) *Substructural Logics.* Oxford.
7. Dov M. Gabbay. *Labelled Deductive Systems; principles and applications. Vol 1: Introduction.* Oxford University Press, 1996.
8. Dov M. Gabbay (ed.) *What is a Logical System?* Studies in Logic and Computation. Clarendon Press, Oxford, first edition, 1994.
9. Dov M. Gabbay. 'I am a logic'. Interview for the *Computational Linguistics Magazine TA* 1995, reprinted in this volume.
10. Dov M. Gabbay and Ruy de Queiroz. Extending the Curry–Howard interpretation to linear, relevant, and other resource logics. *Journal of Symbolic Logic* 57, 1319–1365, 1992.
11. Jean-Yves Girard, Paul Taylor and Yves Lafont. *Proofs and Types.* Cambridge Tracts in Theoretical Computer Science 7, Cambridge, 1989.
12. Rajeev Goré. Substructural Logics on Display. *Logic Journal of the IGPL*, 6(3), 451–504, 1998.
13. Philippe de Groote and Christian Retoré. On the semantic readings of proof-nets. Proceedings *Formal Grammar*, 57–70, ESSLLI Barcelona, 1996.
14. Herman Hendriks. *Studied Flexibility. Categories and Types in Syntax and Semantics.* Ph.D. Dissertation, ILLC, Amsterdam, 1993
15. Mark Hepple. *The Grammar and Processing of Order and Dependency.* Ph.D. Dissertation, Edinburgh, 1990.
16. Mark Hepple. Hybrid Categorial Logics. *Logic Journal of the IGPL*, 3(2,3), 343–355, 1995.
17. Joshua S. Hodas and Dale Miller. Logic programming in a fragment of Intuitionistic Linear Logic. *Information and Computation* 110, 327–365, 1994.
18. Roman Jakobson (ed.) Structure of Language and Its Mathematical Aspects. *Proceedings of the Twelfth Symposium in Applied Mathematics.* Providence, Rhode Island, 1961.
19. Esther König. Parsing as natural deduction. *Proceedings of the 27th Annual Meeting of the ACL*, Vancouver, 272–279, 1991.
20. Natasha Kurtonina. *Frames and Labels. A Modal Analysis of Categorial Inference.* Ph.D. Dissertation, OTS Utrecht, ILLC Amsterdam, 1995.

21. Natasha Kurtonina and Michael Moortgat. *Structural Control.* In Patrick Blackburn and Maarten de Rijke (eds.) *Specifying Syntactic Structures.* CSLI, Stanford, 75–113, 1997.
22. Joachim Lambek. The Mathematics of Sentence Structure, *American Mathematical Monthly* 65, 154–170, 1958.
23. Joachim Lambek. *On the calculus of syntactic types.* In [18], 1961.
24. Joachim Lambek. *Categorial and categorical grammar.* In [35], 297–317, 1988.
25. Joachim Lambek. *Logic Without Structural Rules.* In [6], 179–206, 1993.
26. Michael Moortgat. Labelled Deductive Systems for categorial theorem proving. *Proceedings Eighth Amsterdam Colloquium*, ILLC, Amsterdam, 403–424, 1992.
27. Michael Moortgat Multimodal linguistic inference. *Journal of Logic, Language and Information,* 5(3,4)(1996), 349–385.
28. Michael Moortgat Categorial Type Logics. Chapter Two in Van Benthem & ter Meulen (eds.) *Handbook of Logic and Language*, 93-177, Elsevier, 1997.
29. Michael Moortgat and Richard T. Oehrle. Adjacency, dependency and order. In P. Dekker and M. Stokhof (eds.) *Proceedings Ninth Amsterdam Colloquium*, ILLC, Amsterdam, 447–466, 1994.
30. Richard Moot. Automated Deduction for Categorial Grammar Logics. In Esther Kraak and Renata Wassermann (eds.) *Proceedings Accolade 97*, 50–65, Amsterdam, 1998.
31. Richard Moot. Grail: An Automated Proof Assistant for Categorial Grammar Logics. To appear in Proceedings UITP98, http://www.win.tue.nl/cs/ipa/uitp/proceedings.html.
32. Glyn Morrill. *Type Logical Grammar. Categorial Logic of Signs.* Kluwer, Dordrecht, 1994.
33. Glyn Morrill. Higher-order Linear Logic programming of categorial deduction. *Proceedings of the European Chapter of the Association for Computational Linguistics*, Dublin, 1995.
34. Richard T. Oehrle. Term-labelled categorial type systems, *Linguistics & Philosophy* 17, 633–678, 1995.
35. Richard T. Oehrle, Emmon Bach and Deidre Wheeler (eds.) *Categorial Grammars and Natural Language Structures.* Reidel, Dordrecht, 1988.
36. Dirk Roorda. *Resource Logics. Proof-Theoretical Investigations.* Ph.D. Dissertation, Amsterdam, 1991.
37. Anne S. Troelstra. *Lectures on Linear Logic.* CSLI Lecture Notes. Stanford, 1992.
38. Heinrich Wansing. Formulas-as-types for a hierarchy of sublogics of intuitionistic propositional logic. In D. Pearce & H. Wansing (eds.) *Non-classical Logics and Information Processing.* Springer Lecture Notes in AI 619. Berlin, 1992.
39. Heinrich Wansing. *The Logic of Information Structures.* Ph.D. Dissertation. Berlin, 1992.

FORMALISMS FOR NON-FORMAL LANGUAGES

JULIUS M. MORAVCSIK

In recent decades computer scientists, linguists, and philosophers converged on offering what are called formal representations of natural languages. In this essay I wish to go back to some earlier work that Dov Gabbay and I did in cooperation, show its significance, and tie it to more recent work that I did on lexical semantics, showing that the early joint work and the recent lexical theory of mine are not only compatible but supplement each other. At the end I wish to make some general remarks about the future of formal representations of natural languages, and propose a way in which such representations can tie in with work in artificial intelligence, provided that we give the latter notion a reinterpretation.

The idea that formalisms should be applied to natural languages has a variety of historical sources. Tracing some of these helps to unravel the different senses of 'formal' involved in such proposals, thus distinguishing also different projects, answering different questions. While attending to differences, we should keep in mind certain questions that need be raised about any of these projects. Some of these are: why is such a formal representation valuable? Why and in what sense is it explanatory? What kinds of abstractions and idealizations are involved? Are the presentations meant to have empirical impact?

From a philosophical point of view at least two important sources for 'formalism' should be distinguished. One of these stems from Curry's program, that tried to explain most of mathematics on a purely syntactic basis.[1] Inspired by that effort one might try to construct a formalist – in this sense - representation of the semantics of natural languages. There is, however, also a quite different tradition, originating in the work of Tarski [12]. Tarski defines a formal language in well known technical ways,[2] and shows both how to give a semantics to such languages in terms of the notions of truth and satisfaction. Though Tarski did not believe that this

[1] See e.g. [1].

[2] For a brief summary see [6], pages 103–104. For a more detailed account, see [12].

H.J. Ohlbach and U. Reyle (eds.), Logic, Language and Reasoning, 401–415.

conception could be applied to natural languages as a whole, Richard Montague, considerably later, thought that he could show how just this kind of representation can explicate the semantics of a natural language like English [5].

Formal representations can be associated also with what is in general terms logical analysis of a natural language; i.e. to explicate either certain types of lexical meaning or compositional rules in terms of the notions of symbolic logic and set theory. Showing the utility of such work and the viability of this type of analysis does not commit the researcher either to the success of Curry's program or any analogue of it in the analysis of natural languages, nor to Montague's project. One need not accept either of these schemes, and at the same time one can maintain the utility of set-theoretic notions for the formalization of some aspects of the semantics of natural languages.

Such work – and these rejections – formed the basis of the work on verb semantics that Gabbay and I presented in 1980 at a conference in Stuttgart. One key part of this work was a semantic classification with an interpretation of tense, aspect, and the addition of temporal modifiers [4]. Montague left the representation of lexical meaning open. For he simply posited lexical meanings as functions. This is hardly informative, since just about anything in reality can be represented as functions. In our joint work we wanted to get at the internal anatomy of the meanings of verbs of different sorts. One of our key assumptions was that verbs should be represented as denoting elements of time; thus events or happenings in a broad sense. For example, the denotation of 'eat' is a set of eating events, and not simply eaters. This approach requires in the representation of full sentences quantifying over both things (including agents) and events; thus two basic kinds of particulars. The advantage of this move is twofold. First, it gives a framework that can be tied to empirical work on how children learn to form conceptions of time and temporal configurations. Secondly – as we have shown – it facilitates linking the semantics of the meanings of individual verbs to the operations of aspect and temporal modifiers.

In my recent book, *Thought and Language*, I presented a theory of lexical meaning, based on some of my earlier work [6, chapter 6]. This theory was not presented in formal terms, and – as we shall see – some of its features might be interpreted as claiming that there can be no formal representation of lexical meaning in the sense of 'formal' we invoked. In this essay I wish to show that formal representations of lexical meaning and my theory, abbreviated as AFT (Aitiational Frame Theory), are mutually supportive. The paper has the following four sections. In the first section I shall present some of the relevant aspects of the verb semantics classification of our joint earlier work. In the second I shall present the relevant aspects of AFT, as applied to verbs. In the third section I shall show how the earlier joint work contained seeds for the development of AFT, and how our formal verb classification can be combined with lexical representations within AFT. In the final section I shall attempt to draw some morals about the value of formalizations of the sort Gabbay and I attempted, and link it to speculations about the future of theories concerning natural languages. Before we launch into this project, I shall offer a definition of 'natural language', a phrase often used but rarely defined. As applied both to the earlier joint work and AFT, a natural language is a language that can be learned by a human or sufficiently human like device as his/her or its first language under the normal conditions we can associate with human language acquisition. Admittedly this definition contains some vague phrases. But it has the advantage that it can be used in theoretically significant ways, and it leaves such questions as to whether this or that computer language, or esperanto for that matter, are natural languages as empirical questions [6]. I regard this important even if for ethical reasons at our current stage of knowledge experimentation in this area is not possible.

1 Temporal Intervals, Points, and a Semantic Classification of Verb Types

We started with a framework within which we have both temporal intervals and points, with neither reducible to the other. We saw the need for this, because this enables us to bring out some empirically verifiable differences between verb types, and also the relation of verbs to the aspect of the progressive. This consideration weighed more heavily with us then purely theoretical considerations of economy and parsimony. Our classification centres on a threefold distinction between event verbs, state verbs, and process verbs. The classification covers also verb phrases. The following intuition under lies this structure. Some verbs denote events; i.e. happenings requiring as their minimal parts intervals and not instances. We cannot eat, walk, read, think, etc. instantaneously; it takes time to eat, read, etc. even for the shortest time. The duration of this time can be left indeterminate; its specification does not affect either the formal verb classification or the lexical theory to be sketched. Events have a starting point and a terminal point. We start reading and stop reading. Thus the event occupies a time slice, which we mark as

( )

In the case of events the time slice is filled either by an interval of continuous activity, or by intervals of activity interrupted by temporal gaps.

(__________) or (- - - - - - -)

For example someone might read a piece of fiction through, without interruptions. On the other hand, someone might read a difficult technical article. After a while he gets up and fixes himself a bit of instant coffee, or takes a snack out of the refrigerator. Such gaps are allowed by the semantics, and do not divide the event of reading into distinct events of reading. Likewise, someone might be walking a couple of blocks to the bus station. During the walk she stops briefly to pick up a newspaper, or deposits a soiled tissue into a garbage can. We still call this a walk. How long can the gaps be? What can fill these? Answers to these questions depend not only on the meanings of individual lexical items, but also on contextual considerations. An executive reads a report. He is interrupted by a phone call that he briefly answers. This gap is allowed. But if he reads the report for a part of the morning, then has a conference and lunch, and then reads some more of the report, this counts as two occasions of reading; hence two time slices in our terminology. These distinctions interact with pragmatic factors, and hence need not be very sharply delineated; or still better, have their exactness determined by utilitarian considerations. 'How many observations have you made during that day?' may be an important question in a laboratory, while it may not demand a very specific answer in the context of a school project.

The general specification, then, of the denotation of an event verb will be a set of time slices with the structure indicated; i.e.:

(- - - - - - - - - -) (- - - - - - - - -) (- - - - - - - - -) etc.

For formal purposes, then, within this framework we can define the meaning of an event verb or verb phrase as a function that picks out of possible worlds the relevant sets.

In contrast with event verbs, we find state verbs and verb phrases like 'know' or 'to be ill'. Being in the states denoted by such verbs also takes an interval of time, but the meanings of these terms do not specify any particular activity that would necessarily have to take place during the interval. To be sure, both the event of reading and being in a state of being ill will involve activities and ongoing processes. But while the definition of e.g. 'read' specifies some of these activities, the definition of being ill does not; it specifies the condition under which physical activities take place. Hence we characterize the contents of the time slices during which someone is ill as a set of temporal points - densely organized. We are ill at any moment during the time in which we are in this state. There is no contrast of gaps and actual goings on definitionally specified, as with event verbs. This is reflected in English in the differing relations the two categories have towards the progressive aspect. Event verbs can take the progressive; e.g. 'he is reading right now', something that we would not say if at that moment the subject is taking out his pipe and tobacco before resuming reading. But we cannot say: 'she is knowing', for knowing is a state, and we are in this state regardless of whether or not we have some particular item actually in our conscious thoughts.

Hence the representation of the denotation of e.g. 'to be ill' as a set of time slices filled with densely organized temporal points.

(.............) (...............) (..............) etc.

The meaning of the term will be, accordingly a function selecting these sets across possible worlds.

The third major category of process verbs and verb phrases, such as 'complete', 'build a house'. will be formally represented as having in the denotation a combination of both interval and point. The activity leading up to completion, or specifically a house having been built, has the formal shape of events as we just characterized. The basic units are time slices containing intervals, with gaps, optionally. But the event and its parts are supposed to lead up to a state, i.e. the state of completion, that does not obtain at any stage of the process of building or completing. Hence in our formal representation we must include a temporal point at the end of the time slice, with a star signifying that it is the terminal point of the process indicated by the verb. Hence the denotation has the structure:

(- - - - - - - - - F*) (- - - - - - - - -F*) (- - - - - - - - - F*) etc.

In other words a set of time slices with the structure of process and completion. The meaning is then specified as in the other cases above. There may be cases in which the process is incomplete. People started building a house, but ran out of money; thus the house was never built. Still, something must have been accomplished. If there is nothing there, then they at most planned to build a house, but have not done any building. Thus what we express in the vernacular as 'We were building a house but never built it' can be expressed more precisely as 'We did some house building, but never completed the process of building a house'.

An interesting feature of process verbs or verb phrases is that the object of the verb cannot exist until the process no longer takes place. If we are building a house, the house cannot exist. When the house exists, we are not building it any more. This places such verbs between those verbs that have only an intentional object, and those that have a real spatio temporal object.

Temporal modifiers distinguish process from event verbs by selection rules. We cannot say: 'They were cooking the dinner in an hour', but we can say that 'they cooked the dinner in an hour'. The temporal modifier signifies the time interval within which something has been completed; hence its fitting process verb phrases.

The basic classification has added to it two more distinctions. One of these is between instantaneous verbs like 'stop' and durational verbs like the ones we explored so far. ‘Stop’ in English has many senses. In one sense it is strictly instantaneous; i.e. it refers to an occurrence of a process or motion simply coming to an end. ‘That was when it stopped’ we say, and semantically we indicate a temporal point, though in pragmatic contexts its exact nature may open to variations. In another sense we can construe ‘stop’ as an achievement, hence locution like ‘it took him x minutes to stop the engines’. Within such a context we treat ‘stop’ as a process verb, to be analysed as shown above.

The contrast with instantaneous is ‘durational’, and all of the examples except the last one fall into that category. The other distinction is between repetitious verbs and non-repetitious ones. For example, the usually interpretation of ‘knocking on the door’ is not just one knock, but a series of knocking; hence the denotation having the shape of a set of time slices like:

((.) (.) (.) (.)) etc.

Nailing or hammering things have also a repetitious structure. The non-repetitious can be state, event, process, or instantaneous happening. (These distinctions help to explicate the difference between ‘he knocked on the door’, and ‘he was knocking on the door’.)

These temporal structures, constituting denotations for verbs and verb phrases, need be related to temporal modifiers the structures of which are also subject to analysis via temporal structures of the sort we used. Some modifiers are durationals like ‘for some time’, while others are containers like ‘within an hour’. We can see things happening for a while or for some time, and we can witness events being completed within an hour or a day. Both of these are distinct from simple time reference as ‘on June 26, 1994’.

Working out the various permissible combinations and rules forbidding certain complexes is done in our early work by showing which temporal structures can merge with each others.

Note that such analysis gives rules for constructing certain complexes out of relatively simple elements, but its use does not commit one to any position with regards to whether natural languages are formal languages in Tarski’s sense. In the next section we shall summarize some of the salient features of AFT. Since AFT is committed to denying that natural languages are formal languages in Tarski’s sense, we are left with exploring the possibility of joining such a lexical theory with the earlier semi-formal work on temporal structures.

2 The AFT Lexical Theory

This lexical theory presupposes the thesis that natural languages like English cannot be, and cannot be represented as, formal languages in Tarski’s sense. The evidence and arguments for this thesis have been presented elsewhere

[7, 8]. The lexical theory shows how we can avail ourselves to a framework within which meaning and denotation can be represented in a systematic and rigorous fashion, with empirical implications and hence verifiability, without representing lexical meanings in the uninformative way adopted by Montague as simply functions, and without adopting Carnap's framework of meanings as necessary and sufficient conditions for applications.

Meanings are a set of special necessary conditions for application. This is uninformative, since there are uncountably many necessary conditions for application of any specific lexical item. But behind the qualification of 'special' in the statement above lies the intuition, empirically testable, that meanings of descriptive items in natural languages are explanatory schemes. That is to say, the meaning of a descriptive word w is a scheme within which one would explain what and why some item counts as falling under w. To know in this sense the meaning of, e.g. 'tree' is to know roughly how one would explain why a certain items should be counted as a tree.

There is also a more general intuition underlying this theory, according to which humans are not basically labelling and referring creatures, but explanation seeking creatures. Labelling and reference are construed as elements in continuous explanation seeking and giving abstracted out of such structures when considered in isolation. We cannot explore this view in detail here. It has theoretically important implications both for psycholinguistics, and for the representation of human language use in artificial intelligence [10].

Given the universality of natural languages as Tarski [12] described these – the explanatory schemes and frameworks must have a very general scope. The basic structure of the scheme is made up of a relation tying four factors together; the first two of which is obligatory – i.e. all lexical items must have it – while the last two are elements only in the meanings of some items.

One factor is the specification of the domain within which the denoted items are located. So for most verbs it is time, for many nouns it is either the abstract or the spatio-temporal. For words that function as semantic modifiers the domain specified by what is called the m-factor, is more complicated, and the complications need not concern us here. It is clear that when we explain something, we must know whether it is an abstract entity, or a temporal, or a spatio-temporal one. Having located the domain, we turn to the s-factor that specifies: (i) principle of individuation, (ii) principle of persistence, and (iii) the necessary distinguishing factors that set apart the items falling under the word from other elements. Intuitively, to know the principle of individuation for e.g. 'horse' is to know on what basis we can count a collection of such items as having a certain number. This intuition is linked to the formal mechanism of quantification. If I know what it is to have within an area of 8 horses, then I know also how to handle phrases like 'many horses', 'one horse', etc. An understanding of the nature of individuation linked to the meaning of a given term gives us also the mastery of the distinction between count and mass terms, a distinction that within this theory cuts across the syntactic division of noun, adjective, verb. Principles of persistence specify what it is for an item to persist through time. E.g. it is not enough to understand what it is to know that there are 8 horses in an area, but also what the conditions are for the persistency of any one of these entities. Principles of individuation and persistence differ from kind to kind; hence are parts of lexical meaning.[3]

The distinguishing features attached to a given word in this theory will not define the denotation across all possible worlds, but only for that subset of such worlds within which some general presuppositions hold; specifically, those needed for the spelling out of individuation and persistence conditions; e.g. the timelessness of abstracta, the resemblance of the past to the future for the concrete, the flow of time, etc. The distinguishing characteristics need not form a sharply delineated class.

The third factor and ingredient in this meaning analysis is the a-factor that specifies what are the necessary properties associated with the items in denotation ranges that deal with causal powers, both in terms of necessary antecedents (such as: need to have animate predecessors) or effects. (e.g. what is cleansing must make things – normally – clean.)

Finally, the f-factor specifies the necessary functional elements in the explanatory schemes. The dominance of this element is most obvious in the case of words for artifacts. These are typically defined and explained in terms of what their function is, what they are meant to do. E.g. chairs to sit on, cars to give us transportation, etc.

In the case of each of the factors an adequate specification centers on what we take to be as specific and unique to the kind explained as possible.

Given a structure like $R(m, s, a, f)$, why do these not yield both necessary and sufficient conditions? Note first, the explanatory schemes typically do not accomplish that. Within an explanation we claim only that elements of kind K can be seen as having this or that structure, this or that power, etc. that enable them to be distinct and persist – or in the case of abstracta only the former.

What sorts of things do our scheme explain? Among other items, they explain the denotation of verbs signifying achievement, like 'walk', 'write', 'read'. Achievement is related to agents at different stages of life, different roles, etc. Hence any general specification of an activity involving achievement has to have the form 'activity of ... (constituting reading writing, etc.) in appropriate magnitude, scale of accomplishment, etc.' This scheme needs to be filled in by a variety of descriptions referring to the sick and the

[3]Detailed presentation of AFT in [6] chapter 6.

healthy, the adult and the child, the expert and the layperson, etc. What counts as walking or writing for a child is different from what counts as these activities for healthy adult. How many steps, taken at what speed, count as walking for a patient recovering in a hospital?

These examples show the need for a 3-level lexical semantics. Level 1 is the explanatory scheme in terms of the four factors mentioned (where the third and fourth may be the 0 element). We add to these in the right places conditions referring to appropriate accomplishment, or in the case of terms denoting masses of different kinds (hill, pile, etc.) appropriateness of size and quantity. These elements of the meaning specification on level 1 provide guidelines for specifying contexts of denotation. For example, for 'walk' the different contexts for denotation will include: walk for a baby, for an invalid, for an adult, for an octogenarian, etc. In each context we can ask: 'what counts as an event of walking within this context?' Note that we can provide guidelines for generating contexts, but not a decision procedure or other mechanical process. The class of all possible contexts in which achievement, size, etc. has to be evaluated and determined in an indefinite class. We cannot survey it, the way we survey the class of positive integers. There is nothing analogous here to the successor function. At the same time, not anything goes. An individual cannot simply announce that he determined now a criterion for what counts as writing for 50 year old humans in his village. The aspect of size, achievement, realization of potentialities, etc. that are semantically relevant depend on the semantics of a language, and thus are communal matters, not matters of arbitrary choice and taste. The set of these denotation fixing contexts is Level 2.

Thus we arrive at the 3rd level, a specific denotation range. At this level we do obtain additional conditions that yield sufficiency, not only necessity. These conditions may be more or less precise, depending on pragmatic considerations. Precision may be needed in some medical or chemical experimental context, and not in deciding what is in given contexts a family emergency. A crucial aspect of this lexical theory is that the necessary and sufficient conditions on level 3 are not merely for reference or labelling, but for explanatory contexts as well. 'We understand what F-ness is in this context when we see it as the combination of factors X,Y,Z, etc.' Understanding what a lexical item denotes is more than merely being able to refer or pick out an item in a context.

3 The Complementarity of Formal Temporal Structures and AFT

Let us consider a lexical item such as 'walk'. Informally, its definition is: 'an activity of movement placing one foot in front of the other, keeping one foot on the ground at any time, covering appropriate distance, with possible gaps of appropriate length.' The following is the AFT analysis of this definition. m-factor: time. (AFT leaves it open how time is to be represented and analyzed so as to be suitable as a domain specification for items with s- and a-factors in their definitions.)

s-factor: (i) Principle of individuation: a walking is an event requiring the same agent, and continuous happening (with possible gaps) answering the conditions for walk, distinguishing this from other activities of motion, and having its own spatio-temporal boundary. 2 walks are either two agents walking at the same time or different times, or the same agent walking at different distinct temporal slices. (This leaves open the question of how we represent the temporal structures making up a time slice of walking.)

(ii) Persistence condition. Same agent, spatio-temporally continuous activity meeting descriptive conditions of walking, with appropriate delineation of the difference between mere gaps, and termination of a time slice of walking. (This leaves open how to represent difference between there being a gap, and there being an end to a walking.)

(iii) Distinguishing characteristics: it must be an activity, hence animate subject, the motion has to be what is needed for walking in contrast with running, jogging, sauntering, etc. (See above in informal definition.) a-factor: event requires agent, and in particular animate agent. It has no necessary definitionally linked causal effect. The activity requires in its causal structure an agent with legs.

f-factor: the purposive accomplishment of covering appropriate distance at appropriate speed.

We have seen already the formal representation of temporal structure of a verb like 'walk' as:

(- - - - - - - - -) (- - - - - - - -) (- - - - - - - -) etc.

In the original joint work we took this, as mentioned above, as denotation, and defined meaning in the conventional way in a function across possible worlds. However, reflexion on AFT shows that such a function cannot cover all possible worlds, since the events in question carry presuppositions, and thus denotation can be defined only within a subset of possible worlds. (It is not enough to say that in some possible worlds there are no such events, but that in some possible worlds (all entities constantly fluctuate in and out of existence, etc.) the notion cannot be defined.)

Furthermore, we saw that one cannot define a denotation for 'walk' as such as a set extensionally characterized. For the same event can count as a walk for a recovering invalid, and not as a walk for a healthy adult (but only as 'taking a few slow steps'). Thus if we are to relate the temporal structures of the early work to AFT, at one level, namely level 1, these structures enter into the semantic specification, not as describing actual denotation, but rather as conceptual structures, sketching in precise ways what certain aspects of walking must be.

At the same time we should note that the temporal structure of the early work are needed for AFT in order to present individuation and persistence principles in a precise way. In order to provide a clear principle of individuation for what a verb or verb phrase stands we need to know, e.g. whether it is an event or a state verb. In the latter case, the one state must hold continuously, and at every instant, while in the case of event or process verbs this is not the case. It is not enough to say that individuation gives us distinct spatio-temporally continuous time slices within which a certain activity is taking place. Depending on whether it is a state or event verb, spatio-temporal continuity will have different structures.

The abstractly specified temporal structures are needed also for the persistence criteria, i.e. criteria telling us under what conditions a certain happening persists. Some of these issues are settled by the specification of the nature of the activity. A walk can turn into a trot, reading can turn into musing, etc. But as we saw in the case of event verbs a crucial issue is when a temporal space of non-activity constitutes merely a permissible gap, and when it signals the end of one event and the beginning of another. John may have walked from his house to the building in which a friend of his lives. He might have stopped just to pick up his friend's mail because the friend is on vacation, or he might stop and spend the whole evening discussing philosophy with friends. Let us suppose that in both cases he proceeds from that house to the promenade. In one case we can take what he does as one walk, with gaps, in the other case we construe it as two separate walks. We need contextual and pragmatic information to make these decisions, and at times the decisions are trivial while in other cases these may be very important (was there only one knocking on the door with some interruptions, or two separate knockings? This question may be important for a criminal investigation.) The pragmatic and contextual information is not sufficient to get the right persistence principles. We need to know also the differences between the temporal structures of the denotation of the event verbs and state verbs or verb phrases.

We need the temporal structures also on the third level of analysis. For at that stage we have necessary and sufficient conditions of application. Hence we do have actual ranges of denotation, and these should be - optimally - characterizable in formal and logical terms. This is no mere aesthetic requirement. We need the logical structure so that some structure can be used as underlying and explaining the entailments in which the terms under investigation play key roles. One could do this with analogues of what Carnap and Montague called meaning postulates, but that method is ad hoc, and leads to our missing generalizations that the anatomy of temporal structures enables us to formulate.

The need to specify denotation with the help of formal temporal structures, and our ability to do so should not seduce us into thinking that we can treat the sum of segments of a language on the third level of AFT analysis as a formal language in Tarski's sense. For that treatment requires a specification of the domain of entities over which the interpretations will be given. And for reasons already touched on, such general specification, as in the case of numbers or space-time points, etc. cannot be given in these cases.

The dependency between AFT and the temporal structures is not asymmetrical. If we take the temporal structures arising out of the early joint work as denotation, the we in effect treat a natural language as a formal language, and we are exposed to all of the objections, starting with those of Tarski himself, that show that such a construal is illegitimate.

Practice might suggest to researchers that either we treat English as a formal language or if this is not possible than using set-theoretic notions to illuminate the semantics of such a language will not be possible. Reflecting on the relationship between AFT and the temporal structures of the early work shows that these are not the only alternatives. As we see in this section, interactions between the two systems are possible and in fact desirable on the first and third level of AFT analysis. Set-theoretic notions can help adequate analysis provided that we isolate a certain level of semantics and can concentrate on isolable sets of phenomena. The issue of the utility of such analyses leads us to larger questions to which we turn in the final section.

4 The Semantics of Natural Languages and Human Nature

We have seen how employment of formal analysis helps both with lexical semantics and with some aspects of the use of quantifiers in natural languages, and that such employment of formalism is compatible with the denial that natural languages are formal languages in Tarski's sense. We shall now review briefly other parts of earlier work in which we showed that one can specify semantic complexity for certain parts of the semantics of natural languages, even if this is done negatively, i.e. showing that a certain minimal level of complexity is required.

One such work was on tense iteration in English [2]. It was shown in that work that usual assumptions about how to establish temporal reference in series of tensed sentential parts will not be adequate. The assumption that we need to keep in mind the initial temporal point at which the statement starts being given, and then keep in mind at any stage only the previous point is inadequate to capture English sentences like: 'she regretted that she married the man who was to become an officer of the bank where she had had her account', where the calculation of the last temporal reference point requires jumping back to the last reference to past, thus 'over' the future. This leads to the conclusion that tracking of temporal references in these iterations can require return to points arbitrarily distant from the point that is being calculated. To be sure, there may be grammatical criteria specifying the 'islands' we avoid, but these grammatical structures are not from the grammar of pure logical representations.

Another piece of work centers on branching quantifiers, i.e. sequences of quantifiers representing branching rather than linear dependence [3]. The fact that a natural language like English is permeated by such construction across a wide range of syntactic structures shows that under standard semantic analysis this forces us to construe meaning structures in English as involving fragments of second-order logic. This result too indicates a level of complexity that the semantics of natural languages require as a minimum. Our work contains also the demonstration that the presence of branching quantifiers in English is not a matter of finding a few examples. Rather, we showed the syntactic devices that are constituents of a language like English, that enable us to generate the infinite number of a variety of such sentences. These early results, then, show two things. First, giving up the hypothesis that English is a formal language does not go against the utility of set-theoretic tools being used to show significant aspects of the semantics of natural languages. Secondly that such work says theoretically interesting things about English even if the results are negative; i.e. showing what is not sufficient.

The claim that English cannot be analyzed in the way in which Montague envisaged this emerges already in our work of 1980 [4, page 60]. In connection with the pair of sentences: 'she arrives Tuesday' and 'she arrives on Tuesday every week' we showed that the temporal semantics of sentences cannot be built molecularly; i.e. by assigning a semantic object to each of the linguistic units regardless of context. The addition of a linguistic unit can change the meaning assignments of other units in the sentence; 'arrives' in one sentence denotes a singular event, in the other a series of repetitions.

The analysis of event verbs meanings as allowing 'gaps' speaks also against the Montague hypothesis, since the nature, extent, and legitimacy of these depends on the meanings of lexical items and factors specified later in AFT.

We can combine these observations with more recent work both in AFT, and the lexical semantics developed by James Pustejovsky, employing a framework for lexical analysis overlapping with that of AFT [9]. Pustejovsky points out that a lexicon of a natural language should be construed as productive. For example, the verb 'give' will be analyzed as having different components in its semantic structure in the combinations: to give an apple, to give a recital, to give a lecture, to give thanks. We have no way of delineating all of the combinations in which 'give' can have a role, and thus we cannot specify all of the ways in which 'give' will have a variety of semantic elements in its meaning structure, together with elements that remain constant. On the basis of somewhat different evidence the same conclusion follows from the structuring of what we called above the third level of analysis in AFT.

This productivity of the lexicon provides a challenge for new ways of formalizing the lexical meanings, combining presumably the compositional with the procedural.

In conclusion, let us ponder some general remarks. At CSLI, many researchers have the view that after the big dreams of A.I. (analysis of the complete human understanding process and language mastery; total machine translation down to the last pragmatic stress) come to be seen as beyond the pale of what we can hope for, much of the A.I. work can be used for more practical and more limited projects. Perhaps an analogous conclusion should be drawn from the recent history of formal analysis of natural languages. We will most likely not be able to characterize the semantics of a whole language – especially since a natural language is a diachronic phenomenon – but can use formalism already at hand as well as new ones likely to be developed to analyze significant fragments that show either limits of complexity, or significant features of the human language processing system.

In light of these remarks, let us return to questions raised at the outset: how valuable is this work, what is its explanatory power, and what are viable idealizations that leave us with empirical import in this work?

The value of the work may be in some cases practical. E.g. recent work on the notation of caring in medicine benefitted from AFT analysis. Many of the words used in social/political theory, such as 'freedom', 'equality' can also be given fruitful treatment within this framework ([11]).

But the practical should not be the only criterion for value. The theoretical work can also yield important clues for aspects of cognition. There are two ways of looking at the formal aspects of semantics for natural languages. One is to assume that these form a theoretically and logically interesting 'natural kind'. So far this has not been proven. But this does not mean that work on such semantics is without interest. In our initial definition of natural languages we linked this notion to cognitive capacities such as certain kinds of learning. What may be from the logical point of view a heterogeneous collection may nevertheless acquire unity of a different kind; namely as systems with properties that shed light on various kinds of cognitive processes.

Given this outlook, analyses of natural language semantics should aim at illuminating not only the end product of mastery, but also the processes of acquisition and maintenance. Thus abstract representations with idealization conditions that abstract away from anything that is significantly human in the language processing, aiming only at extensional equivalence between the system proposed and a part of natural language are not within the range of approaches recommended in this essay. But at this point we can hardly have firm views on which idealizations are suitable and which ones make the work uninteresting. This is a matter of degrees of abstractness and of different features of the human cognitive processing, about which we still do not know much, and perhaps never will.[4]

Stanford University, USA.

References

1. Haskell B. Curry. Remarks on the definition and nature of mathematics. In Paul Benacerraf and Hilary Putnam, editors, *Philosophy of Mathematics*, pages 202–206. Cambridge, 1964.
2. Dov M. Gabbay. Tense logic and the tenses of English. In Julius M. Moravcsik, editor, *Logic and Philosophy for Linguists*, pages 177–186. Mouton Publishing Co., 1974.
3. Dov M. Gabbay and Julius M. Moravcsik. Branching quantifiers, English, and montague grammar. *Theoretical Linguistics*, 1:139–155, 1974.
4. Dov M. Gabbay and Julius M. Moravcsik. Verbs, events, and the flow of time. In Christian Rohrer, editor, *Time, Tense, and Quantifiers*, pages 59–83. Niemayer, 1980.
5. Richard Montague. English as a formal language. In Richmond H. Thomason, editor, *Philosophy*. New Haven: Yale University Press, 1974.
6. Julius M. Moravcsik. *Thought and Language*. Routledge, London, 1990.
7. Julius M. Moravcsik. All A's are B's; form and content. *Journal of Pragmatics*, 16:427–441, 1991.
8. Julius M. Moravcsik. Is snow white? In Paul Humphreys, editor, *Patrick Suppes: Scientific Philosopher*, pages 71–87. Kluwer, 1994. vol. 3.
9. James Pustejovsky. The generative lexicon. *Computational Linguistics*, 17:409–441, 1991.
10. Mark Richard. Reference and competence: Moravcsik's thought and language. *Dialogue*, 32:555–563, 1993.
11. Robert Scott, Linda Aiken, David Mechanic and Julius M. Moravcsik. Organizational aspects of caring. *The Milbank Quarterly*, 73:77–95, 1995.
12. Alfred Tarski. Der Wahrheitsbegriff in den formalisierten Sprachen. *Studia Philosophica*, 1:261–405, 1936.

[4] I am indebted to Professors Solomon Feferman and Grigory Mints for helpful suggestions.

Names Index

Index